OPTICAL ROTATORY POWER

BY

T. MARTIN LOWRY

C. B. E., M. A., D. Sc., F. R. S.

LATE PROFESSOR OF PHYSICAL CHEMISTRY IN THE UNIVERSITY OF CAMBRIDGE

WITH 186 ILLUSTRATIONS

DOVER PUBLICATIONS, INC.

NEW YORK NEW YORK

This Dover edition, first published in 1964, is an unaltered and unabridged republication of the work first published by Longmans, Green and Co., London, in 1935.

Library of Congress Catalog Card Number 64-20883

Manufactured in the United States of America

Dover Publications, Inc.
180 Varick Street
New York 14, N.Y.

This new edition is dedicated
to the memory of
T. MARTIN LOWRY
(1874-1936)

PREFACE

DURING the 120 years that have elapsed since Biot's discovery of optical rotatory power was described in the *Memoirs of the Institute of France*, no original book on the subject appears to have been written in the English language. Two editions of Landolt's "Optische Drehungsvermögen," published in Germany in 1879 and 1897 respectively, formed the basis of an English translation by Robb and Veley in 1882, and of an American edition by Long in 1902; but the original book and the translations have been out of print for many years, and no modern work is now available in the English language. The present period is in many respects particularly opportune for such a publication, since experimental methods have been developed to a point at which only those wave-lengths (in the far ultra-violet and in the deep infra-red regions of the spectrum) to which quartz is opaque remain unexplored; and, on the theoretical side, the masterly work of Born has at last provided an adequate physical basis for the interpretation of one of the most difficult of optical phenomena.

In attempting to fill this obvious gap, an endeavour has been made to provide a well-balanced account of the subject, in which the classical work of the pioneers is described as adequately as the researches which happen to be in progress at the present time. With this end in view, the original literature has been surveyed over a wide range, and an account has been given of the work of every important school of research, from Biot to Born. Numerous quotations have been made, especially from the less accessible works of the earlier writers, and in many instances references are given to the actual page or pages on which important statements are made or discoveries announced, in order that the reader may be able to find them immediately, without being obliged to search for them in memoirs which are often printed very much *in extenso*. In order to minimise the risk of inaccuracy, the references have been checked either against the original papers or by means of independent citations, e.g. in the Royal Society's "Catalogue of Scientific Literature." In the same way, nearly all the numerical data have been checked against the papers from which they are cited. The references are accompanied by authors' names, but the following abbreviations have been used in a specially shortened form :

A.C.P. = *Annales de Chimie et de Physique.*
A.C.R. = *Alembic Club Reprints.*
Ann. = *Liebig's Annalen der Chemie.*
Ber. = *Berichte der deutschen chemischen Gesellschaft.*
C.R. = *Comptes rendus de l'Académie des Sciences.*
J. = *Journal of the Chemical Society.*
J.A.C.S. = *Journal of the American Chemical Society.*
P.R.S. = *Proceedings of the Royal Society.*
T.F.S. = *Transactions of the Faraday Society.*
Z.ph.C. = *Zeitschrift für physikalische Chemie.*

The book is divided into four parts, which are arranged in a rough chronological sequence. Part I, *Historical and General*, describes the pioneer researches of Biot and Fresnel ; but the record is continued to include the epoch-making researches of Pasteur on Molecular Dissymmetry, which provided a firm foundation for the science of Stereochemistry ; the development of this science by le Bel, van't Hoff, Pope and Werner is also described in this section. On the physical side, an account is given of Biot's discovery of rotatory dispersion in quartz, of anomalous rotatory dispersion in tartaric acid, and of an artificially-produced anomaly resulting from the superposition of the normal but unequal rotatory dispersions of turpentine and camphor, after the manner of an achromatic doublet in refractometry. Biot's Law of Inverse Squares is described, with its subsequent modifications, culminating in the well-known formula of Drude, which suffices to express the rotatory dispersion of transparent media of all kinds with a precision that is remarkable and unique. Cotton's discovery in absorbing optically-active media of the twin phenomena of circular dichroism and of anomalous rotatory dispersion, which are indissolubly associated with his name, is also included amongst the "classics," although the more recent studies of these phenomena are reserved for a subsequent section of the book. The section closes with an account of Faraday's discovery of Magnetic Rotatory Power and of its application by Perkin to studies of the chemical constitution of organic compounds.

Part II, *Polarimetry*, is a record of the development of polarimetric apparatus, ranging from the earliest polarising prisms of Rochon and Nicol, to the newest saccharimeters of the Bureau of Standards, and the apparatus devised by Descamps, Bruhat and Kuhn for polarimetric work in the ultra-violet.

Part III, *Special Cases*, records the application of polarimetric methods to the study of quartz, of amyl alcohol and *iso*-valeric acid, of tartaric, malic and lactic acids, of the sugars, of camphor and borneol, of nicotine, and of some of Werner's coloured co-ordination-compounds; but the record is limited to those researches in which the polarimeter has been used for some purpose beyond the mere characterisation and identification of optically-active compounds, since a register of the data provided for this purpose has already been made, in collaboration with Dr. E. E. Walker, in Vol. VII of the *International Critical Tables*. In this section, however, some problems of general interest are discussed, e.g. Hudson's "iso-rotation rules," the relative configuration of related compounds, and the phenomenon of mutarotation, discovered by Dubrunfaut in 1846, on which the author presented a report to the *British Association* in 1904 and another to the *Union International de Chimie* in 1930.

Part IV, *Theoretical Considerations*, includes an account of the optical rotatory power of crystals, "liquid crystals," and solutions, and is illustrated by diagrams showing the dissymmetric structure of crystals of sodium chlorate and of α and β quartz. This section is, however, devoted mainly to post-war work on rotatory dispersion in transparent and absorbing media, and to the attempts that have been made to express the absolute magnitude of the former and the relative magnitude of the latter rotations by means of mathematical equations. A fitting climax to this work is provided by a recent paper of Born, which the author had the honour of communicating to the Royal Society for publication in the *Proceedings*.

The author is indebted to several of his colleagues for help rendered in the preparation of the present volume. In particular, Dr. H. Hudson not only drafted the chapters in which his own researches are described, but wrote the whole of the section on "Polarimetry," which the author then revised and edited for publication. Dr. Hudson wishes to acknowledge the help which he received in this work from studying Bruhat's *Traité de Polarimétrie*, and from the writings of Descamps on Ultra-violet Polarimetry; we also wish to thank Dr. Descamps for reading the proofs of this section of the book. In the same way, Dr. C. B. Allsopp, in addition to contributing some of the data cited in the text, wrote the two chapters on Magnetic Rotatory Power; and Dr. H. G. Rule contributed the section on the Rotatory Power of Solutions. The author is also indebted to Mr. H. F. Willis, to Dr. Werner Kuhn, and to Professor Max Born for reading the proofs of the theoretical section, and for contributing many helpful suggestions. The courtesy and consideration shown by the printers and publishers also call for a word of grateful appreciation.

T. M. L.

UNIVERSITY CHEMICAL LABORATORY,
CAMBRIDGE, *March*, 1935.

TABLE OF CONTENTS

PART I—HISTORICAL AND GENERAL

To the Immortal Memory of

LOUIS PASTEUR,

and in grateful appreciation of the teaching and friendship of

PROFESSOR SIR WILLIAM JACKSON POPE

whose pioneer researches, fifty years after those of Pasteur, mark the beginning of a new period in the history of Stereochemistry

PART I.—HISTORICAL AND GENERAL.

CHAPTER I.

ROTATORY POLARISATION.

JEAN BAPTISTE BIOT, 1774–1862.
AUGUSTIN FRESNEL, 1788–1827.

Polarisation by Double Refraction.—The effects produced by the polarisation of light have been well known since CHRISTIAAN HUYGENS published in 1690 his *Treatise on Light*, "in which are explained the causes of that which occurs in reflection and in refraction, and particularly in the strange refraction of Iceland crystals." The Preface of this book (of which an excellent translation was provided by Professor Silvanus Thompson in 1912) [1] concludes with the statement that "there remains much more to be investigated touching the nature of light which I do not pretend to have disclosed, and I shall owe much in return to him who shall be able to supplement that which is here lacking to me in knowledge. The Hague, The 8th January, 1690."

After citing Römer's experiments on the satellites of Jupiter as evidence of the finite velocity of light, and indirectly of its undulatory character, Huygens uses his wave-theory to explain the laws of reflection, refraction and total reflection. He then passes on to a chapter, which occupies approximately two-fifths of the whole treatise, *On the Strange Refraction of Iceland Crystal*, as observed in 1669 by Erasmus Bartholinus. In this chapter the term DOUBLE REFRACTION was introduced (*ib.*, p. 62) as applying both to ordinary (rock) crystal and to Iceland crystal, since both crystals gave a double image of an object which was seen through them (Fig. 1). Of these two images, one was produced by ORDINARY REFRACTION by means of spherical waves, the other was produced by EXTRAORDINARY REFRACTION by means of waves which in Iceland spar were spheroidal, the axis of the spheroid being parallel to the principal axis of the crystal (Fig. 2). When two plates of Iceland spar were superposed in parallel positions only two images were produced, in the same way as by a single plate of equal total thickness. The same phenomenon was noticed when one crystal was rotated through an angle of 90° relatively to the other; but the ray which underwent ordinary

[1] Macmillan & Co., London, 1912.

refraction in the one crystal then suffered extraordinary refraction in the other, and conversely. When, however, one of the plates was rotated through an arbitrary angle, a new pair of images appeared,

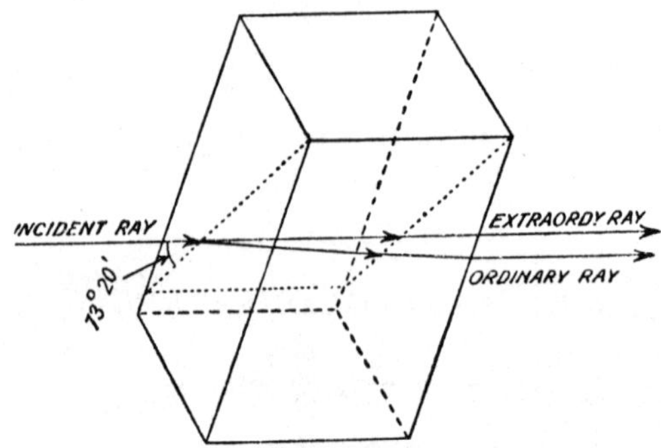

FIG. 1.—DOUBLE REFRACTION IN A RHOMB OF ICELAND SPAR. (After Huygens, 1690.)

When the angle of incidence is 73° 20′, the extraordinary ray passes straight through the crystal without suffering refraction, like a perpendicular ray through a plate of glass.

so that four images of variable intensity were seen. These coincided to give a single pair of images, only when the plates of spar were parallel, or were crossed by rotation through an angle of 90° (Fig. 3).

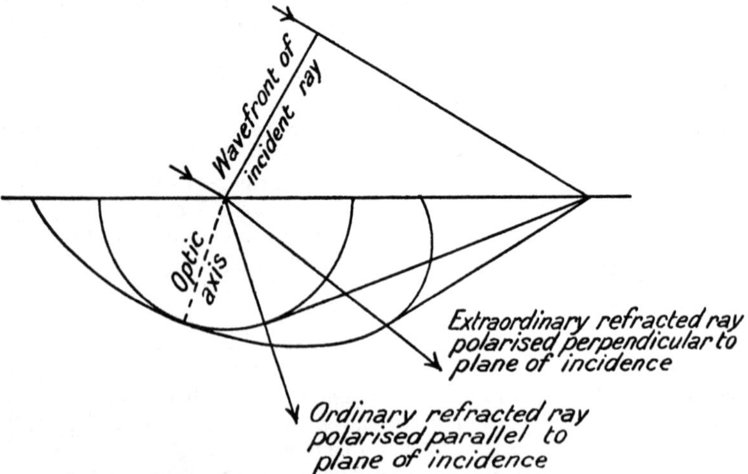

FIG. 2.—HUYGENS' WAVE-THEORY OF DOUBLE REFRACTION.

After describing the phenomena thus produced by rotating one crystal relatively to another, Huygens concludes " that the waves of light, after having passed through the first crystal, acquire a certain form or disposition in virtue of which, when meeting the texture of the second crystal, in certain positions, they can move the two different kinds of matter which serve for the two species of refraction; and when meeting the second crystal in another position are able to move only one of these kinds of matter. But to tell how this occurs, I have hitherto found nothing which satisfies me" (*ib.*, p. 94). Similar observations are recorded by NEWTON.[1]

Polarisation by Reflection.—A second method of imparting to the waves of light "a certain form or disposition" which enables them to give either a single or a double image when viewed through a crystal of Iceland spar, was discovered at the beginning of the nineteenth century by MALUS. His first observations were recorded in two papers published in the *Mémoires de la Société d'Arceuil* for 1809,[2] under the titles *Sur une propriété de la lumière réfléchie* and *Sur une propriété des forces répulsives qui agissent sur la lumière.* In the first of these papers Malus recalls that a beam of light, which has been doubly refracted in one rhomboid of Iceland spar, is not resolved further by a second rhomboid, when this is either parallel to the first or rotated through a right angle, but that in intermediate positions the two rays from the first crystal are each resolved into two rays by the second.

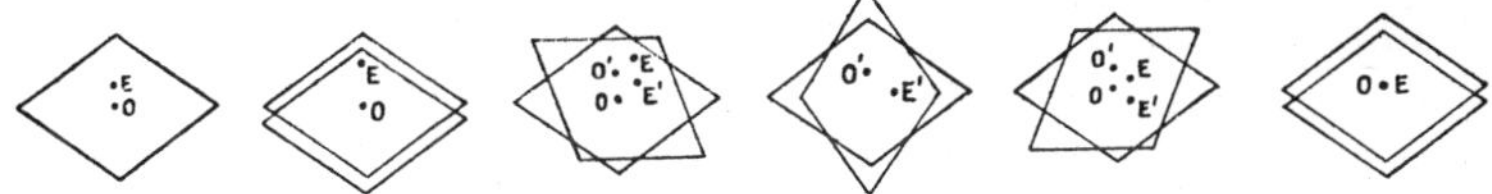

FIG. 3.—PRODUCTION OF DOUBLE IMAGES BY PLATES OF ICELAND SPAR.

> "The direct light differs from that which has been acted on by the first crystal, in that the former possesses constantly the property of being divided into two beams, whilst in the latter this property depends on the angle between the plane of incidence, and the principal section of the crystal" (*ib.*, p. 146).

In this paragraph, the PLANE OF INCIDENCE is one which contains the incident ray and a line perpendicular to the surface of the crystal, whilst the PRINCIPAL SECTION of the crystal is one containing the ray of light and a line parallel to the PRINCIPAL AXIS of the crystal, namely the axis of three-fold symmetry.

The same phenomena were observed when the second crystal was composed of carbonate of lead, or sulphate of barium, or when the first crystal was of sulphur and the second of rock crystal. Malus then records the newly-discovered property of reflected light as follows:

> "Light reflected from the surface of water at an angle of 52° 45′ to the vertical has all the characteristics of one of the

[1] NEWTON, *Opticks* (G. Bell, London, 1931), Book III, Part I, Query 29, p. 373.

[2] MALUS, *Mém. Soc. Arceuil*, 1809, **2**, 143–158, 254–267.

beams produced by the double refraction of a crystal of spar of which the principal section is parallel or perpendicular to the plane containing the incident and reflected rays, which we will call the plane of reflection.

" If this reflected ray falls on any crystal having the property of doubling the images, and of which the principal section is parallel to the plane of reflection, it will not be divided into two rays like a ray of direct light, but will be refracted entirely according to the ordinary law, as if the crystal had lost the property of doubling the images. If, on the other hand, the principal section of the crystal is perpendicular to the plane of reflection, the reflected ray will be refracted entirely according to the extra-

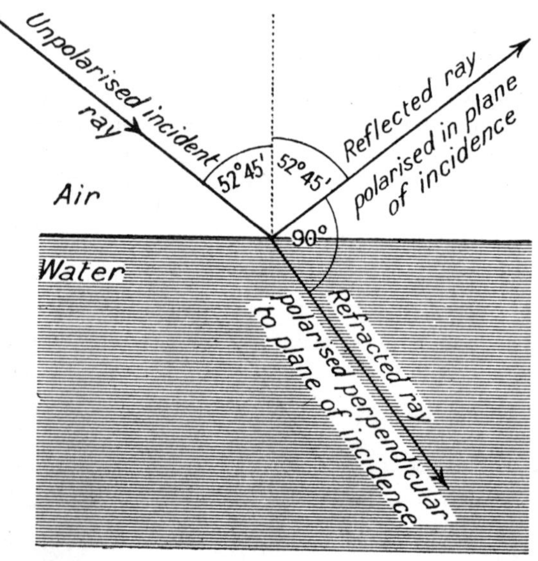

Fig. 4.—Polarisation of Light by Reflection and Refraction at the Surface of Water.

ordinary law. In the intermediate positions it will be divided into two beams according to the same law and in the same proportion as if it had acquired its new character by the influence of double refraction " (*ib.*, pp. 149–150).

The converse experiment was described as follows :—

" To analyse this phenomenon completely, I arranged a crystal with its principal section in a vertical plane and, after having divided a ray of light by double refraction, I received the two emergent beams on the surface of water at an angle of 52° 45′. The ordinary ray . . . was partially reflected, like a beam of direct light, but the extraordinary ray penetrated the liquid

> entirely, none of it escaping refraction. Conversely, when the principal section of the crystal was perpendicular to the plane of incidence, the extraordinary ray alone was partially reflected, and the ordinary ray was entirely refracted" (*ib.*, p. 150).
>
> "The angle at which the light suffers this modification, on reflection at the surface of transparent bodies, varies and in general is greater for those which refract the light more" (*ib.*, pp. 150–151).

A corresponding angle could be found for opaque substances, such as black marble or ebony, but polished metals did not show the phenomenon.

Three more papers, dealing with the polarisation of light by reflection from *glass* surfaces, were read before the *Institut de France*

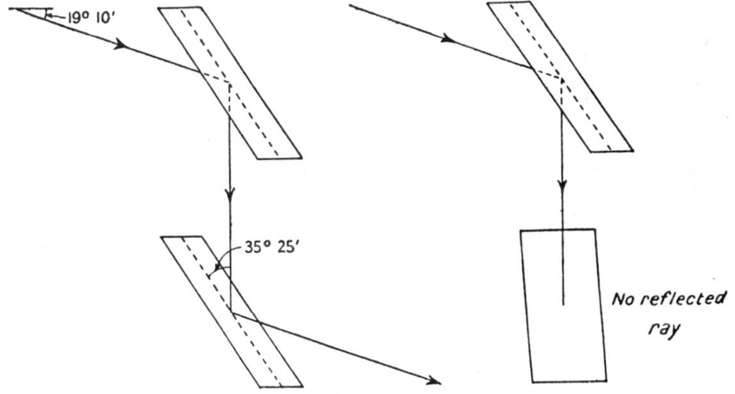

FIG. 5.—POLARISATION OF LIGHT BY REFLECTION FROM GLASS.

on March 11, 1811, May 27, 1811, and August 19, 1811, and published in the *Mémoires de l'Institut* for the year 1810,[1] under the titles:

Mémoire sur de nouveaux Phénomènes d'Optique (*ib.*, pp. 105–111).
Mémoire sur les Phénomènes qui accompagnent la réflexion et la réfraction de la Lumière (*ib.*, pp. 112–120).
Mémoire sur l'axe de réfraction des Cristaux et des Substances organisées (*ib.*, pp. 142–148).

The phenomenon of polarisation by reflection from surfaces of polished glass was described as follows:

> "Let us direct, by means of a heliostat, a ray of sunlight in the plane of the meridian, in such a way that it makes an angle of 19° 10′ with the horizon. Then let us fix an untinned mirror in such a way as to reflect the beam vertically downwards. If we place a second mirror below the first and parallel to it, it will

[1] MALUS, *Mém. Inst.*, 1810, **11**, Part II, 105–111, 112–120, 142–148.

> make an angle of 35° 25′ with the downward ray, which will be reflected again parallel to its first direction. In this case one will not observe anything remarkable; but if this second mirror is turned so that it faces east or west, without changing its inclination to the vertical ray, it will no longer reflect a single molecule of light, either at its first or at its second surface. If, whilst keeping its inclination to the vertical ray unchanged, its face is turned towards the south, it will begin anew to reflect the ordinary proportion of incident light. In intermediate positions, the reflection will be more or less complete, according as the reflected ray approaches more or less to the plane of the meridian. In these circumstances, in which the reflected ray behaves so differently, its inclination to the incident ray is kept constant. Thus, we see a vertical ray of light which, falling on a transparent body, behaves in the same way when the reflecting surface is turned to the north or south, and in a different way when this surface is turned to the east or west, although these surfaces are always inclined at an angle of 35° 25′ to the vertical direction of the ray.
>
> "These observations lead us to conclude that the light acquires in these circumstances properties which are independent of its inclination to the surface which reflects it, but are unique relatively to the sides of the vertical ray. These are the same for the south and north sides, and different for the east and west sides. Giving to these sides the names of poles, I will describe as POLARISATION the modification which gives to the light its properties relatively to these poles" (*ib.*, pp. 105–106).

The plane of the meridian, which included both the incident ray and the ray reflected from the first surface, was selected as the PLANE OF POLARISATION of the light, although this term was not used specifically at this stage.

The second memoir describes the partial polarisation of the ray transmitted through the glass, which was a mixture of unpolarised light with light polarised in a plane at right angles to the plane of polarisation of the reflected ray. It also describes the use of a series of parallel plates to produce more complete polarisation of the transmitted ray.

The third memoir describes the occurrence of double refraction in all crystals except those belonging to the cubic system and in all the vegetable and animal substances which were submitted to the test.

Nature of Polarised Light.—From experiments, made in 1816, in which it was found that the phenomenon of interference could only be observed with polarised light when the beams were polarised in the same plane,[1] AUGUSTIN FRESNEL concluded in 1821 that the vibrations of light were transverse and not longitudinal, and moreover

[1] FRESNEL, *Oeuvres*, **1**, 465; ARAGO and FRESNEL, *ib.*, 509–522; *A.C.P.*, 1819, [ii], **10**, 288–305.

drew from his theory the deduction[1] "that the vibrations of a polarised beam must be perpendicular to what is called its *plane of polarisation.*" Fresnel's theory of the nature of polarised light was set out clearly in a long *Mémoire sur la double Réfraction* read before the *Académie des Sciences* on November 26, 1821, January 22 and April 22, 1822,[2] in which he described the nature and origin of double refraction in biaxial crystals, such as topaz, where both rays of light behaved as extra-ordinary rays. The relevant passage, taken from an abstract of the paper,[3] is as follows :

> "The luminous vibrations take place only in directions parallel to the surface of the waves. . . . It suffices to admit in the ether a sufficient resistance to compression to understand the absence of longitudinal vibrations. . . . Polarised light is that in which the transverse oscillations take place constantly in one direction, and ordinary light is the bringing together and the rapid succession of an infinite number of systems of waves polarised in all directions. The act of polarisation does not consist in creating transverse vibrations, but in decomposing them along two fixed directions at right angles to one another, and separating the two systems of waves thus produced, either merely by their difference of velocity as in crystalline plates, or also by a difference of direction of the waves and of the rays, as in crystals cut into prisms or in thick plates of carbonate of lime; for, wherever there is a difference of velocity between the rays, refraction can make them diverge. Finally, according to the same theory, the plane of polarisation is the plane perpendicular to that in which the transverse vibrations take place" (*ib.*, pp. 265–266).

This type of polarisation was described in a Memoir read before the *Académie des Sciences* on December 9, 1822, as RECTILINEAR POLARISATION.[4]

Depolarisation by a Plate of Quartz.—On August 11, 1811, the astronomer ARAGO presented to the *Institut de France* a Memoir "*Sur une modification remarquable qu' éprouvent les rayons lumineux dans leur passage à travers certains corps diaphanes, et sur quelques autres nouveaux phénomènes d'optique.*"[5] In this Memoir, he described the colours which were produced when plates of mica, gypsum and quartz were inserted in the path of a beam of sunlight which had been polarised by reflection from a pile of glass plates, and was then analysed by viewing the solar images through a plate of spar. A plate of quartz, with surfaces perpendicular to the principal axis of the crystal, differed from the plates of mica or gypsum in that the

[1] FRESNEL, *Bull. Soc. Philomat.*, 1824, p. 150.
[2] FRESNEL, *Oeuvres*, **2**, 479–596.
[3] FRESNEL, *A.C.P*, 1825, [ii], **28**, 263–279; *Oeuvres*, **2**, 465-478.
[4] *Ibid.*, **1**, 744.
[5] ARAGO, *Mém. Inst.*, 1811, Part I, 93–134.

complementary colours of the two solar images were independent of the orientation of the quartz plate, but changed when the analyser was rotated.

In a converse experiment (following Malus) light which had passed through a rhomb of Iceland spar with its principal section vertical was analysed by reflection from a glass mirror. When a plate of quartz was inserted, instead of obtaining a single image, Arago observed the following phenomena :

> "Neither image disappears, since . . . the molecules of light of the two beams are not polarised in a single sense. The green rays of the first image and the red rays of the second, being polarised in the same way, will escape partial reflection simultaneously, and the images in the mirror will be one red and the other green. On turning the rhomboids, without changing the position of the plate, the images reflected by the mirror will pass successively, at each half-turn, through all the prismatic colours, with this peculiarity that the ordinary and extraordinary beams will always have complementary tints" (*ib.*, p. 123).

In these experiments, Arago was evidently studying the effects produced by rotation of the plane of polarisation in a plate of quartz, but without appreciating clearly the nature of the phenomenon. We may therefore concur with the verdict of a modern French worker in the same field who writes as follows :

> "Arago did not distinguish sharply between rotatory polarisation and chromatic polarisation. It is to Biot that we owe a complete study of these two phenomena, a study pursued during many years, with an untiring perseverance, and which enabled him to establish, in a definite way, the laws of polarisation." [1]

Rotatory Polarisation.—The classical experiments of JEAN BAPTISTE BIOT on rotatory polarisation are described in five Memoirs presented to the French *Académie des Sciences* from 1812 to 1838, and in a long series of notes in the *Comptes rendus*, many of them being abstracts of papers which were published *in extenso* in the *Annales de chimie et de physique* and elsewhere. The titles of the five principal Memoirs are as follows :

1. *Mémoire sur un nouveau genre d'oscillations que les molécules de la lumière éprouvent, en traversant certains cristaux.*[2] Read on November 30 and December 7, 1812, February 8, April 5 and May 31, 1813. The page-headings of this Memoir, which was not published until 1814, are *Recherches sur la polarisation de la lumière.*

2. *Sur les rotations que certaines substances impriment aux axes de polarisation des rayons lumineux.*[3] Read on September 22, 1818.

[1] LONGCHAMBON, *Bull. Soc. Fr. Min.*, 1922, **45**, 161.
[2] BIOT, *Mém. Inst.*, 1812, **1**, 1–372.
[3] BIOT, *Mém. Acad. Sci.*, 1817, **2**, 41–136.

3. *Mémoire sur la polarisation circulaire et sur ses applications à la chimie organique.*[1] Read on November 5, 1832.

4. *Methodes mathématiques et expérimentales pour discerner les mélanges et les combinaisons chimiques définies ou non définies qui agissent sur la lumière polarisée, suivies d'applications aux combinaisons de l'acide tartrique avec l'eau, l'alcool et l'esprit de bois.*[2] Read on January 11, 1836.

5. *Mémoire sur plusieurs points fondamentaux de mécanique chimique.*[3] Read on November 27, 1837.

Biot's Discovery of Rotatory Polarisation and Rotatory Dispersion.—Biot's detailed studies of the colours, which Arago had produced with plates of quartz, were described in his first (1812) Memoir,[4] which covers nearly 400 pages, and was read before the Institute in five sections extending over a period of six months.

Biot starts by defining a polarised ray, according to the experiments of Malus, as

> " a ray of which all the molecules [of light] are disposed so that they behave in precisely the same way, when they fall perpendicularly on a natural face of a rhomb of Iceland spar, of which the principal section is oriented in a given direction. Under these conditions the ray, if it is polarised, is not divided into two beams. . . . In addition, if the polarised ray falls upon a glass mirror, polished but not tinned, at an angle of 35° 45′ with the direction of the light, Malus has discovered that the quantity of light reflected is not the same for all positions of the glass relatively to the ray " (*ib.*, p. 5).
>
> " The simplest thing is to consider the molecules of the polarised ray as being all arranged in similar positions to one another " (*ib.*, p. 6).

The earliest parts of the Memoir are concerned with the behaviour of crystals of gypsum and of quartz cut parallel to the axis, and exhibiting only the ordinary phenomena of double refraction. The fifth part, however, read on May 31, 1813, bears the title *Expériences sur les plaques de cristal de roche taillées perpendiculairement à l'axe de cristallization.* The results of these experiments may be summarised as follows :

(i) When a beam of polarised light falls on a rhomb of Iceland spar, placed so that its principal axis coincides with the plane of polarisation of the light, all the light is refracted as an ordinary ray, and no extraordinary ray is developed.

(ii) If, however, a plate of quartz, cut perpendicular to the axis, is interposed in the path of the beam, a coloured extraordinary ray is developed.

(iii) The colour of this ray shows the tints of Newton's rings,

[1] Biot, *Mém. Acad. Sci.*, 1835, **13**, 39–175.
[2] *Ibid.*, 1838, **15**, 93–279.
[3] *Ibid.*, 1838, **16**, 229–396.
[4] Biot, *Mém. Inst.*, 1812, **1**, 1–372.

beginning with a violet tint in a plate 0·4 mm. thick, and finally becoming white when the thickness becomes excessive.

(iv) As Arago found, no change of colour is produced by rotating the plate of quartz in its own plane.

(v) If, however, the rhomb of spar is rotated, the colours change, e.g. to those produced by a thinner plate of quartz.

The angle through which the analyser must be turned to render the extraordinary ray as nearly colourless as possible was found to be proportional to the thickness of the plate, ranging from 9° 45′ for 0·400 mm. to 80° for 3·478 mm. (These numbers would now be described as the rotatory power of quartz for the NEUTRAL TINT, i.e. for the part of the spectrum to which the eye is most sensitive.) It therefore appeared that the plane of polarisation of the light had been rotated by its passage through the plate of quartz to an extent that was proportional to the thickness of the plate; but the simple hypothesis of a uniform rotation, depending only on the thickness of the plate, was rendered untenable by the fact that, when the angle of rotation of the analyser was rather less, the image was dull blue or reddish-violet, whereas on the other side of the minimum the colour became red or orange. Biot therefore concluded that

> "Luminous molecules of different kinds, which have passed through the plate of rock crystal, have, by the action of this plate, turned their axes of polarisation into different azimuths" (*ib.*, p. 247).
>
> "The violet molecules turn faster than the blue, the blue faster than the green, the green faster than the yellow, and so on to the red molecules which will be the slowest of all" (*ib.*, pp. 256–257).

In these terms Biot established simultaneously the phenomena of ROTATORY POLARISATION, i.e. the rotation of the plane of polarisation by passing through a suitable medium, and of ROTATORY DISPERSION, i.e. the unequal rotation of the plane of polarisation of light of different wave-lengths.

Optical Superposition.—In the same Memoir Biot records his discovery of a second form of quartz which rotated the plane of polarisation in the opposite direction to the first form. This discovery was followed by an experiment in which a plate of the second kind, 4·1105 mm. in thickness, was covered by a plate 4·005 mm. in thickness of the first kind. As a result of this superposition the extraordinary ray disappeared, when the analyser was set in its normal position.

> "On turning the rhomb, white images were produced, the intensity of which varied as if the polarised ray had not passed through any crystalline substance" (*ib.*, p. 267).

The compensation of the rotatory power of one form of quartz by the other was therefore complete when equal thicknesses were used.

Biot's Law of Inverse Squares.—In his first (1812) Memoir, Biot recorded an increase in the angle of rotation of the plane of polarisation as the colour of the light changed from red to violet. In his second (1818) Memoir, he describes a more detailed *Recherche de la loi de rotation des différents rayons simples, dans le cristal de roche* (*loc. cit.*, p. 48). In this research, Biot had at his disposal no source of monochromatic light, and his knowledge of wave-lengths was limited to the estimates which Newton had made of the wave-lengths at the limits of the seven simple colours. Biot's values for the rotatory power of quartz at these limits (*ib.*, p. 58) are set out in the following table, together with some more modern data collected from the measurements of Lowry and Coode-Adams.[1]

TABLE I.—ROTATORY DISPERSION IN QUARTZ.

(*a*) *Biot* (1817).		(*b*) *Lowry and Coode-Adams* (1927).	
Désignation du rayon simple.	Arc de rotation en degrés sexagsm.	Wave-lengths.	Rotation per millimetre.
Rouge extrême . . .	17°, 4964	Cd 6438·470	18·0243°
Limite du rouge et de l'orangé	20°, 4798	Na 5895·932	21·7010°
Limite de l'orangé et du jaune	22°, 3138	Cu 5782·158	22·6170°
Limite du jaune et du vert .	25°, 6752	Hg 5460·742	25·5384°
Limite du vert et du bleu .	30°, 0460	Cd 5085·822	29·7323°
Limite du bleu et de l'indigo .	34°, 5717	Zn 4722·164	34·8885°
Limit de l'indigo et du violet	37°, 6829	Cd 4678·163	35·6057°
Violet extrême . . .	44°, 0827	Hg 4358·342	41·5506°

The results of these observations were summarised as follows:

> "On observing successively the rotations of the different simple rays traversing the same plate, I found that they were unequal and increased with the refrangibility. The angles observed with several plates of different thickness were proportional amongst themselves for rays of the same kind, which testified to the accuracy of the results. It only remained then to compare the absolute values of the rotations in each plate, for the different simple rays. On doing this I recognised that they were inversely proportional to the squares . . . of the lengths of their vibrations in the system of waves" (*ib.*, pp. 48–49).

In reference to this law of Inverse Squares, $\alpha = k/\lambda^2$, Biot was able to affirm that

> "If it is not the natural law, it is sufficiently near to be substituted for it in all the observations" (*ib.*, p. 85).

[1] LOWRY and COODE-ADAMS, *Phil. Trans.*, 1927, A. **226**, 399.

Biot illustrated the law of inverse squares by means of a diagram (Fig. 37, p. 125) in which the abscissæ represent the square of the wave-length, while the ordinates show the thickness of quartz required to produce a given rotation. Thus a plate whose thickness is given by the horizontal line NN would produce a rotation of about 180° in the red and 360° in the violet, so that, if the polariser and analyser were "crossed," both ends of the spectrum would be extinguished together. This diagram is of special interest at the present day, in view of the modification of the inverse square law which was introduced some 80 years later by DRUDE (p. 120), who wrote $\alpha = k/(\lambda^2 - \lambda_0^2)$ in place of $\alpha = k/\lambda^2$. Drude's formula, which applies to a large number of organic compounds [1] can be tested most readily [2] by plotting $1/\alpha$ against λ^2 in exactly the same way as in Biot's diagram. There can therefore be little doubt that if observations of sufficient accuracy had been possible, Biot would have discovered immediately that the lines which he plotted do not converge towards the origin, but to a finite value of λ^2, namely the λ_0^2 of Drude's equation, and in this way would have discovered nearly a century earlier the law which is now used to express the rotatory dispersion of the simplest types of optically-active compounds.

Fresnel's Memoirs on the Constitution and Properties of Polarised Light.—Contemporaneously with Biot's Memoirs on Rotatory Polarisation, an important series of Memoirs on the constitution and properties of polarised light was published by AUGUSTIN FRESNEL over the period from 1816 to 1823. This series of Memoirs was reproduced in the first volume of the *Oeuvres complètes d'Augustin Fresnel* published in 1866, where they are numbered XV to XXX; but the Memoirs on Double Refraction, XL and XLVII in Vol. II, belong to the same period and are of equal importance in the development of the theory of polarisation. The Memoirs, which are often of great length, were usually preceded or followed by the publication of independent abstracts, and in certain cases these were the only form in which the Memoirs were at first available. Thus the Memoir (XXX) read on January 7, 1823, was not printed until 1831, when it was rediscovered amongst the papers of Fourier; and two essential Memoirs, XVI and XVII, read on November 10, 1818, and January 19, 1818, were not printed until 1866.

The key to this series of Memoirs is found in the conception of transverse vibrations, as set out in a passage which has already been cited (p. 7). This hypothesis was at first "not . . . regarded as a reality, but only as a manner of representing the facts"; [3] but it was developed as a fully-fledged theory, covering the whole field of polarisation, in the Memoir (No. XXVIII) which was read on December 9, 1822.

[1] LOWRY and DICKSON, *J.*, 1913, **103**, 1067; LOWRY and ABRAM, *J.*, 1919, **115**, 300; LOWRY and RICHARDS, *J.*, 1924, **125**, 1593, 2511; 1925, **127**, 238.

[2] LOWRY and DICKSON, *J.*, 1913, **103**, 1075.

[3] FRESNEL, *Oeuvres*, **1**, 736.

Fresnel's Discovery of Circular Polarisation.—Fresnel's theory of rotatory polarisation, which is still accepted universally at the present day, was developed as a sequel to experiments on the reflection of polarised light, first at exterior surfaces, and then at interior surfaces where total reflection was possible.

According to the LAW OF MALUS, a plane-polarised ray is equivalent to, and can be resolved into, two rays polarised in rectangular planes, their intensities being governed by the ordinary law of resolution of forces. Thus, when reflected at the polarising angle from a surface of glass or of a liquid, a ray polarised in an oblique plane is resolved into two rays, which are polarised parallel and perpendicular to the plane of incidence respectively. Since the former component alone is reflected, and the latter is refracted, the reflected ray is still polarised, but in a different plane to the incident ray, since it is always polarised in the plane of incidence.[1]

Reflection at an interior surface gave rise to similar phenomena, provided that a part of the light was also refracted; but total reflection, instead of giving a reflected ray polarised in the plane of incidence, resulted in a partial depolarisation of the light (*ib.*, p. 452). This apparent depolarisation was not, however, a reversion to ordinary light, since the polarisation was restored by a second reflection, *at the same angle of incidence but in a plane at right angles to the first.* The apparent depolarisation was greatest when the plane of polarisation was inclined at an angle of 45° to the plane of incidence, so that the components parallel and perpendicular to this plane would be of equal intensity. A second reflection, in the same plane of incidence, increased the apparent depolarisation to a maximum, since the ordinary and extraordinary rays, into which the reflected ray was resolved by a rhomb of calcite, were then of equal intensity at all orientations of the calcite, just as with ordinary unpolarised light. A third and fourth reflection in the original plane of incidence resulted, however, in a partial and then a complete repolarisation of the light, but in a plane at right angles to the original plane of polarisation (*ib.*, p. 454).

These phenomena were explained by the supposition "that the polarised light is divided, by total reflection, into two systems of waves, one polarised parallel to the plane of the incidence and the other perpendicularly, and separated by an interval of an eighth of a wave" (*ib.*, p. 473). Complete depolarisation therefore resulted when a *retardation of a quarter of a wave* had been produced by two total reflections; but, when this Memoir was presented on November 10, 1817, the nature of polarised light was still too obscure (*ib.*, p. 484) to permit of a clear exposition of the results of this retardation of phase. After the theory of transverse vibrations had been successfully established at the end of 1821, however, the accuracy of the

[1] FRESNEL, *Mémoire sur les modifications que la réflexion imprime à la lumière polarisée,* presented to the *Institut de France,* November 10, 1817; *Oeuvres,* **1**, No. XVI, 441–485, p. 445.

diagnosis was recognised in a Memoir (No. XXVIII) presented to the *Académie des Sciences* on December 9, 1822, by describing the phenomenon as CIRCULAR POLARISATION.[1]

Production and Properties of Circularly Polarised Light.—The method used to produce CIRCULARLY POLARISED LIGHT by two internal reflections, and the characteristics of this type of polarisation, as elucidated in the Memoirs on the reflection of polarised light, were set out concisely at the beginning of Fresnel's great *Mémoire sur la double réfraction que les rayons lumineux éprouvent en traversant les aiguilles de cristal de roche suivant des directions parallèles à l'axe*[2] as follows:

> " After having first polarised a beam of light, either by passing it through a rhomb of Iceland spar, or by reflection at an untinned mirror inclined at 35°, one introduces it into a parallelopiped of glass,* where it suffers successively at opposite faces two internal total reflections at an incidence of about 50°, and at a plane inclined at 45° to the original plane of polarisation. The angles between the faces of entry and exit of the parallelopiped and the two reflecting faces must be such that the former are almost perpendicular to the incident and emergent rays, in order that they may not exert any polarising action upon them " (*ib.*, p. 733).
>
> " The light leaving the parallelopiped of glass appears to be completely depolarised, since, if it is analysed with a rhomb of Iceland spar, it always gives two white images of equal intensity, in whatever azimuth one turns the principal section of the rhomb. Nevertheless it is not ordinary light, for, if it is passed through a thin plate of gypsum or quartz, and then analysed by a rhomb of Iceland spar, instead of the two white images given by ordinary light, two coloured images are seen, but of different tints from those which would have been developed in the same plates by ordinary polarised light. Yet another remarkable characteristic distinguishes the new modification from the polarisation of Malus and from unmodified light: it is that the light thus modified recovers all the properties of perfect polarisation when it suffers two total reflections at an incidence of 50° in the interior of a parallelopiped of glass: the plane of polarisation of the emergent rays is then inclined at 45° to the plane of reflection, to which one can give any direction whatever. Unmodified light, on the contrary, acquires no new property after two total reflections, whilst polarised light, as has just been stated, appears to be completely depolarised when it is analysed with a rhomb of Iceland spar, provided that the plane of reflection makes an angle of 45° with the original plane of polarisation " (*ib.*, pp. 733–734).

A month later, in a *Mémoire sur la loi des modifications que la réflexion imprime à la lumière polarisée*, read before the *Académie des*

[1] FRESNEL, *Oeuvres*, 1, 744. [2] *Ibid.*, 1, No. XXVIII, 731–751.

* See Fig. 110, p. 238.

Sciences on January 7, 1823,[1] the conditions were deduced under which a retardation of a quarter of a wave-length was to be expected, when light, polarised at an angle of 45° to the plane of incidence, was submitted to multiple internal total reflections in crown glass of refractive index 1·51. Calculation showed that the required retardation would be produced after

	two	total reflections	at an	angle of	48° 37′ or 54° 37′,
	three	"	"	"	43° 11′ or 69° 12′,
or	*four*	"	"	"	42° 20′ or 74° 42′.

Experiment showed that these angles were correct, but that the larger angles of incidence, being further removed from the boundary of the total reflection, gave better results.

Circular polarisation can also be produced by passing plane-polarised light through a plate of a double-refracting crystal, of such a thickness that there is a retardation of a quarter of a wave-length between the ordinary and extraordinary rays. If the plane of polarisation of the incident light bisects the angles between the planes of polarisation of the ordinary and extraordinary rays, these will be of equal intensity and will give rise to circular polarisation (see p. 239).

Fresnel's Theory of Circular and Elliptical Polarisation.* —We have seen that Fresnel regarded a ray of circularly polarised light as the resultant of two rays of plane-polarised light of equal intensity, polarised in rectangular planes, and differing in phase by a quarter-wave. In the Memoir (No. XXVIII, 1822) in which the terms *rectilinear* and *circular polarisation* were first introduced, he reserved the theoretical development of the problem for a supplement which was probably never written.[2] The promised theory was, however, outlined a month later in the concluding paragraph of the *Mémoire sur la loi des modifications que la réflexion imprime à la lumière polarisée* (No. XXX, 1823) as follows:

> "When the incident beam is polarised at 45° to the plane of reflection, the two systems of waves, polarised parallel and perpendicular to this plane, of which the light is composed, are of equal intensity; if by two or more total reflections a difference of path of a quarter-wave (or of an integer and an odd number of quarters) has been established between them, the molecules will describe little circles with a uniform velocity around their positions of equilibrium. If the difference of path is an even number of quarter-waves, they will describe straight lines. Finally, if this difference is not a whole number of quarter-waves, the curves described will be ellipses. When the difference of path is an odd number of quarter-waves, the curves will still be ellipses if the

[1] FRESNEL, *Oeuvres*, 1, No. XXX, 767–799. [2] *Ibid.*, 1, 744.

* For diagrams illustrating this theory see Fig. 46, p. 152, and Fig. 111, p. 239.

two systems of waves have not the same intensity, as would be the case if the incident light were not polarised at 45° from the plane of reflection, or if two systems of waves interfere under any conditions such that their planes of polarisation are not rectangular " (*Oeuvres*, **1**, pp. 798–799).

The theory thus sketched in brief outline was set out fully in the second *Mémoire sur la double réfraction.*[1] The relevant paragraphs were also published separately in 1824 under the title *Considérations théoriques sur la polarisation de la lumière.*[2] After citing evidence that " in polarised light the ethereal molecules cannot have any vibratory movement in the direction of the rays " and therefore " that ordinary light also contains only vibrations perpendicular to the rays " (*ib.*, p. 151), Fresnel passes on to consider the interference of rays originally polarised in a plane PP′ (Fig. 6), which are then resolved along two rectangular planes OO′ and EE′, e.g. by passing through a plate cut from a doubly-refracting crystal. For convenience the lines in the figure may be regarded as the directions of vibration of the ray. The law of Malus follows from the resolution of a velocity v along CP into $v \cos i$ along CO and $v \sin i$ along CE′, since the energies of the two beams are proportional to $\cos^2 i$ and $\sin^2 i$, which add up to unity. The following cases are then considered :

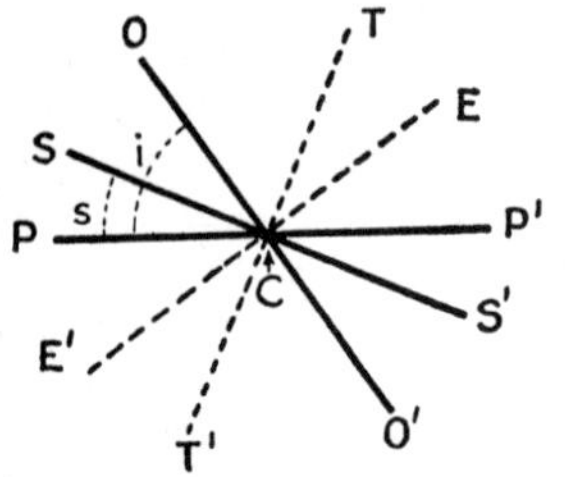

Fig. 6.—Illustrating Fresnel's Theory of Polarisation.

(*a*) " Suppose that the crystalline plate is so thin that there is no appreciable difference of path between the two emergent rays, or that it has a thickness such that the difference of path contains a whole number of waves, which is the same thing . . . ; their resultants will be parallel and will always be projected on PP′, since the components are always in the ratio $\cos i$ to $\sin i$. Thus the light produced by the reunion of two emergent beams will still be polarised, since all its vibrations will take place in parallel directions, and its plane of polarisation will be the same as that of the incident beam."

(*b*) " Suppose now that the difference of path of the ordinary and extraordinary beams, on emerging from the crystal, is a half-wave or an odd number of half-waves ; it is as if, with no difference of path, the sign of the absolute velocities of one of the two systems of waves were reversed ; . . . the resultant, instead of being directed along CP, will be directed along a line situated on the other side of CO, and making with it an angle equal to the

[1] Fresnel, *Oeuvres*, **2**, No. XLVII, §§ 6–16, pp. 496–507.
[2] Fresnel, *Bull. Soc. Philomat.*, 1824, 147–158.

angle i, between CO and CP. . . . Thus, the total light composed of the two emergent beams will still be polarised on leaving the crystal, since all its vibrations will be parallel to a constant direction; but its plane of polarisation, instead of coinciding with the original plane, as in the previous case, will be removed from it by an angle equal to $2i$" (*ib.*, pp. 154–155).

(*c*) "Let us consider now the case where the difference of path is no longer a whole number of half-waves . . . then the reunion of the two systems of waves will no longer present the characters of polarised light.

"If one wishes to calculate the curve described by the molecule . . . one can be certain in advance that the curve can only be an ellipse.

"This curve becomes a circle when, i being equal to 45°, the difference of path contains an odd number of quarter-waves, or in other terms, when the two systems of waves polarised at right angles are of the same intensity and differ in their path by an odd number of quarter-waves.

"Another remarkable peculiarity of the oscillatory movement in the same case, is that the velocity of the molecule is uniform.

"The molecules which, when at rest, are all on a perpendicular through C, . . . , form a spiral, of which the radius is that of the little circles described by the vibrating molecules, and of which the pitch is equal to the wave-length. If one turns this spiral uniformly on its axis . . . one will have a correct idea of the type of luminous vibration which I propose to call *circular polarisation* whilst describing as *rectilinear polarisation* that which was first noticed by Huygens in the double refraction of Iceland spar, and which Malus has reproduced by simple reflection at the surface of a transparent body" (*ib.*, pp. 155–157).

Fresnel's Theory of Rotatory Polarisation.—For Fresnel's theory of rotatory polarisation we must now turn back again to the celebrated Memoir (No. XXVIII) of December 9, 1822, on *Double Refraction along the Axis in Quartz*. In this Memoir he was able to demonstrate the reality of his hypothesis, that a ray of light, travelling along the axis of a crystal of quartz, is resolved into two circularly polarised rays of opposite sign, which travel with unequal velocities and can therefore be separated by a peculiar type of double refraction, in just the same way as the ordinary and extraordinary rays in calcite travel with unequal speeds and can therefore be separated by the more familiar type of double refraction.

"In view of the very feeble double refraction of quartz along its axis, the separation of the light into two distinct beams was effected by cutting a prism of crystal of which the faces of entry and exit were inclined equally to the axis and at an angle of 152° to one another. This prism was achromatised as well as possible by two prisms of glass—of which the refracting angles were much less than the half of 152°, because St. Gobain crown is more dispersive than quartz. . . . Perfect achromatism [was obtained] by replacing the two half-prisms of crown glass by two half-prisms of quartz, of which the double refraction along the axis was of opposite sign to that of the intermediate prism (compare Fig. 7). . . . In this way one obtains a very sensible separation of the two images, which one could increase by multiplying the number of prisms " (*Oeuvres*, **1**, 738–739).

" As a result of this double refraction . . . the rays which are circularly polarised from right to left do not traverse the crystal of quartz with the same speed as the rays which are circularly polarised from left to right, since . . . a difference of refraction between two beams always supposes a difference of speed.

" It results from the laws of interference of polarised rays, that one can always replace a beam of light, with a rectilinear polarisation, by two beams of equal intensity and circularly polarised from left to right and from right to left, since the resultant of these two beams is the equivalent of the incident beam. But since these component beams do not traverse the crystal with the same speed, they will differ in step [marche] in

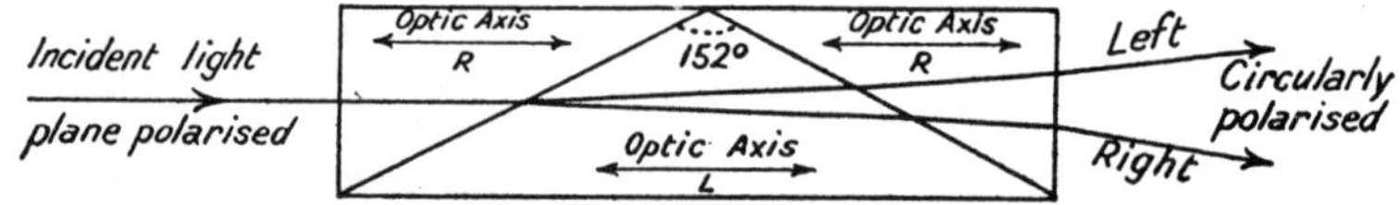

FIG. 7.—ILLUSTRATING FRESNEL'S EXPERIMENT ON CIRCULAR DOUBLE REFRACTION IN QUARTZ.

proportion to the length of the path in the crystal, and also in direction, if the refracting faces are not perpendicular to the beams. It is this divergence that we have rendered manifest by cutting the crystal into a prism " (*ib.*, pp. 748–749).

The manner in which the double refraction of circularly polarised light gives rise to rotatory polarisation is set out in the following paragraph :

" Let us consider what happens when a polarised beam traverses, in the direction of the axis, a plate of quartz cut perpendicular to the axis. Then the two circularly polarised beams, into which one can imagine the incident ray to be divided, will traverse the crystal with different velocities, but will not separate in direction ; only one will fall behind the other to an extent proportional to the length of the path of the crystal. But if the effect of this difference of step [marche] is calculated by the same rules of interference, for any given kind of ray, one will find that the resultant [ensemble] of the two beams must always show the characteristics of rectilinear polarisation, but that its plane of polarisation, instead of coinciding with that of the incident beam, will be separated from it by a certain angle, and that this angle must be proportional to the difference of step divided by the wave-length, and thus to the length of the path in the crystal, for a given kind of ray, as Biot concluded from his observations. The original plane of polarisation is deviated from right to left or from left to right, according as the beam which is circularly polarised from right to left traverses the crystal faster or slower than the beam which is circularly polarised from left to right" (*ib.*, p. 749).

Finally, the conditions under which a difference of phase can be established between two circularly polarised beams, in such a way as to give rise to rotatory polarisation, were described in the part of the later Memoir (No. XLVII) which was reprinted in the *Bull. Soc. Philomat.* for 1824,[1] as follows:

"There are certain refracting media, such as quartz in the direction of its axis, turpentine, essence of lemon, etc., which have the property of not transmitting with the same velocity circular vibrations from right to left and those from left to right. This may result from a peculiar constitution of the refracting medium or of its molecules, which produces a difference between the directions right-to-left and left-to-right; such, for instance, would be a helicoidal arrangement of the molecules of the medium, which would present inverse properties according as these helices were dextrogyrate or lævogyrate.

"The mechanical definition of circular polarisation which we have just given enables us to conceive how the singular double refraction of quartz along its axis may arise; it is that the arrangement of the molecules of the crystal does not present the same aspect from right to left and from left to right, so that the beams of light whose circular vibrations are right to left bring into play an elasticity or force of propagation somewhat different from that which is excited by the other beam, whose vibrations are from left to right" (*ib.*, pp. 157–158).

In this connection it is also of interest to quote the passage referred to by PASTEUR in 1860 when he stated at the conclusion of his first lecture on *Molecular Dissymmetry*[2] that "FRESNEL, with one of those flashes of genius, of which he had so many, had a sort of presentiment of this cause of rotatory dispersion." The passage runs as follows:

"Rock-crystal shows optical phenomena which cannot be reconciled with complete parallelism of the molecular lines, and which would seem to indicate a progressive and regular deviation of these lines in the passage from one layer of the medium to the next."[3]

[1] FRESNEL, *Bull. Soc. Philomat.*, 1824, 147–158.
[2] PASTEUR, *Alembic Club Reprints*, No. 14, p. 25, footnote.
[3] FRESNEL, *A.C.P.*, 1825, [ii], **28**, 279; *Oeuvres*, **2**, 477.

In this passage Fresnel, in general terms, anticipates by nearly a century the spiral or twisted type of structure to which Bragg and his colleagues, by determining the positions of the silicon and oxygen atoms with the help of X-rays [1] have been able to attribute the optical rotatory power of quartz (see Fig. 143, p. 343).

Optical Rotatory Power and Rotatory Dispersion of Organic Compounds.—In the *Bulletin de la Société Philomatique* of December, 1815, BIOT [2] recorded the substance of two communications to the *Institut de France* on October 15 and 30, 1815, in which he disclosed his discovery of rotatory polarisation in oils of turpentine, laurel and lemon, and in alcoholic solutions of camphor. Experiments on the rotatory dispersion of these organic compounds were described in the latter part of his second (1818) Memoir,[3] following the experiments which led to the discovery of the law of inverse squares in the case of quartz. The rotatory dispersions in oil of turpentine and in an aqueous solution of cane sugar were determined by the method of compensation, which he had already applied to the two forms of quartz. The conclusions drawn from these experiments were set out as follows : [4]

> " The evidence which I have just reported seems to me sufficient to prove that the law of rotation of different simple rays is the same for the liquid sugar as for rock-crystal and oil of turpentine. I have obtained the same result in all other substances which I have submitted to similar experiments, and further examples can be seen below. From this one can infer with great probability that the law is a general one, for otherwise the deviations from it should have shown themselves already, and have become obvious in cases so dissimilar as those which we have just examined."

The exact compensation, throughout the visible spectrum, of the rotatory power of cane sugar with that of a quartz plate of opposite sign, forms the basis of a series of technical saccharimeters (Chapter XV), in which the concentration of the sugar is deduced from the thickness of the column of quartz which is required to compensate its optical rotatory power for white light, or for the tinted light transmitted by the sugar solution.

Two later series of experiments on the rotatory dispersion of organic compounds may conveniently be referred to here in inverse chronological order. An appendix to the third Memoir [5] (pp. 174–175) describes the important discovery of the " inversion " of cane-sugar by hydrolysis with acids :

$$\underset{\text{Sucrose.}}{C_{12}H_{22}O_{11}} + H_2O = \underset{\text{Glucose.}}{C_6H_{12}O_6} + \underset{\text{Fructose.}}{C_6H_{12}O_6}$$

[1] BRAGG, *P.R.S.*, 1914, A. **89**, 575 ; BRAGG and GIBBS, *P.R.S.*, 1925 A. **109**, 405–427 ; GIBBS, *P.R.S.*, 1926, A. **110**, 443.
[2] BIOT, *Bull. Soc. Philomat.*, 1815, 190–192.
[3] BIOT, *Mém. Acad. Sci.*, 1817, **2**, 41–136.
[4] *Ibid.*, p. 114.
[5] *Ibid.*, 1835, **13**, 39–175.

Subsequent experiments [1] showed that the rotatory power of cane-sugar and of invert sugar provided an exact compensation for one other, just as in the case of cane-sugar and quartz. When, however, Biot attempted in 1836 to compensate the rotatory powers of lævorotatory turpentine and dextrorotatory oil of lemon, either in two separate columns or mixed together in a single polarimeter tube, he found that "the compensation of the rotations although very close was neither complete nor general for all the rays." [2] The Law of Inverse Squares was therefore only an approximation in at least one of the liquids whose rotatory powers he thus failed to compensate.

Anomalous Rotatory Dispersion of Tartaric Acid.—Biot's third (1832) Memoir, *Sur la polarisation circulaire et sur ses applications à la chimie organique* [3] includes a record of the optical rotatory power of a large number of essential oils and syrups. It also contains, however, a reference to the optical rotatory power of tartaric acid. Thus, in a table facing page 169, Biot records that an aqueous solution of tartaric acid, containing approximately 50 per cent. by weight of the acid, gave a rotation of $+8{\cdot}5°$ (for white light with its optical centre of gravity in the yellow), but that its rotation was "stronger for the less refrangible rays."

In the fourth (1836) Memoir [4] the anomalous rotatory dispersion of tartaric acid is set out as a solitary exception to the Law of Inverse Squares as follows:

> "In different active substances, the rotatory power exerted upon the different simple rays is unequal. In all, with the single exception up to the present of tartaric acid, this inequality follows the same law, rendered evident by the composite colours which appear on analysing white polarised light transmitted through a thickness inversely proportional to the rotatory power of the system. The exception in this respect presented by tartaric acid is the more remarkable, since its compounds with salt-forming bases, and even with boric acid, have rotatory powers which conform to the general law, at least within the limits of accuracy which I have been able to reach on comparing their succession of tints with those produced by all the substances" (*ib.*, p. 99).

In a later part of the same Memoir (*loc. cit.*, p. 236) the following data are tabulated for the rotatory power at 26°, of 1003 mm. of an aqueous solution containing 34·27 per cent. by weight of tartaric acid, for the seven primary colours:

red	violet	orange	indigo	yellow	blue	green
38° 7′ 11″	39° 38′ 3″	40° 29′ 14″	42° 8′ 55″	42° 57′ 29″	44° 39′ 47″	46° 10′ 37″

[1] Biot, *A.C.P.*, 1844, [iii], **10**, p. 35, footnote.
[2] Biot, *C.R.*, 1836, **2**, 542–543.
[3] Biot, *Mém. Acad. Sci.*, 1835, **13**, 39–175.
[4] *Ibid.*, 1838, **15**, 93-279.

These data show clearly the anomaly of a maximum rotation in the green, but were not made the subject of any more complete investigation.

> "I have not pushed this research further, since I propose to determine later the true rotations of the different rays derived from the light of a prism, as well as their relative tendency to be absorbed by the solutions which we are considering. But the close concordance of the preceding values is sufficient to demonstrate the order of the rotations which they express, and that is for the moment my only goal" (*ib.*, p. 237).

It is, however, of interest to recall that, a few months later in the year 1836, Biot produced a similar anomalous sequence of colours by the incomplete compensation of lævorotatory oil of turpentine and dextrorotatory oil of lemon, and in this way provided at least one possible explanation of the origin of anomalous rotatory dispersion, namely, by the superposition of two normal rotations of opposite sign but of unequal dispersion.

Definition of Specific and Molecular Rotations.—In his fourth (1836) Memoir[1] Biot introduced (*loc. cit.*, p. 162), under the title *le pouvoir de rotation moleculaire*, the modern definition of the SPECIFIC ROTATORY POWER $[\alpha]$, of a liquid, or of a mixture or solution, namely:

$$[\alpha] = \frac{\alpha}{l\delta} \quad \text{or} \quad \frac{\alpha}{l\epsilon\delta},$$

where α is the observed rotation in degrees,
l is the length of the column in decimetres,
δ is the density of the liquid, and
ϵ is the fraction by weight of the optically-active compound.

Biot introduced the convention of expressing the length of the column in decimetres "in order that the significant figures may not be uselessly preceded by two zeros," so that our specific rotatory powers are *ten times greater* than in c.g.s. units. On the other hand, the MOLECULAR ROTATORY POWER $[M]$ of a compound, is now defined for similar reasons by the equation $[M] = \frac{M}{100}[\alpha]$ (instead of $M[\alpha]$) where M is the molecular weight; this quantity is therefore *ten times smaller* than in c.g.s. units.* The wave-length and temperature are now indicated by subscript and superscript signs, e.g. $[\alpha]_D^{20^\circ}$, $[M]_{5461}^{20^\circ}$.

Variations of Specific Rotatory Power of Tartaric Acid.—The quantity $[\alpha]$, as defined in the preceding paragraph, was introduced by Biot as a means of expressing the influence on the rotatory power of tartaric acid (for light of different colours) of solvents or reagents,

[1] BIOT, *Mém. Acad. Sci.*, 1838, **15**, 93–279.

* F. M. Jaeger uses the c.g.s. system and gives numbers 10 times larger than those usually deduced; and C. S. Hudson uses $[M] = M[\alpha]$, so that his numbers are 100 times larger.

such as water, methyl alcohol, ethyl alcohol, sulphuric acid and boric acid. He had already established in his third (1832) Memoir [1] that sugar rotates the plane of polarisation both in solution and in the solid amorphous state, as in barley sugar (see below, p. 104) and that the optical rotatory power of turpentine persists in the vapour state (p. 102). Biot therefore held that rotatory power was a specific property of the molecule, which should be independent of its concentration.

> " If the molecular effect of the acid remains the same in the different proportions of water in which it is dissolved, so that it is merely disseminated through it (as is proved by other experiments, in which the mere distance of the active particles did not influence their specific rotatory action), the quantity $\alpha/l\epsilon\delta$ should remain constant for all the different solutions. But, if it changes with the proportion of water, one must conclude that the specific action of the acid is modified by this proportion, either because its molecular constitution is altered, or more probably because it combines with a proportion of the water which is present and thus forms a new molecule with a different rotatory power inherent in its constitution." [2]

This test was applied immediately to two solutions of tartaric acid which had given a strict proportionality between α and l, both for the nearly monochromatic light transmitted by a deep red copper glass, and (less accurately) for observations of the minimum of colour in a beam of white light (i.e. for the ." sensitive tint "), but which had given a different ratio between the colours in the two solutions.

	t.	δ.	ϵ.	l.	α.	$[\alpha]$.
Solution A.	26°	1·16919	0·3427	1515	58·125°	9·55402°
Solution B.	25°	1·32307	0·59992	912·5	44·809°	6·18666°

The wide variation in $[\alpha] = \alpha/l\epsilon\delta$ is immediately obvious.

The general results of the investigation of aqueous solutions of tartaric acid are shown diagrammatically in Fig. 28, page 91, where the abscissæ show the proportion of water in the solutions and the ordinates show the rotations of the solutions for the seven primary colours. In the text corresponding with this series of figures, Biot claims to have established a linear relationship between specific rotatory power and concentration (which, however, was afterwards shown not to be precise), namely,

$$[\alpha] = \frac{\alpha}{l\epsilon\delta} = A + Be,$$

where e is the proportion of water in the solution. Since the rotatory power increases as e increases, Biot concluded that a molecular group of higher rotatory power is formed by the union of the acid and water,

[1] Biot, *Mém. Acad. Sci.*, 1835, **13**, 39-175. [2] *Ibid.*, 1838, **15**, 161.

and his observations showed further that the rotatory power of this group must increase with rising temperature, provided that the heating was not sufficient to decompose it. Conversely, since A was already negative at 6·8° C., it could be predicted that a REVERSAL OF SIGN might be observed in concentrated solutions of tartaric acid at low temperatures.

Action of Acids and Bases on the Rotatory Power of Tartaric Acid.—In his fifth (1837) Memoir,[1] Biot made a detailed study of the action of acids and alkalis on the rotatory power of tartaric acid. His prediction then took a more precise form, since it was shown that the limiting rotatory power for red light when $e = 0$ should be positive above 23° C., and negative below this temperature. The preparation of a glassy amorphous form of tartaric acid was realised some years later by LAURENT, and Biot was then able (see below, p. 104) to demonstrate the accuracy of his forecast, as well as to record the fact that the rotatory power of tartaric acid is increased by the addition of boric acid when in the glassy state as well as in solution.

[1] BIOT *Mém. Acad. Sci.*, 1838, **16**, 229–396.

CHAPTER II.

MOLECULAR DISSYMMETRY.

LOUIS PASTEUR, 1822–1895.

Hemihedrism and Rotatory Power of Quartz.—A typical crystal of quartz consists of a hexagonal prism surmounted at each end by a hexagonal pyramid. This simple figure (Fig. 8) is highly symmetrical. In addition to a CENTRE OF SYMMETRY, it has seven PLANES OF SYMMETRY, one perpendicular to the principal axis of the crystal, and six others passing through it. It has an AXIS OF SIX-FOLD SYMMETRY coinciding with the principal axis of the crystal, since the figure is superposed upon itself six times during a complete revolution about this axis; and there are also six AXES OF TWO-FOLD SYMMETRY, perpendicular to the six-fold axis, since there are six lines about which the crystal can be rotated so as to be superposed on itself by merely turning it upside down.

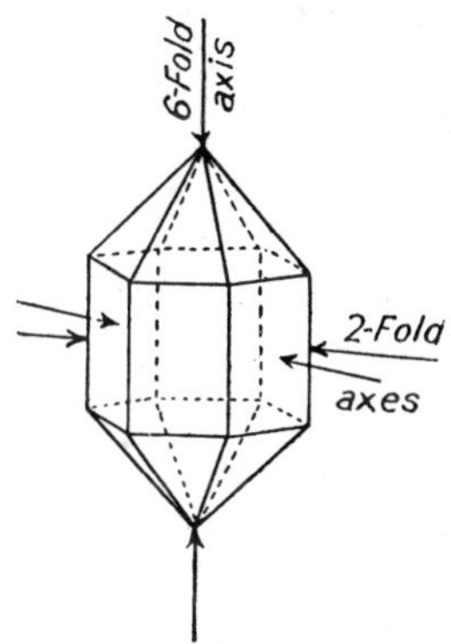

FIG. 8.—HOLOHEDRAL HEXAGONAL CRYSTAL.

Three 2-fold axes pass through the *edges* of the crystal and three through the centres of the *faces*.

At the beginning of the nineteenth century, however, HAUY noted the presence of certain small facets which reduced the symmetry of the crystals very considerably (Fig. 9). These facets were found only on alternate corners of the crystal and were therefore described as HEMIHEDRAL. These hemihedral crystals possess no *plane* or *centre* of symmetry; but they still possess four *axes* of symmetry, since the main crystallographic axis is now an AXIS OF THREE FOLD SYMMETRY, and there are three AXES OF TWO-FOLD SYMMETRY perpendicular to it. The hemihedral facets are distributed in such a way as to give rise to two forms of quartz, which cannot be superposed on one another, but are related in the same way as an object and its image in a mirror. The two hemihedral forms of quartz were distinguished as left and right, but it was not until 1820 that HERSCHEL, in a paper *On the rotation impressed by plates of rock-crystal on the planes of polarisation of the rays of light, as connected with certain peculiarities in its crystallisation*[1] correlated Hauy's crystallographic observation with Biot's physical observations and showed that the two crystalline forms of quartz rotated the plane of polarisation to the right and to the left respectively.

Until recently two opposite conventions existed as to the sign of the rotatory power of quartz. In the case of organic compounds such as *d*-camphor, chemists defined a dextrorotatory substance as producing a clockwise rotation of the plane of polarisation, *as viewed from the eye-piece of the polarimeter.* Physicists, however, defined dextrorotatory quartz as producing a clockwise rotation *when viewed along the beam of light*, so that *d*-camphor and dextro-quartz were opposite in sign. The chemists' convention has, however, now been adopted universally. It applies to circular polarisation as well as to optical rotation, so that dextro-rotation always implies a left-handed spiral in the light.

Dissymmetry and Asymmetry.—The term DISSYMMETRY was used by PASTEUR in 1848 to describe hemihedral crystals of a tartrate

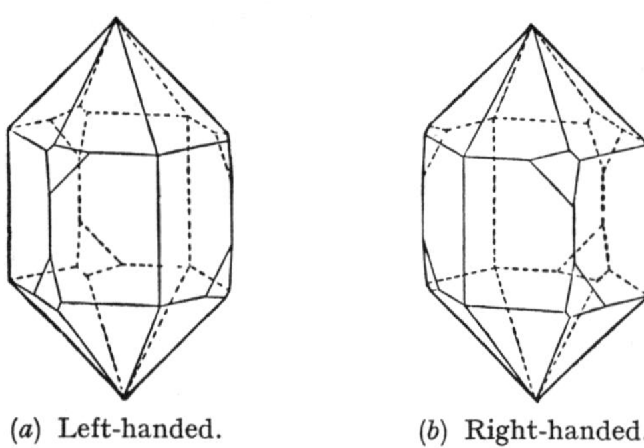

(*a*) Left-handed. (*b*) Right-handed.

FIG. 9.—HEXAGONAL CRYSTALS OF QUARTZ.

Two enantiomorphous forms of the crystal are shown, which are " mirror-images " of one another.

" which differ only as an image in a mirror differs in its symmetry of position from the object which produces it." [2] Hemihedral crystals are, however, not necessarily dissymmetric, since the latter term is restricted to those systems in which the symmetry has been degraded * to a point at which the figure is no longer superposable on its image in a mirror (*ib.*, p. 451). Pasteur therefore distinguishes sharply between SUPERPOSABLE HEMIHEDRISM, which is observed in Iceland spar, but is not associated with any rotatory polarisation in the crystal, and NON-SUPERPOSABLE HEMIHEDRISM, as observed in rock-crystal, where it is linked indissolubly with the rotatory power of the crystal, according to the relationship established by Herschel.

[1] HERSCHEL, *Trans. Camb. Phil. Soc.*, 1822, **1**, 43–52.
[2] PASTEUR, *Oeuvres*, p. 78 ; *A.C.P.*, 1848, [iii], **24**, 457.
* Compare the use of the prefix *dis* in words such as discoloured and displaced.

Dissymmetric figures are not necessarily ASYMMETRIC, since they may possess a complex system of *axes* of symmetry ; but dissymmetry, as defined by the formation of a non-superposable image in a mirror, is only possible in figures which possess *no plane or centre of symmetry.* It is also necessary that the figure should have *no alternating axis of symmetry*, a property which depends (like a knight's move in chess) on two operations, since a rotation and reflection are needed to bring each facet to the position occupied by a similar facet.

Molecular Dissymmetry.—Pasteur's views on dissymmetry are expressed very clearly in the second of two lectures "*Recherches sur la Dissymétrie Moléculaire des Produits Organiques Naturels*," delivered on January 20 and February 3, 1860, before the *Société Chimique de Paris*, and published by them in a volume entitled "*Leçons de chimie professées en* 1860." * The relevant passage is as follows :

> "When we study material things of whatever nature, as regards their forms and the repetition of their identical parts, we soon recognise that they fall into two large classes of which the following are the characters. Those of the one class, placed before a mirror, give images which are superposable on the originals ; the images of the others are not superposable on their originals, although they faithfully reproduce all the details. A straight stair, a branch with leaves in double row, a cube, the human body—these are of the former class. A winding stair, a branch with the leaves arranged spirally, a screw, a hand, an irregular tetrahedron—these are so many forms of the other set. The latter have no plane of symmetry" (*A.C.R.*, **14**, p. 26).

In the case of quartz, which loses its optical rotatory power when dissolved or fused, the dissymmetry depends on the structure of the crystal and not on the molecules of silica.

> "Imagine a spiral stair whose steps are cubes, or any other objects with superposable images. Destroy the stair and the dissymmetry will have vanished. The dissymmetry of the stair was simply the result of the mode of arrangement of the component steps" (*ib.*, p. 30).

In the case of organic substances, which are optically active in solution, this explanation cannot be given, and Pasteur attributes their optical activity to MOLECULAR DISSYMMETRY. This phrase, which is introduced in the preface to the 1848 paper in the form "une dissymétrie dans les molécules"[1] has been selected as the title of the first volume of his collected works, and is inscribed as a key-word in his Mausoleum in the Pasteur Institute in Paris. The general proposition that molecules, like all other geometrical figures, must fall into two classes—those with superposable images and those with

* An English translation has been issued by the Alembic Club as No. 14 of its series of reprints ; but the distinction between asymmetry and dissymmetry has been ignored in the translation.

[1] PASTEUR, *Oeuvres*, p. 66 ; *A.C.P.*, 1848, [iii], **24**, 443.

non-superposable images—was shown to be in accord with actual fact, and was demonstrated in a remarkable way by the experiments on tartaric acid and the tartrates which are summarised below.

Isomerism of Tartaric and Racemic Acids.—(*a*) *Tartaric acid* was discovered by SCHEELE in 1769.[1] Its optical activity was observed in 1815 by Biot, who recorded its anomalous rotatory dispersion in 1832, and in 1835 and 1837 made a detailed study of the influence of water, of alcohol, and of various acids and bases upon its rotatory power.

(*b*) *Racemic acid* was separated from argol in 1820 by KESTNER, and investigated in detail in 1826 by GAY LUSSAC [2] who showed that it was identical in composition with tartaric acid, although it differed widely from it in its physical properties. It thus provided one of the earliest examples of ISOMERISM to be established, and was cited by BERZELIUS when defining this term in 1830. It was examined by BIOT [3] and shown to be optically inactive.

Crystalline Form of Tartrates and Paratartrates.—The crystalline forms of tartaric and racemic acids and their salts were examined by DE LA PROVOSTAYE in 1841. They were studied in fuller detail by PASTEUR, who discovered the existence of hemihedral facets in the tartrates, but—with one vital exception—not in the *racemates* or *paratartrates*, as he called them, in view of their close relation to the tartrates. The experiments were described in outline in the *Comptes rendus* for May 15, 1848,[4] and in detail in a Memoir *Recherches sur les relations qui peuvent exister entre la forme cristalline, la composition chimique et le sens de la polarisation rotatoire.*[5] His lectures, delivered in 1860, include the following references to this work:

> "When I began to devote myself to special work, I sought to strengthen myself in the knowledge of crystals, foreseeing the help that I should draw from this in my chemical researches."
>
> "I crystallised tartaric acid and its salts, and investigated the forms of the crystals. All the tartrates which I examined gave undoubted evidence of hemihedral faces."
>
> "This observation would probably have remained sterile without the following one. . . . There is a kind of semi-isomorphism among all the tartrates. . . . Now if we compare the disposition of the hemihedral faces on all the prisms of the primitive forms of the tartrates, when they are oriented in the same manner, this disposition is found to be the same. These results, which have been the foundation of all my later work, may be summed up in two phrases: the tartrates are hemihedral, and that in the same sense."

[1] See KOPP, *Geschichte der Chemie*, IV, 349.
[2] GAY LUSSAC, *Schweigger's J.*, 1826, **48**, 381.
[3] BIOT, *A.C.P.*, 1838, [ii], **69**, 22.
[4] PASTEUR, *C.R.*, 1848, **26**, 535–538.
[5] PASTEUR, *A.C.P.*, 1848, [iii], **24**, 442–459.

"I also occupied myself with the examination of the crystalline forms of paratartaric acid and its salts. These substances are isomeric with the tartaric compounds, but had all been found by BIOT to be inactive towards polarised light. None of them exhibited hemihedry" (*A.C.R.*, **14**, pp. 14, 15, 17).

Pasteur's Resolution of Sodium Ammonium Racemate into Optically-active Forms.—Pasteur encountered a remarkable anomaly in the case of *sodium ammonium tartrate*, where the active and inactive salts were reported by MITSCHERLICH to be identical in crystalline form and other properties.

"I must first place before you a very remarkable note by MITSCHERLICH which was communicated to the *Académie des Sciences* by BIOT. [C.R., 1844, **19**, 719-725.]

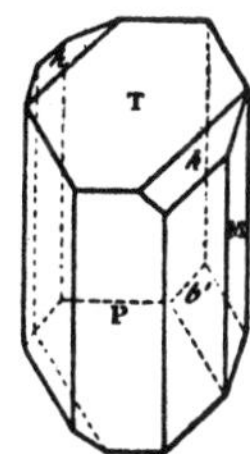

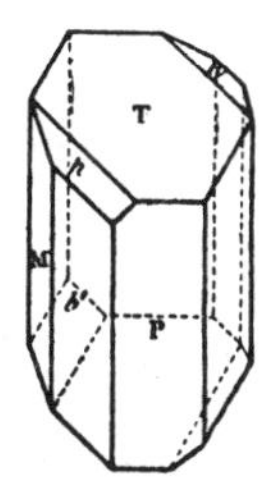

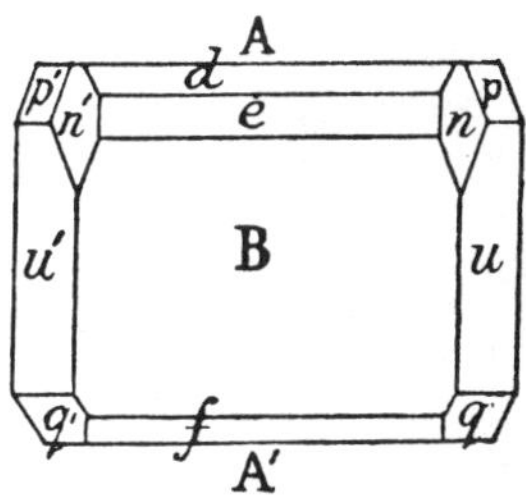

(*a*) Dextrorotatory. (*b*) Lævorotatory. (*c*) Inactive.

FIG. 10.—HEMIHEDRAL CRYSTALS OF SODIUM AMMONIUM TARTRATE (Pasteur). HOLOHEDRAL CRYSTAL OF SODIUM AMMONIUM RACEMATE (Scacchi[1]).

'The double paratartrate and the double tartrate of soda and ammonia have the same chemical composition, the same crystalline form with the same angles, the same specific weight, the same double refraction, and consequently the same inclination in their optical axes. When dissolved in water their refraction is the same. But the dissolved tartrate deviates the plane of polarised light, while the paratartrate is indifferent, as has been found by BIOT for the whole series of those two kinds of salts.'

"I hastened therefore to re-investigate the crystalline form of MITSCHERLICH's two salts. I found, as a matter of fact, that the tartrate was hemihedral, like all the other tartrates which I had previously studied, but, strange to say, the paratartrate was hemihedral also. Only the hemihedral faces, which in the tartrate were all turned the same way, were inclined in the paratartrate sometimes to the right and sometimes to the left. I carefully separated the crystals which were hemihedral to the right from those hemihedral to the left, and examined their solutions separately in the polarising apparatus. I then

[1] SCACCHI, *Rendiconti Accad. Sci. Fis. Mat., Napoli*, 1865, **4**, 250–263.

saw with no less surprise than pleasure that the crystals hemihedral to the right deviated the plane of polarisation to the right, and that those hemihedral to the left deviated to the left" (*A.C.R.*, **14**, pp. 17, 19).

In these epoch-making experiments, Pasteur succeeded in resolving a salt, which was really a mixture of equal parts of dextrorotatory and lævorotatory sodium ammonium tartrate, into its optically-active components, by the simple device of picking out by hand the hemihedral crystals of opposite types. The experiment was repeated in BIOT's laboratory and under his close supervision.[1] The Memoir then formed the subject of a long report by Regnault, Balard, Dumas and Biot, who recommended its approval to the Academy. It is noteworthy that the sodium ammonium salt is (with the possible exception of the sodium potassium salt [2]) the only inactive tartrate which can be resolved in this way, since the two forms of tartaric acid and the two forms of all its other salts unite to form crystalline racemic acid and racemates, which exhibit no hemihedrism or optical activity, and may be classified as "double salts" of the *d* and *l* forms, comparable with the alums or with the familiar double sulphates of the Schönite series. Even more remarkable is the fact that the two sodium ammonium tartrates unite to form a racemate at all temperatures above 26° C., so that Pasteur's discovery would probably not have been made if he had been working in a tropical climate.

Pasteur [3] records the want of a word in chemical language to express the concealment of optical activity by the neutralisation of opposite forms. The products thus formed are now generally known as RACEMIC MIXTURES or RACEMIC COMPOUNDS.

Relationship between Tartaric and Racemic Acids.—The relationship between tartaric acid and paratartaric or racemic acid, which is described in the preceding paragraph, was established by PASTEUR in a second Memoir, *Recherches sur les propriétés spécifiques des deux acides qui composent l'acide racémique.*[4] The following extracts are taken from the 1860 lectures:

"Let us return to the two acids furnished by the two sorts of crystals deposited in so unexpected a manner in the crystallisation of the double paratartrate of soda and ammonia."

"One of them deviates to the right, and is identical with ordinary tartaric acid. . . . The other deviates to the left, like the salt which furnishes it. The deviation of the plane of polarisation produced by these two acids is rigorously the same in absolute value. The right acid follows special laws in its deviation, which no other active substance had exhibited. The

[1] PASTEUR, *A.C.R.*, **14**, 21.
[2] WYROUBOFF, *Bull. Soc. Chim.*, 1884, [ii], **41**, 212; 1886, [ii], **45**, 52; JUNGFLEISCH, *ib.*, 1884, [ii], **41**, 222.
[3] PASTEUR, *A.C.R.*, **14**, 32.
[4] PASTEUR, *A.C.P.*, 1850, [iii], **28**, 56–99; *C.R.*, Sept. 17, 1849, **29**, 297–300.

left acid exhibits them, in the opposite sense, in the most faithful manner, leaving no suspicion of the slightest difference."

"That paratartaric acid is really the combination, equivalent for equivalent, of these two acids, is proved by the fact that, if somewhat concentrated solutions of equal weights of each of them are mixed, their combination takes place with disengage-

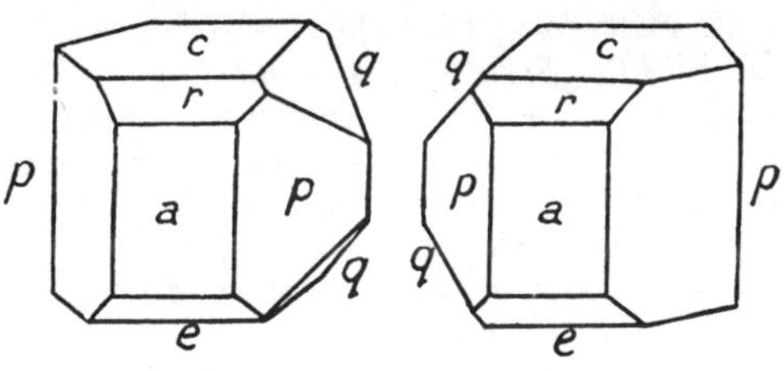

(*a*) Dextrorotatory. (*b*) Lævorotatory.

FIG. 11.—HEMIHEDRAL CRYSTALS OF TARTARIC ACID (Pasteur).

ment of heat, and the liquid solidifies immediately on account of the abundant crystallisation of paratartaric acid, identical with the natural product" (*A.C.R.*, **14**, p. 22).

Conversion of Tartaric Acid into Racemic Acid and into Mesotartaric Acid.—The origin of racemic acid presented a very difficult problem. In 1820 KESTNER had obtained several hundred kilograms of the acid, but no further supplies became available during the next thirty years, and the material used by Pasteur was a part of the original sample supplied to him by Kestner himself.[1] The quest for the origin of racemic acid is described by Pasteur in a *Notice sur l'origine de l'acide racémique.*[2] Statements that the inactive acid had been produced commercially in England proved to be unfounded; but needles of racemic acid were found in the interstices of large, clear crystals of tartaric acid in big lead-lined tanks in a factory in Saxony; and tiny crystals of racemic acid (which had been mistaken for potassium sulphate) were found on stocks of tartaric acid in a Vienna factory, but in such minute quantities that several hours were required to collect a few decigrams. Further investigation showed (i) that racemic acid only appeared when using crude instead of semi-refined tartar, (ii) that even then it only began to appear in mother liquids that had been in service for a year or more, and lastly (iii) that the acid was not produced at all from

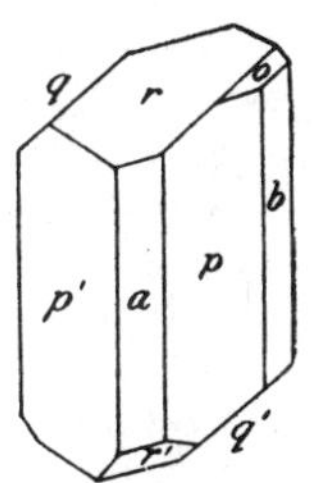

FIG. 12.—HOLOHEDRAL CRYSTAL OF RACEMIC ACID.

[1] PASTEUR, *Oeuvres*, p. 87; *A.C.P.*, 1850, [iii], **28**, 56-99.

[2] PASTEUR, *C.R.*, Jan. 3, 1853, **36**, 19–26; *Oeuvres*, pp. 242–249.

tartar derived from Alsace or Burgundy, although it was present in crude tartars from Austria, Hungary, Italy and Styria. A crop of several kilograms was finally obtained in December, 1852, by Kestner, from the calcium tartrate precipitated from the mother liquors of a factory which had gone into liquidation, and by Seybel, who precipitated as calcium salt the acid contained in 1400 kilos of the last mother liquors from three years of manufacture. The natural origin of the acid, as well as its rarity, was thus clearly established.

In the meanwhile the *Société de Pharmacie de Paris* in 1857 had offered a prize for the solution of the following problems :—

(i) Do tartars exist which contain pre-formed racemic acid ?

(ii) Determine the conditions under which tartaric acid can be transformed into racemic acid.

The prize was awarded to Pasteur [1] in view of the note cited above, which provided a solution to the first problem, and of the experiments described below, which solved the second problem, in addition to providing another inactive form of tartaric acid. The experiments were published under the title *Transformation des acides tartriques en acide racémique. Découverte de l'acide tartrique inactif. Nouvelle methode de séparation de l'acide racémique en acides tartriques droit et gauche.*[2]

The transformation of the active into the inactive acid was effected by heating dextro-tartaric acid to 170° in combination with an alkaloid. Thus when cinchonine tartrate was heated progressively, the alkaloid first underwent isomeric change, and the salt was converted into cinchonidine tartrate.

> " On continuing to heat the salt, the cinchonidine changes ; it loses water, becomes discoloured and is converted into quinoidine. The tartaric acid also undergoes important changes, and, after maintaining the temperature at 170° for five or six hours, a part has become racemic acid. One breaks the flask. The black resinous mass which it contains is extracted repeatedly with boiling water, the liquid is cooled and filtered and an excess of calcium chloride is added, which precipitates immediately all the racemic acid as calcium racemate, from which racemic acid can easily be extracted " (*Oeuvres*, p. 258).
>
> " The racemic acid, thus prepared artificially, is completely identical in all its physical and chemical properties with natural racemic acid. Above all it possesses the supremely important characteristic of being resolvable into dextro and lævo tartaric acids, which show equal and opposite rotatory powers in their compounds with bases " (*ib.*, p. 259).

The preceding experiments were made in the hope of destroying the optical activity of tartaric acid, and not of effecting a conversion

[1] Pasteur, *Oeuvres*, pp. 468–472.
[2] Pasteur, *C.R.*, Aug. 1, 1853, **37**, 162–166 ; *Oeuvres*, pp. 258–262.

of the dextro acid into the lævo acid, as was actually done in producing racemic acid from it.

> " In searching for inactive tartaric acid I obtained racemic acid. But, by a curious and fortunate circumstance, the same operation also gave me notable quantities of inactive tartaric acid. In other words, I obtained, at the same time as racemic acid, tartaric acid without any action on polarised light, and incapable of being resolved like racemic acid into right and left tartaric acid, a very curious acid, crystallising perfectly, and giving salts which in the beauty of their forms were inferior neither to the tartrates nor to the racemates " (*Oeuvres*, p. 260).

Thus after filtering off the calcium racemate in the preparation described above, the liquor deposited after 24 hours a fresh crop of crystals, consisting of pure inactive calcium tartrate, from which the inactive tartaric acid can easily be extracted.

This inactive acid, which cannot be resolved into optically-active forms, since the molecules are no longer dissymmetric, is now known as MESOTARTARIC ACID. On the basis of this experiment Pasteur concluded, wrongly, that *all* optically-active compounds can exist in *four* forms, namely, *dextro*, *lævo*, *dextro* + *lævo* and *inactive*, e.g that an acid like malic acid can be untwisted, so that if " the natural acid is a spiral stair as regards the arrangement of its atoms," the inactive acid is " the same stair made of the same steps, but straight instead of being spiral."[1] The special structure which leads to four varieties in the case of tartaric acid, but not of malic acid, is discussed in the next chapter.

Resolution with the help of Alkaloids.—Pasteur laid much stress on the absolute identity of the physical and chemical properties of dextro and lævo tartaric acids and of their derivatives:

> " There is nothing that one can do with one of these acids that one cannot do with the other in the same circumstances, and the products have invariably the same solubility, the same density, the same double refraction, the same interfacial angles. In a word, everything is identical, but for the impossibility of superposing the crystalline forms, and that the rotatory power is to the right in one case, to the left in the other, but rigorously the same in numerical value. And it need not be supposed that this identity is only manifest in important properties like solubility and density; it is all-pervasive. Suppose that a dextrotartrate, or in general a derivative of dextrotartaric acid, is deposited in crystals which are voluminous, pointed, clear or cloudy, with plain or striated faces, grouped together in some special way, with a ready cleavage; in fact that they present a thousand details which cannot be put on record: one

[1] PASTEUR, *A.C.R.*, **14**, 36.

can be certain of recognising them with the same characteristics in the lævo derivative of the same name."

"This absolute identity in everything but hemihedrism and sign of rotatory power, exists, however, only so long as the two tartaric acids are combined with compounds which are inactive towards polarised light. But, put the acids or their derivatives in the presence of products which have an action on the plane of polarisation and all identity then ceases. The corresponding compounds have neither the same composition, nor the same solubility; they no longer behave in the same way when heated. If by chance their composition is the same, their crystalline forms are incompatible, their solubilities are widely different. In fact, it will often happen that combination is possible with the dextro body and impossible with the lævo body."[1]

In the 1860 lectures a similar statement is summarised by saying:

"Towards the two tartaric acids, quinine does not behave like potash, simply because it is dissymmetric and potash is not" (*A.C.R.*, **14,** p. 40).

These phenomena were illustrated by noting the contrast between the two tartaric acids when homogeneous and when mixed with the acid of opposite sign to form racemic acid,[2] but they were especially conspicuous in the relationship of dextro and lævo tartaric acids towards the alkaloids.

"The neutral dextrotartrate of cinchonine contains eight equivalents of water, the lævotartrate contains only two. The dextrotartrate dissolves easily in absolute alcohol, the lævotartrate is very slightly soluble in it. The dextrotartrate already loses its water and becomes discoloured at 100°; the lævotartrate also loses its water of crystallisation at 100°, and thus becomes strictly isomeric with the dextrotartrate, but it can be heated to 140° without becoming discoloured" (*Oeuvres*, p. 201).

Similar experiments were made with brucine and strychnine, but the sharpest contrast was obtained with two artificial alkaloids, quinicine and cinchonidine, isomers of quinine and cinchonine, which Pasteur himself had prepared by heating the two natural alkaloids with tartaric acid.[3] The resolution of racemic acid into *d* and *l* tartaric acid with the help of an alkaloid was described in the 1860 lectures as follows:

"I prepare the paratartrate of cinchonidine by neutralising the base; then, after adding as much of the acid as was necessary for the neutralisation, I allow the whole to crystallise.

[1] Pasteur, *C.R.*, Aug. 2, 1852, **35,** 176–183; *Oeuvres*, p. 200; compare *A.C.P.*, 1853, [iii], **38,** 437–483.

[2] Pasteur, *Oeuvres*, p. 103; *A.C.P.*, 1850, [iii], **28,** 56–99.

[3] *Ibid.*, pp. 252 and 254; *C.R.*, 1853, **38,** 110–114.

The first crystallisations consist of perfectly pure left tartrate of cinchonidine. All the right tartrate remains in the mother liquor because it is more soluble. Finally this itself crystallises, with an entirely different aspect, since it does not possess the same crystalline form as the left salt. We might almost believe that we were dealing with the crystallisation of two distinct salts of unequal solubility " (*A.C.R.*, **14,** p. 41).

Resolution with the Help of Living Organisms.—A third (and last) method of preparing an optically-active tartaric acid from racemic acid was described in his 1860 lectures as follows:

> " It had long been known, from the observation of a German manufacturer of chemical products, that the impure tartrate of lime of the works, mixed with organic matters, when left under water in summer, could ferment, giving various products " *ib.*, pp. 43–44).
>
> " Knowing this, I set the ordinary right tartrate of ammonia to ferment in the following manner. . . . So far there is nothing peculiar; it is a tartrate fermenting. The fact is well known " (*ib.*, p. 44).
>
> " But let us apply this method of fermentation to paratartrate of ammonia, and under the above conditions it ferments. The same yeast is deposited. Everything shows that things are proceeding absolutely as in the case of the right tartrate. Yet, if we follow the course of the operation with the help of the polarising apparatus, we soon discover profound differences between the two operations. The originally inactive liquid possesses a sensible rotatory power to the left, which increases little by little and reaches a maximum. At this point the fermentation stops. There is no longer a trace of the right acid in the liquid. When it is evaporated and mixed with an equal volume of alcohol it gives immediately a beautiful crystallisation of left tartrate of ammonia " (*ib.*, p. 44).
>
> " The yeast which causes the right salt to ferment leaves the left salt untouched, in spite of the absolute identity in physical and chemical properties of the right and left tartrates of ammonia, as long as they are not subjected to dissymmetric action " (*ib.*, p. 45).

Pasteur had already been impressed by the fact that optical rotatory power was an attribute of natural organic products.

> " All artificial products of the laboratory and all mineral species are superposable on their images. On the other hand, most natural organic products . . . the essential products of life, are dissymmetric in such a way that their image cannot be superposed upon them " (*ib.*, p. 29).
>
> " Cellulose, starches, gums, sugars—tartaric, malic, quinic, tannic acids—morphine, codeine, quinine, strychnine, brucine

—oil of turpentine and of lemon—albumen, fibrine, gelatine —all of these primary products are molecularly dissymmetric. All these materials exhibit optical rotatory power in solution, a character necessary and important to establish their dissymmetry, even when, crystallisation being impossible, hemihedrism is useless for the recognition of this property" (*ib.*, p. 32).

In the experiments on fermentation, molecular dissymmetry "intervenes in a phenomenon of a physiological kind"[1] and these experiments may therefore be regarded as the bridge by which Pasteur crossed over from chemistry to the masterly biological researches which formed his life-work.

[1] Pasteur, *A.C.R.*, 14, 45.

CHAPTER III.

CHEMISTRY IN SPACE.

Joseph Achille le Bel, 1847–1930.
Henricus Jacobus van't Hoff, 1852–1911.
Alfred Werner, 1866–1919.

The Tetrahedral Model of the Carbon Atom.—Since molecular dissymmetry is impossible in figures which are confined to two dimensions, Pasteur's work was based on the fundamental postulate that *molecules are three-dimensional figures*, in which the atoms are linked together according to a definite plan, and in such a way as to preserve their orientation, as well as the mere sequence of atoms of carbon, oxygen, hydrogen, etc. Pasteur's elucidation of the phenomenon of molecular dissymmetry therefore marks him out as a pioneer in the study of CHEMISTRY IN SPACE. When Pasteur was at work on this subject, however, say from 1846 to 1853, the study of structural chemistry had not yet progressed to a point at which the sequence of the atoms, even in the simplest organic compounds, could be specified; and, indeed, it was not until 1861, a year after Pasteur had made his masterly survey of his work as a chemist and crystallographer, that Kekulé introduced the crude symbols which are shown in Fig. 13.

Pasteur's views as to the structure of tartaric acid were therefore put in the form of a question:

> "Are the atoms of the dextro-acid grouped on the spirals of a dextrogyrate helix, or placed at the summits of an irregular tetrahedron, or disposed according to some particular dissymmetric grouping or other? We cannot answer these questions. But it cannot be doubted that the atoms are grouped in some dissymmetric order having a non-superposable image. It is not less certain that the atoms of the lævo acid have precisely the opposite dissymmetric grouping" (*A.C.R.*, **14,** p. 24).

In this passage Pasteur puts forward the irregular tetrahedron, which even at the present day provides the simplest and commonest illustration of molecular dissymmetry as a basis of optical rotatory power; and he also foreshadows the helical or spiral structure, which was detected in the structure of crystalline tartaric acid sixty years later by Astbury[1] with the help of X-ray analysis. It is therefore not surprising that, when the quadrivalency of carbon was clearly established, the tetrahedral model was adopted almost immediately as a correct representation of the carbon atom in space of three dimensions, and was used as a basis for explaining isomerism

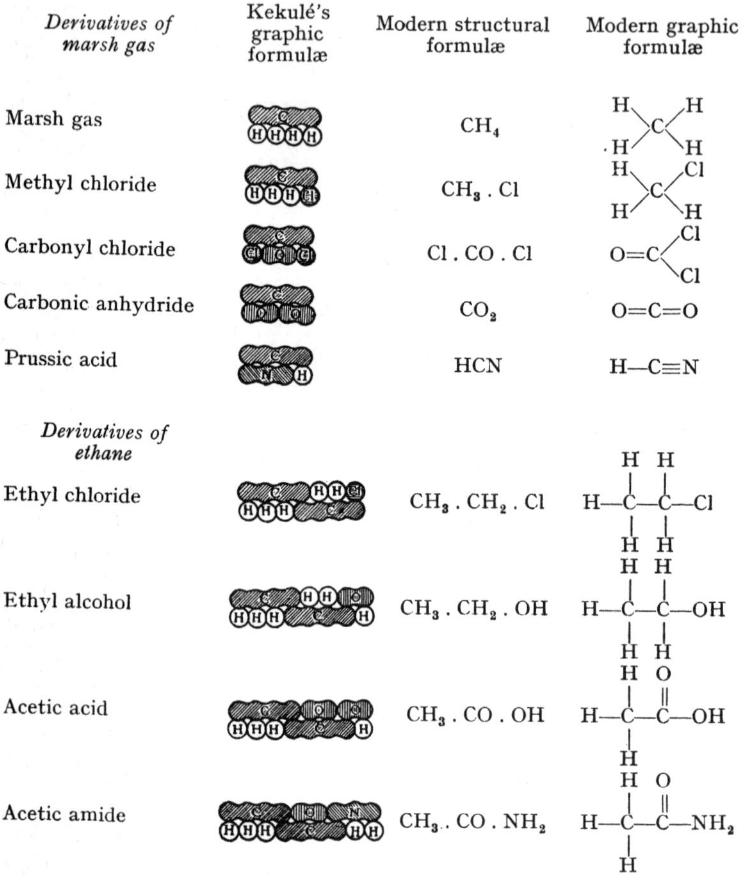

Derivatives of marsh gas	Kekulé's graphic formulæ	Modern structural formulæ	Modern graphic formulæ
Marsh gas		CH_4	H H C H H
Methyl chloride		$CH_3 . Cl$	H Cl C H H
Carbonyl chloride		Cl . CO . Cl	Cl O=C Cl
Carbonic anhydride		CO_2	O=C=O
Prussic acid		HCN	H—C≡N
Derivatives of ethane			
Ethyl chloride		$CH_3 . CH_2 . Cl$	H H H—C—C—Cl H H
Ethyl alcohol		$CH_3 . CH_2 . OH$	H H H—C—C—OH H H
Acetic acid		$CH_3 . CO . OH$	H O H—C—C—OH H
Acetic amide		$CH_3 . CO . NH_2$	H O H—C—C—NH_2 H

FIG. 13.—KEKULÉ'S GRAPHIC FORMULÆ.

(or the absence of isomerism) amongst carbon compounds. Thus, as early as 1869, PATERNO[2] made use of the diagram shown in Fig. 14 to represent three isomeric forms of the compound $C_2H_4Cl_2$ which were then supposed to exist. Paterno explained the supposed

[1] ASTBURY, *P.R.S.*, 1923, A. **102,** 506; *Nature*, July 26, 1924, **114,** 122.

[2] PATERNO, *Gior. Sci. Palermo*, 1869, **5,** 117; compare *Gaz. chim. Ital.*, 1919, **49,** 341.

existence of two isomeric forms of ethylene chloride, Cl . CH_2 . CH_2 . Cl, by postulating a rigid orientation of two halves of the molecule, relatively to one another, giving rise (in modern terms) to *cis* and *trans* forms of the dichloride. His priority in deducing the facts of organic chemistry from the tetrahedral model for carbon was therefore lost when this isomerism was proved to be non-existent; but some restitution is called for at the present day, in view of the experiments of Debye,[1] who has established with the help of X-ray analysis the existence in the vapour of ethylene chloride of two isomeric forms, in which the halogens are separated by 3·4 and 4·4 A.U. respectively.

Relationship between Chemical Constitution and Optical Rotatory Power.—Pasteur's question as to the grouping of the atoms in the molecule of tartaric acid found an almost complete answer when W. H. Perkin, senr., in 1867[2] assigned to it the structural formula of a dihydroxy-dicarboxylic acid, which we should now write as HO . CO . CH(OH) . CH(OH) . CO . OH. In the meanwhile

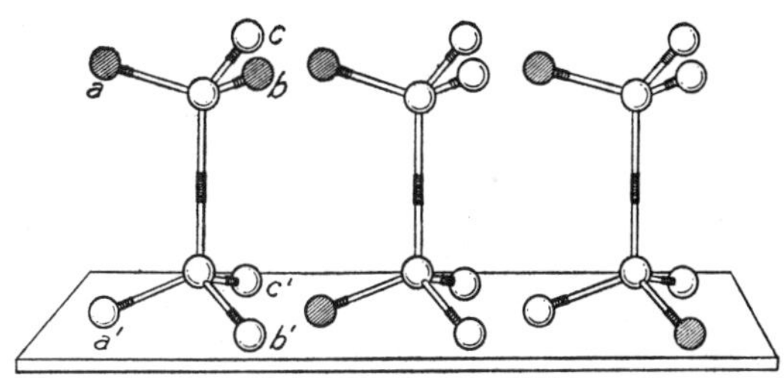

Fig. 14.—Tetrahedral Models of Dichloroethane, $C_2H_4Cl_2$ (Paterno, 1869).

trustworthy structural formulæ had been assigned by Wislicenus and others to the molecules of a number of other simple compounds which were endowed with the property of optical rotatory power. The time was therefore ripe, if not indeed over-ripe, for the disclosure of the elements of structure which gave rise to molecular dissymmetry in these compounds, on the lines envisaged so clearly by Pasteur. It is, therefore, probably not a mere coincidence that the ripe fruit was plucked, independently and almost simultaneously, by young chemists working in France and in Holland.

The main principles, which govern the development of optical activity in organic compounds, were set out in November, 1874, by J. A. le Bel, in a Memoir *Sur les relations qui existent entre les formules atomiques des corps organiques et le pouvoir rotatoire de leurs dissolutions.*[3] Le Bel's two general principles were as follows:

[1] Debye and Ehrhardt, *Physikal. Zeitsch.*, 1930, **31**, 142.
[2] W. H. Perkin, senr., *J.*, 1867, **20**, 138–160.
[3] Le Bel, *Bull. Soc. Chim.*, 1874, [ii], **22**, 337–347.

First General Principle.—" Let us consider a molecule of a chemical compound having the formula MA_4, where M is a simple or complex radical, combined with four monovalent atoms A, capable of being replaced by substitution. Let us replace three of these atoms by simple or complex monovalent radicals, differing from each other and not identical with M. The compound thus obtained will be dissymmetric."

" There are two exceptional cases, which are quite distinct. 1st. If the molecule under consideration possesses a plane of symmetry containing the four atoms A, the substitution of these by radicals (which we must consider as being non-oriented) cannot alter the symmetry relatively to this plane, and the whole series of substituted compounds will then be inactive. 2nd. The last radical substituted for A may be composed of the same atoms as the remainder of the group into which it enters. The effect of these two equal groups on the polarised light may then be either a compensation or an addition. If it is a compensation, the compound will be inactive " (*ib.*, p. 338).

Second General Principle.—" If in our fundamental type we introduce only two substituent radicals R and R′, either symmetry or dissymmetry may appear, depending on the constitution of the molecule MA_4. If this molecule had originally a plane of symmetry passing through the two atoms A which have been replaced by R and R′, this plane will remain a plane of symmetry after the substitution. The resulting compound will then be inactive."

" In particular, if a single derivative is formed, not only by a single substitution, but by two or even three substitutions . . . we are obliged to admit that the four atoms A occupy the apices of a regular tetrahedron of which the planes of symmetry are identical with those of the whole molecule MA_4. In this case no disubstituted compound will possess optical rotatory power " (*ib.*, pp. 338–339).

Application to Saturated Compounds of the Aliphatic Series.—As examples of dissymmetric molecules produced by replacing three of the four atoms of hydrogen of marsh gas, le Bel cited the following groups :

```
                    H
                    |
Lactic group  CH3—C—CO . OH
                    |
                   HO

                            H
                            |
Malic group   HO . CO—C—CH2—CO . OH
                            |
                           HO
```

Tartaric group $\left[HO.CO-\overset{H}{\underset{HO}{C}}- \right] -\overset{H}{\underset{OH}{C}}-CO.OH$

Amyl group including

$$C_2H_5-\overset{H}{\underset{CH_3}{C}}-CH_2-OH \quad \text{and} \quad C_2H_5-\overset{H}{\underset{CH_3}{C}}-CO-OH$$

Amyl alcohol. Valeric acid.

Other optically-active compounds of similar type were found in the alcohols, aldehydes and ketones of the sugar series.

The first exception to the "first general principle" that molecules of the type MARR′R″ are dissymmetric, refers to planar or two-dimensional molecules, which necessarily have a plane of symmetry if we are content to consider the four univalent radicals as "equivalent to spheres or material points, equal if the atoms or radicals are equal, and different if they are different" (*ib.*, p. 338). The second exception refers directly to Pasteur's inactive or *meso*-tartaric acid, where the fourth radical, shown in square brackets in the formula cited above, consists of "the same atoms as all the rest of the group into which it enters"; the fourth radical thus exerts a neutralising effect on the rotatory power of the acid, since the whole molecule acquires a plane of symmetry in view of the symmetrical disposition of the radicals inside and outside the square bracket.

Application to Unsaturated Aliphatic Compounds.*—The second general principle was illustrated by reference to the case of iso*amylene* or *methylethylethylene*. In general, the symbol M does not necessarily refer to a single atom; and in this case it corresponds with two atoms of carbon in compounds of the type $C_2H_2(CH_3)(C_2H_5)$. Le Bel discusses the question whether the four hydrogen atoms of ethylene, C_2H_4 (or the four radicals of a substituted ethylene), are or are not in a plane, and suggests that, if they are not, it should be possible to prepare optically active forms of one of the isomeric amylenes

$$\begin{matrix} CH_3 \\ C_2H_5 \end{matrix}\!\!>\!C{=}C\!<\!\!\begin{matrix} H \\ H \end{matrix} \quad \text{or} \quad \begin{matrix} CH_3 \\ H \end{matrix}\!\!>\!C{=}C\!<\!\!\begin{matrix} C_2H_5 \\ H \end{matrix}$$

Compounds of this type are in fact inactive, and the conclusion, which follows logically from le Bel's argument, that *the four hydrogen atoms of ethylene are in a plane*, is now admitted universally to be

* In the original *Corps gras à deux atomicités libres.*

correct. The force of this argument can perhaps be realised most easily by noticing that *methylethylallene*,

$$\begin{matrix} CH_3 \\ H \end{matrix}\!\!>\!C{=}C{=}C\!<\!\!\begin{matrix} C_2H_5 \\ H \end{matrix}$$

in which the four groups are generally regarded as not being in one plane, is an excellent example of van't Hoff's "second case of optical activity" (see below, p. 44) and is obviously capable of existing in optically-active forms. If, therefore, the four radicals in the corresponding derivative of ethylene were not coplanar, optical activity should be possible in that compound also.

Chemistry in Space.—In 1874, Dr. J. H. VAN'T HOFF, "an official of the Veterinary School at Utrecht" (with "no taste for exact chemical investigation," according to Kolbe's description of him), published in Dutch a pamphlet under the title *A treatise on a system of atomic formulæ in three dimensions and on the relation between rotatory power and chemical constitution.* This treatise first became known, after the appearance of le Bel's paper, in the form of a French translation, published in May, 1875, under the title *La Chimie dans l'Espace* or CHEMISTRY IN SPACE. A German translation, *Die Lagerung der Atome im Raume*, was also issued in 1877. In spite of the criticisms with which the German translation was assailed by KOLBE,[1] the space-chemistry of carbon compounds, as set out in van't Hoff's pamphlet, has held good, without correction and almost without addition, during a period of at least half a century; and even to-day it is so sound that modern electronic formulæ may be judged very largely by their conformity to the principles laid down by van't Hoff in 1874.

In his memoir, van't Hoff urged that structural formulæ in two dimensions were not only inadequate but also misleading. Thus, a difference of function should already exist amongst the three hydrogen atoms of mono-derivatives of methane of the type CH_3R, according as the hydrogen atom is adjacent or opposite to the radical R; and actual isomerism should be found in di-derivatives of the type CH_2R_2. By arranging the atoms in three dimensions, however (e.g. at the corners of a regular tetrahedron containing the carbon atom at its mass centre), isomerism would only appear in asymmetric derivatives of the type CRR′R″R‴. Moreover, the isomerism would then be accompanied by optical activity.

Van't Hoff then proceeded to work out the number of isomers which would be expected in compounds containing 1, 2, 3 and 4 ASYMMETRIC CARBON ATOMS, each linked to four different radicals in the manner postulated above, when joined together by single bonds. In the case where all the groups were different the number of isomers should be 2, 4, 8 and 16; but if similar substituents were used, so that the two-dimensional structural formulæ were symmetrical, these numbers would be reduced, e.g. from 4 to 3 and from 16 to 10.

[1] KOLBE, *J. prakt. Chem.*, 1877, **15**, 473.

He also investigated the number of isomers which would be formed when the carbon atoms were linked by double bonds, as represented by tetrahedra united by a common edge, and showed that derivatives of ethylene of the type RHC=CHR, should yield two isomers, devoid of optical activity, whilst the corresponding derivatives of allene, RHC=C=CHR, should yield two optically-active isomers.

Some illustrations of van't Hoff's tetrahedra are reproduced in Fig. 15.

(*a*) Two carbon atoms united by a *single bond* are represented by two tetrahedra joined at an apex. In the first two drawings the upper and lower tetrahedra are identical, when viewed from the centre of the model, but cannot be rotated to produce a plane of symmetry. These figures are therefore dissymmetric and are used to represent *d*- and *l*-tartaric acid. In the third drawing the upper and lower tetrahedra are "mirror images" of one another, so that the figure has a horizontal plane of symmetry. It therefore corresponds with an optically-inactive molecule and is used to represent *meso*-tartaric acid.

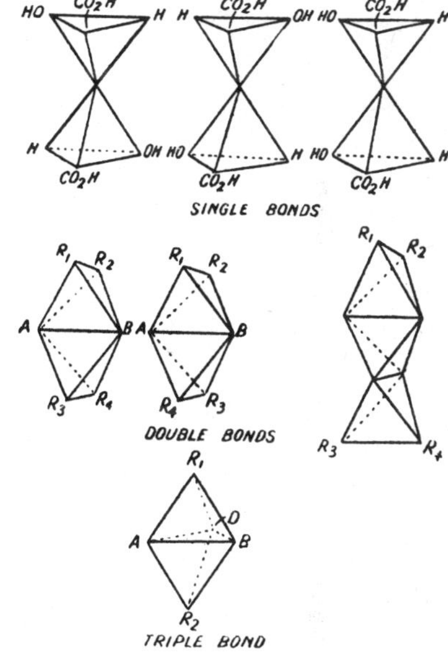

FIG. 15.—TETRAHEDRAL MODELS OF CARBON COMPOUNDS.

(*b*) Two carbon atoms united by a *double bond* are represented by two tetrahedra with a common *edge*. This model gives rise to *cis* and *trans* isomerism (p. 45), in compounds of the type $C_2RR'R''R'''$, but both isomers have a plane of symmetry and neither of them is optically active. When, however, *three* carbon atoms are united in this way, the model $C_3RR'R''R'''$ has no plane of symmetry and is dissymmetric. Optical activity is therefore predicted in compounds of this type, as well as in compounds of the simpler type, RR'C=C=CRR', since these also exhibit no plane of symmetry.

(*c*) Two carbon atoms united by a *triple bond* are represented by two tetrahedra with a common *face;* but the models do not enable us to represent the formation of a *quadruple* bond between two carbon atoms. The deduction that a quadruple covalence between two atoms of carbon is impossible is still universally adopted, since it is in strict accord with modern theory and experiment.

Optical Isomerism and Stereoisomerism of Carbon Compounds.—On comparing his theory with experiment, VAN'T HOFF was able to make the following assertions :—

(*a*) *All carbon compounds which in solution rotate the plane of polarisation, possess an asymmetric carbon atom.* This rule he illustrated in the case of active lactic, aspartic, maleic, valeric, caproic and saccharic acids, of amyl alcohol and its derivatives, of the sugars and glucosides, and finally of borneol, camphor and camphoric acid.

(*b*) *Derivatives of compounds which are active in solution lose their rotatory power when the asymmetry of all the carbon atoms disappears ; in the opposite case they often retain it.* Thus, malic acid gives inactive malonic, fumaric and maleic acids ; tartaric acid gives inactive succinic and tartronic acids ; whilst camphor gives inactive cymene. On the other hand, active lactose gives active tartaric acid ; the glucosides yield active glucose ; mannitol gives active nitromannitol ; and in general optically active acids yield optically active salts.

(*c*) The converse rule that *compounds which contain an asymmetric carbon atom rotate the plane of polarised light* could not be maintained, since

(i) Many such compounds are obtained (as Pasteur pointed out) as inactive mixtures of optical isomers, which, on account of the close similarity of their properties, are difficult to separate.

(ii) Small rotatory powers are difficult to observe, as in the case of mannitol, for which $[\alpha]_D^{20} = -0{\cdot}25°$ in water, although this compound gives derivatives of high optical activity, e.g. $[\alpha]_D = +28{\cdot}3°$ in presence of $N/2$ H_3BO_3.[1]

(iii) The presence of asymmetric carbon atoms does not suffice to produce optical activity, since internal compensation may occur, as in *meso*-tartaric acid.

Nevertheless, it was possible to assert that *the principal, although not the only, cause of optical activity is to be found in the fact that the compound contains asymmetrical carbon atoms.*

(*d*) van't Hoff describes the dissymmetric derivatives of the allene type as the SECOND CASE OF OPTICAL ACTIVITY, but was not able to quote a genuine example of this type. He was, however, very successful in explaining the isomerism of derivatives of ethylene, where inactive isomers are to be expected. As examples of this form of isomerism he quotes the cases of

fumaric and *maleic acids* $CO_2H\,.\,\underbrace{CH{=}CH}\,.\,CO_2H$;

the two *bromomaleic acids* $CO_2H\,.\,\underbrace{CBr{=}CH}\,.\,CO_2H$;

citra- ita- and *mesa-conic acids*,

$$CH_2{=}C(CH_2\,.\,CO_2H)\,.\,CO_2H \quad\text{and}\quad CO_2H\,.\,\underbrace{C(CH_3){=}CH}\,.\,CO_2H\ ;$$

the two *crotonic acids* $CH_3\,.\,\underbrace{CH{=}CH}\,.\,CO_2H$,

[1] IRVINE and STEEL, *J.*, 1915, **107**, 1229.

and their monochloro-derivatives. He was, however, in error in adopting Kekulé's formula CHOH=C=C=CH . CO_2H for *mucic acid*, since this acid actually contains a ring system.

Compounds which differ only in the sign of their molecular dissymmetry are described as OPTICAL ISOMERS. Those which differ only in the CONFIGURATION or arrangement in space of the atoms are described as STEREOISOMERS. Stereoisomers derived from ethylene are distinguished as *cis* and *trans*.[1] Thus maleic acid with two carboxyl groups on the same side of the plane containing the double bond is the *cis* acid, whilst fumaric acid with the two carboxyls on opposite sides of this plane is the *trans* acid:

H . C . CO . OH	HO . CO . C . H
()	()
H . C . CO . OH	H . C . CO . OH
Maleic acid (*cis*).	Fumaric acid (*trans*).

Stereochemistry of the Metals.—The arguments used by le Bel and van't Hoff in deducing the tetrahedral configuration of quadrivalent carbon compounds of the type MA_4 were developed and used nearly twenty years later by ALFRED WERNER to establish the configuration of metallic radicals of the types MA_4 and MA_6. Werner's paper, *Beitrag zur Konstitution anorganischer Verbindungen*,[2] was issued from the laboratory of Professor Hantzsch at Zürich at the end of 1892, and should be read on account of its exceptional clearness, as well as its historical interest, before passing on to the much more difficult text of his *Neuere Anschauungen auf dem Gebiete der anorganischen Chemie*.

Composition of Ammines and Hydrates.—Werner's paper begins with a discussion of the composition of the metal-ammonia salts of bivalent, tervalent and quadrivalent metals, which he classified under two principal headings as follows:

	Bivalent.	*Tervalent.*	*Quadrivalent.*
Type MA_6 . .	$Ni(NH_3)_6Cl_2$	$Co(NH_3)_6Cl_3$	$Pt(NH_3)_6Cl_4$
Type MA_4 . .	$Pt(NH_3)_4Cl_2$		

In each of these types the molecules of ammonia could be replaced by molecules of water, but the total of *six* or *four* molecules was usually maintained, e.g.

$$Co(NH_3)_6Cl_3 \rightarrow Co(NH_3)_5(H_2O)Cl_3$$

Luteo salt. Roseo salt.

It was also possible to remove ammonia without replacing it, e.g.

$$Co(NH_3)_6Cl_3 \rightarrow Co(NH_3)_5Cl_3 \rightarrow Co(NH_3)_4Cl_3 \rightarrow Co(NH_3)_3Cl_3$$

Luteo salt. Purpureo salt. Praseo salt. Hexammine salt.

$$Pt(NH_3)_4Cl_2 \rightarrow Pt(NH_3)_3Cl_2 \rightarrow Pt(NH_3)_2Cl_2.$$

[1] VON BAEYER, *Ann.*, 1888, **245**, 137.
[2] A. WERNER, *Zeitschr. anorg. Chem.*, 1893, **3**, 267–330.

In that case, however, there was an alteration of function in the acid radical, since, for each vacancy that was created in the total of four or six associated molecules, one atom of halogen ceased to act as an ion. This change of function was indicated by transferring the negative radical to the inside of the bracket * which was used to mark out the complex ion of the salt, thus :

$$\left[Co(NH_3)_6\right]Cl_3 \rightarrow \left[Co{}^{(NH_3)_5}_{Cl}\right]Cl_2 \rightarrow \left[Co{}^{(NH_3)_4}_{Cl_2}\right]Cl \rightarrow \left[Co{}^{(NH_3)_3}_{Cl_3}\right]$$

$$\left[Pt(NH_3)_4\right]Cl_2 \rightarrow \left[Pt{}^{(NH_3)_3}_{Cl}\right]Cl \rightarrow \left[Pt{}^{(NH_3)_2}_{Cl_2}\right]$$

In the case of the most stable complex ions, the halogen " inside the bracket " was not precipitated by silver nitrate, and was not acted on even by concentrated sulphuric acid ; but this extreme inertness was not observed in the less stable complex ions.

The process of replacing ammonia by negative radicals could be carried still further in " complex salts " as in the series :

$$\underset{\text{Luteo-salt.}}{\left[Co(NH_3)_6\right]Cl_3} \rightarrow \underset{\text{Xantho-salt.}}{\left[Co{}^{(NH_3)_5}_{NO_2}\right]Cl_2} \rightarrow \underset{\text{Croceo-salt.}}{\left[Co{}^{(NH_3)_4}_{(NO_2)_2}\right]Cl} \rightarrow$$

$$\underset{\substack{\text{Triammino-}\\\text{cobaltic trinitrite.}}}{\left[Co{}^{(NH_3)_3}_{(NO_2)_3}\right]} \rightarrow \underset{\substack{\text{Potassium diammino-}\\\text{cobaltic tetranitrite.}}}{\left[Co{}^{(NH_3)_2}_{(NO_2)_4}\right]K} \rightarrow \underset{\substack{\text{Potassium}\\\text{cobaltinitrite.}}}{\left[Co(NO_2)_6\right]K_3}$$

The valency of the complex ion in these compounds could always be deduced by subtracting the sum of the valencies of the negative radicals " inside the bracket " from the positive valency of the metallic ion. Thus it was $+3$ in $[Co(NH_3)_6]$, 0 in $[Co(NH_3)_3(NO_2)_3]$ and -3 in $[Co(NO_2)_6]$.

Stereochemistry of Radicals of the Type MA_6.—In reply to the question as to the configuration in space of complex radicals of the type MA_6, Werner suggested that " if we place the metal atom at the centre of the system, the six radicals can be arranged most simply at the corners of an octahedron " (*ib.*, p. 298). When one molecule of ammonia in the symmetrical complex $[M(NH_3)_6]$ is replaced by a negative radical to give the complex $[M(NH_3)_5X]$, as in Fig. 16, the model indicates that only one form of the complex should be obtained ; and this result is in agreement with experiment, since the isomerides of this type described by Gibbs in 1856 are now written as $[Co . 5NH_3 . H_2O]Cl_3$ and $[Co . 5NH_3 . Cl]Cl_2 + H_2O$;

* In Werner's original paper *round* brackets were used ; in the present text they have been replaced by his later *square* brackets.

compare $[Co . 4NH_3 . H_2O . Cl]Cl_2$ and $[Co . 4NH_3 . Cl_2]Cl + 1H_2O$.[1] If, however, a second molecule of ammonia is displaced by a negative radical, as in Figs. 17 and 18, the model indicates that two isomeric forms of the complex $[M(NH_3)_4X_2]$ should be possible, according as the two radicals X_2 are placed at the ends of an edge or of a main diagonal of the octahedron. An isomerism which could be explained in this way had already been observed by JÖRGENSEN in the tetram-

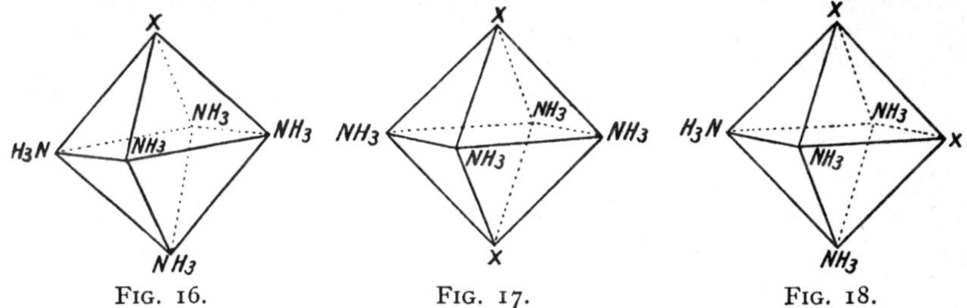

FIG. 16. FIG. 17. FIG. 18.

WERNER'S OCTAHEDRAL MODELS.

mines of the cobaltic series $[Co . 4NH_3 . X_2]X$, which were known as green "praseo" salts and as violet "violeo" salts, and these compounds were cited by Werner as evidence of the correctness of his theory.

This argument can be set out more fully as follows. The simplest symmetrical arrangements of six similar radicals in space are (i) the hexagon, (ii) the prism, (iii) the octahedron (Fig. 19). These can all be drawn on the same hexagonal outline as (i), since the conventional

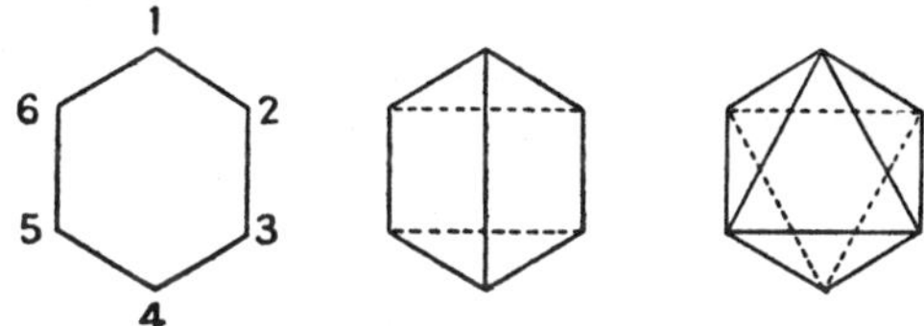

(i) Hexagon. (ii) Prism. (iii) Octahedron.

FIG. 19.—SYMMETRICAL ARRANGEMENTS OF SIX SIMILAR RADICALS IN SPACE.

"prism-formula" (ii) suggests an exaggerated perspective view from a point close to one vertical edge of the prism, whilst (iii) is a perfectly correct projection of an octahedron, of which the top and bottom faces are parallel to the plane of the paper. This projection, which was introduced by the author in 1923,[2] is used repeatedly in the present chapter in place of the setting selected by Werner, which

[1] JÖRGENSEN, *Zeitschr. anorg. Chem.*, 1898, **17**, 465; WERNER and KLEIN, *ibid.*, 1897, **14**, 32; WERNER and GUBSER, *Ber.*, 1901, **34**, 1587.

[2] LOWRY, *Chem. and Ind.*, 1923, **42**, 224.

(in its modern simplified form) represents two of the apices of the octahedron in a different manner from the other four.

The hexagon and the prism have both been used to represent the configuration of benzene, which yields *three* series of isomeric di-derivatives, *ortho*, *meta* and *para*. The octahedral configuration, on the other hand, can only account for *two* series of di-derivatives of the type $[MA_4B_2]$ or $[MA_4BC]$, which Werner described as *cis* and *trans*, following von Baeyer's use of these terms to describe the stereo-isomerism of derivatives of ethylene such as maleic and fumaric acids. Thus if the hexagon is numbered in the usual way, the 1 : 2, 1 : 3, 1 : 5 and 1 : 6 di-derivatives have the *cis* configuration, whilst the

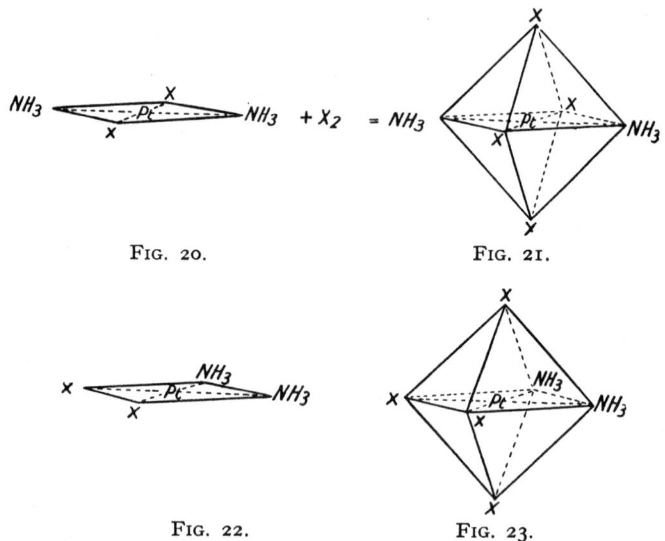

FIG. 20. FIG. 21.

FIG. 22. FIG. 23.

SHOWING THE RELATION BETWEEN THE TYPES MA_2X_2 AND MA_4X_2 (A. Werner, 1892).

1 : 4 di-derivative is of the *trans* type. A typical case is that of the yellow *flavo*-cobaltic ion $[2NO_2 . Co . 4NH_3]$ to which a *cis* configuration is assigned, since it is formed by displacing the bivalent carbonato-radical by two contiguous nitrito groups in the kation of $[CO_3 . Co . 4NH_3]NO_3$.* The orange-coloured *croceo*-cobaltic chloride $[2NO_2 . Co . 4NH_3]Cl$, to which a *trans* configuration is assigned, is prepared directly by oxidising a mixture of cobaltous chloride and sodium nitrite by a current of air in presence of ammonium chloride.

* This is prepared by oxidising cobalt chloride in presence of ammonium carbonate and ammonium nitrate. The non-ionised carbonate radical can be displaced by the action of sodium nitrite and nitric acid, but is so disguised that the salt does not effervesce with acids nor precipitate chalk from calcium chloride.

Stereochemistry of Radicals of the Type MA_4.—To the radicals of the type MA_4, whether in the form of ions as in [Pt . $4NH_3$]Cl_2 or neutral molecules such as [Pt . $2NH_3$. Cl_2], Werner assigned the structure of an incomplete octahedron, Figs. 20–23.

> "The molecules [MA_4]X_2 are incomplete molecules [MA_6]X_2: the radicals MA_4 result from the octahedrally-conceived radicals MA_6 by loss of two groups A, but with no function-change of the acid residue" (*ib.*, p. 310). "They behave as if the bivalent metallic atom in the centre of the octahedron could no longer bind all six of the groups A and lost two of them, leaving behind the fragment [MA_4]" (*ib.*, p. 303).

To this fragment a square or planar configuration was assigned.

On this basis the compounds of the type MA_2X_2 must exist in two forms just like the compounds of the type MA_4X_2. The existence of such an isomerism had been established many years before by the preparation of isomeric forms of the diammine [$2NH_3$. Pt . Cl_2],* and at a more recent date, of isomeric forms of the corresponding dipyridine compound [$2C_5H_5N$. Pt . Cl_2]. The addition compound [$2NH_3$. Pt . $2C_5H_5N$]Cl_2 had also been obtained in α and β forms, of which the former could be prepared from the preceding *cis* compounds, by the addition of two molecules of pyridine or of ammonia, whilst the latter could be prepared in the same way from corresponding *trans* compounds.†

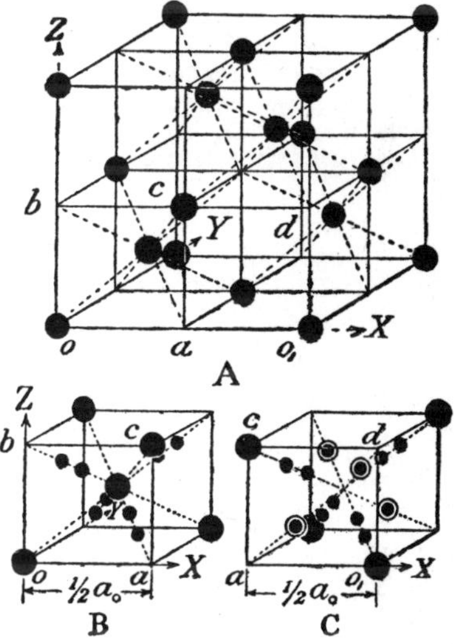

FIG. 24.—CUBIC LATTICE OF $K_2MC_4N_4$ (Dickinson, 1922).

The potassium ions form a large face-centred cube A, in which the complex anions are embedded, as in fluor-spar. The tetrahedral structure of these anions is indicated in B and C, where the small dots represent the atoms of carbon and nitrogen.

Werner pointed out that "The geometrical isomerism of

* The *polymer* [Pt . $4NH_3$]$PtCl_4$ was prepared in 1828 (G. MAGNUS, *Ann. Physik*, **14**, 204) and is still known as "MAGNUS' GREEN SALT": the *cis* and *trans* forms were prepared in 1844 (PEYRONE, *Ann.*, 1844, **51**, 1; REISET, *C.R.*, 1844, **18**, 1103).

† When decomposed the α-compound lost NH_3 and C_5H_5N from *trans* positions and gave *trans* [Pt . NH_3 . C_5H_5N . Cl_2], the β-compound lost two molecules either of ammonia or of pyridine and gave a mixture of the two *trans* compounds cited above.

the platinum compounds does not fall into line with the geometrical isomerism of the carbon and nitrogen compounds." Nevertheless the planar structure which he assigned to the quadrivalent radical of the co-ordinated platinous salts has survived to the present day,

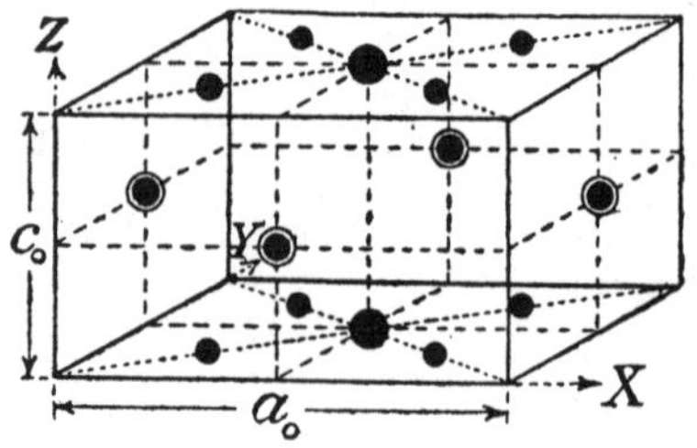

FIG. 25.—TETRAGONAL LATTICE OF K_2PtCl_6 ETC. (Dickinson, 1922).

The top and bottom surfaces of the tetragonal prism are occupied by the *planar* anions $PtCl_4$, etc., whilst the potassium atoms (represented by ringed circles) occupy the centres of the vertical sides of the prism.

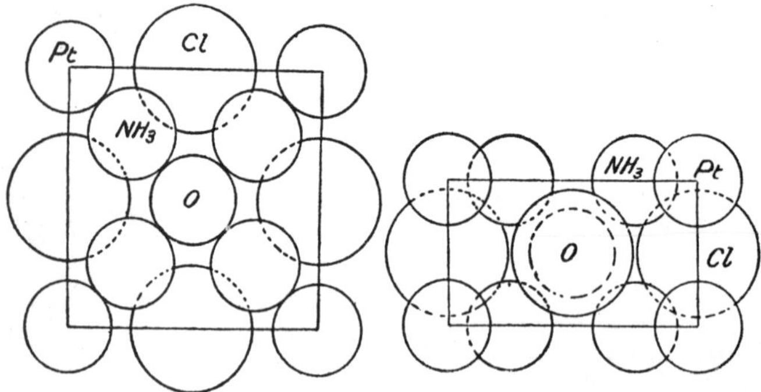

FIG. 26.—TETRAGONAL LATTICE OF TETRAMMINOPLATINOUS CHLORIDE (Cox, 1932).

The planar kations [Pt . $4NH_3$] occupy the corners of a tetragonal prism, whilst the chloride ions occupy the centres of the sides. A molecule of water occupies a hollow in the centre of the prism. Each platinum atom is surrounded by 4 ammonia groups in a square, and at a greater distance by 8 equidistant chlorine ions. Each ammonia group is at the centre of 4 coplanar chlorine ions, while each chlorine ion is surrounded by 8 ammonias.

and appears likely to persist indefinitely. In particular, whilst the X-ray analysis of Dickinson [1] has indicated (Fig. 24) a *tetrahedral* configuration of the quadrivalent anions in the *cubic* crystals of

$$K_2[ZnC_4N_4], \qquad K_2[CdC_4N_4], \qquad K_2[HgC_4N_4],$$

it has revealed [2] (Fig. 25) a *planar* configuration of the quadrivalent

[1] DICKINSON, *J.A.C.S.*, 1922, 44, 774. [2] *Ibid.*, 1922, 44, 2404.

anions in the *tetragonal* crystals of

$$K_2[PtCl_4], \quad K_2[PdCl_4], \quad Am[PdCl_4].$$

Confirmation of this planar configuration has been obtained from the experiments of Cox on

$[Pt.4NH_3]Cl_4$ (Fig. 26)[1] and on $[Pt.4NH_3][PtCl_4]$ (Fig. 27).[2]

Optical Activity in Metallic Compounds.—Werner's views in reference to the octahedral configuration of metallic radicals of the type MA_6 were supported by the fact that more than 30 series of isomer-pairs of the type MA_4B_2 were discovered amongst the co-ordination compounds of cobalt, in addition to those which are found amongst the derivatives of chromium, platinum, etc., but without a single instance of the triple isomerism which is characteristic of benzene derivatives of the type $C_6H_4X_2$. Nevertheless they were received with widespread scepticism, until the preparation of optically-active forms of compounds such as *potassium cobaltioxalate*,

$$K_3Co(C_2O_4)_3,$$

provided evidence of a type which is always admitted to be decisive, in view of the firm foundations on which it rests. The application to metallic compounds of Pasteur's principle, which provided this new evidence, is described in Chapter VI.

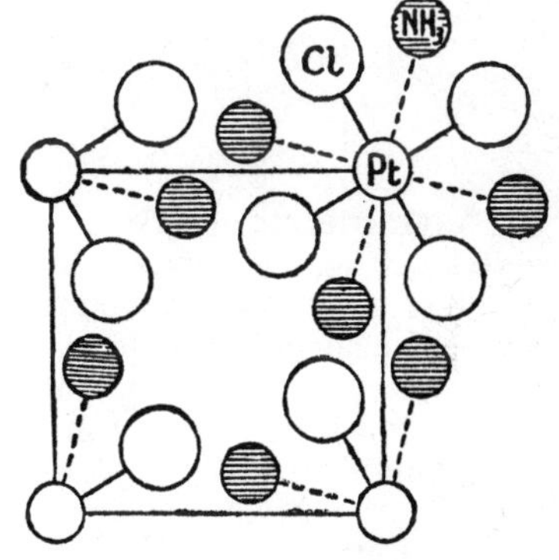

FIG. 27.—TETRAGONAL LATTICE OF "MAGNUS' GREEN SALT" (Cox, 1932).

The planar ions $[Pt.4NH_3]^{++}$ and $[Pt\,Cl_4]^{--}$ are piled alternately on one another along the corners of a tetragonal prism.

[1] Cox, *J.*, 1932, p. 1917. [2] *Ibid.*, 1932, p. 2530.

CHAPTER IV.

APPLICATIONS OF PASTEUR'S PRINCIPLE.

(*a*) NON-METALS.

Asymmetric Carbon Compounds.—PASTEUR'S PRINCIPLE states that optical rotatory power in structureless media is conclusive proof of molecular dissymmetry. This principle became of direct value as evidence of molecular structure when le Bel and van't Hoff picked out the ASYMMETRIC CARBON ATOM as the characteristic feature which gave rise to molecular dissymmetry in natural organic compounds, so many of which (as Pasteur observed) possess the optical activity which is lacking in laboratory products. The quarter of a century which followed was devoted to the verification of the main principles of "chemistry in space" as applied to carbon compounds. This verification included (*a*) the detection of asymmetric carbon atoms in natural compounds which exhibited optical activity, (*b*) the preparation of optically-active forms of compounds of which the structural formula included one or more asymmetric carbon atoms.

The former task was relatively easy. Asymmetric carbon atoms had already been detected by le Bel and van't Hoff in the simpler types of optically-active compounds. As the structure of more complex compounds became elucidated, it was then an easy matter to point to the atom or atoms whose asymmetry was responsible for the optical rotatory power of the compound. Indeed, the rule, that the formulæ of optically-active compounds *must* include at least one asymmetric atom, soon became so well established that formulæ which did not show this characteristic were at once repudiated as impossible in the case of an optically-active compound, such as camphor.

The converse problem, namely, the RESOLUTION of asymmetric compounds into their component OPTICAL ISOMERS, was much more difficult. The methods used were those which Pasteur had developed, and two of these were of very limited application. Thus the separation of optical isomers by picking out two types of hemihedral crystals can only be effected when the isomers do not unite to form a racemic compound (p. 30); and even then the process can only be carried out by a trained crystallographer. The method was applied to ammonium zinc lactate, by Purdie and Walker[1] in 1892, and by Jaeger and Thomas (p. 86) to potassium cobaltioxalate; but it has only been used about a dozen times since Pasteur first developed it in 1848. The method of separation with the help of ferments is also limited in application, since synthetic laboratory products are not often fermentable, and if fermentable may be equally so in both forms. More commonly, ferments and moulds have been used as a means of procuring optically active raw materials, but these do not differ in any important respect from those derived from other vegetable or animal sources.

In practice, then, resolution must be effected in nearly every case by combining the racemic material with a compound which is already optically active, as in Pasteur's resolution of racemic acid with the help of the alkaloid cinchonidine. Thus LADENBURG[2] resolved his synthetic coniine into its optically-active components by combining the base with tartaric acid, and then taking advantage of the fact that *d*-coniine *d*-tartrate is less soluble than *l*-coniine *d*-tartrate, just as *l*-cinchonidine *l*-tartrate is less soluble than *l*-cinchonidine *d*-tartrate. Alkaloids have also been used very extensively by Pickard and Kenyon[3] in the resolution of asymmetric alcohols into their optically-active components. For this purpose the alcohol is combined with the anhydride of a dibasic acid and the acid ester is resolved as a salt of the type $RO . CO . C_6H_4 . CO . \bar{O}\overset{+}{X}$, where R is the alkyl radical and X is the ion of the alkaloid. Important advances were made possible, however, when POPE in 1899[4] introduced a new technique for the resolution of optically-active bases, and thus inaugurated a new era in the study of optical rotatory powers, especially in synthetic laboratory products.

Improved Methods for the Resolution of Racemic Mixtures.—The improvements which Pope made in Pasteur's method for the resolution of racemic mixtures were as follows:

(*a*) *The use of non-ionising solvents* in order to diminish the risk of racemisation.

(*b*) *The use of camphor sulphonic acids* in resolving racemic bases. These acids have the advantages (i) of being much stronger than tartaric acid, and therefore less likely to give basic products by partial hydrolysis of the salts, (ii) of being monobasic, and therefore able to form only one series of salts, (iii) of giving in many cases a magnificent crystallisation, (iv) of being practically free from all risk of themselves undergoing racemisation.

(*c*) *The use of half an equivalent of an optically-active acid with half the equivalent of a mineral acid* for resolving racemic bases. Under these conditions four salts may be formed, e.g.,

[1] PURDIE and WALKER, *J.*, 1892, **61**, 754; 1893, **63**, 1143. The active salt crystallises with $2H_2O$, but the racemate with $3H_2O$. The crystallisation was induced by "seeding."

[2] LADENBURG, *Ber.*, 1886, **19**, 2578.

[3] PICKARD and KENYON, *J.*, 1912, **101**, 620 *et seq.*

[4] W. J. POPE, *The Application of Powerful Optically-active Acids to the Resolution of Externally Compensated Basic Substances*, *J.*, 1899, **75**, 1105–1109.

d-base *d*-sulphonate
l-base *d*-sulphonate
d-base hydrochloride
l-base hydrochloride.

On evaporation or cooling, the least soluble of these four salts begins to crystallise, e.g. the *d*-base *d*-sulphonate. This then carries down the whole of the active sulphonic acid, leaving the *l*-base in solution in the form of a hydrochloride. Instead, therefore, of having to separate two isomeric salts, it is only necessary to be able to separate the sulphonate of one base from the hydrochloride of the other, so that much larger differences of solubility may be looked for than in Pasteur's original method of separation.

The new methods thus developed were immediately fruitful in disclosing a new type of molecular dissymmetry, the first really new experimental development since the work of Pasteur half a century before. Up to this point all optically-active organic compounds, whether natural or synthetic, had belonged to the same type as tartaric acid, and had contained one or more asymmetric carbon atoms. By the application of his new methods Pope was able to prepare in the course of a few years asymmetric derivatives of *nitrogen*, *sulphur*, *selenium* and *tin*, and the list of elements from which optically-active compounds have been prepared has now been extended until it includes twenty-one elements as follows :

Non-metals.				*Metals.*							
B	C	N		Be							
	Si	P	S		Al						
		As	Se			Cr	Fe	Co	Ni	Cu	Zn
			Te				Ru	Rh			
								Ir	Pt		

In addition, a number of optically-active compounds have been prepared in which (as in van't Hoff's "second case of optical activity" p. 44) no one atom can be picked out as a cause of molecular dissymmetry. The optically-active compounds of elements other than carbon can be described conveniently in groups as follows :

A. Optically-active Derivatives of Nitrogen, Phosphorus and Arsenic.

(*a*) *Quaternary Ammonium salts.*

$$\text{I.}\quad \begin{matrix} C_6H_5 & & CH_2\,.\,C_6H_5 \\ & \overset{+}{\mathbf{N}} & \\ CH_3 & & CH_2\,.\,CH : CH_2 \end{matrix} \quad \bar{I}$$

Phenylbenzylallylmethylammonium iodide.

Resolved by Pope and Peachey (1899) [1] by crystallising the *d*-camphorsulphonate from acetone and ethyl acetate and then precipitating the iodide or bromide.

[1] POPE and PEACHEY, *J.*, 1899, **75**, 1127.

(*b*) *Amine-oxides and analogous compounds.*

$$\text{II.}\quad \begin{matrix} C_6H_5 \searrow & \overset{+}{\mathbf{N}} & \nearrow CH_3 \\ C_2H_5 \nearrow & & \searrow \overset{-}{O} \end{matrix}$$

Methylethylaniline oxide.

Resolved by Meisenheimer (1908).[1]

$$\text{III.}\quad \begin{matrix} C_6H_5 \searrow & \overset{+}{\mathbf{P}} & \nearrow CH_3 \\ C_2H_5 \nearrow & & \searrow \overset{-}{O} \end{matrix}$$

Methylethylphenylphosphine oxide.

Resolved by Meisenheimer and Lichtenstadt (1911).[2]

$$\text{IV.}\quad \begin{matrix} HO\,.\,CO\,.\,C_6H_4 \searrow & \overset{+}{\mathbf{As}} & \nearrow CH_3 \\ C_2H_5 \nearrow & & \searrow \overset{-}{S} \end{matrix}$$

Methylethylcarboxyphenylarsine sulphide.

Resolved by Mills and Raper (1925).[3]

At the time when Pope and Peachey first resolved methylallylphenylbenzylammonium iodide into its two optically-active components, the clear distinction which is now made between covalence and electrovalence had not yet been recognised. It was therefore necessary to assign positions in space to all the five radicals of an ammonium salt. Pickering [4] and others found a plausible solution of this problem by representing the molecule of ammonia by an equilateral triangle, and converting this into a triangular bipyramid, by the addition of hydrogen on one side and chlorine on the other, in order to represent the quinquevalency of nitrogen in ammonium chloride. Bischoff,[5] on the other hand, proposed a model in which the four hydrogens occupied similar positions at the four corners of a square pyramid, with the halogen at the apex and the nitrogen somewhere in the centre of the figure, and this model was adopted by H. O. Jones [6] when reviewing the work done on the resolution of quaternary ammonium salts.

Modern physics has greatly simplified this problem by asserting that the ammonium ion is an independent entity, which persists even in the crystalline state, since the X-ray analysis of crystals of ammonium chloride represents them as an aggregate of alternating ammonium and chloride ions.[7] It therefore indicates, in accordance with a conclusion already foreshadowed by Werner,[8] that *the stereo-*

[1] MEISENHEIMER, *Ber.*, 1908, **41**, 3966.

[2] MEISENHEIMER and LICHTENSTADT, *Ber.*, 1911, **44**, 356.

[3] MILLS and RAPER, *J.*, 1925, **127**, 2479.

[4] PICKERING, *J.*, 1893, **63**, 1069. [5] BISCHOFF, *Ber.*, 1890, **23**, 1197.

[6] H. O. JONES, *J.*, 1905, **87**, 1721 ; *B.A. Report*, 1904, 188.

[7] WYCKOFF, *Amer. J. Sci.*, 1922, **3**, 177 ; **4**, 469. See especially p. 473, where a diagram of the structure of AmCl crystals is given.

[8] WERNER, *New Ideas on Inorganic Chemistry*, English translation, 1911, p. 53.

chemistry of the ammonium radical should be identical with that of methane, the only difference being that one is a positively charged ion, whilst the other is an electrically-neutral molecule.

A direct experimental vindication of the tetrahedral configuration of the ammonium ion has been given by Mills and Warren,[1] who have prepared a compound which would yield a pair of optically-inactive stereoisomers if the four hydrogens of the ammonium radical were coplanar, as in ethylene, whereas, if they are arranged at the corners of a tetrahedron, a pair of optically-active isomerides would be formed. The optical activity of the compound formulated below is therefore a clear vindication of the tetrahedral configuration of the ammonium ion, as contrasted with the alternative pyramidal configuration.

$$\text{V.}\quad \left[\begin{matrix} C_6H_5 \\ H \end{matrix} \!>\! C \!<\! \begin{matrix} CH_2 \cdot CH_2 \\ CH_2 \cdot CH_2 \end{matrix} \!>\! \overset{+}{N} \!<\! \begin{matrix} CH_2 \cdot CH_2 \\ CH_2 \cdot CH_2 \end{matrix} \!>\! C \!<\! \begin{matrix} CO \cdot OEt \\ H \end{matrix} \right] \overset{-}{Br}$$

4-Phenyl-4′-carbethoxy*bis*piperidinium-l : l′-spirane bromine.

Meisenheimer's resolution of methylethylaniline oxide, II, conforms strictly to le Bel's first principle (p. 40), since the nitrogen atom is associated with four different radicals and is therefore asymmetric in the same sense as a carbon atom associated with four dissimilar radicals. The valency problem has been solved by writing the compound with a " mixed " or " semipolar " double bond,[2] so that four of the valencies are covalencies and the fifth is an electrovalence, exactly as in the case of Pope's asymmetric ammonium salt :

$$\begin{matrix} C_6H_5 \\ C_2H_5 \end{matrix} \!>\! \overset{+}{N} \!<\! \begin{matrix} CH_3 \\ \overset{-}{O} \end{matrix} \quad \text{compare} \quad \left[\begin{matrix} C_6H_5 \\ C_7H_7 \end{matrix} \!>\! \overset{+}{N} \!<\! \begin{matrix} CH_3 \\ C_3H_5 \end{matrix} \right] \overset{-}{I}$$

The preparation of an optically-active derivative of phosphorus was effected by making use of an analogous phosphine oxide ; but in the case of arsenic it was necessary to make use of a sulphide in order to secure sufficient stability for effective resolution. An optically-active arsenic compound of a totally different type is described below (p. 68). The preparation of an optically-active compound of boron is also described in a later section, on account of the different considerations which are involved.

The problem of preparing optically-active derivatives of quinquevalent *antimony* and *bismuth* still remains unsolved. Attempts to resolve compounds of nitrogen linked to three different radicals have also been unsuccessful, in spite of the fact that sulphonium ions of similar type yield optical isomers of a high degree of stability.

[1] MILLS and WARREN, *J.*, 1925, **127**, 2507.
[2] LOWRY, *T.F.S.*, 1923, **18**, 285 ; *Phil. Mag.*, 1923, [vi], **45**, 1105.

B. **Optically-active Sulphur, Selenium and Tellurium Salts.**

$$\text{I.}\quad \overset{+}{S}(C_2H_5)(CH_3)(CH_2\,.\,CO\,.\,OH)\ \overset{-}{Br}$$

Methylethylthetine bromide.

Pope and Peachey (1900) [1] crystallised the *d*-camphorsulphonate and the *d*-bromocamphorsulphonate from alcohol and ether, and then precipitated the bromide. Platinochloride $[M]_D = +31°$.

$$\text{II.}\quad \overset{+}{S}(C_2H_5)(CH_3)(CH_2\,.\,CO\,.\,C_6H_5)\ \overset{-}{Br}$$

Methylethylphenacylsulphine bromide.

Smiles (1900) [2] crystallised the *d*-bromocamphorsulphate from alcohol. Picrates $[M]_D = -39{\cdot}3°$ and $+34{\cdot}2°$.

$$\text{III.}\quad \overset{+}{Se}(C_6H_5)(CH_3)(CH_2\,.\,CO\,.\,OH)\ \overset{-}{Br}$$

Methylphenylselentine bromide.

Resolved by Pope and Neville (1902),[3] compare I. Chloroplatinates $[M]_D +55°$ and $-54{\cdot}3°$ in acetone.

$$\text{IV.}\quad \overset{+}{Te}(CH_3\,.\,C_6H_4)(C_6H_5)(CH_3)\ \overset{-}{I}$$

Phenyl-*p*-tolylmethyltelluronium iodide.

Resolved by Lowry and Gilbert (1929).[4] Precipitated from the α-bromocamphor π-sulphonate, $[M]_{5461} > -70°$. Precipitated from the camphor β-sulphonate $[M]_{5461} > +30°$.

The resolution of a pair of sulphonium salts by Pope and Peachey and by Smiles in 1900 followed quickly after the resolution of an unsymmetrical ammonium salt by Pope and Peachey in 1899, and was effected by the same methods. When this resolution was effected, the sulphur in the sulphonium salts was regarded as a quadrivalent element, strictly analogous with carbon, and with its valencies distributed according to the same tetrahedral scheme. The later developments of the theory of valency, which simplified the problem of the asymmetric nitrogen atom, by assigning the tetrahedral configuration of the carbon atom also to the ammonium ion, introduced a fresh complication in the case of the asymmetric sulphonium salts, since it limited the number of bonds in the sulphonium radical to *three* and attributed optical activity to the *ion* $[SRR'R'']^+$. Experiment had shown that this optically-active ion is quite

[1] Pope and Peachey, *J.*, 1900, **77**, 1972.
[2] Smiles, *J.*, 1900, **77**, 1174.
[3] Pope and Neville, *J.*, 1902, **81**, 1552.
[4] Lowry and Gilbert, *J.*, 1929, 2867.

stable—a result that is now attributed to the fact that the sulphur atom carries a complete *octet* of valency-electrons, of which three pairs are shared with the three univalent radicals of the sulphonium salt, whilst the fourth is present as a "lone pair," and may be regarded as the equivalent of a fourth radical, occupying the fourth apex of the tetrahedron. On these lines, however, we might expect also to obtain optically-active forms of a substituted ammonia NRR′R″, cf. $\overset{+}{S}$RR′R″, since the central atom again carries three pairs of shared valency-electrons and one "lone pair." The difference between a molecule and an ion has been seen to be unimportant when comparing CRR′R″R‴ with $\overset{+}{N}$RR′R″R‴; but the invariable failure of attempts to resolve compounds of tervalent nitrogen indicates that in this case there may be a real difference between the two elements, arising perhaps from the greater flexibility of structure in elements of the first short period as compared with the second. Thus, non-polar double bonds are easily formed in the first period, but not in the second period (contrast the properties of CO_2 and SiO_2).

The selenium salt resolved by Pope and Neville in 1902 is closely analogous with the sulphonium salt resolved by Pope and Peachey in 1900, but the resolution of an asymmetric telluronium salt introduces us to a new problem. In 1920, Vernon,[1] working in Professor Pope's Laboratory at Cambridge prepared a second form of dimethyltelluronium diiodide, $TeMe_2I_2$, by a mechanism which he formulated as follows:

$$\underset{\substack{\alpha\text{-Diiodide}\\ \text{(trans).}}}{\overset{CH_3\ \ \ \ I}{\underset{I\ \ \ \ CH_3}{>Te<}}} \rightarrow \underset{\substack{\alpha\text{-Base}\\ \text{(trans).}}}{\overset{CH_3\ \ \ \ OH}{\underset{HO\ \ \ \ CH_3}{>Te<}}} \rightarrow \underset{\substack{\beta\text{-Base}\\ \text{(cis).}}}{\overset{CH_3}{\underset{CH_3}{>Te{=}O}}} \rightarrow \underset{\substack{\beta\text{-Diiodide}\\ \text{(cis).}}}{\overset{CH_3\ \ \ \ I}{\underset{CH_3\ \ \ \ I}{>Te<}}}$$

It was therefore believed that tellurium had the same planar configuration which Werner had assigned in 1893 to platinum and palladium in MA_4 (p. 49). This view, which made all attempts at resolution appear as futile as they would be disagreeable, was rendered untenable when Drew, in 1929,[2] showed that the "isomerism" involved the migration of a methyl-group during the preparation of the β-base, whereas Vernon had observed this transformation only when he had acted on the β-salts with an alkali. The essential stages could therefore be formulated as follows:

$$\underset{\alpha\text{-Diiodide.}}{TeMe_2I_2} \rightarrow \underset{\alpha\text{-Base.}}{TeMe_2(OH)_2} \rightarrow \underset{\beta\text{-Base.}}{[TeMe_3]^+\,[TeMeO_2]^-} \rightarrow \underset{\beta\text{-Diiodide.}}{[TeMe_3]^+\,[TeMeI_4]^-}$$

The evidence on which the planar configuration of the quadrivalent tellurium atom had been based was thus destroyed. It was there-

[1] VERNON, *J.*, 1920, **117**, 86, 889; 1921, **119**, 105, 687.
[2] DREW, *J.*, 1929, 560.

fore likely that quadrivalent tellurium had the same tetrahedral configuration as quadrivalent sulphur, but this conclusion was still equally unproved. An asymmetric telluronium salt, which Lederer had described in 1916,[1] was, however, available for experiments on resolution, which soon resulted in the preparation by Lowry and Gilbert[2] of optically-active forms of the iodide $CH_3 . C_6H_4 . Te(C_6H_5)(CH_3)I$. The three elements S, Se, Te of the sulphur family have thus been shown to possess the same tetrahedral configuration; but the preparation of an optically-active oxonium salt is a problem that still awaits solution, and may prove to be as difficult as the preparation of an optically-active ammonia.

C. **Optically-active Sulphoxides, etc.**—

$$\text{I.}\quad \begin{matrix} CH_3 . C_6H_4 \diagdown \\ C_2H_5O \diagup \end{matrix} \overset{+}{\mathbf{S}}—\overset{-}{O}$$

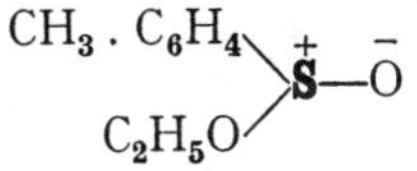

Ethyl *p*-toluene sulphinate.

Resolved by Phillips (1925).[3]

$$\text{II.}\quad \begin{matrix} CH_3 . C_6H_4 \diagdown \\ NH_2 . C_6H_4 \diagup \end{matrix} \overset{+}{\mathbf{S}}—\overset{-}{O}$$

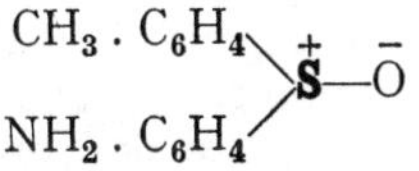

4-Methyl-4′-amino-diphenyl sulphoxide.

Resolved by Harrison, Kenyon and Phillips (1926).[4]

$$\text{III.}\quad \begin{matrix} HO . CO . C_6H_4 \diagdown \\ CH_3 \diagup \end{matrix} \overset{+}{\mathbf{S}}—\overset{-}{O}$$

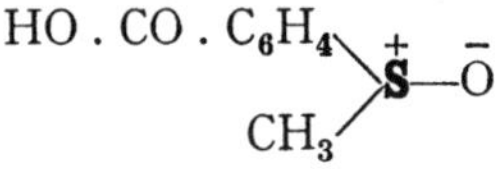

m-Carboxyphenylmethyl sulphoxide.

Resolved by Harrison, Kenyon and Phillips (1926).[4]

Towards the end of 1922 it was suggested by the author[5] that, whilst a single bond must be either a covalence or an electrovalence, a double bond might very well be a "mixed" (or "semi-polar") bond, composed of one covalence and one electrovalence. Subsequent experiments have shown that this type of bond is unstable in compounds such as ethylene and acetone, where its development implies the formation of a *sextet* of valency-electrons round a carbon atom, and that the non-polar double bond in compounds of this type only becomes semi-polar when the compound is activated, e.g. by the catalytic action of the polar walls of a vessel containing ethylene and bromine. With the help of the parachor, however, Sugden was able to show that the semi-polar double bond is stable in compounds such as $SOCl_2$, SO_2Cl_2 and $POCl_3$, where the formation of a non-

[1] LEDERER, *Ber.*, 1916, **49**, 1615.
[2] LOWRY and GILBERT, *J.*, 1929, 2867.
[3] PHILLIPS, *J.*, 1925, **127**, 2552.
[4] HARRISON, KENYON and PHILLIPS, *J.*, 1926, 2079.
[5] LOWRY, *T.F.S.*, 1923, **18**, 285; *J.*, 1923, **123**, 822.

polar double bond would require a *decet* of electrons. This classification of double bonds received a remarkable vindication, when Phillips and his colleagues in 1925 prepared an optically active sulphinate of the type R . SO . OEt,[1] and in 1926 a pair of optically-active sulphoxides of the type R . SO . R′.[2] Unlike the aniline oxide and phosphine oxide which Meisenheimer had resolved, these compounds carried only *three* radicals instead of *four*. There was therefore no reason for supposing, on the basis of le Bel's principle, that they would be capable of existing in optically-active forms; indeed, if a tetrahedral configuration is assigned to the quadrivalent sulphur atom, the sulphoxide $\begin{matrix} R_1 \\ R_2 \end{matrix}\rangle S{=}O$ would be just as incapable of optical resolution as methyl ethyl ketone $\begin{matrix} C_2H_5 \\ CH_3 \end{matrix}\rangle C{=}O$. If, however, the valency of the sulphur atom in these compounds is limited (as we have seen in the case of the sulphonium salts) to *three* co-valencies, and *one* electrovalence, the two valencies of the oxygen atom must be of different types, and the " double bond " which links it to sulphur must be unsymmetrical and semi-polar. The optical activity of the sulphoxides provides clear proof that the three radicals are not coplanar with the sulphur atom and thus establishes the semi-polar character of the double bond.

D. **Optically-active Derivatives of Silicon and Tin.**

I. $\begin{matrix} C_3H_7 \\ C_2H_5 \end{matrix}\rangle \mathbf{Sn} \langle\begin{matrix} CH_3 \\ I \end{matrix}$

Methylethylpropylstannonium iodide.

Resolved by Pope and Peachey (1900).[3] The *d*-camphorsulphonate gave $[M]_D + 95°$, of which about 50° is due to the camphor-radical. Iodide $[\alpha]_D + 23°$ in ether.

II. $\overset{+}{Na}\overset{-}{O} . SO_2 . C_6H_4 . CH_2 - \begin{matrix} C_2H_5 \\ C_3H_7 \end{matrix} \rangle \mathbf{Si}{-}O{-}\mathbf{Si} \langle \begin{matrix} C_2H_5 \\ C_3H_7 \\ CH_2 . C_6H_4 . SO_2 . \overset{-}{O}\overset{+}{Na} \end{matrix}$

Sodium salt of sulphobenzylethylpropylsilicyloxide.

Resolved by Kipping (1907)[4] $[\alpha]_D + 3{\cdot}3°, - 4{\cdot}5°$.

III. $\begin{matrix} C_3H_7 \\ C_2H_5 \end{matrix}\rangle \mathbf{Si} \langle\begin{matrix} CH_2 . C_6H_5 \\ CH_2 . C_6H_4 . SO_2 . ONa \end{matrix}$

Sodium ethylpropyldibenzylsilicanesulphonate.

Resolved by Challenger and Kipping (1910)[5] $[M]_D - 3{\cdot}3°$.

[1] Phillips, *J.*, 1925, **127**, 2552.
[2] Harrison, Kenyon and Phillips, *J.*, 1926, 2079.
[3] Pope and Peachey, *Proc. Chem. Soc.*, 1900, **16**, 42 and 116.
[4] Kipping, *J.*, 1907, **91**, 209, 717.
[5] Challenger and Kipping, *J.*, 1910, **97**, 755.

In spite of the great experimental difficulties which were encountered and overcome by Kipping and his colleagues in the attempt to prepare an optically-active derivative of *silicon*,[1, 2] the stereochemistry of the element appears to be comparable in all essential respects with that of carbon. In the case of *tin*, however, an important difference of behaviour was noticed in the readiness with which optical inversion took place.[3] This gave rise to variable readings for the optical rotatory power of the iodide; but a more striking result was the complete conversion into a salt of the dextro-base which took place when an aqueous solution of the racemic base was evaporated to dryness with *d*-camphorsulphonic acid. In this process the least soluble salt, namely, *d*-base *d*-acid, crystallises out first, and (as in all other balanced actions) this is followed by a complete conversion of the equilibrium-mixture, *d*-base $\rightleftharpoons$ *l*-base, into this form.

The racemisation may perhaps be attributed to a thermal dissociation of the salt,

$$\mathbf{Sn}\text{MeEtPrI} \rightleftharpoons \text{SnEtPr} + \text{MeI},$$

taking the place of an electrolytic dissociation which we can represent as follows:

$$\mathbf{Sn}\text{MeEtPrI} + \text{H}_2\text{O} \rightleftharpoons [\mathbf{Sn}\text{MeEtPr(OH}_2)]^+ + \text{I}^-.$$

The intervention of a molecule of water is postulated, since otherwise the stannic ion $\overset{+}{\text{Sn}}\text{RR}'\text{R}''$ would only have a sextet of valency electrons and might be expected to racemise immediately; this risk is eliminated if the ion is an aquo-compound, and ionisation would then probably serve to stabilise the optical activity of the salt. This stabilisation is actually observed in the iodide

$$\overset{+}{\mathbf{N}}(\text{CH}_3)(\text{C}_3\text{H}_5)(\text{C}_6\text{H}_5)(\text{C}_7\text{H}_7)\overline{\text{I}},$$

which is racemised in chloroform but not in water.

[1] KIPPING, *J.*, 1907, **91**, 209, 717.
[2] CHALLENGER and KIPPING, *J.*, 1910, **97**, 755.
[3] POPE and PEACHEY, *Proc. Chem. Soc.*, 1900, **16**, 42 and 116.

CHAPTER V.

APPLICATIONS OF PASTEUR'S PRINCIPLE.

(*b*) Dissymmetry without Asymmetry.

The "Second Case of Optical Activity."—The validity of Pasteur's principle has been demonstrated in a very striking way by the preparation of optically-active forms of a number of compounds which contain no asymmetric atom in the sense of van't Hoff's definition, but which owe their activity to the fact that the *molecule* is *dissymmetric*, even in the absence of an *asymmetric atom*.

The existence of this type of optical activity was foreshadowed by van't Hoff in the case of derivatives of allene (compare Fig. 15, p. 43). Thus it was pointed out by van't Hoff that, if the four hydrogen atoms of *methane* are represented as lying at the corners of a regular tetrahedron, the four hydrogen atoms of *ethylene* $H_2C : CH_2$ must be represented as coplanar. In the case of the hydrocarbon *allene*, however, the four hydrogen atoms, which lie on two parallel straight lines in ethylene, again become crossed as a result of the interposition of a third atom of carbon, between the two methylene radicals, as shown in the formula $H_2C : C : CH_2$. Optical activity, therefore, once more becomes possible, just as in the case of methane; but, whereas, in order to develop optical activity in derivatives of methane, all the four radicals must be different, it is sufficient in the case of allene that one hydrogen atom of each pair should be replaced by some other radical as in

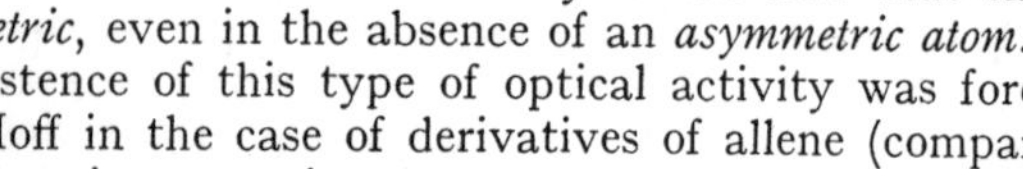

$$\begin{matrix}H\\X\end{matrix}\!>C=C=C<\!\begin{matrix}H\\X\end{matrix} \qquad \begin{matrix}H\\X\end{matrix}\!>C=C=C<\!\begin{matrix}H\\X\end{matrix}$$

Dissymmetric compounds of the allene type, in which it is not possible to pick out an asymmetric atom or atoms as being responsible for the molecular dissymmetry, were described by van't Hoff as the *second case of optical activity*. No example of this second case has been discovered amongst natural organic compounds, and Lapworth and Wechsler[1] were unable to resolve a compound of this type, namely,

[1] Lapworth and Wechsler, *J.*, 1910, **97**, 38.

diphenylnaphthyl allenecarboxylic acid, I, which they had prepared for this purpose.

$$\text{I.}\quad \begin{matrix} C_6H_5 \\ C_{10}H_7 \end{matrix} \!\!>\! C{=}C{=}C \!<\!\! \begin{matrix} C_6H_5 \\ COOH \end{matrix}$$

Optically-active derivatives of allene are therefore still unknown, but ring-compounds of analogous type have been resolved by Pope and others as described in the following paragraphs.

Centro-Asymmetry.—Pope has suggested that dissymmetric compounds which contain no asymmetric atom should be described as CENTRO-ASYMMETRIC.

Success in the preparation of an optically-active compound of this type was first achieved in 1909 by Perkin, Pope and Wallach,[1] who resolved 1-*methyl* cyclo *hexylidene-4-acetic acid*, II, by fractional crystallisation of the brucine salt. Marckwald and Meth[2] had already prepared and resolved a compound to which they assigned this formula, but this was in reality the isomeric compound III.

$$\text{II.}\quad \begin{matrix} H \\ CH_3 \end{matrix} \!\!>\! C \!<\!\! \begin{matrix} CH_2{-}CH_2 \\ CH_2{-}CH_2 \end{matrix} \!\!>\! C \frown C \!<\!\! \begin{matrix} CO\,.\,OH \\ H \end{matrix}$$

$$\text{III.}\quad \begin{matrix} H \\ CH_3 \end{matrix} \!\!>\! \overset{(1)}{\mathbf{C}} \!<\!\! \begin{matrix} \overset{(6)}{CH_2}{-}\overset{(5)}{CH} \\ \underset{(2)}{CH_2}{-}\underset{(3)}{CH_2} \end{matrix} \!\!\geqslant\! \overset{(4)}{C}\,.\,CH_2\,.\,CO_2H$$

It contains an asymmetric carbon atom (1) of the ordinary type, and has therefore no bearing on the problem of realising experimentally van't Hoff's second case of optical activity.

The compound II differs from the derivatives of allene mainly in that one of the double bonds is replaced by a six-atom ring. Pope directs attention to the fact that the optical activity of each of these acids is maintained when it is converted into a pair of stereoisomeric dibromides, each of which contains an asymmetric carbon atom, as shown in heavy type in the formula IV.

$$\text{IV.}\quad \begin{matrix} H \\ CH_3 \end{matrix} \!\!>\! C \!<\!\! \begin{matrix} CH_2.CH_2 \\ CH_2.CH_2 \end{matrix} \!\!>\! CHBr.\mathbf{C}HBr.CO.OH$$

$$\xrightarrow{NaOH}\ \text{V.}\quad \begin{matrix} H \\ CH_3 \end{matrix} \!\!>\! C \!<\!\! \begin{matrix} CH_2.CH_2 \\ CH_2.CH_2 \end{matrix} \!\!>\! C \frown C \!<\!\! \begin{matrix} Br \\ CO.OH \end{matrix}$$

$$\xrightarrow{Na_2CO_3}\ \text{VI.}\quad \begin{matrix} H \\ CH_3 \end{matrix} \!\!>\! C \!<\!\! \begin{matrix} CH_2.CH_2 \\ CH_2.CH_2 \end{matrix} \!\!>\! C \frown C \!<\!\! \begin{matrix} Br \\ H \end{matrix}$$

[1] PERKIN, POPE and WALLACH, *J.*, 1909, **95**, 1789.
[2] MARCKWALD and METH, *Ber.*, 1906, **39**, 1171, 2035, 2404.

Moreover, the optical activity still persists when these dibromides are reconverted into compounds of the original type, (i) by eliminating hydrogen bromide under the influence of caustic soda, as in V, or (ii) by eliminating hydrogen bromide and carbon dioxide, by warming with a solution of sodium carbonate, as in VI. The remarkable stability of the dissymmetric structure can be attributed to the fact that the ketonic form of the acid, II, is a conjugated compound, whilst the enolic form is no longer conjugated, but is instead a derivative of the allene type. On esterification with ethyl alcohol and sulphuric acid, however, the ester of the acid III is obtained in an inactive racemic form. This result follows at once from the fact that the double bond can wander indifferently either to the 4 : 3 or to the 4 : 5 position, and thus give rise to the two enantiomorphous forms of the acid III.

Centro-Asymmetric Oximes, Hydrazones and Semi-carbazones.—The preparation of optically-active forms of the centro-asymmetric acid described above was followed in the next year by the resolution of a centro-asymmetric oxime I. This was the first of a series of five compounds, each containing a doubly-bound atom of tervalent nitrogen $>C{=}NR$, which Mills and his colleagues have used to test the hypothesis (put forward by Hantzsch and Werner in 1890) that the three bonds of doubly-linked nitrogen do not lie in a plane, and can therefore give rise to isomerism just as in the case of ethylene.

$$\begin{array}{c} C_6H_5 . C . H \\ \| \\ HO . N \end{array} \quad \text{and} \quad \begin{array}{c} C_6H_5 . C . H \\ \| \\ N . OH \end{array}$$

Benzaldoxime.

compare

$$\begin{array}{c} CH_3 . C . H \\ \| \\ HO . CO . C . H \end{array} \quad \text{and} \quad \begin{array}{c} CH_3 . C . H \\ \| \\ H . C . CO . OH \end{array}$$

Crotonic acid.

$$\text{I.}\quad \begin{matrix} H \\ HO . CO \end{matrix}\!>\!C\!<\!\begin{matrix} CH_2 . CH_2 \\ CH_2 . CH_2 \end{matrix}\!>\!C{=}N{-}OH$$

$$\text{II.}\quad \begin{matrix} H \\ HO . CO \end{matrix}\!>\!C\!<\!\begin{matrix} CH_2 . CH_2 \\ CH_2 . CH_2 \end{matrix}\!>\!C{=}N{-}NH . CO . NH_2$$

$$\text{III.}\quad \begin{matrix} H \\ HO . CO \end{matrix}\!>\!C\!<\!\begin{matrix} CH_2 . CH_2 \\ CH_2 . CH_2 \end{matrix}\!>\!C{=}N{-}N(C_6H_5) . CH_2 . C_6H_5$$

Oxime, semi-carbazone and phenylbenzylhydrazone of 1-*cyclo*-hexanone-4-carboxylic acid.

Resolved by Mills and Bain (1910 and 1914).[1]

[1] MILLS and BAIN, *J.*, 1910, **97**, 1866; 1914, **105**, 64.

$$\text{IV.}\quad \begin{array}{l} CH_2 . CH_2 . CH . S \\ | \qquad\qquad\quad | \\ CH_2 . CH_2 . CH . S \end{array} \!\!\!> C \!\!=\!\! N \diagup NH . C_5H_4N$$

Pyridylhydrazone of *cyclo*-hexylenedithiocarbonate.
Resolved by Mills and Schindler (1923).[1]

$$\text{V.}\quad CH_3 . CH \!<\! \begin{array}{l} CH_2 . S \\ CH_2 . S \end{array} \!\!\!> C \!\!=\!\! N \diagup NH . C_6H_4 . CO . OH$$

o-Carboxyphenylhydrazone of methyltrimethylene dithiocarbonate
Resolved by Mills and Saunders (1931).[2]

The ketones from which these compounds are derived possess a plane of symmetry, perpendicular to the plane of the ring, as in the formula

$$\begin{array}{r} H \\ HO . CO \end{array} \!\!\!> C \!<\! \begin{array}{l} CH_2 . CH_2 \\ CH_2 . CH_2 \end{array} \!\!\!> C \!\!=\!\! O$$

This plane of symmetry is destroyed by the non-planar configuration of the radical $=N-R$, where R may be $-OH$, $-NH . CO . NH_2$, $-NRR'$, etc. The compound IV was prepared[1] in order to avoid the risk that the double bond in I, II and III might have migrated into the ring $>C=NR \rightarrow \ >C-NHR$, in which case the carbon atom carrying the carboxyl-group would become asymmetric, exactly like the corresponding atom in Marckwald and Meth's isomeride of the dissymmetric acid of Pope, Perkin and Wallach. The compound, V, was prepared[2] in order to eliminate the risk that the dissymmetry of IV might depend on the buckling of the ring, giving rise to a possible *cis-trans* isomerism comparable with that of the decalones.

The optical activity of the *oxime*, I, is not very stable, and only one salt is obtained when it is evaporated with an alkaloid (compare p. 61), but both forms could be prepared by using different alkaloids. Thus morphine gave an ammonium salt with $[M]_D + 50°$ to $60°$, whilst quinine gave an ammonium salt with $[M]_D - 70°$ to $- 80°$. The ammonium salt racemised to the extent of 50 per cent. in 13 minutes, but in presence of half an equivalent of ammonia this half-change occupied 8·5 hours. In the same way, half the N/5 sodium salt had racemised at the end of 24 minutes, but in presence of N/10 sodium hydroxide the half-change occupied 22 hours. The racemisation is therefore checked by the addition of alkalis, and perhaps proceeds through the enolisation of the free acid. The sodium salt of the *hydrazone* when liberated from the quinine salt, gave $[M]_D +238·6°$. The ammonium salt of the *semicarbazone*, when liberated from the

[1] Mills and Schindler, *J.*, 1923, **123**, 312.
[2] Mills and Saunders, *J.*, 1931, 537.

morphine salt, gave $[M]_D + 30°$ to $+ 40°$. In both cases only a single salt with the alkaloid could be obtained, and, when this was decomposed with sodium hydroxide or with ammonia, rapid auto-racemisation occurred.

Dissymmetry without Asymmetry.—The centro-asymmetric compounds resolved by Pope, Perkin and Wallach, and by Mills and his colleagues, do not contain an asymmetric carbon atom in the sense of van't Hoff's definition. Thus the radicals attached to the carbon atom (1) in cyclohexylidene-acetic acid may be dissected out as follows:

$$\begin{matrix} CH_3 \diagdown \overset{(1)}{} \diagup CH_2 . CH_2 . \overset{|}{C}{=}CH . CO . OH \\ \qquad C \\ H \diagup \qquad \diagdown CH_2 . CH_2 . \underset{|}{C}{=}CH . CO . OH \end{matrix}$$

Two of them are identical in constitution, although they differ in configuration. On this basis Pope suggests [1] that the atom (1) is of the type CHRR′R′ and not CHRR′R″. A more detailed study shows that the two large radicals (which are twisted into different shapes when they are blended into a single ring-system) are, in fact, *cis* and *trans* isomers of the type of acrylic acid $CH_3 . CH : CH . COOH$, so that the problem is quite different in type from that encountered in the tartaric acids. Similar considerations apply to the oximes resolved by Mills and his colleagues.

The compounds cited above are not merely dissymmetric but asymmetric, since, in addition to being without a plane or centre of symmetry, they contain no axis of symmetry. In this respect they are different from the simplest dissymmetric compounds of the allene type, e.g.

$$\begin{matrix} H \diagdown \\ \quad C{=}C{=}C \\ X \diagup \end{matrix} \begin{matrix} \cdot\cdot H \\ \diagdown X \end{matrix}$$

These are relatively symmetrical, since they possess an axis of two-fold symmetry, similar to that of a two-bladed propeller in which the forward edges of the blades are unlike the backward edges. They are therefore dissymmetric without being asymmetric.

The phenomenon of dissymmetry without asymmetry is very common amongst the optically-active metallo-organic compounds which are described in the following chapter, but has also been established in compounds of carbon, nitrogen, boron and arsenic.

Dissymmetric Carbon Atoms.—We have seen that a symmetrically disubstituted allene is dissymmetric in spite of the fact

[1] POPE, *Report of the First Solvay Conference*, 1922, p. 145: "An asymmetric carbon atom is one which is attached separately to four atoms or radicals which differ in composition or in constitution."

that it has a two-fold axis of symmetry, so that the complex $\begin{matrix}H \\ X\end{matrix}\!\!>C_3\!<\!\!\begin{matrix}H \\ X\end{matrix}$ can be obtained in optically-active forms. The same result can be secured in derivatives of methane if pairs of bonds are " ear-marked " by forming part of a ring. Thus a spiro compound of the type $\begin{matrix}A \\ | \\ B\end{matrix}\!\!>C<\!\!\begin{matrix}A \\ | \\ B\end{matrix}$ will have a two-fold axis of symmetry in virtue of the identity of the two " propeller blades," but will still be dissymmetric. A compound of this type, namely, the *keto-dilactone*, I, of *benzophenone-2 : 4 : 2′ : 4′-tetracarboxylic acid*, II,

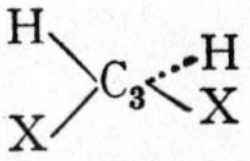

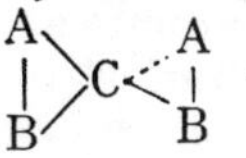

I. HO . OC . C_6H_3—C(—O—CO—)(—CO—O—)—C_6H_3 . CO . OH

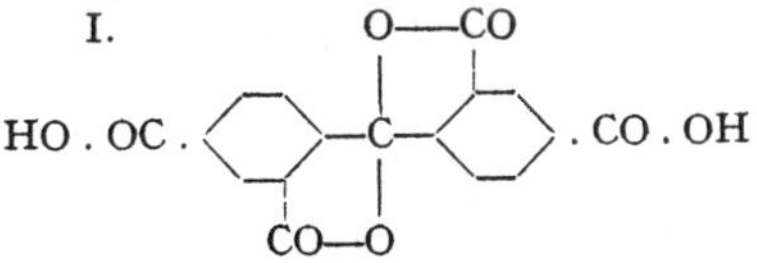

II. $(HO.CO)_2C_6H_3 . CO . C_6H_3(COOH)_2$

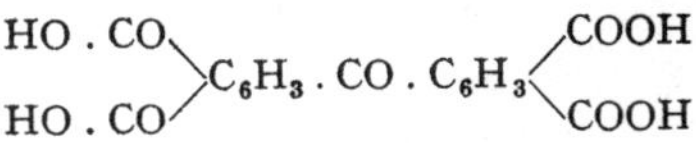

was resolved by Mills and Nodder in 1920.[1] A closer approach to the original type of a symmetrically disubstituted allene is found, however, in the compound III, resolved by Backer and Schurink in 1928,[2] since this compound differs only in the expansion of the double bonds of allene to a pair of 4-carbon rings and thus preserves all the essential characteristics of the type.*

III. 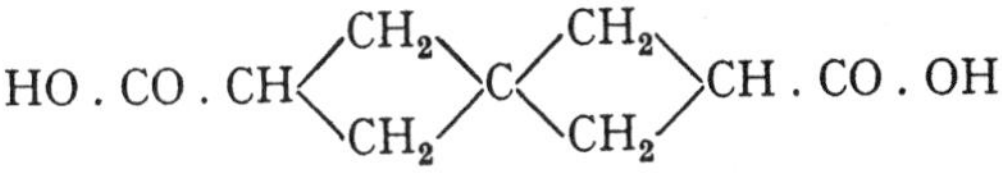 HO . CO . CH<$(CH_2)_2$>C<$(CH_2)_2$>CH . CO . OH

cyclo Butane*spirocyclo*butane-1 : 1′-dicarboxylic acid $[M]_D + 1{\cdot}9°$.

A number of spiro compounds, in addition to those cited above, have been resolved into optically-active forms, some of them with identical bivalent radicals and therefore possessing axial symmetry, and others with different bivalent radicals attached to the central carbon atom and therefore belonging to the original type of centro-asymmetry.[3]

Dissymmetric Compounds of Nitrogen, Boron and Arsenic.—

IV. 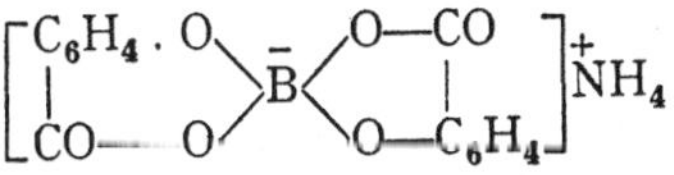 $\left[\begin{matrix}C_6H_4 . O \\ | \\ CO—O\end{matrix}\!\!>\bar{B}<\!\!\begin{matrix}O—CO \\ | \\ O—C_6H_4\end{matrix}\right]\overset{+}{N}H_4$

Ammonium borosalicylate.

Resolved by Boeseken and Meuienhof (1924).[4]

* The resolution of the corresponding diamine is described on p. 391.

1 Mills and Nodder, *J.*, 1920, **117**, 1407; 1921, **119**, 2094.

2 Backer and Schurink, *Proc. Akad. Wet., Amsterdam*, 1928, **31**, 370.

3 W. H. Mills, *T.F.S.*, 1930, **26**, 431.

4 Boeseken and Meulenhoff, *Proc. Akad. Wet., Amsterdam*, 1924, **27**, 174–177.

V. $H[As(O_2C_6H_4)_3]$, $5H_2O$
Pyrocatecholarsinic acid.

Resolved by Rosenheim and Plato (1925).[1]

Werner pointed out in 1893 that the "valency number" of carbon, as measured by its combining power for hydrogen or chlorine separately, e.g. in CH_4 and CCl_4, is identical with its "co-ordination number" as measured by its combining power for the two other elements taken together, e.g. in CH_3Cl, CH_2Cl_2 and $CHCl_3$. In the case of nitrogen and boron, the "valency number" as measured by combination with hydrogen or a halogen separately is only three in NH_3 and BF_3, but the "co-ordination number" is the same as in carbon, since the combining power for hydrogen or fluorine can be increased to four by making use of elements of both types as in $(\overset{+}{N}H_4)\overset{-}{F}$ and $\overset{+}{K}(\overset{-}{B}F_4)$.

We have already seen that the conditions for producing an asymmetric atom are the same for an ammonium *ion* as for a *molecule* derived from methane. A precisely similar statement can now be made in reference to the production of optical activity in compounds exhibiting axial symmetry round a central atom of nitrogen, in place of carbon. Thus the compound (formula V, p. 56) which was resolved by Mills and Warren [2] in an endeavour to establish the tetrahedral configuration of the quadrivalent nitrogen in an ammonium ion only fails to belong to this type because the COOEt radical in one ring is balanced by C_6H_5 in the other. In the borosalicylates, which were resolved by Böeseken and Meulenhof in 1924 (formula IV), the four univalent radicals of the ion BF_4^- are replaced by two identical but unsymmetrical bivalent acid radicals. The type is therefore exactly the same as in the diketolactone of Mills and Nodder

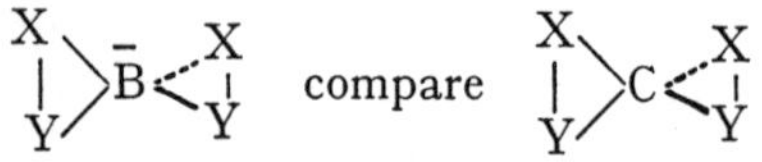

X⟩$\overset{-}{B}$⟨X ... compare X⟩C⟨X
Y ... Y ... Y ... Y

The symmetry is again that of a two-bladed propeller with the front and back edges of the blades dissimilar to each other, and the resolution of the ion into optically-active forms is a perfectly normal phenomenon.

The *borosalicylates* can be regarded as derived from an overhydrated form of orthoboric acid, $H_3BO_3 + H_2O = H_5BO_4$. This acid must, however, be written as $H[B(OH)_4]$ in order to bring out the anomaly that four of five hydrogens are *basic*, whilst the fifth is strongly *acidic*, at least when the others have been replaced by acid radicals. Similar considerations apply to the *pyrocatecholarsinic acid*, V, $H[As(O_2C_6H_4)_3]$, $5H_2O$, resolved by Rosenheim and Plato in 1925. This can be derived

[1] Rosenheim and Plato, *Ber.*, 1925, **58**, 2000.
[2] Mills and Warren, *J.*, 1925, **127**, 2507.

from ordinary arsenic acid by the addition of two molecules of water, and again contains one molecule of water in excess of the "ortho" acid

$$H_3AsO_4 + 2H_2O = H_7AsO_6 \text{ or } H[As(OH)_6].$$

In pyrocatecholarsinic acid, the basic hydrogens of the six hydroxyl groups are replaced by three bivalent acid radicals, and the arsenic ion then has the symmetry of a three-bladed propeller. In this case it is no longer necessary that the front and back edges of the blades should be distinguished from one another, since the increase in the number of blades from two to three eliminates a potential centre of symmetry; it is therefore possible to make use of the symmetrical catechol radical, —O . C_6H_4 . O—, instead of the unsymmetrical salicylic radical, —O . C_6H_4 . CO . O—, which was used in the case of boron. The substituted arsenic ion has then the well-developed axial symmetry which we have already found in crystals of quartz (p. 26), namely, one three-fold axis of symmetry—the main axis of the propeller—and three two-fold axes, running down the centres of the three propeller blades and emerging midway between the two blades on the other side of the main axis.

Molecular Dissymmetry depending on Restriction of Rotation about a Single Bond.—Another case in which optical activity has afforded evidence of dissymmetry in compounds that appeared to be symmetrical has been provided by the derivatives of diphenyl. The *o*-dinitrodicarboxylic acid was known in two isomeric forms, which behaved in certain respects as if they were *cis* and *trans* isomers of the formula I. This isomerism could be explained either (i) by assuming that free rotation of the single bond between the two phenyl groups was prevented by the interlocking of the *ortho*-substituents, or (ii) by supposing that the diphenyl molecule was folded like the wings of a butterfly. On the first hypothesis, both the *cis* and the *trans* forms of the acid might exist in optically-active form; on the second hypothesis this should occur only in the *trans* form. Actually, Kenner and his colleagues [1] obtained optically-active forms of both acids, and only discovered afterwards [2] that they had the structures I and II and were therefore not stereoisomers after all. In the meanwhile, however, they had also obtained optically-active forms of the acids III [3] and IV [4] as well as of the dichloro-acid [5] corresponding to I.

[1] CHRISTIE and KENNER, *J.*, 1922, **121**, 614.
[2] CHRISTIE, HOLNERNESS and KENNER, *J.*, 1926, 671.
[3] CHRISTIE and KENNER, *J.*, 1923, **123**, 779.
[4] *Ibid.*, 1922, **121**, 614.
[5] CHRISTIE, JAMES and KENNER, *J.*, 1923, **123**, 1948.

I. HO . CO, NO_2 / NO_2, CO . OH

II. NO_2, HO . CO / NO_2, CO . OH

III. NO_2, HO . CO / NO_2, CO . OH, NO_2

IV. NO_2, HO . CO, NO_2 / NO_2, CO . OH, NO_2

The fixed dissymmetry of these acids, which appear to be too stable to undergo racemisation, has been attributed [1] to the inhibition by the *ortho*-substituents of free rotation about the median bond of the diphenyl molecule. A similar dissymmetry has been found by Mills and Elliott [2] in the naphthalene series, where the compound V has been obtained in optically-active forms :

V. $C_6H_5SO_2$, CH_2 . CO . OH, N, NO_2

This compound, however, racemises very quickly, the " half-life period " in chloroform being only 17 minutes at 15°. The brucine salt therefore crystallised from *acetone* as a pure product, derived entirely from the *l*-acid ; but the brucine salt of the *d*-acid was obtained by crystallising the same material from *methyl alcohol*, in which it is less soluble than the *l*-isomeride.

[1] See, for example, MILLS, *Chem. and Ind.*, 1926, **45**, 884.
[2] MILLS and ELLIOTT, *J.*, 1928, 1291.

CHAPTER VI.

APPLICATIONS OF PASTEUR'S PRINCIPLE.

(*c*) Molecular Dissymmetry in Metallic Compounds.

Tetrahedral and Octahedral Models.—The tetrahedral model, which van't Hoff used as a basis for predicting both stereoisomerism and optical isomerism in carbon compounds, has proved to be equally valid when used to represent the elements, silicon and tin, of the same group, which are definitely quadrivalent in their optically-active compounds. The facts recited in Chapter IV show, however, that it has a wider application than van't Hoff supposed, since it is equally applicable to asymmetric ammonium salts, which were then regarded as containing quinquevalent nitrogen, but are now formulated as containing a quadricovalent ion, comparable with the molecule of methane. We have seen further in Chapter V that, in the borosalicylates, the boron atom, which was formerly regarded as tervalent or quinquevalent, can also be formulated as containing a quadricovalent ion, which may be represented by the same tetrahedral model as carbon. Finally, the model has been found to be valid in the case of sulphur and its homologues, which were formerly regarded as bivalent, quadrivalent or sexavalent, but are now formulated as only tercovalent in the ions of the salts in which optical activity has been observed. In these compounds the tetrahedral model has been retained, but one of its apices is assigned to a lone pair of electrons, whilst the others are occupied in the usual way by the three radicals of the asymmetric ion. In general then, the tetrahedral model has served to explain the configuration of all the compounds of the non-metals which have been obtained up to the present in optically-active forms, with the single exception of the pyrocatecholarsinic acid $H[As(O_2C_6H_4)_3]$, which was resolved by Rosenheim and Plato in 1926. When, however, we pass on to consider the configuration of optically-active compounds of the *metals*, we find that a similar predominant importance must be assigned to the octahedral model.

Vindication of the Octahedral Model.—The octahedral model was introduced by A. Werner in 1892[1] as an aid to the interpretation of the configuration in space of metallic radicals of the type MA_6, as in the hexammines $Ni(NH_3)_6Cl_2$, $Co(NH_3)_6Cl_3$, $Pt(NH_3)_6Cl_4$.

[1] Werner, *Zeitschr. anorg. Chem.*, 1893, **3**, 267–330.

Its correctness was demonstrated by the existence of long series of pairs of isomers of the type MA_4B_2, and has since been established by X-ray analysis of salts such as Am_2SiF_6, Am_2SnCl_6, K_2PtCl_6 and $[Ni . 6NH_3]Cl_2$; but the whole theory of co-ordination was received with widespread scepticism,* in spite of the vast array of stereochemical evidence by which it was supported, until, after a period of nearly 20 years, it finally received immediate and almost universal recognition, as the result of the preparation of an optically-active salt, the dissymmetry of which could not have been foreseen apart from the hypothesis of co-ordination.† Just as in the case of Hantzsch and Werner's theory of stereoisomeric oximes (p. 64) the prediction and detection of optical activity then sufficed to place the fundamental doctrine of co-ordination beyond the reach of further controversy.

Molecular Dissymmetry in Co-ordination Compounds.— The arguments which Werner applied in 1911 in his paper on *The Asymmetric Cobalt Atom*,[1] to the co-ordinated compounds of the metals, are very similar to those used by van't Hoff in 1874 in reference to the configuration of the carbon atom in his memoir on *Chemistry in Space*. Thus, starting with the symmetrical complex MA_6, we have seen (p. 46) that no isomerism appears in the mono-derivatives of the type MA_5B, since the isomerism which Gibbs described in the complex ions of the series $[Co . 5NH_3 . H_2O]Cl_3$ is now attributed to a change in the composition and not merely in the configuration of the ion. In the di-derivatives of the type $[MA_4B_2]$ or $[MA_4BC]$, *cis* and *trans* isomerism has been observed in a large range of cases, but none of these compounds is dissymmetric, since a plane of symmetry can be drawn through the central atom M and the four radicals AABC. It is, however, clear that, on diminishing the symmetry of the molecule still further, non-superposable isomers must be produced and that these should exhibit optical activity. The simplest cases in which optical activity could appear were set out by Werner as follows:

(i) *Complex radicals of the type* $[MeA_3BCD]$. If three dissimilar radicals B C D occupy the three corners of a triangular face, enantiomorphous forms of the compound can be obtained by giving to the groups B C D either a clockwise or a counter-clockwise sequence on the face of the octahedron, thus:

* Thus the index of Stewart's *Stereochemistry* (Longmans, 1919) has one entry in reference to the *Co-ordination hypothesis*, and six entries under the heading *Co-ordination hypothesis, defects of*.

† An excellent summary of Werner's work in this direction is given in a lecture *Sur les Composés Métalliques à Dissymétrie Moléculaire* delivered before the Société Chimique de France on May 24, 1912 (*Bull. Soc. Chim.*, 1912, [iv], **11**, i-xxiv). Later work on the same subject is summarised in lectures by Professor P. Job, *Les Complexes du Cobalt* (*Bull. Soc. Chim.*, 1923, [iv], **33**, 6–21), and by Professor F. M. Jaeger, *Sur le Pouvoir Rotatoire des Composés Chimiques* (*Bull. Soc. Chim.*, 1923, [iv], **33**, 853–889).

[1] Werner, *Ber.*, 1911, **44**, 1887, *Zur Kenntnis des asymmetrischen Kobaltatom*; continued in *Ber.*, 1911, **44**, 2445, 3272, 3279; 1912, **45**, 121, 3281, 3287, 3294; 1913, **46**, 3674; 1914, **47**, 1961, 2171, 3087.

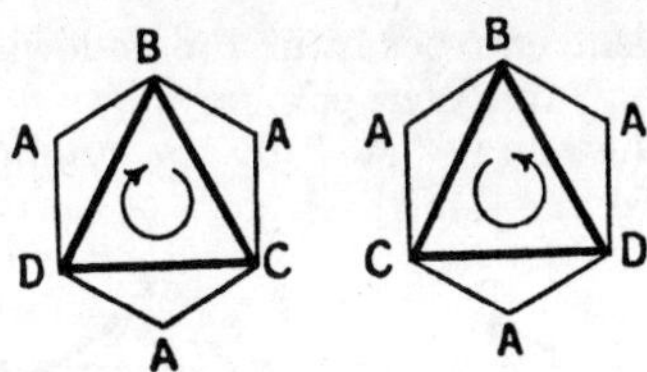

If, however, this condition is not fulfilled, enantiomorphism will not occur in compounds of this type.

(ii) *Complex radicals of the type* [$MeABC_2D_2$]. If the groups B C D do not lie at the corners of a face of the octahedron, they must necessarily occupy three corners of one of the squares which form the principal planes of symmetry of the octahedron. The four groups A B C D can then be arranged either in a clockwise or in a counter-clockwise direction round the square, as viewed from an outlying apex of the octahedron. But since this sequence would be reversed by viewing the square from the opposite apex, the arrangement now suggested would not suffice to give rise to enantiomorphism. If, therefore, optical activity is to appear in this case, the remaining plane of symmetry of the molecule must be destroyed by assigning different radicals to the apices on either side of the square section of the octahedron. These two radicals must differ from one another, but need not differ from the other radicals already present in the molecule (compare the case of allene, p. 43). Thus, if the groups A B C D be represented as occupying four corners of a square, it would be sufficient to introduce the radical A at one apex and B at the opposite apex, or C at one apex and D at the other. The two optical isomers could then be represented as follows:

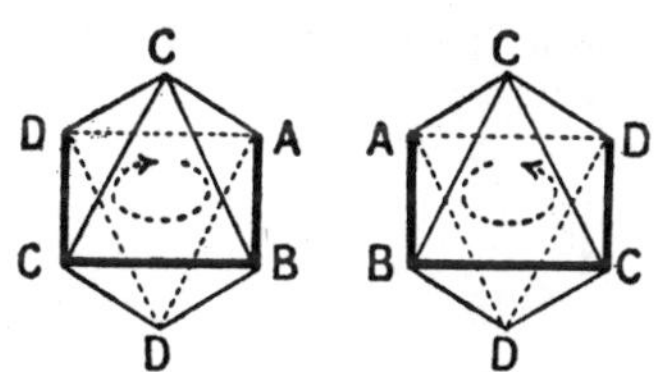

Enantiomorphism in Bridged Compounds.—Just as in the case of the dilactone resolved by Mills and Nodder (p. 67), the number of different groups that are required to produce molecular dissymmetry in co-ordination compounds is reduced very greatly when the groups are united by bridges of atoms. This bridging can be effected (*a*) by using a bivalent base, such as ethylene diamine ("en"), $NH_2 . CH_2 . CH_2 . NH_2$, as a substitute for two molecules of ammonia in compounds of the type of the cobaltammines, or (*b*) by using a dibasic acid, such as oxalic acid HO . CO . CO . OH, in place of the nitrous acid of compounds such as the cobaltinitrites. An examination of the octahedral model shows that, when bivalent radicals are used, enantiomorphism can occur in the following cases, where the bivalent radical "en" of Werner's formula is represented by the symbol E—E and the metal "Me" by the symbol M.

(i) *Radicals of the type* [MeAB en_2].

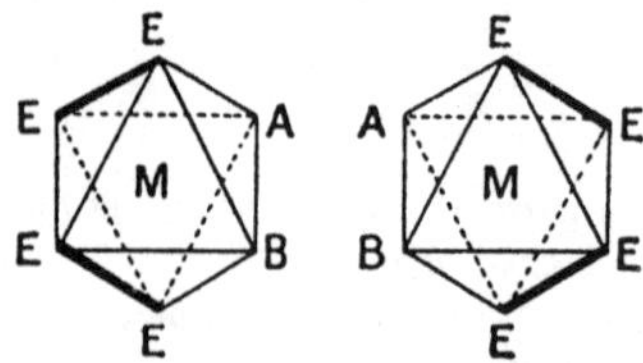

Werner distinguishes this type from (ii) and (iii) on the ground that it contains two non-superposable tetrahedra (A, B, en, Me), and may therefore be regarded as containing an asymmetric metallic atom.

(ii) *Radicals of the type* [MeB_2 en_2].

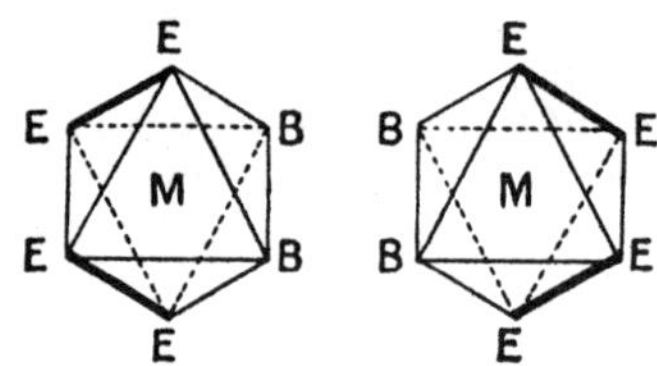

When A and B are identical the octahedron cannot be analysed into two non-superposable tetrahedra, and the molecule cannot therefore be regarded as containing an asymmetric cobalt atom; Werner[1] therefore distinguishes this type as "*Molecular dissymmetry* * *of type I.*"

In both of the above cases, the univalent radicals A and B, or B and B, must be in the *cis* (or "ortho") position to one another, since the *trans* (or "para") isomers have obviously a plane of symmetry as in the formulæ

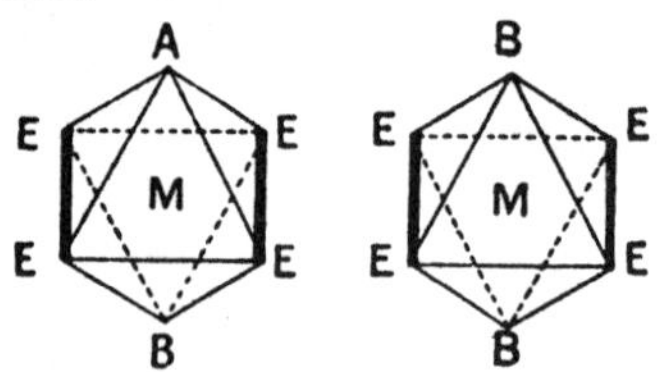

(iii) *Radicals of the type* [Me en_3].

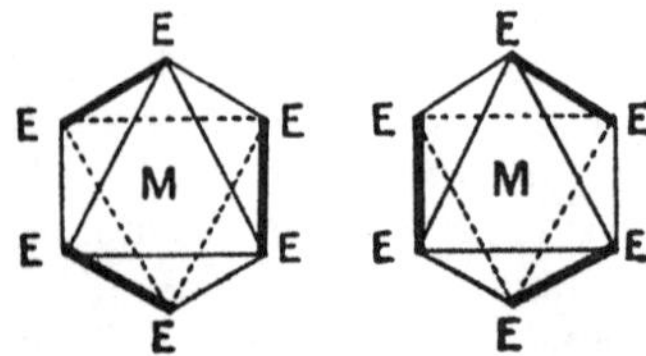

[1] *Ber.*, 1911, 44, 2446.

* See footnote on p. 75.

This type, which Werner describes[1] as "*Molecular dissymmetry* * *of type II*," is unique in that only one metallic atom and one kind of bivalent radical are required to produce molecular dissymmetry. The preparation of optically-active forms of compounds of such extremely simple composition may be regarded as a final proof of the truth of Werner's hypothesis, and its realisation places his octahedron on the same firm basis of fact as the tetrahedral carbon atom of le Bel and van't Hoff.

It should be noted that the method of projection used above to represent the asymmetric octahedron has the important advantage of showing clearly the existence of a three-fold axis of symmetry (with 3 two-fold axes of symmetry perpendicular to it) in each figure. The thick lines representing the three bivalent radicals are arranged round this three-fold axis like the blades of a propeller, but the arrangement is of opposite signs in two cases. Thus if the figures were rotated in a clockwise direction about the three-fold axis, the blades would tend to drive the first figure towards the observer, but the second figure would be driven away from him.

Methods of Resolving Dissymmetric Co-ordination Compounds.†—The methods used by Werner for resolving co-ordination compounds were in the main those modifications of Pasteur's third method which Pope had already introduced and used successfully for the resolution of asymmetric compounds of nitrogen, sulphur, selenium, etc. Although co-ordinated complexes can be obtained either as acid or as basic radicals, nearly all the resolutions described by Werner were of basic radicals, to which alone these special devices can be applied.‡ Their application to co-ordinated compounds may be summarised as follows:

(*a*) *Fractional Crystallisation of Salts.* Werner stated in 1912 [2] that with the help of Reychler's *d*-camphorsulphonic acid he had resolved two series of dissymmetric compounds, whilst with Kipping and Pope's *d*-bromocamphor-π-sulphonic acid he had resolved eight series of dissymmetric compounds. Thus the bivalent complex ions

$$\left[\begin{matrix}Cl \\ NH_3\end{matrix}\ Co\ en_2\right]^{++} \quad \text{and} \quad \left[\begin{matrix}Br \\ NH_3\end{matrix}\ Co\ en_2\right]^{++}$$

were resolved with the help of bromocamphorsulphonic acid, although the univalent and tervalent ions

* The term "asymmetry" used in the German text is here translated as "dissymmetry" in accordance with the nomenclature of Pasteur.

† Most of the information contained in this and the following sections is derived from a lecture by Werner,[2] but references to the original papers have been added in the majority of cases.

‡ The resolution by Werner of the acidic radical of the chromioxalates and rhodioxalates is described on pp. 84–85.

[1] *Ber.*, 1912, **45**, 122. [2] *Bull. Soc. Chim.*, 1912, [iv], **11**, i–xxiv.

$$\left[CO_3\,Co\,en_2\right]^{+} \quad \text{and} \quad \left[\begin{matrix}NH_3\\ \\ NH_3\end{matrix}\;Co\;en_2\right]^{+++}$$

could not be resolved by means of either of these acids. The univalent ion $\left[\begin{matrix}NO_2\\ \\ NO_2\end{matrix}\;Co\;en_2\right]^{+}$ was, however, resolved with the help of both acids.[1]

In most cases fractional crystallisation gave the less soluble isomer in a fairly pure form, whilst the more soluble isomer was troublesome to purify. In order to get over this difficulty, the sulphonates in the mother liquor were in many cases reconverted into chlorides, and reprecipitated with *sulphonates derived from lævocamphor.* Since, however, lævocamphor is not easy to procure, it was satisfactory to find that the same result could often be achieved by the alternate use of the β- and π-sulphonic acids derived from *d*-camphor and *d*-bromocamphor respectively.

Tartaric acid proved of comparatively little value in these resolutions. It was, however, found that the tervalent ions

$$[Co\;en_3]^{+++} \quad \text{and} \quad [Rh\;en_3]^{+++}$$

could be resolved in the form of chloro-tartrates,[2] whilst the former ion could also be resolved by making use of the bromotartrate [3]

$$[Co\;en_3]\;\begin{matrix}Br\\ C_4H_6O_4.\end{matrix}$$

(*b*) *Precipitation.* An important variant on these processes was the direct precipitation of insoluble sulphonates. Thus, the following complex ions were precipitated in optically-active forms by means of ammonium *d*-bromocamphorsulphonate,[4]

$$d\left[\begin{matrix}Cl\\ \\ NH_3\end{matrix}\;Co\;en_2\right]^{++} \quad d\left[\begin{matrix}Br\\ \\ NH_3\end{matrix}\;Co\;en_2\right]^{++} \quad l\left[\begin{matrix}Cl\\ \\ Cl\end{matrix}\;Co\;en_2\right]^{+}$$

$$d\left[\begin{matrix}Cl\\ \\ NO_2\end{matrix}\;Co\;en_2\right]^{+} \quad l\left[\begin{matrix}Cl\\ \\ CNS\end{matrix}\;Co\;en_2\right]^{+} \quad l\left[\begin{matrix}Cl\\ \\ Cl\end{matrix}\;Cr\;en_2\right]^{+}$$

In a similar way ammonium camphorsulphonate was used to precipitate the two optically-active ions [4]

$$l\left[\begin{matrix}Cl\\ \\ NO_2\end{matrix}\;Co\;en_2\right]^{+} \quad l\left[\begin{matrix}Br\\ \\ NO_2\end{matrix}\;Cr\;en_2\right]^{+}$$

[1] *Ber.*, 1911, **44**, 2450. [2] *Ber.*, 1912, **45**, 121.
[3] *Bull. Soc. Chim.*, 1912, [iv], **11**, i–xxiv.
[4] *Ber.*, 1911, **44**, 3275–3276.

This method was of particular advantage in the case of certain chloro-compounds which were of a low order of stability towards water, so they could not be crystallised from aqueous solutions without loss. Thus salts of the univalent ion $\left[\begin{matrix} \mathrm{Cl} & \\ & \mathrm{Co\ en_2} \\ \mathrm{Cl} & \end{matrix}\right]^{+}$, which were usually converted by the action of water into aquo-compounds containing the bivalent ion $\left[\begin{matrix} \mathrm{Cl} & \\ & \mathrm{Co\ en_2} \\ \mathrm{H_2O} & \end{matrix}\right]^{++}$, could be resolved quite successfully by precipitation.[1] The method worked best, however, with the salts of the series $\left[\begin{matrix} \mathrm{Cl} & \\ & \mathrm{Co\ en_2} \\ \mathrm{CNS} & \end{matrix}\right]^{+}$, since the lævorotatory form of this ion gives an almost insoluble *d*-bromocamphorsulphonate, whilst the dextrorotatory ion can be precipitated in combination with the ion of *l*-bromocamphorsulphonic acid.

The method of resolving co-ordinated salts by precipitation was extended still further by using the sodium derivative of nitrocamphor to precipitate the nitrocamphorates of certain co-ordinated compounds. This method was used successfully in the case of the two tervalent ions [2, 3]

$$[\mathrm{Cr\ en_3}]^{+++} \quad [\mathrm{Rh\ en_3}]^{+++}.$$

In one case it was even possible to use ammonium tartrate for this purpose, since an insoluble *d*-tartrate was obtained by double decomposition with a salt of the dissymmetric ion $[\mathrm{Fe\ dp_3}]^{++}$ where " dp " represents the α-dipyridyl complex $C_5H_4N\,.\,NC_5H_4$.[4] The structure of this remarkable ion is set out in the opposite formula, where the dipyridyl groups are seen to be arranged like propeller blades around the central axis of the molecule.

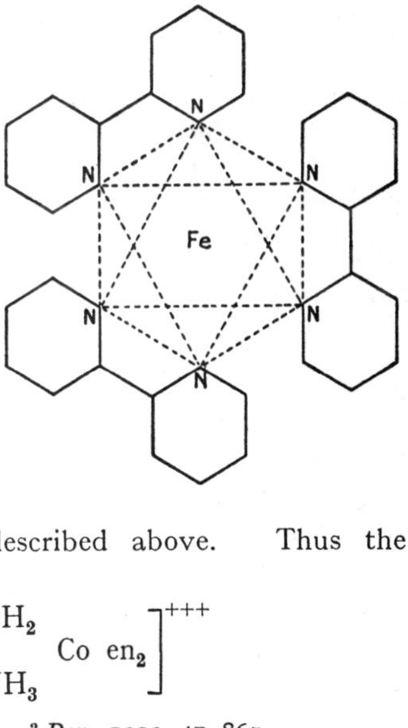

(*c*) *Chemical Methods.* In a large number of cases Werner prepared fresh series of optically-active salts by chemical processes from compounds which he had already resolved by the methods described above. Thus the hydrolysis shown in the scheme

$$\left[\begin{matrix} \mathrm{Br} & \\ & \mathrm{Co\ en_2} \\ \mathrm{NH_3} & \end{matrix}\right]^{++} \rightarrow \left[\begin{matrix} \mathrm{OH_2} & \\ & \mathrm{Co\ en_2} \\ \mathrm{NH_3} & \end{matrix}\right]^{+++}$$

[1] *Ber.*, 1911, **44**, 3279.
[2] *Ber.*, 1912, **45**, 867.
[3] *Ibid.*, p. 1231.
[4] *Ibid.*, p. 433.

could be effected without racemisation, by the action of silver nitrate on aqueous solutions of the bromo-salt, and gave rise to an optically active aquo-salt of the type formulated above. Again, by the action of liquid ammonia, the bromo-compound could be converted into an optically-active diammine [1]

$$\left[\begin{matrix}Br & \\ & Co\ en_2 \\ NH_3 & \end{matrix}\right]^{++} \rightarrow \left[\begin{matrix}NH_3 & \\ & Co\ en_2 \\ NH_3 & \end{matrix}\right]^{+++}$$

The two molecules of ammonia in the optically-active product occupy the *cis* position; but some of the inactive *trans* compound was also produced by molecular rearrangement.

Racemisation of Optically-Active Co-ordination Compounds.—(*a*) *Racemisation in Solution.* The stability of optically-active co-ordination compounds varies very greatly. Thus, the complex ion, $\left[\begin{matrix}NO_2 & \\ & Co\ en_2 \\ NO_2 & \end{matrix}\right]^{+}$, retains its optical activity indefinitely even in aqueous solutions since the salts can be recrystallised from hot water,[2] whilst the optical rotatory power of the ion $\left[\begin{matrix}NO_2 & \\ & Co\ en_2 \\ CNS & \end{matrix}\right]^{+}$ falls to about a quarter in four months. Again, the complex ion $\left[\begin{matrix}Cl & \\ & Co\ en_2 \\ Cl & \end{matrix}\right]^{+}$ lost its rotatory power completely in the course of a few hours;[3] the corresponding carbonato-compound $[CO_3\ Co\ en_2]^+$ retained its activity for some days in the cold, but lost it on heating to 90°.[4]

A remarkable instance of "mutarotation" (p. 271) was recorded [5] in the salts of the complex univalent ion $\left[\begin{matrix}Cl & \\ & Co\ en_2 \\ NO_2 & \end{matrix}\right]^{+}$. The rotatory power of this ion rose first to practically double its initial value, but then fell to zero in the course of two days. The initial increase of rotatory power was found to be due to hydrolysis to the bivalent aquo-compound $\left[\begin{matrix}H_2O & \\ & Co\ en_2 \\ NO_2 & \end{matrix}\right]^{++}$, which, however, rapidly underwent racemisation.

(*b*) *Racemisation during Chemical Change.* (i) *Univalent ions.* In some cases chemical changes can be effected without producing any racemisation, as when Cl is replaced by NO_2 in the action set out below [6]

$$\left[\begin{matrix}Cl & \\ & Co\ en_2 \\ NO_2 & \end{matrix}\right]^{+} + NaNO_2 \rightarrow \left[\begin{matrix}NO_2 & \\ & Co\ en_2 \\ NO_2 & \end{matrix}\right]^{+} + NaCl.$$

[1] *Ber.*, 1912, **45**, 3287.
[2] *Ber.*, 1911, **44**, 2445.
[3] *Ibid.*, p. 3282.
[4] *Ber.*, 1912, **45**, 3285.
[5] *Ber.*, 1911, **44**, 3273.
[6] *Ibid.*, p. 3277.

Partial racemisation occurs, however, when the halogen is replaced by CNS as in the equation[1]

$$\left[\begin{matrix}Cl \\ NO_2\end{matrix}\ Co\ en_2\right]^+ + KCNS \rightarrow \left[\begin{matrix}CNS \\ NO_2\end{matrix}\ Co\ en_2\right]^+ + KCl.$$

Very extensive racemisation also took place in the preparation of the carbonato-compound[2]

$$\left[\begin{matrix}Cl \\ Cl\end{matrix}\ Co\ en_2\right]^+ + K_2CO_3 \rightarrow \left[CO_3\ Co\ en_2\right]^+ + 2KCl.$$

The reverse action, expressed by the equation

$$\left[CO_3\ Co\ en_2\right]^+ + \underset{\text{(alcoholic)}}{2HCl} \rightarrow \left[\begin{matrix}Cl \\ Cl\end{matrix}\ Co\ en_2\right]^+ + H_2O + CO_2,$$

was accompanied by complete racemisation. Complete racemisation also occurred when the carbonate was decomposed by aqueous hydrochloric acid, giving rise to a tervalent diaquo-compound, as shown in the equation

$$\left[CO_3\ Co\ en_2\right]Cl + \underset{\text{(aqueous)}}{2HCl} + H_2O \rightarrow \left[\begin{matrix}H_2O \\ H_2O\end{matrix}\ Co\ en_2\right]Cl_3 + CO_2.$$

(ii) *Bivalent Ions.* The two bivalent ions

$$\left[\begin{matrix}Cl \\ NH_3\end{matrix}\ Co\ en_2\right]^{++} \quad \text{and} \quad \left[\begin{matrix}Br \\ NH_3\end{matrix}\ Co\ en_2\right]^{++}$$

do not undergo racemisation when converted into aquo-compounds by the prolonged action of water or by the action of silver nitrate (see above, p. 77); but total racemisation occurs when the hydrolysis is effected by means of an alkali.

The action of ammonia on the bromo-compound is also of interest in that the optically-active *cis* diammine, which is formed by direct displacement of the halogen by ammonia (see above, p. 78) is accompanied by some of the isomeric *trans* diammine. The latter compound, as required by Werner's theory, is completely inactive.

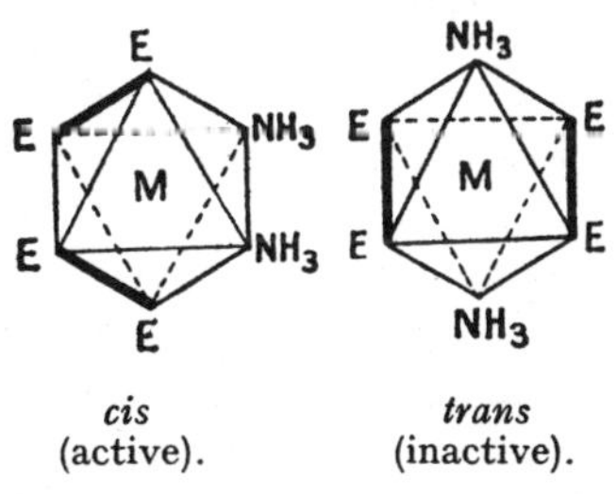

cis (active). *trans* (inactive).

[1] *Ber.*, 1911, **44**, 3278.

[2] *Ber.*, 1912, **45**, 3284.

The bivalent ferrous salts[1] such as [Fe $dp_3]^{++}$ are of interest on account of the fact that they possess a red colour and contain iron in a form that is completely masked, recalling the properties of hæmoglobin. The bromide [Fe $dp_3]Br_2$, prepared through the tartrate [Fe $dp_3]C_4H_4O_6$, has a molecular rotatory power [M] = 4000°; but the salts racemise very rapidly in solution, the rotatory power diminishing to one half in half an hour. Werner suggests that racemisation is due to the reversible separation of one dipyridyl radical, giving rise to a compound with a co-ordination number of only 4. It is important to notice that this optically-active compound contains bivalent ferrous iron, whilst in all the other cases now cited optical activity is associated with a valency of at least three in the central metallic atom.

(*c*) *Behaviour of Tervalent Ions.* Tervalent ions of the type [Co $en_3]^{+++}$ are characterised by a remarkable degree of stability, which may perhaps be attributed to the high degree of symmetry which they possess, as well as to the fact that each molecule of the diammine is held at two points and cannot therefore escape so easily, e.g. by dissociation from the central metallic atom. Thus, whilst salts of the series $\left[\begin{matrix}Cl & \\ & Cr\ en_2 \\ Cl & \end{matrix}\right]^+$ become completely inactive in a few hours, the salts of the series [Cr $en_3]^{+++}$ are almost completely stable in solution,[2] although their optical activity diminishes when the solution is evaporated. The rhodium salts of the series [Rh $en_3]^{+++}$, which were resolved by precipitation with a sodium derivative of nitrocamphor or by crystallising out the chlorotartrate,[3] were also stable in solution.

Co-ordination Compounds Containing more than one Dissymmetric Atom..—Special interest attaches to co-ordination compounds containing two dissymmetric metallic atoms, since these may be expected to show a close analogy to the properties of substances, such as tartaric acid, which contain two asymmetric carbon atoms.

(*a*) Compounds containing *two* cobaltic atoms in a complex ion (compare the two platinum atoms in Magnus' green salt [Pt $4NH_3]PtCl_4$) had been prepared many years previously, but it was not until a comparatively late date that their structure was established. Thus it was in 1907[4] that Werner proposed to double the formula of a compound $Co(OH)SO_4$, $4NH_3$, prepared by Gentele in 1856, and to write it as

$$\left[4NH_3\ Co\begin{matrix}\diagup OH \diagdown \\ \diagdown HO \diagup\end{matrix}Co\ 4NH_3\right](SO_4)_2$$

Octammine-μ-diol-dicobaltic sulphate.

[1] *Ber.*, 1912, **45**, 433.
[2] *Ibid.*, p. 867.
[3] *Ibid.*, p. 1231.
[4] *Ber.*, 1907, **40**, 4434.

In the same way an insoluble red sulphate, prepared by Vortmann in 1885[1] by oxidising ammoniacal cobaltous nitrate by a slow current of air and then adding sulphuric acid, was formulated by Werner in 1907[2] as

$$\left[4NH_3\ Co\begin{matrix} NH_2 \\ \\ HO \end{matrix}Co\ 4NH_3\right](SO_4)_2$$

Octammine-μ-amino-ol-dicobaltic sulphate.

These compounds are of interest as containing radicals, OH or NH_2, which are shared by two atoms of cobalt; the radicals which act as bridges or links between two metallic atoms are indicated by the symbol μ.

(*b*) The octammines formulated above must obviously be represented by two octahedra sharing a common edge as in the scheme

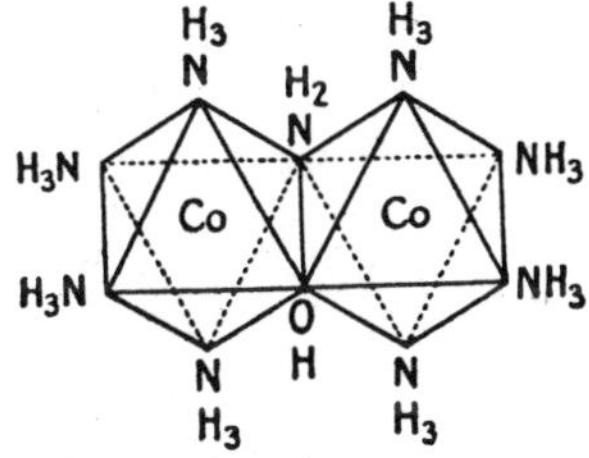

Since, however, two octahedra can also share either a corner or a face, it should be possible to prepare a complete series of compounds in which one, two and three radicals are shared by two cobaltic atoms as follows:

$$\textit{Decammine}\quad \left[5NH_3\ Co\ \begin{matrix} OH \\ or \\ NH_2 \end{matrix}\ Co\ 5NH_3\right]^{\substack{+++\\++}}$$

$$\textit{Octammine}\quad \left[4NH_3\ Co\ \begin{matrix} OH \\ \\ HO \end{matrix}\ Co\ 4NH_3\right]^{\substack{++\\++}}$$

$$\textit{Hexammine}\quad \left[3NH_3\ Co\ \begin{matrix} OH \\ HO \\ OH \end{matrix}\ Co\ 3NH_3\right]^{+++}$$

(i) The tervalent HEXAMMINE ion formulated above is produced[3] by hydrolysis of aquo-triammines of the type

$$\left[\begin{matrix} 3NH_3 & & Cl \\ & Co & \\ H_2O & & Br \end{matrix}\right]^{+}$$

[1] VORTMANN, *Monatshafte*, 1885, **6**, 412.
[2] WERNER, *Ber.*, 1907, **40**, 4605.
[3] *Ibid.*, p. 4434.

Thus the hexammine chloride is obtained by the action of sodium hydroxide on the univalent dichloroaquo-triammine

$$\left[\begin{array}{ccc}3NH_3 & & Cl\\ & Co & \\ H_2O & & Cl\end{array}\right]Cl \xrightarrow{NaOH} \left[\begin{array}{ccccc} & & OH & & \\ 3NH_3 & Co & HO & Co & 3NH_3\\ & & OH & & \end{array}\right]Cl_3,$$

whilst the tervalent hexammine bromide is formed by the action of water at 60° on the chlorobromoaquo-triammine bromide

$$\left[\begin{array}{ccc}3NH_3 & & Cl\\ & Co & \\ H_2O & & Br\end{array}\right]Br \xrightarrow{H_2O} \left[\begin{array}{ccccc} & & OH & & \\ 3NH_3 & Co & HO & Co & 3NH_3\\ & & OH & & \end{array}\right]Br_3.$$

(ii) The hexammine formulated above is converted by the successive action of nitric acid and of liquid ammonia into a quadrivalent OCTAMMINE, which can then be separated as a bromide by the action of ammonium bromide

$$\left[\begin{array}{ccccc} & & OH & & \\ 3NH_3 & Co & HO & Co & 3NH_3\\ & & OH & & \end{array}\right]^{+++} \xrightarrow{HNO_3} \left[\begin{array}{ccccc}3NH_3 & & OH & & 3NH_3\\ & Co & & Co & \\ H_2O & & HO & & NO_3\end{array}\right]^{+++}$$

$$\xrightarrow{NH_3} \left[\begin{array}{cccc} & & OH & \\ 4NH_3\ Co & & & Co\ 4NH_3\\ & & HO & \end{array}\right]^{++++}$$

(iii) Similar treatment converts the octammines into quinquevalent DECAMMINES, e.g.

$$\left[\begin{array}{ccc} & NH_2 & \\ 4NH_3\ Co & & Co\ 4NH_3\\ & HO & \end{array}\right]^{++++} \xrightarrow{HNO_3} \left[\begin{array}{ccc}4NH_3 & & 4NH_3\\ & Co\ NH_2\ Co & \\ H_2O & & NO_3\end{array}\right]^{++++}$$

$$\xrightarrow{NH_3} \left[5NH_3\ Co\ NH_2\ Co\ 5NH_3\right]^{+++++}$$

(*c*) These compounds are of interest on account of the analogy which the octahedral models show with van't Hoff's tetrahedral models for compounds containing single, double and triple bonds. It is, moreover, not difficult to work out the conditions under which optical activity may appear in compounds of this class, and to recognise that these conditions are essentially similar to those which prevail in the case of tartaric acid. As in the case of complex ions containing only one atom of cobalt (p. 73), molecular dissymmetry is produced most readily with the help of bivalent radicals. These can be introduced quite readily by the action of ethylene diamine on ammines such as those formulated above. Thus the compound I (which contains two cobalt atoms linked together by a peroxide group O_2, but also holding an amino group, NH_2, in the co-ordination spheres of both atoms) gives with ethylene diamine the compound II which contains two dissymmetric atoms of cobalt

$$\text{I.}\left[\begin{array}{ccc} & NH_2 & \\ 4NH_3\,Co & & Co\,4NH_3 \\ & O_2 & \end{array}\right]^{++++} \longrightarrow \left[\begin{array}{ccc} & NH_2 & \\ en_2\,Co & & Co\,en_2 \\ & O_2 & \end{array}\right]^{++++}\text{II.}$$

Octammino-μ-amino-peroxo-dicobaltic salts. Tetraethylenediamino-μ-amino-peroxo-dicobaltic salts.

With the help of bromocamphorsulphonic acid this compound was resolved into *d* and *l* components of remarkably high optical activity, the molecular rotatory power of the bromide being as high as $\pm 6855°$.[1] The corresponding aminonitro compound,

$$\left[\begin{array}{ccc} & NH_2 & \\ en_2\,Co & & Co\,en_2 \\ & O_2N & \end{array}\right]^{++++}$$

prepared by the action of nitrous acid on the aminoperoxo-compound, gave *three* isomeric bromocamphorsulphonates, and these yielded dextro-, lævo-, and meso- (inactive non-resolvable) forms of the basic radical.[2] No difficulty is experienced in formulating the meso-compound as III :

III.

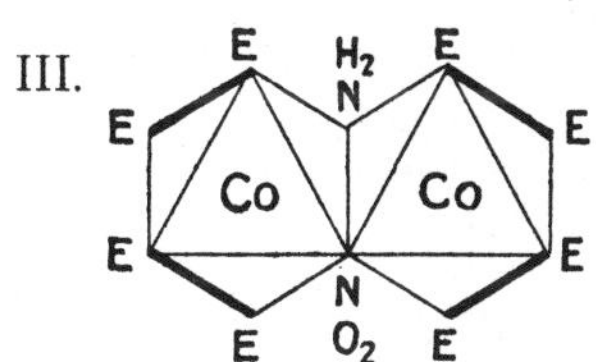

but in order to represent the optically-active forms of the substance it is necessary to show the ethylene diamine molecules in one of the two co-ordinated systems by means of " meta " linkages, as in IV, although these are in no way different from " ortho " linkages, since in each case the two nitrogens of ethylene diamine are in a *cis* position.

IV.

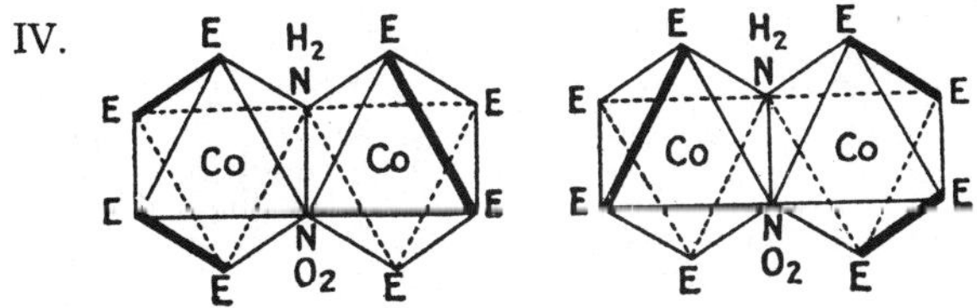

It is noteworthy that, when the optically-active salts are boiled strongly, a partial conversion to the non-resolvable meso-form takes place, just as when tartaric acid is heated with hydrochloric acid

[1] *Ber.*, 1914, **47**, 1961. [2] *Ber.*, 1913, **46**, 3674 ; 1914, **47**, 1961.

at 200°. The inactivity of the meso-form is used by Werner as evidence that the μ-amino- and nitro-groups are situated symmetrically in reference to the two cobalt atoms. Thus, if either of these groups were linked to one atom of cobalt in a different manner to the other, the symmetry of the molecule would disappear, and optical activity might be expected to persist even when the compound is converted into the meso-form.

Optically-active Compounds containing no Carbon.—After resolving a series of compounds containing the group $[Co\ en_3]^{+++}$, Werner[1] varied the experiment by replacing the ethylene diamine by the complex $[4NH_3 . Co(OH)_2]^+$. In this way he obtained salts of a sexabasic radical, e.g.

$$[Co . 3\{4NH_3 . Co . (OH)_2\}]Br_6, 2H_2O.$$

This radical, which is of the same type as $[Co\ en_3]^{+++}$, was resolved by converting it into a π-bromocamphorsulphonate, giving salts of very high rotatory power, containing no carbon at all, but so unstable that they were racemised completely in solution in two hours.

An optically-active inorganic salt has also been prepared by Mann,[2] who was able to resolve *sodium diaquorhodiumdisulphamide* $Na[(H_2O)_2Rh(N_2H_2SO_2)_2]$, i.e. the sodium salt of the univalent kation represented by the formula

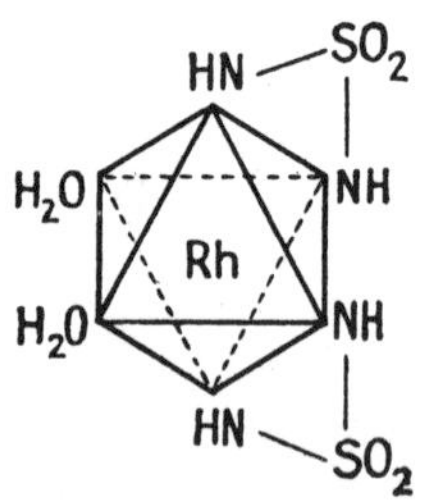

Optically-active Oxalates.—(*a*) *Chromioxalates.* A new chapter in the history of optically-active co-ordination compounds was opened when Werner in 1912[3] succeeded in preparing optically-active forms of the double oxalate $K_3Cr(C_2O_4)_3$. This was resolved by converting *potassium barium chromioxalate*, $KBa[Cr(C_2O_4)_3]$, $2H_2O$ into $KH_2[Cr(C_2O_4)_3]$, and then precipitating the *potassium distrychnine salt* by the addition of strychnine. From this strychnine salt the dextrorotatory *tripotassium* d-*chromioxalate* $K_3[Cr(C_2O_4)_3]$, H_2O was readily prepared, with $[\alpha]_G + 1300°$ in water; but it lost its rotatory power in rather more than an hour. The *d* and *l* salts are therefore very easily interconvertible, and this would account for the fact that the whole of the chromioxalate was obtained in a dextrorotatory form, without any corresponding accumulation of

[1] *Ber.*, 1914, **47**, 3087. [2] MANN, *J.*, 1933, 412.
[3] WERNER, *Ber.*, 1912, **45**, 3061–3070.

the lævo-acid in the mother liquors. The *tri*strychnine salt, on the other hand, consisted entirely of the *l*-chromioxalate, so that the enantiomorphous *tripotassium* l-*chromioxalate* could readily be prepared from it; but this also racemised with great rapidity.

These oxalates are of special interest, not only by reason of their rapid autoracemisation, but also because of their simple chemical composition. Apart from Werner's theory, no one would have thought of looking for optical activity in a salt such as $K_3Cr(C_2O_4)_3$, which appears on ordinary scrutiny to be no more promising than a dichromate or an alum. The octahedral arrangement of three bridge groups round a central metallic atom, however, leads inevitably to two non-superposable structures as set out in the formulæ

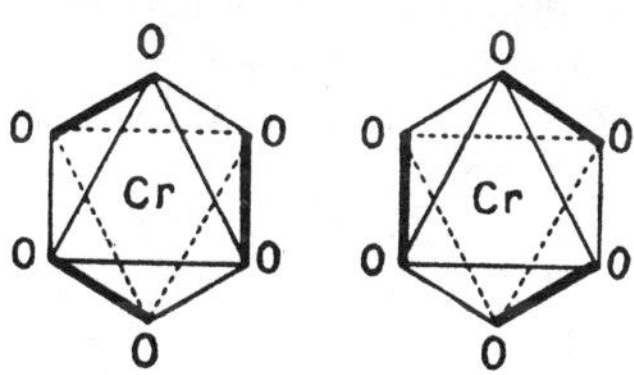

where the symbol O—O is used as an abbreviation for the bivalent radical $\bar{O}$. CO . CO . $\bar{O}$. The discovery of optical activity in the complex oxalates affords a decisive proof of the correctness of these structures, and a dramatic vindication of the theory on which they were based.

(*b*) *Rhodioxalates.* Werner also prepared optically-active forms of the corresponding rhodium oxalates such as $K_3Rh(C_2O_4)_3$, H_2O and $Na_3Rh(C_2O_4)_3$, $4H_2O$.[1] These have a smaller rotatory power than the chromioxalates but are much more stable, since the solutions can be warmed on the water-bath without appreciable change. The two strychnine salts can also be separated by fractional crystallisation, without undergoing isomeric change as a result of auto-racemisation of the acid radical. Werner stated that the enantiomorphous potassium salts could be separated by Pasteur's method of crystallising from solution and picking out mechanically the two opposite types of crystals; but this has been denied by Jaeger,[2] who was unable to repeat this observation, and identified Werner's crystals as having the typical form of the racemate and not of the optically-active salts.

(*c*) *Iridioxalates.* The iridioxalates, e.g. $K_3Ir(C_2O_4)_3$, prepared in optically-active forms by Delepine,[3] by fractional crystallisation of the *tri*strychnine salts, are also stable, the rotations diminishing only by a few hundredths of a degree per hour at 100°.

(*d*) *Cobaltioxalates.* Jaeger and Thomas[4] in 1919 resolved *potassium cobaltioxalate* $K_3Co(C_2O_4)_3$ into its optically-active components

[1] WERNER, *Ber.*, 1914, **47**, 1954.
[2] JAEGER, *Proc. Akad. Wet., Amsterdam*, 1917, **20**, 264.
[3] DELEPINE, *C.R.*, 1914, **159**, 239.
[4] JAEGER and THOMAS, *Proc. Akad. Wet., Amsterdam*, 1919, **21**, 693.

by spontaneous evaporation at temperatures above 13·2°, just like sodium ammonium tartrate below 26° (p. 30), and claimed that this was the first example of its kind amongst complex metallic salts. As an alternative method of resolution, the strychnine salt of the lævo-acid was precipitated directly by the addition of strychnine sulphate to the potassium salt, and was then converted into *potassium* l-*cobaltioxalate* by the addition of potassium iodide, which threw down the strychnine as iodide.

The racemic form of potassium cobaltioxalate crystallises in the triclinic (or anorthic) system with 3½ H_2O, i.e. as *d* $K_3Co(C_2O_4)_3$, *l* $K_3Co(C_2O_4)_3$, $7H_2O$. The active forms have the crystal symmetry of quartz and contain only 1 H_2O. They differ widely in stability from the chromioxalates, which racemise in about an hour, but do not possess quite the same degree of stability as the rhodioxalate and iridioxalate, which are only racemised slowly even at 100°.

(*e*) *Ferrioxalates.* Thomas in 1921 [1] resolved *potassium ferrioxalate*, $K_3Fe(C_2O_4)_3$, by crystallising out the optically-active α-phenylethylamine salt, and converting it quickly through the barium salt into the potassium salt.[2] It had a high optical rotatory power, but lost one half of its optical activity in 15 minutes. It is therefore even less stable than the chromioxalate. The rapid racemisation of the salt was attributed to a secondary ionisation of the complex as shown in the scheme

$$[Fe\,.\,3C_2O_4]^{---} \rightleftharpoons [Fe\,.\,2C_2O_4]^{-} + [C_2O_4]^{--}.$$

This can also be expressed in a more commonplace way as a dissociation of the tripotassium salt, thus

$$K_3Fe(C_2O_4)_3 \rightleftharpoons KFe(C_2O_4)_2 + K_2C_2O_4.$$

(*f*) A number of *optically active malonates* have also been prepared by Jaeger and Thomas.[3]

Co-ordination Compounds of Beryllium, Zinc and Copper.—An entirely different type of optically-active co-ordination compound is seen in the derivatives of beryllium, zinc and copper resolved by Mills and Gotts.[4] These differ from the metallic compounds, which were resolved by Werner and his successors in this field, not only because they have a tetrahedral in place of an octahedral configuration, but also because only two of the apices of the tetrahedron are occupied by "principal valencies," formed (as in the co-ordinated oxalates) by replacement of hydrogen by a metal, whilst the other two apices are occupied merely by "subsidiary

[1] THOMAS, *J.*, 1921, **19**, 1140.

[2] C. H. JOHNSON (*T.F.S.*, 1932, **28**, 845) was unable to repeat this resolution, and could not confirm the resolution of the aluminium compound, $K_3Al(C_2O_4)_3$, $3H_2O$, reported by Wahl in 1927 (*Ber.*, 1927, **60**, 399).

[3] JAEGER and THOMAS, *Proc. Akad. Wet., Amsterdam*, 1918, **21**, 215, etc.; *J.*, 1921, **119**, 1140.

[4] MILLS and GOTTS, *J.*, 1926, 3121.

valencies," linking the metal in some less-defined way to the oxygen of a carbonyl group, as in the scheme

$$>C{=}O \;.\;.\;.\; Be{-}O{-}.$$

The nature of this linkage has formed the subject of much discussion, and it is by no means clear that a final interpretation has yet been reached ; but its reality is clearly proved by the experiments described below, since, if no such linkage existed, the compounds in question would all be of the type MA_2B_2, where A is a univalent radical and B is a lone pair of electrons, and would be too symmetrical to be resolved into optically-active forms.

In the case of beryllium, the existence of optical activity was first deduced from observations of mutarotation in *beryllium benzoylcamphor.*[1]

I. $C_8H_{14}<$ $C{=}C(C_6H_5)\,.\,O$ | $C{=}O$ . . . . **Be** . . . $O{=}C$ | $O\,.\,(C_6H_5)C{=}C$ $>C_8H_{14}$

This phenomenon was explained most readily as due to a reversal in sign of the dissymmetric atom of beryllium shown in the centre of the formula I. Decisive proof that beryllium compounds of this type are dissymmetric was obtained when Mills and Gotts [2] resolved the *beryllium derivative* II of *benzoylpyruvic acid* into its optically-active components.

II. C_6H_5, CH, $HO\,.\,CO$ $>C{-}O$ / $>C{=}O$ $>$ **Be** $<$ $O{=}C<$ / $O{-}C<$ $CO\,.\,OH$, CH, C_6H_5

The molecular dissymmetry of this compound is determined by the tetrahedral configuration of the central atom, and by the unsymmetrical character of the bivalent radicals associated with it, which carry a phenyl group at one end and a carboxyl group at the other. The distribution of valencies in the six-atom ring is then seen to be irrelevant, since the molecular dissymmetry of the molecule would persist even if the distribution of the double and single bonds were reversed, or if they were replaced by some completely symmetrical system of linkages, as in the familiar case of benzene.

Similar results were obtained with the analogous derivatives of copper and zinc. This extension is important, not merely because it adds to the list of elements which have been proved to give rise to molecular dissymmetry, but also because it establishes a tetra-

[1] Lowry and Burgess, *J.*, 1924, **125**, 2081.
[2] Mills and Gotts, *J.*, 1926, 3121.

hedral configuration for the co-ordinated derivatives of two elements which have always been classified as bivalent. Reference should also be made to the platinum compound III, which has been resolved by Mann.[1] In this case molecular dissymmetry depends on the difference between an amino group which is co-ordinated to the metal and one which is not co-ordinated. Its optical resolution, therefore, proves the reality of the link between the metal and the amino group, just as the resolution of the beryllium compound of benzoyl pyruvic acid proves the reality of the link between the metal and the carbonyl group.

$$\text{III.}\quad \left[Cl_4Pt \begin{array}{l} \cdots NH_2 \,.\, CH_2 \\ \qquad\qquad | \\ \diagdown NH_2 \,.\, CH \,.\, CH_2 \,.\, \overset{+}{N}H_3 \end{array} \right] \overset{-}{Cl}.$$

[1] Mann, *J.*, 1927, 1224.

CHAPTER VII.

SPECIFIC AND MOLECULAR ROTATIONS.

Constancy of Specific or Molecular Rotatory Power.— When Biot in 1836 introduced his definition of specific or molecular rotatory power, $[\alpha] = \frac{\alpha}{l\epsilon\delta}$ (p. 22), his experiments led him to believe that this function was a constant property of the molecule. The circumstances in which this view was adopted he afterwards described [1] as follows :

> " In the numerous experiments which I made, between the years 1815 and 1832, to determine the laws according to which molecular rotatory power acts, I worked on organic compounds of mobile structure, sugars, gums, camphors, essential oils, which I was anxious not to alter. This obliged me to dissolve them in inactive and neutral solvents, water, alcohol, ether, fatty oils, which could not modify them chemically, at least during the period covered by the optical observations.
>
> " In these circumstances the specific rotatory power, reduced to unit thickness and unit weight of the active substance, was found to be always in the same sense and of the same intensity, whatever the proportion of inactive solvent which was associated with it. The active molecules thus seemed merely to disperse themselves amongst the inactive molecules, as if in free space, without suffering from them any action which modified their rotatory power appreciably. As a result, I was able to establish the physical law governing the phenomena in these simple conditions ; and the mathematical formula, by which I expressed them, fitted the experiments so well that one could calculate the results in every detail as accurately as they were given by the observations.

Detailed experiments carried out under these conditions showed that $[\alpha]$ was constant within the limits of experimental error, since it was independent of variations in the thickness l of the column of the liquid, of the concentration ϵ of the active material, and of the nature of the indifferent solvents in which it could be dissolved without fear of chemical action. These observations, therefore, appeared to establish beyond question the validity of the hypothesis

[1] Biot, *A.C.P.*, 1852, [iii], **36**, 258.

which underlies the conception of a constant specific or molecular rotatory power, namely, that "the molecules of the active substance are scattered in the inactive medium in which they are dissolved, as if in empty space, without influencing or being influenced by them" (*ib.*, p. 260).

Variable Rotatory Power of Tartaric Acid.—Up to this point optical rotatory power had appeared in the guise of a physical property by which molecules could be distinguished and their identity established. Further experiments showed, however, that tartaric acid, dissolved in water, alcohol or wood spirit, not only exhibited an abnormal sequence of spectral colours, but failed to maintain the rigorous constancy of specific rotatory power that had been observed hitherto. In particular, the rotatory power increased markedly and progressively as the proportion of solvent was increased, but to an unequal extent for different rays, so that the sequence of colours was changed continuously.

The increase of rotatory power on dilution could be expressed by a linear formula

$$[\alpha] = A + Be$$

where e is the proportion by weight of solvent and A and B are arbitrary constants.[1] Thus for red light the equation for aqueous solutions at about 12° was

$$[\alpha]_r = -\ 1{\cdot}17987^\circ + 14{\cdot}3154e.$$

When plotted in the form of a diagram (Fig. 28),[2] the lines showing the influence of dilution at different temperatures were parallel to one another, showing that the increment B was the same, but the limiting value A was different. Thus the data for red light at 5° could be expressed by the equation [3]

$$[\alpha]_r = -\ 2{\cdot}39340 + 14{\cdot}3154e,$$

and at 26° the corresponding equation [4] was

$$[\alpha]_r = +\ 0{\cdot}31739 + 14{\cdot}3154e.$$

The lines corresponding with these equations were, however, unequally inclined for light of different wave-lengths, and might therefore be expected to intersect.

This linear equation was found to be valid over the whole range from the solid amorphous acid, where $e = 0$, to the final stage of dilution, when the rotations were too small for accurate measurement in tubes 1 metre long.[5]

The same linear law of variation of rotatory power, with varying concentration but at constant temperature, was maintained when the rotatory power of tartaric acid was observed in aqueous solutions

[1] Biot, *Mém. Acad. Sci.*, 1838, **15**, 207.
[2] *Ibid.*, p. 280. [3] *Ibid.*, p. 244. [4] *Ibid.*, **16**, 269.
[5] Biot, *A.C.P.*, 1844, [iii], **10**, 5, 175, 307, 385.

of mineral acids such as sulphuric and hydrochloric acids,[1] which appeared to have no immediately perceptible action on tartaric acid, but were supposed to exercise a more or less powerful affinity on the water in which the acid was dissolved.[2] The linear law was also established for aqueous solutions of the alkalis,—potash, soda and ammonia.[3] In these new experiments, however, the coefficient A could assume very different values according to the nature of the

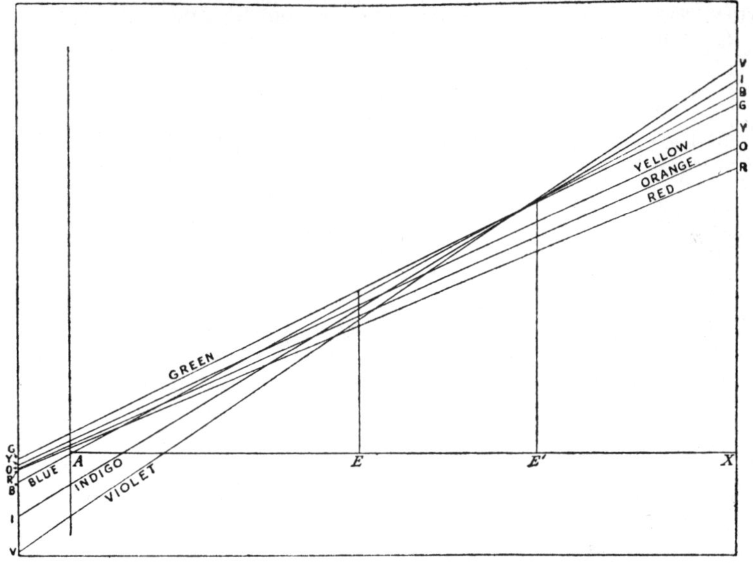

FIG. 28.—SPECIFIC ROTATORY POWER OF AQUEOUS SOLUTIONS OF TARTARIC ACID (Biot, 1838).

The proportions of water, e, are plotted as abscissæ, whilst the specific rotation of the acid for a given concentration and colour is plotted as the corresponding ordinate. The ordinate through A, where $e = 0$, shows the value of the constant A of the equation $[\alpha] = A + Be$, whilst the slope of the line corresponds with the magnitude of the coefficient B for a given colour. The ordinate through X shows the extrapolated value of the specific rotation at infinite dilution. The sequence of colours at X is that of the natural spectrum, since the specific rotation increases continuously as the wave-length decreases in dilute solutions, i.e. for all concentrations on the right of E'. In more concentrated solutions, however, e.g. at E, the specific rotation rises to a maximum in the green.

solvent; and the "coefficient of proportionality" B showed a very wide range of variation and could be "increased, decreased, cancelled or even inverted at the whim of the experimenter."[4]

These variations of rotatory power were incompatible with the idea that the molecules of tartaric acid were merely scattered, without

[1] BIOT, *Mém. Acad. Sci.*, 1838, **16**, 271–304.
[2] *Ibid.*, p. 271.
[3] *Ibid.*, pp. 304–377.
[4] *A.C.P.*, 1852, [iii], **36**, 266.

mutual interaction, amongst the molecules of the solvent. Biot therefore regarded them as evidence of the formation of new chemical systems, e.g. hydrates and alcoholates of tartaric acid, and described the memoir in which they were first recorded as a study in "chemical mechanics." "This variability of rotatory power," in Biot's opinion, "extended greatly the ideas that one had formed of the importance of this property, since, instead of providing merely a physical property for purposes of identification, it proved to be also a chemical property, by which the formation of compounds amongst neighbouring molecules could be disclosed."[1] Experience has shown that optical rotatory power is of even greater value in disclosing changes of molecular structure which are not accompanied by changes of chemical composition as, for instance, the reversible isomeric changes which give rise to the phenomenon of mutarotation (p. 271).

Variable Rotatory Power in Borotartaric Acids.—The utility of optical rotatory power as a means of studying the formation of elusive chemical compounds was established in a more dramatic manner by experiments on ternary mixtures of tartaric acid and boric acid with water.[2]

> "Boric acid, alone, has no rotatory power. But, if one adds it in the cold to an aqueous solution of tartaric acid, even in the proportion of a few thousandths, the optical properties of the solution are suddenly changed. The whole mass, although remaining liquid, forms immediately a new molecular system, which is distinguished from the former system by its greater rotatory power, by its wider range of variation as the proportions are changed, and finally by its different action on lights of unequal refrangibility."[3]

In particular, the characteristic dispersion of tartaric acid, with "the largest deviations in the green, and the smallest in the violet," disappears, and the law of dispersion conforms once more to "the general law, for quartz, sugars, gums, essential oils, according to which the rotations impressed on the planes of polarisation of the simple rays, are approximately in inverse proportion to the squares of their wave-lengths."[4]

As in the case of the strong bases, which had a similar influence in destroying the characteristic rotatory dispersion of tartaric acid, these "inexplicable and quite unforeseen phenomena" could only be explained by chemical combination between the two acids; but this was regarded as a "contact action" (*action de presence*) in which the whole of each component took part, instead of as an equilibrium, corresponding with the laws of mass-action discovered by Guldberg and Waage in 1867.

[1] *A.C.P.*, 1852, [iii], **36**, 266.
[2] Biot, *A.C.P.*, 1844, [iii], **11**, 82.
[3] *A.C.P.*, 1852, [iii], **36**, 267.
[4] *Ibid.*, p. 268.

Variable Rotatory Power of Cane Sugar, Turpentine and Camphor.—In order to determine whether the molecular interactions, which he had discovered in systems containing tartaric acid, were a general characteristic of the mechanism of solutions, Biot [1] decided to repeat his early experiments in which active molecules had appeared to be merely scattered amongst the molecules of an inactive solvent. The simplest method of experiment was to calculate the value of $[\alpha]$ at different concentrations of the active solute; but a more sensitive method was to compare the rotations produced when the length of the column was increased in inverse ratio to the concentrations, so that the total rotation would be constant if there were no change in specific rotatory power.

When this method was applied to aqueous solutions of cane sugar, it appeared that the specific rotatory power of the sugar for light of the neutral tint (*teinte de passage*) was *increased* by approximately 1 part in 237, when the solution was diluted in the ratio 1 : 3.[2] Oil of turpentine showed a similar immediate increase of nearly 1 per cent. when diluted to three times its volume with alcohol, but the numerical value of the increment was regarded as doubtful; when diluted, however, with olive oil in the ratio 1 : 3, there was *no immediate change* of rotatory power, although an increase of nearly 10 per cent. was observed in a mixture which had been kept for about five weeks.

It will be seen that in these two cases, where the rotatory dispersion was similar to that of quartz, and not very different from that corresponding with the requirements of Biot's Law of Inverse Squares, the specific rotatory power was almost independent of the concentration of the solution and of the nature of the diluent. In the case of camphor, however, where the dispersion was considerably greater,[3] changes of rotatory power on dilution were also much larger, and could be expressed by the same linear formula as in the case of tartaric acid. Thus the rotatory power for red light of solutions of camphor in acetic acid decreased on dilution in accordance with the equation

$$[\alpha]_r = + 42{\cdot}5391 - 14{\cdot}2357e.$$ [4]

Solutions of camphor in absolute alcohol showed a similar *decrease* as represented by the equation

$$[\alpha]_r = + 45{\cdot}2467 - 13{\cdot}6877e.$$ [5]

A small *decrease* was also observed on diluting an alcoholic solution of camphoric acid in the ratio 2 : 3, but the effect was much smaller than in the case of camphor and indeed too small to be established with certainty by a single experiment.[6]

Specific Rotatory Power of Turpentine, Nicotine, Ethyl Tartrate and Camphor.—The influence of solvent and concentration

[1] Biot, *A.C.P.*, 1852, [iii], **36**, 275–320.
[2] *Ibid.*, p. 289.
[3] *Ibid.*, p. 307.
[4] *Ibid.*, p. 306.
[5] *Ibid.*, p. 309.
[6] *Ibid.*, p. 315

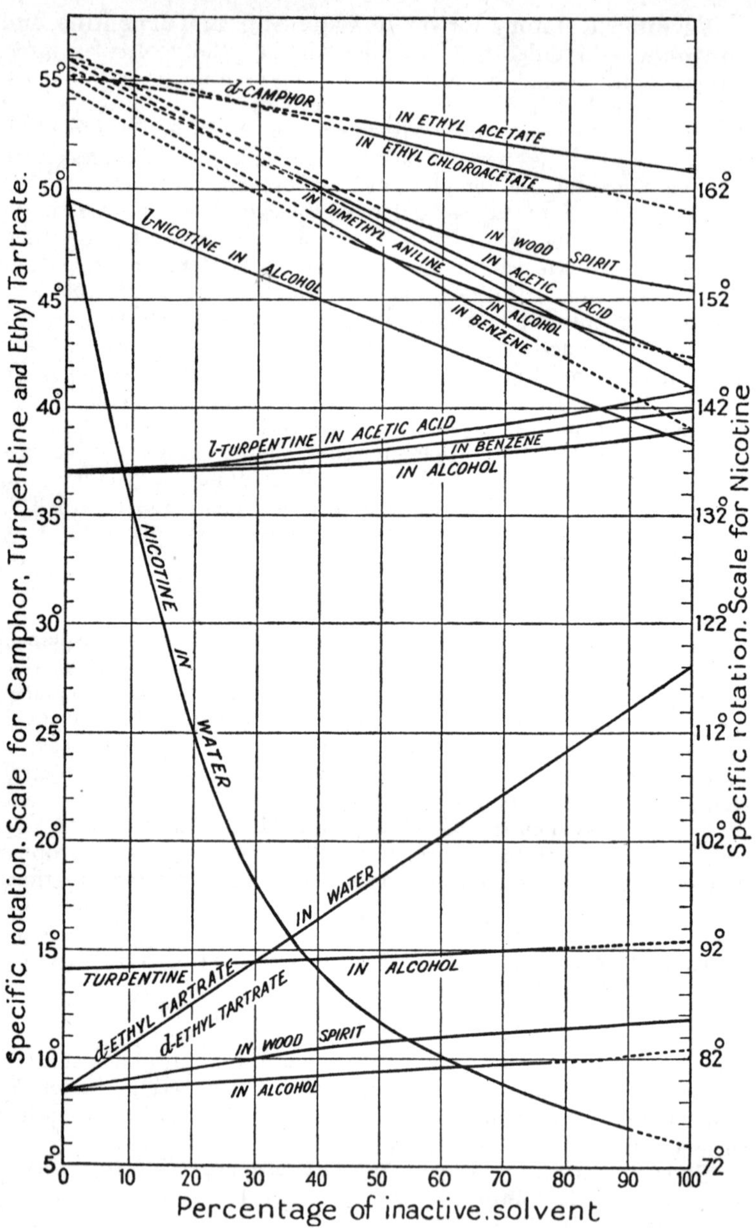

Fig. 29.—Influence of Solvent and Concentration on Specific Rotatory Power (Landolt, 1877).

on the specific rotation of *d*- and *l*-turpentine, *d*-nicotine and ethyl *d*-tartrate when diluted with solvents such as water, methyl and ethyl alcohols, benzene and acetic acid was determined by LANDOLT in 1877 [1] for a series of concentrations from 100 per cent. to a lower limit (never less than 10 per cent.) below which the observed rotations were too small to give accurate values for $[\alpha]$. The values obtained for the pure liquid ($q = 0$), and, by extrapolation, for infinite dilution ($q = 100$) were as follows :

TABLE 2.—SPECIFIC ROTATION OF CAMPHOR.

Solvent.	$[\alpha]$ $q = 0$.	$[\alpha]$ $q = 100$.	Difference.
Acetic acid	55·5°	41·8°	13·7°
Ethyl acetate . . .	55·2°	50·8°	4·4°
Ethyl monochloroacetate .	55·7°	49·0°	6·7°
Benzene	55·2°	38·9°	16·3°
Dimethylaniline . . .	55·8°	40·9°	14·9°
Wood spirit . . .	56·2°	45·3°	10·9°
Alcohol	54·4°	41·9°	12·5°

The results of Landolt's experiments are also shown graphically in Fig. 29, but without regard to the *sign* of the rotation. The graphs show clearly the increase, according to an approximately linear law, of the rotatory power of turpentine and the decrease of rotatory power of camphor on dilution ; but the most striking feature of the diagram is the remarkable influence of *water* in *increasing* the rotatory power of ethyl tartrate in direct proportion to the dilution, and in *decreasing* the rotatory power of nicotine to a limiting value which is less than half that of the rotatory power of the anhydrous base.

Influence of Concentration on the Specific Rotatory Power of Aqueous Solutions of Tartaric Acid and of Concentration and Solvent on the Specific Rotatory Power of Ethyl Tartrate.—From a large mass of more recent data, on the effect of solvent and concentration on specific rotatory power, it will suffice to select two series for reproduction.* (*a*) Fig. 30 [2] shows the molecular rotations of tartaric acid in aqueous solutions at 20° for six different wave-lengths, whilst Fig. 31 [2] shows a set of derived curves, illustrating the deviations of the observed values from a linear relation of the type postulated by Biot and from a simple parabolic formula. The principal diagram is similar to that given by Biot in 1836 (Fig. 28), but is based on accurate measurements instead of being merely illustrative of general principles. The derived

* A third series of curves, for nicotine in different solvents, is reproduced in Fig. 139, p. 330.

[1] LANDOLT, *Ann.*, 1877, **189**, 241–337.

[2] LOWRY and AUSTIN, *Phil. Trans.*, 1922, A. **222**, 267 and 269.

curves intersect the axis at two and at three points respectively, thus showing that the influence of water cannot be expressed by a linear law, nor even by a parabolic formula, but is a much more complex function of the water-concentration.[1]

(*b*) The data for ethyl tartrate (Fig. 32) are given by Patterson.[2] The diagram shows an increase of specific rotation from 7·8° to 26·5° on dissolving ethyl tartrate in a large excess of *water*. This incre-

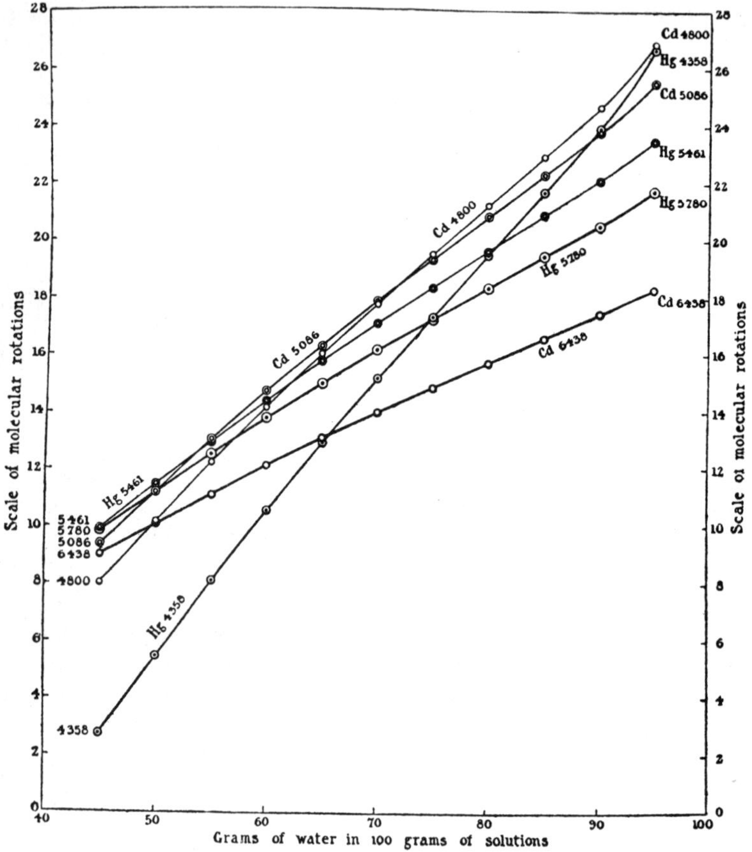

FIG. 30.—SPECIFIC ROTATIONS OF AQUEOUS SOLUTIONS OF TARTARIC ACID (Lowry and Austin, 1922).

ment diminishes in the homologous series of alcohols until a slight decrease occurs in *propyl alcohol*. Later experiments disclosed an

[1] BIOT (*A.C.P.*, 1844, **10**, 385) himself introduced a hyperbolic formula

$$[\alpha] = A + Be/(1 + Ce)$$

when he found that the linear formula was not exact.

[2] PATTERSON, *J.*, 1901, **79**, 180.

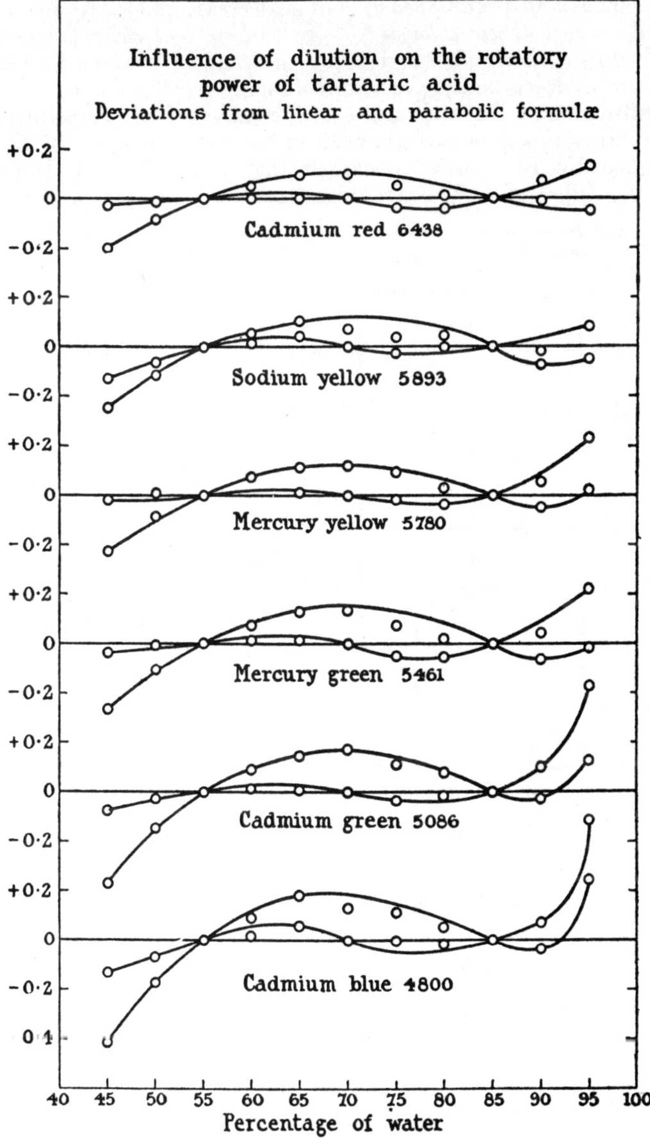

FIG. 31.—SPECIFIC ROTATIONS OF AQUEOUS SOLUTIONS OF TARTARIC ACID (DERIVED CURVES) (Lowry and Austin, 1922).

increase to + 35° in *nitrobenzene*, + 48·5° in *phenol*,[1] + 17° and + 75° in o- and p-*nitrophenol*,[1] but a decrease to + 2·5° in *anisol*,[1] + 1·9° in *carbon tetrachloride*,[2] — 4·2° in *ethylene chloride*,[2] — 16·6° in *acetylene tetrachloride*,[2] and — 20° in *acetylene tetrabromide*,[2] whilst aromatic hydrocarbons gave values ranging from about 0° to + 6°.[3]

Influence of Temperature.—The influence of temperature on the rotatory power of tartaric acid, in the range from 5° to 26°, was examined by Biot, who concluded that the coefficient B in the equation $[\alpha] = A + Be$ was independent of the temperature, but

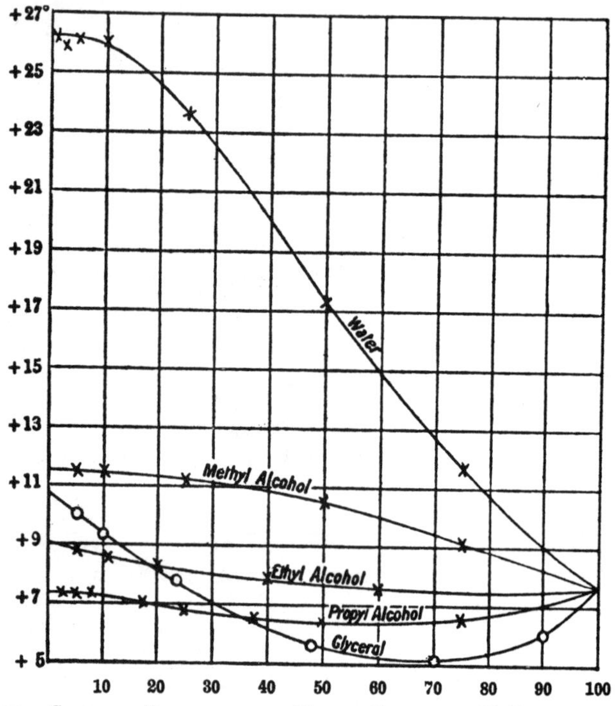

FIG. 32.—SPECIFIC ROTATIONS OF ETHYL TARTRATE (Patterson, 1901).

that the coefficient A, representing the rotatory power of the anhydrous acid, varied rapidly with changing temperature, becoming negative in sign below + 23° C. for the light transmitted by a deep-red glass. The negative rotation thus predicted was realised experimentally with the help of LAURENT and described to the *Académie des Sciences* on December 10, 1849.[4] Thus a rotation of — 3° was

[1] PATTERSON, *J.*, 1910, **97**, 2118. [2] *Ibid.*, 1908, **93**, 355–371.

[3] *Ibid.*, 1902, **81**, 1097–1133.

[4] BIOT, *A.C.P.*, 1844, [iii], **10**, 5, 175, 307, 385. Full details of these experiments are given in a memoir, *Sur la manifestation du pouvoir rotatoire moléculaire dans les corps solides* (*A.C.P.*, 1850, [iii], **28**, 215–240, 351–381).

observed with white light in a 70 mm. column of the solid amorphous acid at + 15° C., as compared with + 16° for 76 mm. of the liquid acid. Three series of observations with a red glass at 3·5° C. gave rotations of − 2·9°, − 3·75°, − 3·2°, mean − 3·28°, corresponding with a specific rotation $[\alpha]_r = -2{\cdot}787°$ at 3·5° C.[1] as compared with the value $A = -2{\cdot}752°$ extrapolated by a parabolic formula from the values of *A* deduced from the study of aqueous solutions from + 5° to + 26° C.

Experiments on the rotatory powers of three essential oils, over the range from 0° to 160°, for 5 Fraunhofer lines, were described by GERNEZ in 1864.[2] They showed a *decrease* of specific rotatory power which was almost constant for the five different wave-lengths in the ratio 1·47 : 1 for oil of orange and oil of bigarade, but in the ratio 1·02 : 1 only in the case of oil of turpentine, when a correction was made for the expansion of the liquids between 0° and 160° C. (p. 102).

A research by KRECKE in 1872[3] *Sur l'influence que la température exerce sur le pouvoir rotatoire de l'acide tartrique et les tartrates* included observations of the rotatory power from 0° to 100° C. of aqueous solutions containing 50, 40, 20 and 10 per cent. of tartaric acid, as well as observations on a series of metallic tartrates. The rotatory power of the acid was increased approximately three-fold in the range from 0° to 100°, as is shown by the data cited in Table 3. It is interesting to notice that the maximum rotation disappears into the ultra-violet on heating the 50 per cent. solution to about 50°, or on diluting the solution to a concentration of 40 per cent. at 0°.

TABLE 3.—INFLUENCE OF TEMPERATURE ON THE SPECIFIC ROTATORY POWER OF TARTARIC ACID.

Fraunhofer Line =	*C.*	*D.*	*E.*	*b.*	*F.*
Tartaric acid, 50 % at					
0° C. [α] =	5·641°	6·425°	5·771°	5·746°	5·635°
100° C. [α] =	13·457°	15·253°	17·819°	18·467°	19·671°
Ratio	2·3	2·5	2·7	3·1	3·4
Tartaric acid, 40 % at					
0° C. [α] =	4·570°	5·459°	6·702°	—	7·082°
100° C. [α] =	15·392°	17·506°	19·924°	20·558°	22·689°
Ratio	3·2	3·2	3·0	—	3·2

For the purpose of comparison with free tartaric acid, Krecke calculated the *molecular* as well as the *specific* rotatory powers of the salts, and summarised his results as in Table 4.

and particularly in Section II, *Propriétes rotatoires manifestées par les molécules de l'acide tartrique, mis en fusion par la chaleur, puis solidifié à l'état amorphe.*

[1] BIOT, *A.C.P.*, 1850, [iii], **28**, 366.

[2] GERNEZ, *Ann. École Normale*, 1864, **1**, 1–38.

[3] KRECKE, *Arch. néer.*, 1872, **7**, 97–116.

TABLE 4.—MOLECULAR ROTATIONS OF THE TARTRATES.

Fraunhofer Line =	*C.*	*D.*	*E.*	*b.*	*F.*
Potassium tartrate, M = 235	51·43°	62·66°	77·61°	81·78°	94·32°
Sodium tartrate, M = 230	48·49	59·31°	73·47°	75·84°	89·27°
Ammonium tartrate, M = 194	57·18°	68·28°	79·21°	83·24°	98·91°
Potassium sodium tartrate, M = 282 0°	51·49°	60·03°	73·21°	76·83°	89·03°
50°	53·76°	64·38°	77·13°	79·59°	91·62°
100°	55·72°	67·66°	79·63°	81·72°	94·05°
Tartar emetic, M = 341 0°	407·4°	503·8°	626·4°	655·8°	767·4°
50°	367·9°	457·1°	597·3°	623·4°	722·6°
100°	358·1°	444·4°	562·6°	585·9°	712·1°
Tartaric acid (Arndtsen), *c* = 0	18·291°	22·470°	26·501°	27·472°	30·569°

The rotations quoted for the first two salts are the means of five series of readings from 0° to 100°; those for the ammonium salt were for 25° only and, for the other two salts, the values for 0°, 50° and 100° have been selected from Krecke's table.

Krecke records that potassium and sodium tartrate conform to Biot's Law of Inverse Squares (pp. 11 and 106) and that the specific rotatory power of the potassium salt *decreases* slightly, whilst that of the sodium salt *increases* slightly between 0° and 100°; actually, however, the *decrease* in the case of the potassium salt was limited to the red and yellow, changing to an insignificant *increase* in the green and blue. Rochelle salt, which conformed only approximately to the Law of Inverse Squares, showed a definite *increase* of rotatory power with rise of temperature. Tartar emetic conformed almost as closely as the sodium and potassium salts to the requirements of the Law of Inverse Squares, but showed a definite *decrease* of rotatory power with rising temperature.

Krecke compared the molecular rotations of the metallic tartrates with those deduced for tartaric acid at infinite dilution from the values of the coefficient A of ARNDTSEN's equation,[1] and showed that comparable numbers were obtained if the molecular rotations of the acid were *multiplied by three.* The molecular rotation of tartar emetic, however, is approximately *seven times* as great as that of the other salts and twenty times greater than that of the free acid.

Optical Rotatory Power of Vapours.—The last section of Biot's second large Memoir [2] bears the title *Sur l'existence de la faculté rotatoire dans les particules même des corps, indépendamment de leur état d'aggregation.* After having shown that the rotatory power of turpentine was unchanged when diluted so as to increase the length

[1] ARNDTSEN, *A.C.P.*, 1858, [iii], **54**, 412.
[2] BIOT, *Mem. Acad. Sci.*, 1817, **2**, 114–133.

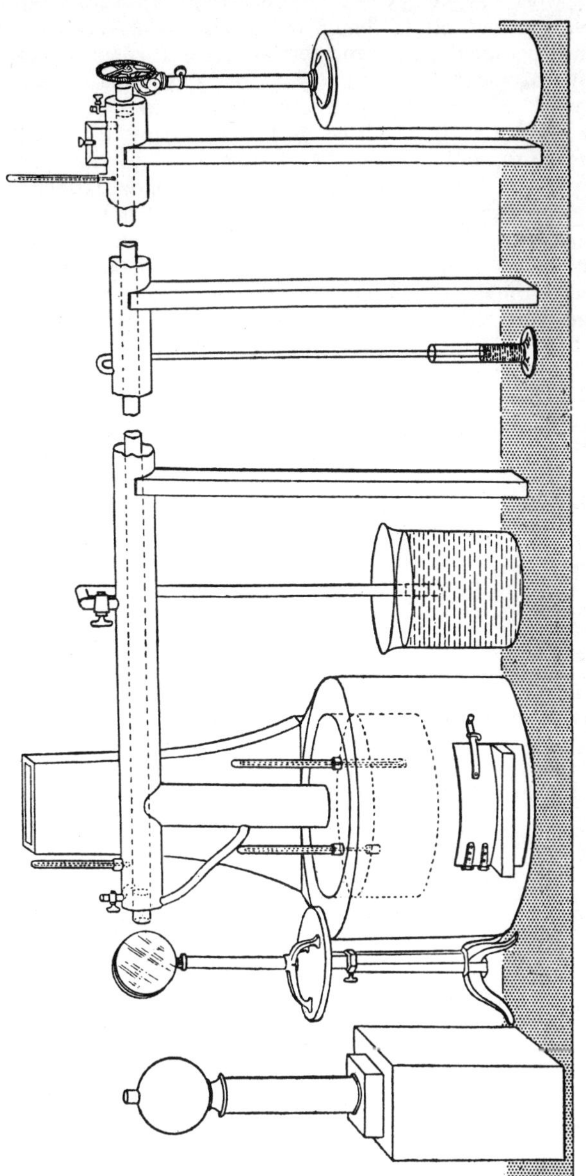

FIG. 33.—APPARATUS FOR OBSERVING THE OPTICAL ROTATORY POWER OF VAPOURS (Biot, 1817).

The light from a lamp was polarised by reflection from a mirror, shown at the left-hand side of the diagram, and was analysed by an achromatised prism of Iceland spar mounted on a graduated circle, shown on the right-hand side of the diagram.

of the column from 163·5 to 338·5 mm., and that the rotatory powers, e.g. of turpentine and oil of lemon, were approximately additive, Biot made an experiment on the rotatory power of turpentine vapour, using the apparatus shown in Fig. 33. This apparatus was set up in the cloister of an ancient church, which was then the orangery of the house of peers. It consisted of a double tube of sheet-iron, 30 metres in length, and had double glass ends, which were heated by hot air passing between them to prevent condensation. The vapour from a boiler filled the outer jacket, which sloped slightly upwards from the boiler and was provided with two water-seals, one of which carried a tap, to act as safety-valves. The vapour was then admitted to the interior, which was also provided with an escape-valve, opening directly to the air at the far end. As the vapour entered the tube, the light passing through it (which was formerly polarised in a single plane and therefore gave only a single image in the analyser) began to give a second coloured image when viewed through the calcite analyser, and finally showed two images with the normal tints observed with liquid turpentine. At this moment, however, the boiler exploded, the vapour and liquid caught fire and ignited a wooden beam, so that outside help was needed to extinguish it.

Biot's experiment established the existence of optical rotatory power in the vapour, but without giving any numerical value for its specific rotatory power. This omission was filled by a series of observations made by Gernez[1] in extension of his experiments on the effect of temperature on optical rotatory power (p. 99). A comparison of the rotation produced by 4 metres of vapour, heated by an oil-bath over a series of gas jets, with that of the liquid condensate, gave the following results:

TABLE 5.—OPTICAL ROTATORY POWER OF VAPOURS.

Substance.	State.	Temp.	Molecular rotation.
d-Essence of orange	Liquid	12·5°	114·97
,, ,, ,,	,,	99·6°	103·20
,, ,, ,,	,,	154°	95·15
,, ,, ,,	,,	176°	88·35
,, ,, ,,	Vapour	195°	77·09
d-Essence of Bigarade	Liquid	11·5°	92·87
,, ,, ,,	,,	100°	83·80
,, ,, ,,	,,	156°	78·06
,, ,, ,,	,,	172°	75·70
,, ,, ,,	Vapour	190°	70·22
l-Essence of Turpentine	Liquid	11°	36·53
,, ,, ,,	,,	98°	36·04
,, ,, ,,	,,	154°	35·81
,, ,, ,,	Vapour	168°	35·49
d-Camphor	Liquid	204°	70·33
,,	Vapour	220°	70·31

[1] GERNEZ, *Ann. École Normale*, 1864, **1**, 1–38; *C.R.*, 1864, **58**, 1108–1111.

The rotatory power of camphor was identical in the liquid and vapour states. Turpentine, with its very small temperature coefficient, showed a negligible decrease of rotatory power on vaporisation. The other two essential oils, however, with their large temperature coefficients, showed an unmistakable further decrease on vaporisation, as if this process served to complete some change

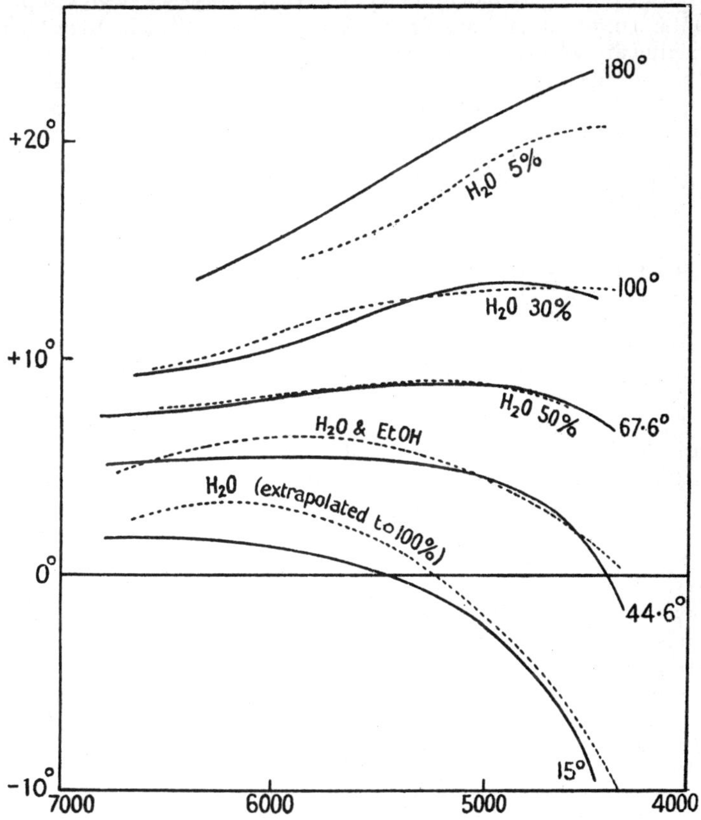

FIG. 34.—ROTATORY DISPERSION OF AMORPHOUS TARTARIC ACID (Bruhat, 1914).

of molecular configuration, which was already in progress when the liquid was being heated.

From the theoretical point of view, the determination of the molecular rotations of vapours is one of the most essential steps in unravelling the problem of optical rotatory power; but after 1864 no important further progress in this direction was made until 1932, when Lowry and Gore[1] observed the rotatory powers of the vapours

[1] LOWRY and GORE, *P.R.S.*, 1932, A. **135**, 13.

of camphor and of camphorquinone, not only in the region of transparency, but also in the region of absorption. Their results are reproduced in Figs. 131 and 133 (pp. 308 and 309).

Optical Rotatory Power of Glassy Solids.—The fact that amorphous quartz shows no optical rotatory power was used by Herschel [1] as evidence against the view that optical rotatory power is an inherent molecular property, since no case was known of optical rotation in an amorphous solid. This gap was filled in 1832 by the experiments which Biot made with barley sugar in his *Memoire sur la polarisation circulaire.*[2] This section of his Memoir bears the title *Invariabilité de la force rotatoire moléculaire d'une même substance dans les deux états de solidité et de simple dissolution* (pp. 126–132). The rotations observed in a block of barley sugar 37·3 × 24·6 mm. were + 24·225° and + 15·5417° for red light. The specific rotatory power of another portion of the same sugar in solution was $[\alpha] = 70{\cdot}975/160$, giving calculated values + 24·97° and 16·47° for the solid block. The rotatory power of the solid was less than 1° above the calculated values, a small difference which could be attributed to errors of sampling or the like. Biot therefore regarded this experiment "as establishing the constancy of molecular rotatory power in the liquid and solid states, when the method of liquefaction of the active material does not alter its molecular constitution." [3]

Nearly twenty years later [4] similar experiments were carried out with glassy tartaric acid, prepared according to the method devised by Laurent, and these were extended to glassy mixtures of tartaric and boric acids. A modern series of experiments on similar lines was described in 1914 by Bruhat,[5] whose curves for the rotatory dispersion of amorphous tartaric acid at temperatures of 15°, 44·6°, 67·6°, 100° and 180° are reproduced in Fig. 34. The diagram also shows the general similarity of these curves to the dotted curves representing the rotatory dispersion (i) of the acid as deduced by extrapolation from that of its aqueous solutions, (ii) of a solution in a mixture of alcohol and water, (iii) of a series of aqueous solutions of decreasing concentration to keep step with the increasing temperature of the amorphous acid.

[1] Herschel, *Light*, p. 55.
[2] Biot, *Mém. Acad. Sci.*, 1832, **13**, 39–173.
[3] *Ibid.*, p. 132.
[4] Biot, *A.C.P.*, 1850, [iii], **28**, 351.
[5] Bruhat, *T.F.S.*, 1914, **10**, 84–90.

CHAPTER VIII.

ROTATORY DISPERSION.

Rotatory Power and Rotatory Dispersion.—Biot's discovery of rotatory polarisation was a sequel to Arago's observations of the colours produced by the rotatory dispersion of a plate of quartz. The phenomenon of rotatory dispersion has therefore been familiar as long as that of optical rotatory power itself. For this reason it is not surprising that, during Biot's life, nearly all measurements of optical rotatory power, including those made by his pupil Pasteur, were associated with measurements of optical rotatory dispersion. It was only after Biot's death in 1862 that the invention in 1866 of the Bunsen burner inhibited the more laborious study of rotatory dispersion, by making it almost too easy to work with the nearly monochromatic light of the sodium flame, and in this way brought to an end the fertile era which Biot had inaugurated half a century before. Thus, when interest in optical rotatory power had been intensified by the work of le Bel and van't Hoff on "Chemistry in Space," Landolt published a long review of the subject under the title *Untersuchungen über optisches Drehungsvermögen*;[1] but the experimental work in this important paper was limited to measurements of the D-line; and this fault was repeated in almost all the measurements which were made during the next thirty years.

This limitation was particularly unfortunate in view of the fact that, during this period, experiments to determine the influence of solvent, concentration, temperature and chemical constitution on optical rotatory power were usually made with derivatives of tartaric acid, which was known to exhibit anomalous rotatory dispersion. This choice of material was based on the fact that (as Biot pointed out) these compounds are also abnormal in their sensitiveness to changes of external conditions. Experiments on these lines, however, when limited to observations of a single wave-length on a dispersion curve of an anomalous type, were necessarily infertile, and it was only when measurements of rotatory dispersion again became customary that the significance of these observations could be effectively discussed. In the meanwhile, deductions of a quantitative character were only possible when the substances under investigation happened to exhibit a simple and constant type of rotatory dispersion,

[1] LANDOLT, *Ann.*, 1877, **189**, 241–337.

as in the sugar derivatives whose rotations were analysed by Simon,[1] C. S. Hudson [2] and Maltby.[3]

Biot's Two Types of Rotatory Dispersion.—Biot's earliest quantitative measurements showed that the rotatory power of quartz varied inversely as the square of the wave-length, $\alpha = k/\lambda^2$. The same law was established by the principle of compensation for solutions of cane-sugar and invert-sugar; oil of turpentine also appeared to obey the law, but dextrorotatory oil of lemon did not obey it exactly, since its rotatory power could not be compensated by oil of turpentine of opposite sign. In general, however, the majority of organic compounds could be cited as conforming, at least approximately, to the Law of Inverse Squares, as established in the case of quartz. This conclusion is stated clearly in a paper read before the *Académie des Sciences* on August 16, 1852, and published under the title:

Expériences ayant pour but d'établir que les substances douées de pouvoir rotatoire, lorsq'elles sont dissoutes dans les milieux inactifs, qui ne les décomposent pas chimiquement, contractent avec eux un combinaison passagère, sans proportions fixes, mais variable avec le dosage, laquelle impressionne toute leur masse, et subsiste, tant que le système mixte conserve l'état de fluidité.

> "All substances other than the two tartaric acids dextro and lævo, in which molecular rotatory power has been recognised up to the present, impress on the planes of polarisation of the simple rays which compose white light, deviations which are unequal and always increasing with the refrangibility. Thus the dispersion of these planes is always of the same character as that which is observed in spectra produced by prismatic refraction." [4]

Tartaric acid, on the other hand, produced an entirely different type of rotatory dispersion. The first recorded measurement of its optical rotatory power was accompanied by a note to the effect that its rotation was "greater for the more refrangible rays," and more detailed measurements showed that its rotatory power was greatest in the green. It therefore exhibited the anomaly of a *maximum* rotatory power, in contrast with the compounds of the preceding group, where the rotatory power always increased with the refrangibility of the light. As we have already seen, the acid was also exceptional in that its specific rotatory power was influenced in a remarkable degree by changes of temperature, solvent and concentration, as well as by the action of various chemical agents such as sulphuric acid, boric acid and the alkalis. This variability

[1] Simon, *C.R.*, 1901, **132,** 487–490.
[2] C. S. Hudson, *J.A.C.S.*, 1909, **31,** 66; 1924, **46,** 462 *et seq.*
[3] Maltby, *J.*, 1922, **121,** 2608; 1923, **123,** 1404.
[4] Biot, *A.C.P.*, 1852, [iii], **36,** 267–268.

made it possible for Biot to predict another anomaly, realised by Biot himself in 1850, and by Arndtsen in 1858, namely, a *reversal of sign* of the rotatory power, which could assume a negative value for certain wave-lengths when the temperature and concentration also were suitably adjusted. Tartaric acid, however, remained for many years the only known example of the remarkable type of rotatory dispersion of which it provided the first illustration.

Normal and Anomalous Dispersion.—Biot, then, divided all optically-active substances into two classes—an orthodox class which obeyed his law of inverse squares, at least approximately, and an heretical class which made no pretence of obedience.

The orthodox class is now generally described as showing NORMAL ROTATORY DISPERSION. The characteristic property of this class was defined by Arndtsen in 1858 [1] in terms of a "general law that the rotation increases with the refrangibility of the rays." In more modern times it has been defined in almost identical terms by Tschugaeff,[2] who specified that in cases of normal rotatory dispersion the *rotatory power increases progressively with diminishing wave-length.* A more precise definition by the author [3] specifies that α, $d\alpha/d\lambda$ and $d^2\alpha/d\lambda^2$ must remain constant in sign throughout the range of wave-lengths to which the medium is transparent. This definition asserts (i) that there must be no *reversal of sign*, which would give $\alpha = 0$ at the point of reversal, (ii) that there must be no *maximum*, which would give $d\alpha/d\lambda = 0$, and (iii) that there must be no *point of inflexion* or *reversal of curvature*, corresponding with $d^2\alpha/d\lambda^2 = 0$, which would make the curve concave as viewed from the axis of wave-lengths during a part of its course, instead of remaining always convex.

The anomalies recorded above have been regarded, ever since they were first discovered, as characteristics of ANOMALOUS ROTATORY DISPERSION, although this term was not used in a formal way until a much later date. Thus Krecke in 1872 [4] merely referred to the "remarkable anomalies" in the optical properties of tartaric acid, and it was left to Landolt,[5] five years later, to make the first formal use of the term.

Later experience has shown that these three anomalies are characteristic of all curves of anomalous rotatory dispersion in transparent media, in which the optically-active absorption bands (p. 393) are in the ultra-violet. Moreover, the inflexion, maximum and reversal of sign always follow the same sequence of decreasing wave-lengths. Thus a strong solution of tartaric acid (Fig. 35),[6] with a

[1] ARNDTSEN, *A.C.P.*, 1858, [iii], **54**, 409.
[2] TSCHUGAEFF, *T.F.S.*, 1914, **10**, 70.
[3] LOWRY, *J.*, 1924, **125**, 2511–2524; compare also an earlier paper on *An Exact Definition of Normal Rotatory Dispersion, J.*, 1915, **107**, 1195.
[4] KRECKE, *Arch. néerland.*, 1872, **7**, 98, 110 and 114.
[5] LANDOLT, *Ann.*, 1877, **189**, 274.
[6] LOWRY and AUSTIN, Bakerian Lecture, *Phil. Trans.*, 1922, A. **222**, 249–308.

maximum rotation in the green, as recorded by Biot, can be shown to give *a reversal of sign* in the ultra-violet ; and the equation of the curve discloses an *inflexion* in the red, near to the limit of the visible spectrum. By reducing the concentration of the aqueous solution, the *inflexion* can be brought to the middle of the visible spectrum, but the *maximum* is displaced towards the limit of visibility in the violet. Conversely by increasing the concentration to the point of supersaturation, or by making use of glassy tartaric acid, the *reversal of sign* can be displaced from the ultra-violet into the blue or green region of the spectrum, but when this is done, the *maximum* is displaced into the red or infra-red region. It is therefore reasonable to regard the discovery of any one of the more obvious anomalies as evidence for the existence of the others, either in the visible or in an

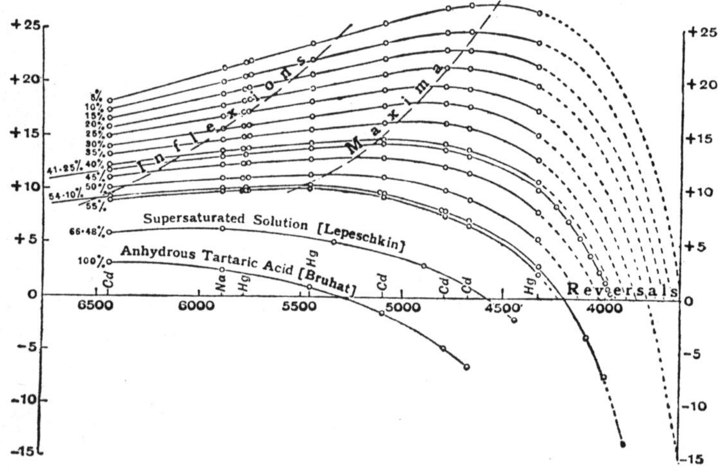

FIG. 35.—ANOMALOUS ROTATORY DISPERSION OF AQUEOUS SOLUTIONS OF TARTARIC ACID (Lowry and Austin, 1922).

invisible region of the spectrum. The three requirements for normal rotatory dispersion, as set out above, are therefore all satisfied, or all violated simultaneously, but in different regions of the spectrum, as is found to be the case in compounds of the type of tartaric acid. Dispersion curves of other types may occur when one of the optically-active absorption bands is of longer wave-length than the region of transparency, but this kind of anomaly is still almost unknown. Anomalous dispersion in absorbing media is described in Chapters XI, XXXIII, and XXXIV.

Origin of Anomalous Rotatory Dispersion.—In order to preserve a correct historical sequence it is necessary to refer again at this point to Biot's synthesis of an anomalous rotatory dispersion by the optical superposition of two normal dispersions of opposite sign, namely, by the imperfect compensation of lævorotatory oil of turpentine and dextrorotatory oil of lemon.[1] Biot was very much interested in the analogy between the unequal rotatory dispersions of organic compounds and the unequal dispersions of the glasses used in making achromatic lenses, and in 1852 published a paper *Sur l'application de la théorie de l'achromatisme, à la compensation des mouvements angulaire que le pouvoir rotatoire imprime aux plans de polarisation des rayons lumineux d'inégale réfrangibilité.*[2] In this investigation he made use of *l*-turpentine and a strong (55 per cent.) solution of *d*-camphor * in acetic acid. These nearly compensated one another when examined with white light in separate polarimeter tubes, but the system showed increasing dextrorotations at wave-lengths shorter than the "neutral tint" (*ib.*, p. 432). Similar results were obtained with a solution of camphor (42 per cent.) in alcohol. By using a much weaker solution of camphor (16 per cent.) in acetic acid, a lævorotatory combination was obtained, which gave almost the same rotation for red and yellow light and showed practically no dispersion colours when the analyser was set at $-20°$ (*ib.*, p. 439). When a more dilute (20 per cent.) solution of camphor was used a lævorotatory combination was obtained, with the normal sequence reversed, since the rotations *decreased* in the series *red, yellow, blue, violet;* but the normal sequence was restored when the lævorotation was increased by adding a second (short) tube of turpentine to the system.

A second series of experiments extended the observations to "the achromatic compensation of rotatory powers, by *active substances, dissolved simultaneously* in the same liquid medium." The first system, containing 34 per cent. of camphor dissolved in lævorotatory turpentine, was practically achromatic for the red, yellow and green, with a total rotation of about $-13°$, but was dextrorotatory in the blue and violet; the second system, containing only 31 per cent. of camphor, was dextrorotatory throughout, and had a maximum rotation in the green, where it was almost achromatic, with a rotation of about 28° over a considerable range of the spectrum (*ib.*, p. 462).

Some years later a further application of Biot's principles was made by G. H. von Wyss in a paper, *Ueber eine neue Methode zur Bestimmung der Rotationsdispersion einer activen Substanz, und über einen Fall von anomaler Dispersion.*[3] Von Wyss applied his improved methods of measurement to a sample of turpentine which had a very small lævorotation, rising to a maximum at $\lambda = 565\mu$ (Fig. 36). Following the precedents set by Biot, von Wyss attributed this anomaly to the presence in the liquid of two components of opposite rotatory powers and unequal dispersions, and verified this

* The use of *d*-camphor to compensate the rotation of *l*-turpentine was first referred to in 1818 (*Mém. Acad. Sci.*, 1817, **2,** 117).

[1] Biot, *C.R.*, 1836, **2,** 543. Cf. p. 21.

[2] Biot, *A.C.P.*, 1852, [iii], **36,** 405–489.

[3] von Wyss, *Pogg. Ann. Phys. Chem.*, 1888, **33,** 554.

opinion by preparing a series of artificial mixtures of two turpentines, namely, T_1, which was lævorotatory and T_2, which was more strongly dextrorotatory. The rotations of these mixtures are also shown in Fig. 36. The mixture M_1, containing only 20·184 per cent. of T_2, already showed a maximum lævorotation, but only in the violet; M_2 and M_3, containing 29·724 and 32·277 per cent. of T_2, gave maxima at 551 μ and 581 μ respectively, and enveloped the curve for the natural turpentine closely on either side; M_4, with 39·839 per cent. T_2, just reached the axis of zero rotation at 514·5μ, so that the two turpentines were exactly compensated at this wave-length, whilst M_5, with 50·233 per cent. of T_2, was dextrorotatory for all wave-lengths. In this way the origin of the anomalous rotatory dispersions of the natural turpentine was established by direct synthesis of the phenomenon with the help of samples which probably consisted predominantly of *l*-pinene and *d*-limonene.[1]

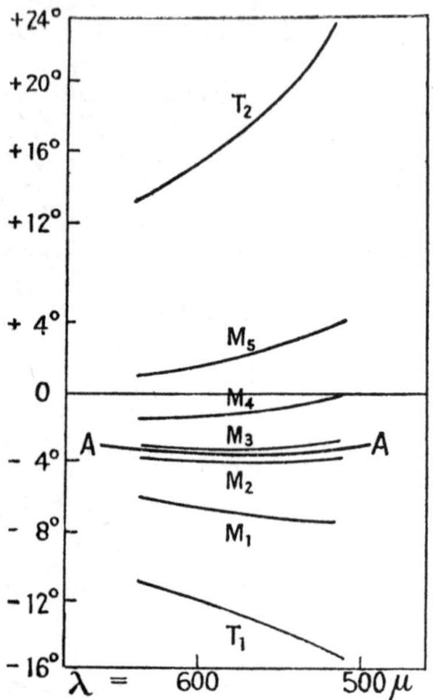

FIG. 36.—ANOMALOUS ROTATORY DISPERSION OF TURPENTINE (von Wyss, 1888).

Mixtures of T_1 and T_2 in various proportions gave the curves M_1, M_2, M_3, M_4, M_5, two of which approximate closely to the curve A of a natural turpentine.

Origin of the Anomalous Dispersion of Tartaric Acid.—Arndtsen, who spent a summer in Biot's laboratory in Paris, suggested in 1858 that Biot's observations could be used to account for the anomalous optical properties of tartaric acid:

> "If one should imagine two active substances which do not act chemically upon one another, of which one turns the plane of polarisation to the right, the other to the left, and, in addition, that the rotation of the first increases (with the refrangibility of the light) more rapidly than that of the other, it is clear that, on mixing these substances in certain proportions, one would have combinations which would show optical phenomena precisely similar to those of tartaric acid, as M. Biot has already proved by his researches on different mixtures of turpentine and natural camphor. One could then regard tartaric acid as a mixture of two substances differing only in their optical pro-

[1] See also DARMOIS, *A.C.P.*, 1911, [viii], **22**, 247–281, 495–590.

perties, of which one would have a negative, and the other a positive, rotatory power, and of which the rotations would vary in different proportions with the refrangibility of the light." [1]

Arndtsen's hypothesis, that aqueous solutions of tartaric acid contain two optically-diverse modifications of the acid, has been thrown into a more precise form by Longchambon,[2] who showed that the crystals of *d*-tartaric acid, when viewed along an optic axis, are lævorotatory, although they yield dextrorotatory solutions. He therefore proposed to describe the lævorotatory molecules of the crystals as α-tartaric acid, whilst the dextrorotatory molecules which are produced from them in solution were described as β-tartaric acid. On this basis, the properties of the acid can be accounted for by postulating a condition of equilibrium between the α and β acids in solution or in the fused state, followed by a complete conversion into the α-form when the crystalline acid separates.

$$\begin{array}{ll} \text{Liquid} & \alpha \rightleftharpoons \beta \\ & \upharpoonleft\downharpoonright \\ \text{Solid} & \alpha \end{array}$$

This type of equilibrium was first recorded in isomeric compounds when Butlerow [3] published his observations on the *iso*dibutylenes, and showed that two isomeric olefines were interconvertible in presence of sulphuric acid; but a closer analogy with Arndtsen's hypothesis is found in the case of prussic acid, which Butlerow regarded as an equilibrium mixture of hydrogen cyanide and isocyanide $H \cdot C \equiv N \rightleftharpoons H \cdot N = C$. In this case he supposed that the two isomers were interconvertible in the absence of a catalyst; equilibrium could therefore be attained without any obvious lag, just as in the case of tartaric acid and its esters, the properties of which do not show any peculiarities when freshly dissolved, or after recent fusion.

Two Optically-active Radicals in One Molecule.—We have seen that anomalous rotatory dispersion can be produced by superposing two normal dispersions in a variety of ways. In Biot's experiments the opposite partial rotations were provided by two different types of molecule, either in different polarimeter tubes, when their superposition was purely optical, or mixed in the same solution, but without forming a chemical compound. Arndtsen suggested yet another mechanism, namely, the presence of two forms of a compound in a solution from which only a single form can be separated; but this hypothesis, which is in perfect harmony with the laws of chemical equilibrium as revealed by subsequent studies of "dynamic isomerism," again calls for the presence of molecules of opposite rotatory powers. Many years later, however, in a paper

[1] ARNDTSEN, *A.C.P.*, 1858, [iii], **54**, 421.
[2] LONGCHAMBON, *C.R.*, 1924, **178**, 951.
[3] BUTLEROW, *Ann.*, 1877, **189**, 44–83.

on *Anomalous Rotatory Dispersion*, Tschugaeff[1] showed that this condition is not necessary, since he was able to prove, in the case of *l*-menthyl *d*-camphorsulphonate, that "anomalous dispersion may be produced by the *superposition of the partial rotations produced by two asymmetric complexes within the molecule of an active body*, the necessary conditions being that these partial rotations should be of *opposite sign* and should possess *different dispersion ratios*." This case differs only from those investigated by Biot in that the two optically-active components have been linked together to form a single complex molecule, instead of remaining free and independent, although dissolved in the same solution. Thus, by using *l*-menthol instead of the hydrocarbon *l*-pinene, and *d*-camphorsulphonic acid instead of *d*-limonene or *d*-camphor, an ester was prepared which exhibited the same anomalies that might have been expected in mixtures of the alcohol and acid.

Anomalous rotatory dispersion, then, cannot be attributed to one universal set of conditions, but in general it appears to be caused by the superposition of normal rotations of opposite sign and unequal dispersion, either (i) in the same molecule, (ii) in different molecules, but in the same solution, or (iii) in different molecules contained in different solutions. The confirmation of this conclusion by quantitative measurements is described below (p. 136), together with suggestions as to yet another mechanism by which anomalous rotatory dispersion may be produced in unsaturated compounds. It should also be noticed that under the second heading, the "different molecules in the same solution" may be (i) identical in composition and interconvertible, like Longchambon's α and β tartaric acid, (ii) identical in composition, but not interconvertible under the conditions of experiment, as in the observations of Biot and of von Wyss on *l*-pinene and *d*-limonene, or (iii) different in composition, as in Biot's experiments on *l*-pinene and *d*-camphor.

[1] TSCHUGAEFF, *T.F.S.*, 1914, **10**, 70.

CHAPTER IX.

SIMPLE ROTATORY DISPERSION.

Quantitative Measurements of Rotatory Dispersion.—In a paper published in the *Comptes rendus* of June 6, 1836,[1] BIOT stated that he "had always preserved some doubt as to the absolute rigour" of the identity of rotatory dispersion in different media, and justified these doubts by experiments which showed that the opposite rotatory powers of oil of turpentine and oil of lemon could not be balanced against one another simultaneously for different wave-lengths. Similar experiments, of an even more striking character, with oil of turpentine and camphor, were described in 1852.[2]

Accurate quantitative measurements of rotatory dispersion were first made in 1846[3] with a polarimetric apparatus (Fig. 82, p. 200) described by FIZEAU and FOUCAULT.[4] In this apparatus, sunlight from a heliostat was passed through Nicol prisms to serve as polariser and analyser, and was then resolved into its spectral components with the help of a refracting prism. When a plate of quartz was set up between the polariser and analyser, dark bands were seen in the spectrum, corresponding with the wave-lengths for which the polarised beam was extinguished. These bands moved across the spectrum when the analyser was rotated, and could be set to coincide with the black Fraunhofer lines of the solar spectrum. In this way Broch[3] determined the rotatory power of quartz for a series of wave-lengths as follows:

Fraunhofer Line.	*B.*	*C.*	*D.*	*E.*	*F.*	*G.*
Rotation α .	15·30° per mm.	17·24° per mm.	21·67° per mm.	27·46° per mm.	32·50° per mm.	42·20° per mm.
Product $\alpha\lambda^2$.	7238	7429	7511	7596	7622	7841

These values were afterwards used by von Lang, Stefan and Boltzmann as evidence of the need for adding another term as a correction to Biot's equation.

1 BIOT, *C.R.*, 1836, **2**, 540–547.
2 BIOT, *A.C.P.*, 1852, [iii], **36**, 405–489.
3 BROCH, *Repert. Physik*, 1846, **7**, 113; *A.C.P.*, 1852, [iii], **34**, 119–121.
4 FIZEAU and FOUCAULT, *C.R.*, 1845, **21**, 1155.

The same method was used by WIEDEMANN in 1851 [1] to compare the natural and magnetic rotatory dispersion of two typical organic liquids, namely, oil of turpentine and oil of lemon. His values for the natural rotations in degrees per millimetre were as follows:

Fraunhofer Line.	*B.*	*C.*	*D.*	*E.*	*F.*	*G.*
Turpentine . α =	0·215°	0·234°	0·293°	0·368°	0·436°	0·559° per mm.
Oil of lemon . α =	0·340°	0·379°	0·485°	0·633°	0·675°	1·060° per mm.

When these rotations were multiplied by the squares of the wave-lengths of the Fraunhofer lines, the product $\alpha\lambda^2$ showed a progressive increase with the refrangibility of the light.

Similar results followed from the more precise measurements made by the same method by ARNDTSEN in 1858.[2] Arndtsen's data for quartz, turpentine and gum arabic are set out below:

Fraunhofer Line.		*B.*	*D.*	*E.*	*F.*	*G.*
Quartz (1 mm.) .	α	—	21·62°	27·40°	32·65°	42·57°
	$\alpha\lambda^2$	—	75,004	75,809	76,800	78,346
Turpentine . .	α	+6·15°	+8·51°	+11·80°	+15·24°	+23·11°
	$\alpha\lambda^2$	26,465	29,522	32,647	35,848	42,531
Aqueous gum-arabic	α	−5·47°	−6·78°	−8·63°	−10·47°	—
	$\alpha\lambda^2$	23,539	23,521	23,877	24,628	—

> "One sees that for all these substances the product ($\alpha\lambda^2$) is far from constant. On the contrary it always increases with the refrangibility of the light." [3]

In the case of camphor the increase was even more striking, since the ratio of the rotations for the red and violet lines, *C* and *E*, was 1 to 4·012, as an average for four solutions, as compared with 1 to 2·635 for sugar.

Modification of Biot's Law of Inverse Squares: von Lang's Equation.—The experiments recorded in the preceding paragraph showed clearly that Biot's Law of Inverse Squares is only approximately true. Thus, even in the case of quartz, the value of $\alpha\lambda^2$ increases by 2·4 per cent. between the *D* and *F* lines of the solar spectrum. A similar increase is observed in aqueous cane-sugar and in oil of turpentine, whilst camphor shows a much larger increase. Some sort of correction is needed, therefore, if the law is to be made to conform accurately to the experimental data, even for

[1] WIEDEMANN, *Pogg. Ann. Phys. Chem.*, 1851, **82**, 222.
[2] ARNDTSEN, *A.C.P.*, 1858, [iii], **54**, 403–421. [3] *Ibid.*, p. 409.

those substances which approximate most closely to the requirements of that law in its original form. Moreover, this correction must be such as to *increase* the calculated dispersion of the medium.

The first modification of this kind was effected by VON LANG in a short theoretical paper, *Zur Theorie der Circular-polarisazion,*[1] written in London in 1863, but published in German. In this paper von Lang developed a formula for optical rotatory dispersion of the same type as that used by CAUCHY [2] for refractive dispersion, thus:

$$\text{Cauchy} \quad \mu = A + B/\lambda^2. \qquad \text{von Lang} \quad \alpha = A + B/\lambda^2.$$

This formula was applied to Broch's data for the rotatory dispersion of quartz for six Fraunhofer lines, which were expressed by the formula

$$\alpha = -3{\cdot}40 + 8{\cdot}5706/10^6\lambda^2.$$

Stefan's Experiments.—A more elaborate series of experiments on rotatory dispersion was described by STEFAN in 1864 in a paper, *Ueber die Dispersion des Lichtes durch Drehung der Polarisationsebene in Quarz.*[3] In this paper Stefan described two methods for developing a ROTATION SPECTRUM comparable with those produced by a prism or by a diffraction grating. (i) In the first method the polarised light emerging from a plate of quartz was analysed with the help of a mirror which could be rotated round the axis of the incident ray; the colour of the reflected ray was then determined by the wave-lengths of the light that was polarised in the plane of incidence, for a given setting of the mirror.* (ii) In the second method, he made use of the familiar figure, given by a uni-axial crystal of calcite in convergent polarised light, consisting of a black cross intersected by concentric coloured circles. By using strongly convergent light, the rings can be reduced to a small central part of the field of view, which is then occupied almost exclusively by the black cross. If now a plate of quartz, cut perpendicularly to the axis, is inserted in the parallel beam of polarised light between the polariser and the condensing lens, the black cross of the calcite is resolved into a series of coloured crosses with different orientation, so that the field of view is occupied by a wedge with coloured sectors. Unlike the spectra produced by a prism or a grating, however, these rotation spectra are not composed of pure colours, but depend on the development of maxima at different wave-lengths as the plane of polarisation is rotated.

In order to compare rotatory dispersion with refractive dispersion, Stefan made use of a series of quartz plates, the thicknesses of which were 70·08, 61·33, 44·80 and 25·28 mm. One of the plates was placed

[1] VON LANG, *Pogg. Ann. Phys. Chem.*, 1863, **119**, 74.

[2] CAUCHY, *C.R.*, 1842, **15**, 1082.

[3] STEFAN, *Sitz. Akad. Wiss., Wien*, 1864, **50**, 88–124.

* A similar demonstration can be given by passing a beam of polarised light upwards through a column of aqueous cane-sugar, when the light scattered by the cloudy solution forms a coloured spiral.

between the polariser and analyser of a polarimeter, and the emergent light was analysed by a spectrometer. Since the plates of quartz were thick, the observed rotations were large, and differed by a considerable number of right angles in the range of wave-lengths covered by the visible spectrum, so that for every position of the analyser several different colours were extinguished simultaneously. The spectrum was therefore crossed by a series of dark bands, increasing in number with the thickness of the quartz plate. By means of the spectrometer, the refractive index of the glass could be determined for those wave-lengths which were extinguished at a given setting of the analyser, e.g. for 9 extinctions with the thickest plate of quartz, and for 8, 6 and 4 when the thinner plates were used.

The successive extinctions corresponded with increments of two right angles in the total rotation by the plate of quartz and therefore with equal increments in the rotation per millimetre. Experiment showed that these equal increments of rotatory power corresponded with equal increments of refraction by the crown-glass prism. Since this refraction could be expressed by a Cauchy formula $\mu=P+Q/\lambda^2$, it was possible to use a similar formula $\alpha = R + S/\lambda^2$ to represent the rotatory dispersion of quartz. This two-fold calculation led to the formula

$$\alpha = -1{\cdot}753 + 8{\cdot}1624/10^6\lambda^2.$$

This formula was valid to tenths of a degree, whereas Biot's formula included an error of nearly two degrees, as represented by the initial constant term of the new formula. A second series of experiments with a prism of flint glass gave

$$\alpha = -1{\cdot}749 + 8{\cdot}1774/10^6\lambda^2$$

in excellent agreement with the constants deduced from the experiments with crown glass. Finally, a series of more exact measurements by Broch's direct method gave

$$\alpha = -1{\cdot}697 + 8{\cdot}1088/10^6\lambda^2.$$

By using a grating instead of a prism, it was possible to make a direct comparison of rotatory power and wave-length, since the deviations by the grating depend directly on the wave-length of the light instead of on the refractive dispersion of an arbitrary sample of glass. When this was done it was found that equal increments of rotation did not correspond to equal increments in $1/\lambda^2$. The latter increments were almost constant from red to green, but then decreased progressively, so that the increments in the violet were about 10 per cent. smaller than at the other end of the spectrum. The deviations from Biot's law were therefore greater in the blue and violet than in the red, yellow and green, and a larger correction was required to express the rotatory dispersion at this end of the spectrum.

When applied to Wiedemann's data, von Lang's formula gave for undistilled oil of turpentine

$$\alpha = -0{\cdot}0064 + 0{\cdot}10392/10^6\lambda^2$$

and for oil of lemon

$$\alpha = -0{\cdot}1254 + 0{\cdot}214567/10^6\lambda^2.$$

In the first case the experimental numbers were too small for accurate comparison with those deduced from the formula, but the second equation showed systematic errors of the order of 2 to 3 per cent. and was obviously incapable of representing the experimental data accurately.

It will be seen that in all these equations the initial constant term is negative, or opposite in sign to the term which depends on the square of the wave-length. This is necessary in view of the fact that the rotatory dispersion of all these substances is greater than that given by Biot's law; but even when this correction has been made the equation is not capable of representing the experimental data within the limits of accuracy which could be obtained by the method of Fizeau and Foucault.

Modification of Biot's Law of Inverse Squares: Boltzmann's Equation.—The equation used by von Lang and by Stefan was criticised in a paper by L. Boltzmann, *Ueber den Zusammenhang zwischen der Drehung der Polarisationsebene und der Wellenlänge der verschiedenen Farben*, which was published in 1874.[1] This paper includes the following statement:

> "It seems to me to be probable on theoretical grounds, that the angle of rotation of the plane of polarisation could be expressed better by a formula of the form $B/\lambda^2 + C/\lambda^4$ than by the formula $A + B/\lambda^2$" (*ib.*, p. 129).

This deduction was based on the view that

> "The rotation of the plane of polarisation is one of those phenomena which depend on the fact that the wave-lengths are no longer very large compared with the sphere of action of a molecule" (*ib.*, p. 134).

The rotatory power of a medium must therefore tend towards zero at infinite wave-length. In the formula used by von Lang and Stefan, however, the rotatory power would have a finite value even if the wave-lengths were infinite, namely, that corresponding with the constant term A. Boltzmann, in harmony with more recent theories, makes $A = 0$; he then corrects Biot's formula by adding a term of higher order. A graphical comparison of the equations of Biot, von Lang and Boltzmann, by plotting α against $1/\lambda^2$, showed that the systematic errors recorded by Stefan were of the kind that would be observed on attempting to represent a parabolic curve, corresponding with Boltzmann's equation, by a straight line corresponding with von Lang's equation. Numerical calculations then showed that the mean error could be reduced to one-fifth and the mean squared error to one-eighteenth of its former value, if Broch's

[1] Boltzmann, *Pogg. Ann. Phys. Chem., Jubelband,* 1874, pp. 128–134.

data for the rotatory power of quartz were represented by the equation

$$\alpha = 7{\cdot}07018/10^6\lambda^2 + 0{\cdot}14983/10^{12}\lambda^4$$

instead of by the equation

$$\alpha = -\ 2{\cdot}303 + 8{\cdot}32777/10^6\lambda^2,$$

which was the best equation of the earlier type. Boltzmann therefore concluded that the equations for rotatory dispersion do not include a constant term, but might be extended if necessary by the addition of more terms of higher order, $\alpha = B/\lambda^2 + C/\lambda^4 + D/\lambda^6$, etc.

Normal and Anomalous Dispersion in Absorbing Media.—In a text-book on *The Theory of Optics* dated Leipzig, January, 1900, PAUL DRUDE made the first satisfactory theoretical study of natural and magnetic rotatory power, and developed "some new dispersion formulæ in the natural and magnetic rotation of the plane of polarisation, formulæ which are experimentally verified."

Drude's equations were based upon the hypothesis that absorbing media contain charged particles or ions, which possess natural periods of vibration. These particles are set into more or less violent vibration according as their natural periods agree more or less closely with the periods of the light-vibrations which fall upon them. This conception was used for the first time by MAXWELL in 1869 [1] as the basis of a theory of anomalous dispersion in absorbing media, but was afterwards developed independently, as the basis of a general theory of dispersion, by SELLMEIER, VON HELMHOLTZ and KETTELER. Thus, in the case of transparent substances, which have no appreciable absorptive power because their natural periods of vibration differ widely from that of the incident light, the NORMAL DISPERSION OF THE MEDIUM can be expressed by SELLMEIER'S EQUATION

$$n^2 = 1 + \sum \frac{D_m\lambda^2}{\lambda^2 - \lambda_m{}^2}$$

or by means of KETTELER'S EQUATION

$$n^2 = n_0{}^2 + \sum \frac{M_m}{\lambda^2 - \lambda_m{}^2},$$

where n is the refractive index of the medium,
$n_0{}^2$ can be identified with its dielectric constant,
λ is the wave-length of the incident light,
and λ_m is the wave-length corresponding with the natural period of vibration of the molecule.

If the molecule has more than one natural period, there will be a series of values for λ_m and a series of values for the constants D_m and M_m.

These equations represent the refraction as decreasing with increasing wave-length to a limiting value, which in one case is

[1] MAXWELL, *Cambridge Calendar*, 1869, Math. Trip. Examination; compare Rayleigh, *Phil. Mag.*, 1889, [v], **48**, 151.

$\sqrt{1+D}$ (and usually not very different from unity), but in the other is equal to the square root of the dielectric constant of the medium. They agree in making the refractive index increase asymptotically to an infinite value at the wave-length corresponding with the natural frequency of the molecule, so that $n = \infty$ when $\lambda^2 = \lambda_m^2$. At still shorter wave-lengths the sign of the term is reversed. The equations indicate an infinite negative value immediately beyond the asymptote, but, since this has no physical meaning, it will be sufficient to note the possible occurrence in this region of refractive indices which are less than unity, or less than the square root of the dielectric constant. If the wave-length is decreased still further, however, the influence of other terms, with a smaller characteristic frequency, will usually give rise once more to larger values of the refractive index.

These phenomena have been realised by R. W. Wood[1] with the help of a non-homogeneous cylinder of sodium vapour acting like a refracting prism. The refractive index of the vapour increased progressively from 1·000117 in the red at 7500 A.U. to 1·002972 at 5916 A.U. just before reaching the D-lines at 5895·932 and 5889·965 A.U. Immediately beyond them, the refractive index was as low as 0·614 at 5889·6 A.U., but rose quickly to 0·9908 at 5882 A.U. Since there is no other important absorption, the refractive index remained just below unity throughout the visible and ultra-violet spectrum, in close agreement with values calculated from Sellmeier's formula.

The absorption by sodium vapour is very intense, giving rise to the two sharply-defined Fraunhofer lines D_1 and D_2 of the solar spectrum, between which the helium line was first seen. In liquid and solid media, on the other hand, the absorption is usually much less complete, and is distributed over a range of wave-lengths, giving rise to an ABSORPTION BAND. The maximum of absorption is then described as the *head* of the band and its *width* can be defined by the separation of the two wave-lengths, on either side of the "head," at which the absorption falls to half its maximum value. Equations for the dispersion in a medium with this imperfect type of absorption were given by Helmholtz as follows:

$$n^2 - \kappa^2 - 1 = -\Sigma P\lambda^2 + \Sigma Q\frac{\lambda^4(\lambda^2 - \lambda_m^2)}{(\lambda^2 - \lambda_m^2)^2 + \alpha^2\lambda^2},$$

$$2n = \Sigma Q\frac{\alpha\lambda^5}{(\lambda^2 - \lambda_m^2)^2 + \alpha^2\lambda^2},$$

where, in addition to the symbols used above, $2\pi\kappa$ is the fraction of the light which is absorbed by a column of the medium of thickness λ, α is the amplitude of the vibration, and P, Q are constants depending on the nature of the band. In a region where there is no absorption, so that $\kappa = 0$, the first equation becomes

$$n^2 = 1 - P\lambda^2 + Q\frac{\lambda^4}{\lambda^2 - \lambda_m^2}$$

[1] R. W. Wood, *Phil. Mag.*, 1904, [vi], **8**, 293–324.

for a medium with only a single absorption band. When $P = Q$ this equation becomes identical with Sellmeier's formula.

These equations can be used to calculate the refractive index, either from the known positions of the bands or by a process of trial and error. In the latter case the refractive indices can be used to predict the positions of absorption bands which have not yet been observed experimentally.

The sinuous curve of refraction in the neighbourhood of an absorption band has been realised by R. W. Wood in nitrosodimethylaniline, which is transparent between the red and blue, but has a very intense absorption in the violet. The refractive index then rose to 2·140 at 4970 A.U., but fell nearly to unity on the other side of the band at 3000 A.U.

Saturated organic compounds are usually transparent throughout the visible and ultra-violet regions up to the limit (about 1850 A.U.) above which further measurements of absorption are rendered difficult by the absorptive power of air and quartz; they therefore exhibit only normal dispersion. Unsaturated organic compounds, on the other hand, often show absorption bands in the accessible ultra-violet region, or even in the visible spectrum, if they contain suitable groupings of "conjugated" double bonds. They may therefore be expected to provide many examples of anomalous dispersion, provided that the refractive index can be measured on both sides of the absorption band.

Drude's Equation for Optical Rotatory Dispersion.— Drude's analysis depends on the consideration of a DISSYMMETRICALLY ISOTROPIC MEDIUM.

> "A dissymmetrically isotropic medium would result if all the molecules were irregular tetrahedra of the same kind, the tetrahedra of the opposite kind (that which is the image of the first) being altogether wanting. The same would be true if one kind existed in smaller numbers than the other. A graphical representation may be obtained by conceiving that, because of the molecular structure, the paths of the ions are not short straight lines, but short helices twisted in the same direction and whose axes are directed at random in space."

In such a medium, the rotation of the plane of polarisation is given by the equation

$$\alpha = \sum \frac{k_m}{\lambda^2 - \lambda_m^2},$$

where λ_m is the wave-length corresponding to a characteristic frequency of the vibration, and

k_m is a constant depending on the number of vibrators in unit volume and other constants of the medium.

This equation, which is closely analogous to Sellmeier's equation for refractive dispersion, represents the optical rotation as increasing asymptotically to infinity when $\lambda^2 = \lambda_m^2$, and then showing a reversal of sign on crossing the wave-length of absorption. No examples have been found of optically-active bodies with linear absorption, to correspond with the sodium vapour of which the refractive dispersion was investigated by R. W. Wood, but his investigation of *magnetic* rotatory dispersion in sodium vapour (p. 461) gave rise to precisely similar phenomena in the neighbourhood of the *D*-lines. In the case of all naturally-active bodies, such as quartz or an alcoholic solution of camphor, the optical properties of the medium are determined by absorption *bands* of finite width, often extending over a considerable range of the spectrum. Drude's equation is then only valid in a region of complete transparency, and cannot be used to predict the course of the curve within the region covered by the absorption band (see below, p. 154).

Experimental Test of Drude's Equation.—We have seen how accurate measurements of rotatory dispersion revealed the inadequacy first of Biot's Law of Inverse Squares, $\alpha = k/\lambda^2$ and then of the formula $\alpha = -A + B/\lambda^2$ used by von Lang and by Stefan. If all the natural periods of the active ions lie in the ultra-violet, Drude's equation can be developed in ascending powers of $1/\lambda^2$ and put into the form

$$\alpha = B/\lambda^2 + C/\lambda^4 + D/\lambda^6 + \text{etc.}$$

which becomes identical with Boltzmann's equation if all the terms but the first two are ignored. Drude established the validity of his equation by showing that the rotatory dispersion of quartz, which had been measured over the range from 21,400 A.U. to 2193·5 A.U. by Gumlich[1] could not be deduced accurately from Boltzmann's equation, but was correctly represented by the equation

$$\alpha = \frac{12{\cdot}200}{\lambda^2 - 0{\cdot}010627} - \frac{5{\cdot}046}{\lambda^2}.$$

The observed and calculated values are set out in Table 6.

In deducing this equation Drude assumed that the characteristic frequencies of quartz could be deduced from its refractive dispersion, together with measurements of its dielectric constant, and observations by Rubens of the selective reflection of "residual rays" in the infra-red. Of the frequencies thus deduced, namely, $\lambda_1^2 = 0{\cdot}010627$, $\lambda_2^2 = 78{\cdot}22$, $\lambda_3^2 = 430{\cdot}6$, only the first appears in the final equation, since it was found, by calculation of the corresponding numerators, that $k_2 = k_3 = 0$. Drude therefore concluded *that the kinds of ions whose natural periods lie in the infra-red are inactive.* On the other hand, the term which included the ultra-violet frequency at $\lambda^2 = 0{\cdot}010627$ ($\lambda = 1030$ A.U.) had to be supplemented by a second term of a similar kind in order to represent the experimental data accurately, even when two infra-red terms were available for use if necessary. The second ultra-violet term does not depend on a

[1] GUMLICH, *Wied. Ann.*, 1898, **64**, 333–359.

TABLE 6.—ROTATORY DISPERSION IN QUARTZ (Gumlich, 1898; Drude, 1900).

λ (in μ).	α Obs.	α Calc.
2·140	1·60	1·57
1·770	2·28	2·29
1·450	3·43	3·43
1·080	6·18	6·23
0·67082	16·54	16·56
0·65631	17·31	17·33
0·58932	21·72	21·70
0·57905	22·55	22·53
0·57695	22·72	22·70
0·54610	25·53	25·51
0·50861	29·72	29·67
0·49164	31·97	31·92
0·48001	33·67	33·60
0·43586	41·55	41·46
0·40468	48·93	48·85
0·34406	70·59	70·61
0·27467	121·06	121·34
0·21935	220·72	220·57

characteristic frequency, since the experimental data then available were satisfied completely with the help of *two* arbitrary constants in the equation

$$\alpha = A/(\lambda^2 - \lambda_1^2) + B/\lambda^2.$$

There were therefore no errors to be corrected by introducing an additional arbitrary constant λ_2^2 in the second term, i.e. by writing $B/(\lambda^2 - \lambda_2^2)$ instead of B/λ^2. The more accurate measurements which are now available (p. 257) have, however, made it possible to deduce from the rotations themselves arbitrary values for λ_1^2 and λ_2^2, which are independent of measurements of refraction, and can be used instead to calculate the refractive dispersion of the crystal.

Application of Drude's Equation to Organic Compounds.—At the time when Drude developed his equation for optical rotatory dispersion, optical activity had been recorded in many hundreds of organic compounds, but the study of their rotatory dispersion had not advanced beyond the point represented by the measurements of Wiedemann (1851), Arndtsen (1858), Krecke (1872) and von Wyss (1888). These included a very limited number of liquids and solutions, and covered a range of four to six wave-lengths in the visible spectrum only. The first serious attempts to investigate the law of rotatory dispersion in organic compounds were therefore made a dozen years after the publication of Drude's book, and are described in a paper on *The Form of the Rotatory-dispersion Curves* published in 1913.[1] In this paper data are given for the *magnetic* rotatory dispersion of some 50 simple organic compounds, and for the *natural* rotatory dispersion of quartz, *iso*valeric acid and a series

[1] LOWRY and DICKSON, *J.*, 1913, **103**, 1067–1075.

of 10 optically-active alcohols prepared by Pickard and Kenyon, together with values calculated with the help of Drude's equation, for a series of 8 wave-lengths in the visible spectrum. The *natural* rotations, expressed in terms of the DISPERSION RATIO, α/α_{5461}, are reproduced in Table 7 below.

TABLE 7.—FORM OF THE ROTATORY DISPERSION CURVE

	Red.		Yellow.	Green.		Blue.		Violet.
	Li 6708.	Cd 6438.	Na 5893.	Hg 5461.	Cd 5086.	Cd 4800.	Cd 4678.	Hg 4358.
Quartz	0·648	0·706	0·851	1·000	1·164	1·319	1·394	1·627
$k = 0·2809$ $\lambda_0^2 = 0·0173$	0·649	0·707	0·851	1·000	1·164	1·318	1·394	1·627
sec. Amyl alcohol $CH_3 . CH(OH) . C_3H_7$	0·642	—	0·846	1·000	—	—	—	1·652
,, Hexyl alcohol $CH_3 . CH(OH) . C_4H_9$	0·639	—	0·847	1·000	1·169	1·331	—	1·653
,, Heptyl alcohol $CH_3 . CH(OH) . C_5H_{11}$	0·638	—	0·847	1·000	—	—	—	1·648
,, Octyl alcohol $CH_3 . CH(OH) . C_6H_{13}$	0·640	0·699	0·847	1·000	1·168	1·329	1·407	1·653
,, Nonyl alcohol $CH_3 . CH(OH) . C_7H_{15}$	0·637	—	0·846	1·000	—	—	—	1·651
,, Decyl alcohol $CH_3 . CH(OH) . C_8H_{17}$	0·641	—	0·848	1·000	—	—	—	1·649
,, Undecyl alcohol $CH_3 . CH(OH) . C_9H_{19}$	0·643	—	0·850	1·000	—	—	—	1·651
,, Dodecyl alcohol $CH_3 . CH(OH) . C_{10}H_{21}$	0·645	—	0·848	1·000	—	—	—	1·653
Average	0·641	—	0·847	1·000	—	—	—	1·651
$k = 0·2745$ $\lambda_0^2 = 0·0237$	0·644	0·702	0·848	1·000	1·168	1·329	1·406	1·651
*iso*Valeric acid . . .	0·631	0·695	0·841	1·000	1·178	1·355	—	1·710
$k = 0·2607$ $\lambda_0^2 = 0·0375$	0·632	0·6915	0·842	1·000	1·179	1·352	—	1·710
Phenylmethylcarbinol $C_6H_5 . CH(OH) . CH_3$	0·629	0·687	0·839	1·000	1·184	1·361	—	1·736
$k = 0·2553$ $\lambda_0^2 = 0·0429$	0·627	0·687	0·839	1·000	1·183	1·362	1·450	1·735

These observations showed clearly that

> " both the optical and the magnetic dispersions can be expressed by the simple equation $\alpha = k/(\lambda^2 - \lambda_0^2)$. This equation has the merit of separating completely and easily the rotatory power of a substance from its dispersive power, so that these can now be measured and discussed as separate and independent properties " (*ib*., p. 1068).

Simple and Complex Rotatory Dispersion.—In a paper published under the above title[1] it was proposed to describe as

[1] LOWRY and DICKSON, *T.F.S.*, 1914, **10**, 96.

SIMPLE ROTATORY DISPERSION all those cases which can be expressed by one term of Drude's equation $\alpha = k/(\lambda^2 - \lambda_0^2)$. The constant k was described as the ROTATION CONSTANT, and λ_0^2 as the DISPERSION CONSTANT of the equation. Any dispersion which could not be expressed by a simple equation of this type, with only two arbitrary constants, was described as a COMPLEX ROTATORY DISPERSION. This classification is not identical with the older method of distinguishing between "normal" and "anomalous" rotatory dispersion, since a "normal" curve need not necessarily conform to the equation of simple rotatory dispersion and may, therefore, be "complex." On the other hand, a simple dispersion must necessarily be normal, and an anomalous dispersion will obviously be complex.

If it were required to compare the two systems of classification, it might be pointed out that the distinction between simple and complex dispersion in doubtful cases depends on measuring the rotations up to the limit of transparency in the ultra-violet, since the existence of a second term with a different dispersion constant can be demonstrated most readily as the asymptote of the theoretical curve is approached, i.e. when readings are taken in close proximity to the region of absorption, within which Drude's equation ceases to be valid. On the other hand, the distinction between normal and anomalous dispersion in doubtful cases depends on finding out whether the dispersion curve does or does not cut the axis of zero rotation when prolonged into the infra-red. At the present time there are several methods available for measuring optical rotations in the ultra-violet, with an ease and accuracy that are not much less than in the visible region of the spectrum. Measurements in the infra-red are, however, still difficult to make and are relatively inexact. Whilst, therefore, it is little more than a matter of routine to test the simple dispersion equation over the range of its validity, and up to the limits of accuracy which are experimentally practicable, there is no easy experimental method of deciding whether a rotation is normal or anomalous, unless the anomalies happen to fall within the limits of visual or photographic observation. In practice, therefore, doubtful cases can only be classified as normal or anomalous by a process of extrapolation, i.e. by calculating a two-term equation of the Drude type and comparing the relative magnitudes of the two rotation constants with those of the corresponding dispersion constants (see below, p. 139) in order to determine whether the calculated curve would show a reversal of sign in the infra-red.

Graphical Test for Simple Rotatory Dispersion.—Biot in 1818 introduced the device of plotting $1/\alpha$ against λ^2 as a method of demonstrating his Law of Inverse Squares in the case of quartz (Fig. 37).[1] A precisely similar method can be used to demonstrate the validity of a dispersion formula containing only a single term of Drude's equation.

[1] BIOT, *Mém. Acad. Sci.*, 1818, **2**, 41–136 (Figs. 15 and 16, Plate 3).

"A very simple method of testing the form of the dispersion curve is to plot the reciprocals of the rotatory powers (or of the dispersion ratios) against the squares of the wave-lengths, using for this purpose the following data :

	Li 6708	Cd 6438	Na 5893	Hg 5461	Cd 5086	Cd 4800	Cd 4678	Hg 4358
$\lambda^2 =$	0·4500	0·4145	0·3473	0·2982	0·2587	0·2304	0·2189	0·1900

If the simple dispersion formula is valid, the observations will then plot out to a straight line." [1]

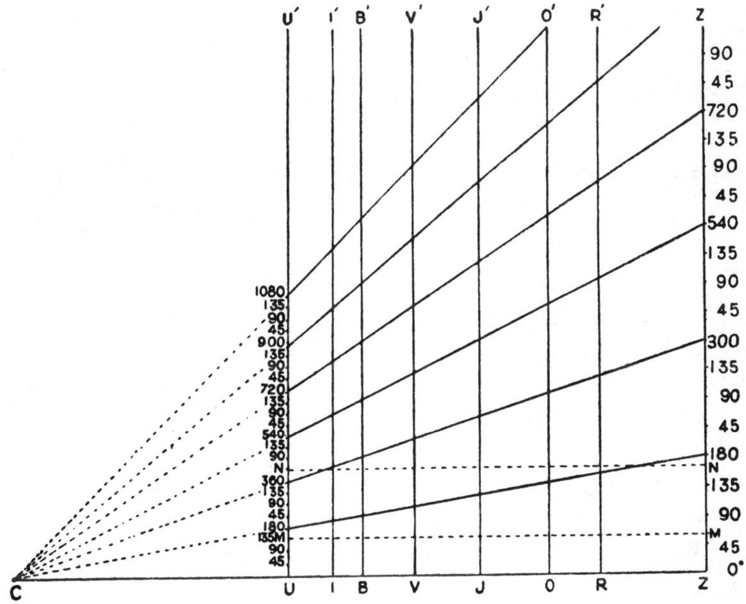

FIG. 37.—DIAGRAM ILLUSTRATING THE ROTATORY DISPERSION OF QUARTZ (Biot, 1818).

The squares of the wave-lengths are plotted as abscissæ ; the ordinates show the length of the column required to produce a rotation of 180°, 360°, etc., and are therefore proportional to $1/\alpha$, where α is the rotation in degrees per millimetre.

This graphical method provides only a rough test of the validity of the equation, and therefore serves to disclose only the grosser deviations from the law of simple rotatory dispersion ; but it can be used to demonstrate the fact that this law approximates more closely to the experimental data than any other formula that has yet been proposed. The simple dispersion law can therefore be used as a standard from which deviations may be measured, even when

[1] LOWRY and DICKSON, *J.*, 1913, **108**, 1075.

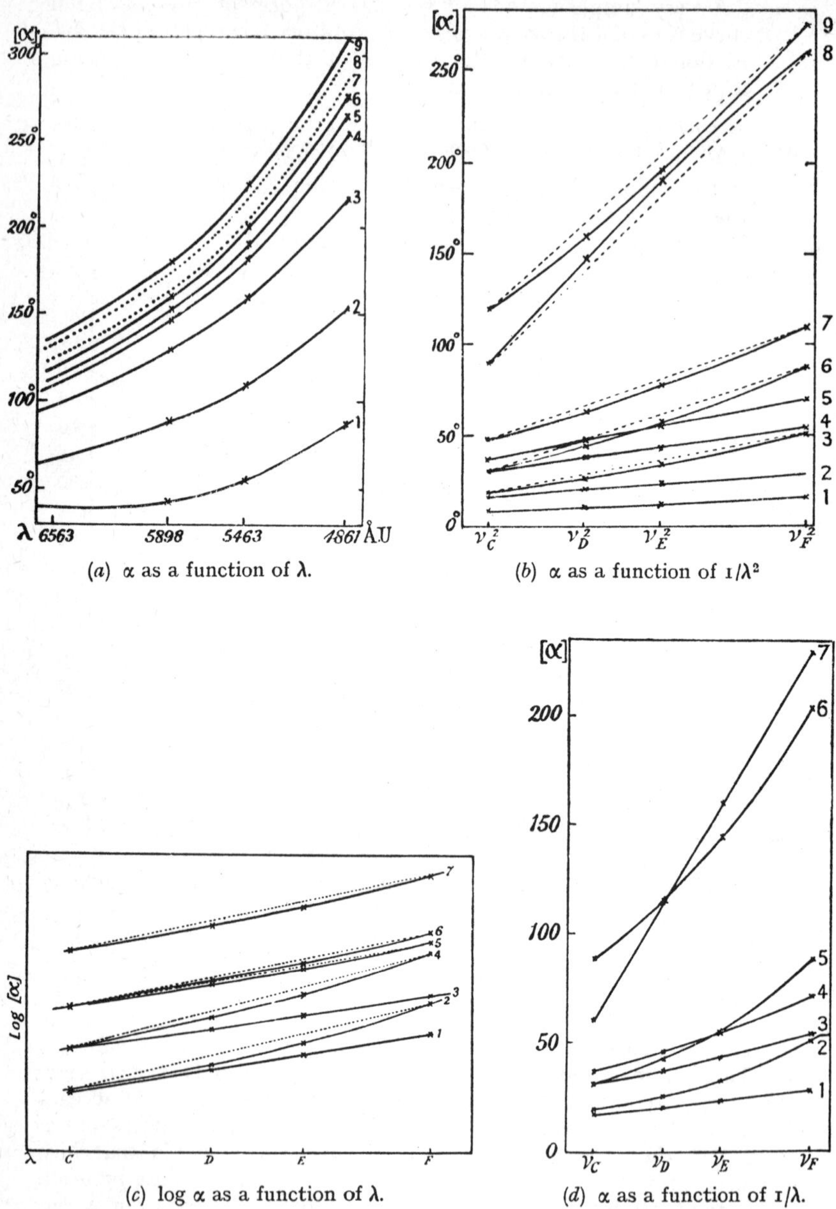

(*a*) α as a function of λ.

(*b*) α as a function of $1/\lambda^2$

(*c*) log α as a function of λ.

(*d*) α as a function of $1/\lambda$.

FIG. 38.—ROTATORY DISPERSION OF TERPENE DERIVATIVES (Rupe, 1915).

it is not rigorously exact, and in particular it provides an ideal method for deciding whether the deviations between the observed and calculated values are systematic or merely casual.

A vindication of Drude's equation on these lines can be found in a paper by H. Rupe, *Uber normale und anormale Rotations-dispersion*,[1] and in a paper by his colleague, Hagenbach, *Uber die Rotations-dispersion homologer Reihen*.[2] Hagenbach's graphical analysis was vitiated by an error as regards the wave-length of the line described by the letter E, so that his graphs were a series of broken lines; but Figs. 38 (*a*) to (*d*) show the curves which Rupe obtained * by plotting:

(*a*) α against λ. (*b*) α against $1/\lambda^2$.
(*c*) $\log \alpha$ against λ. (*d*) α against $1/\lambda$.

On the other hand, Figs. 39 to 42 [3] show the lines obtained by plotting $1/\alpha$ against λ^2, according to the method first suggested by Biot.

It is particularly noteworthy that, whereas *curves* were obtained almost invariably when α was plotted against $1/\lambda^2$, in order to test Stefan's equation, the points obtained by plotting $1/\alpha$ against λ^2 (i.e. by interchanging the reciprocals) could be represented by *lines* drawn with the help of a ruler, with only three obvious exceptions amongst forty-two cases. It cannot be claimed that the rotatory dispersion was "simple" in the remaining thirty-nine cases, since the method is only an approximate one; but it is at least clear that these straight lines, and the formulæ to which they correspond, provide an ideal starting-point for the investigation of the possible complexity of the dispersions.

Simple Rotatory Dispersion of Octyl Alcohol and of Sucrose.—In order to find out whether the simple dispersion formula is valid up to the limits of experimental accuracy, or would be found to be a mere approximation like Biot's Law, or the formulæ of von Lang and Stefan, two series of critical observations were made by

[1] Rupe, *Ann.*, 1915, **409**, 327–357.

[2] Hagenbach, *Z. ph. C.*, 1915, **89**, 570.

[3] Lowry and Abram, *Simple Rotatory Dispersion in the Terpene Series*, *J.*, 1919, **115**, 300–311.

* The substances used were as follows:—

(*a*) Solutions in benzene of (1) Camphor. (2) Oxymethylenecamphor. (3) *cyclo*Hexyl-. (4) *iso*-Butyl-. (5) Propyl-. (6) Ethyl-. (9) Methyl-methylenecamphor. (7) and (8) Propyl- and Ethyl-methylenecamphor without a solvent.

(*b*) (1) Dimethylnonadiene. (2) Benzyldihydrocarvone. (3) Pulegone. (4) Menthyl β-phenylcinnamate. (5) Menthol. (6) Camphor. (7) Dimethylphenyloctadiene. (8) Diphenylmethylenecamphor. (9) Ethylmethylenecamphor.

(*c*) (1) Diphenylmethylenecamphor. (2) Pulegone. (3) Menthyl β-phenylcinnamate. (4) Camphor. (5) Carvone. (6) Dimethylphenyloctadiene. (7) Benzylmethylenecamphor.

(*d*) (1) Benzyldihydrocarvone. (2) Pulegone. (3) Menthyl β-phenylcinnamate. (4) Menthol. (5) Camphor. (6) Benzylmethylenecamphor. (7) Diphenylmethylenecamphor.

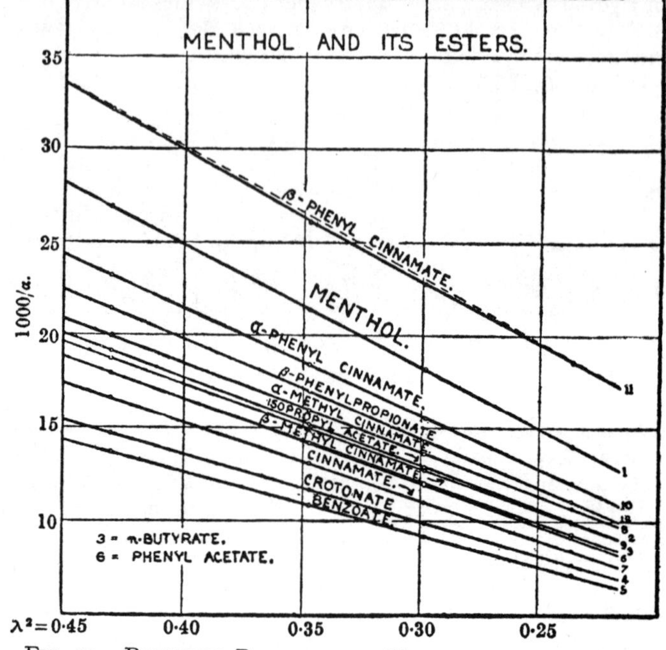

FIG. 39.—ROTATORY DISPERSION IN MENTHOL AND ITS ESTERS.

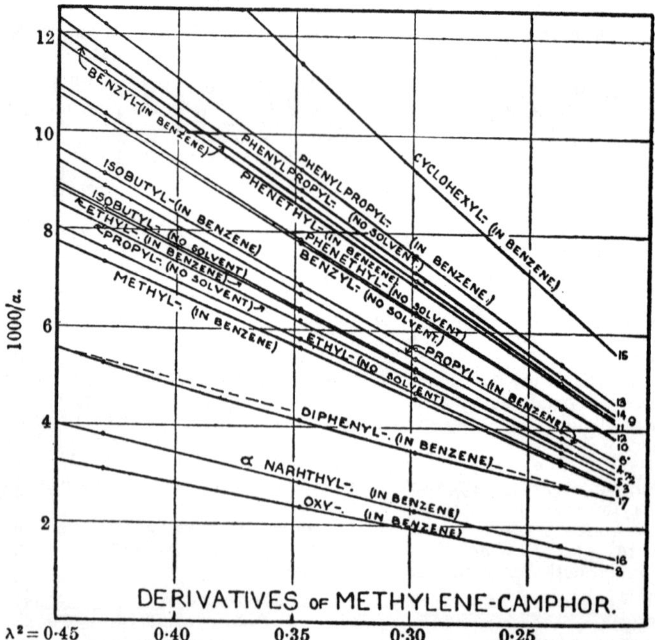

FIG. 40.—ROTATORY DISPERSION IN DERIVATIVES OF METHYLENE-CAMPHOR.

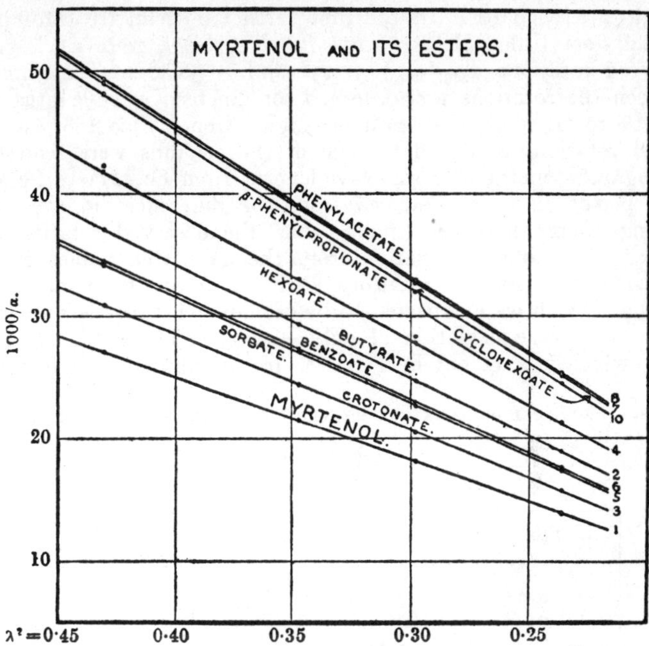

FIG. 41.—ROTATORY DISPERSION IN MYRTENOL AND ITS ESTERS.

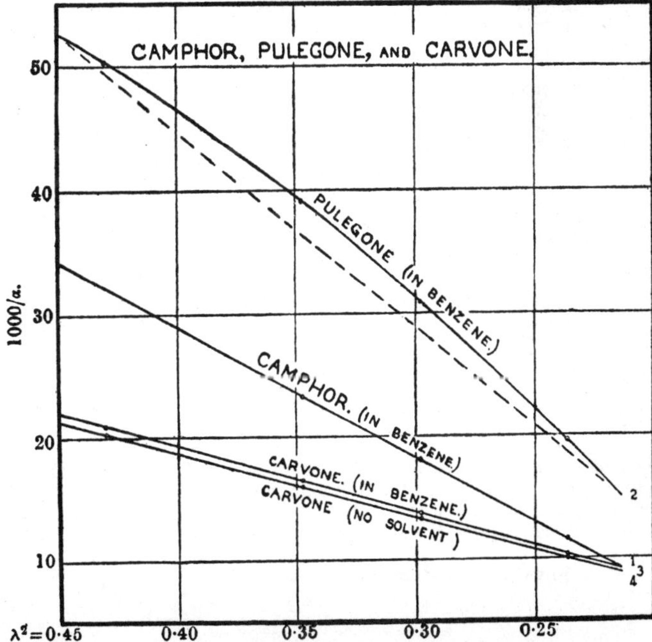

FIG. 42.—ROTATORY DISPERSION IN CAMPHOR, PULEGONE AND CARVONE.

Lowry and Richards. In the first series[1] 6-decimetre columns of *d*- and *l*-octyl alcohol, or *methyl hexyl carbinol*, were used, giving observed rotations $\alpha_{5461} = +57{\cdot}13^\circ$ and $-56{\cdot}40^\circ$. The *difference* between the rotations was observed for nineteen wave-lengths from Li 6708 to Hg 4358, the readings ranging from 73° to 188°, and the actual rotations produced by one of the columns were read by a photographic method for six wave-lengths from Fe 4271 to Fe 3814. The specific rotations deduced from the differences in the visual readings were ten times smaller than the observed rotations and showed a maximum deviation from the calculated values of only $\pm 0{\cdot}01^\circ$. The specific rotations deduced from the photographic readings, which were expected to be perhaps ten times less accurate, gave an average deviation of only $\pm 0{\cdot}03^\circ$. These minute deviations, which are set out in Table 8, are distributed in a fortuitous

TABLE 8.—SPECIFIC ROTATORY POWER OF *sec*.-OCTYL ALCOHOL AT 20°.

$$[\alpha] = 3{\cdot}176/(\lambda^2 - 0{\cdot}0244).$$
$$\alpha_{5161} = +57{\cdot}13^\circ \text{ and } -56{\cdot}40^\circ.$$

Line.	$[\alpha]_{obs.}$	$[\alpha]_{calc.}$	$[\alpha]_{obs.} - [\alpha]_{calc.}$
	Degrees.	Degrees.	Degrees.
Li 6708	7·46	7·47	− 0·01
Cd 6438	8·14	8·14	±
Zn 6362	8·35	8·35	±
Na 5893	9·84	9·85	− 0·01
Cu 5782	10·25	10·25	±
Hg 5780	10·25	10·26	− 0·01
Cu 5700	10·57	10·57	±
Hg 5461	11·61	11·60	+ 0·01
Cu 5218	12·82	12·81	+ 0·01
Ag 5209	12·86	12·85	+ 0·01
Cu 5153	13·18	13·18	±
Cu 5106	13·44	13·44	±
Cd 5086	13·56	13·56	±
Zn 4811	15·36	15·35	+ 0·01
Cd 4800	15·43	15·42	+ 0·01
Zn 4722	16·00	15·99	+ 0·01
Zn 4680	16·33	16·32	+ 0·01
Cd 4678	16·34	16·34	±
Hg 4358	19·19	19·19	±
Fe 4271	*20·10*	20·10	±
Fe 4261	*20·25*	20·21	+ 0·04
Fe 4251	*20·35*	20·32	+ 0·03
Fe 4046	*22·80*	22·80	±
Fe 3860	*25·44*	25·50	− 0·06
Fe 3814	*26·16*	26·23	− 0·07

Note.—The observed difference between the two alcohols for Hg 5461 was 113·53°. The factor for reducing the readings to specific rotations was 0·1022 for the visual and 0·2057 for the photographic readings. Values determined by the photographic method are shown in italics.

[1] LOWRY and RICHARDS, *J.*, 1924, **125**, 1593–1597.

manner, with no indications of systematic error. They, therefore, prove conclusively that the one-term equation is not a mere approximation, but provides a perfect expression of the experimental data, as obtained from an exceptionally careful and accurate series of measurements. The possibility of deviations being found at shorter wave-lengths, by using shorter columns of liquid, is irrelevant to the question which is here at issue, since the rotations were followed up to the limit imposed by the incipient absorption of light by the medium, and any further extension would violate the essential condition which limits the validity of the equation to the region of complete transparency, in which the absorptive power of the medium is negligible.

In the second series of experiments[1] a concentrated aqueous solution of highly purified *sucrose* was used, containing 26 grams of sugar in 100 c.c. The rotations produced by a 6-decimetre column were measured for 18 wave-lengths from Li 6708 to Hg 4358 and by a photographic method for 13 wave-lengths from Fe 4384 to Fe 3826. The specific rotations in this case are much larger fractions of the observed rotations, and for this reason are from three to six times less accurate than in the case of octyl alcohol. Nevertheless the deviations, which are set out in Table 9, are insignificant in magnitude and fortuitous in distribution, so that once again the equation is shown to provide an accurate expression of the

TABLE 9.—ROTATORY DISPERSION OF SUCROSE IN AQUEOUS SOLUTION AT 20°.

$$[\alpha] = 0{\cdot}641\alpha = 21{\cdot}648/(\lambda^2 - 0{\cdot}0213).$$

Line.	$[\alpha]_{obs.}$	$[\alpha]_{calc.}$	$[\alpha]_{obs.} - [\alpha]_{calc.}$	Line.	$[\alpha]_{obs.}$	$[\alpha]_{calc.}$	$[\alpha]_{obs.} - [\alpha]_{calc.}$
	Degrees.	Degrees.	Degrees.		Degrees.	Degrees.	Degrees.
Li 6708	50·51	50·50	+ 0·01	*Fe 4384*	*126·5*	126·7	− 0·2
Cd 6438	55·04	55·05	− 0·01	*Fe 4376*	*127·2*	127·2	±
Zn 6362	56·51	56·45	+ 0·06	Hg 4358	128·49	129·37	+ 0·12
Na 5893	66·45	66·44	+ 0·01	*Fe 4353*	*128·5*	128·7	− 0·2
Cu 5782	69·10	69·16	− 0·06	*Fe 4337*	*129·8*	129·8	±
Hg 5780	69·22	69·21	+ 0·01	*Fe 4315*	*130·7*	131·3	− 0·6
Cu 5700	71·24	71·30	− 0·06	*Fe 4282*	*133·6*	133·6	±
Hg 5461	78·16	78·18	− 0·02	*Fe 4272*	*134·2*	134·3	− 0·1
Cu 5218	86·21	86·25	− 0·04	*Fe 4261*	*134·9*	135·1	− 0·2
Cu 5153	88·68	88·63	+ 0·05	*Fe 4191*	*140·0*	140·2	− 0·2
Cu 5106	90·46	90·44	+ 0·02	*Fe 4144*	*144·2*	143·9	+ 0·3
Cd 5086	91·16	91·20	− 0·04	*Fe 3889*	*166·7*	166·7	±
Zn 4811	103·07	103·03	+ 0·04	*Fe 3833*	*171·8*	172·3	− 0·5
Cd 4800	103·62	103·53	+ 0·09	*Fe 3826*	*173·1*	173·2	− 0·1
Zn 4722	107·38	107·33	+ 0·05				
Zn 4680	109·49	109·48	+ 0·01				
Cd 4678	109·69	109·58	+ 0·11				

The photographic readings are shown in italics.

[1] LOWRY and RICHARDS, *J.*, 1924, **125**, 2511–2524.

experimental data, instead of a mere approximation. Incidentally, these observations show that simple rotatory dispersion is compatible with the presence of *nine* asymmetric carbon atoms in the molecule, presumably because the *dispersion* is equal or similar throughout the nine centres of asymmetry.

Dispersion-ratios and Dispersion-constants of Saturated Compounds.—In the preceding paragraphs it has been shown that the rotatory dispersion of a number of secondary alcohols can be expressed by means of a single term of Drude's equation. It is noteworthy that the dispersion-ratio $\alpha_{4358}/\alpha_{5461}$ remains constant at 1·651, and the dispersion-constant at 0·0237, in the homologous series of methylalkylcarbinols (Table 7). Similarly, in the case of cane sugar, the dispersion-ratio $\alpha_{4358}/\alpha_{5461}$ is 1·644, and the dispersion-constant is 0·0213. These numbers correspond with a characteristic frequency at about 1500 A.U. in the SCHUMANN REGION of the spectrum, which can only be studied with the help of a vacuum spectrograph and specially prepared photographic plates. This frequency appears to be characteristic of *saturated* organic compounds, which show no selective absorption in the accessible ultra-violet, but may be expected to give very intense absorption bands in the Schumann region.

The approximate constancy of the characteristic frequency in *saturated* compounds also provides an explanation of the relative ease with which numerical relationships can be detected in the rotatory powers of compounds of this type. On the other hand, *unsaturated* and *aromatic* secondary alcohols give rise to larger dispersions, e.g. the dispersion-ratio $\alpha_{4358}/\alpha_{5461} = 1{\cdot}673$, $\lambda_0^2 = 0{\cdot}0293$ for *phenylethylcarbinol*, or 1·735 and 0·0429 for *phenylmethylcarbinol*. The corresponding numbers for iso*valeric acid*, 1·710 and 0·0375, are intermediate between those for the two aromatic alcohols. It therefore appears that unsaturation *increases* the rotatory dispersion of a compound, but that there is a limit beyond which the rotatory dispersion cannot *decrease* even in the simplest saturated molecules.

> "Biot's Law appears to express the lower limit of dispersive power as it would be observed in an absolutely transparent medium. Such a medium does not appear to exist amongst optically-active compounds, and if it did its optical rotatory power would probably be reduced to the vanishing point. In all the actual cases that have been studied the dispersion is increased by the presence of absorption in the ultra-violet region, whereby λ^2 is converted into $\lambda^2 - \lambda_0^2$." [1]

Pseudo-simple Rotatory Dispersion.—(*a*) The distinction between simple and complex rotatory dispersion is essentially a practical one. Its main purpose is to discriminate between (i) compounds which conform to the requirements of a one-term Drude's equation within the limits of experimental accuracy and up to the

[1] LOWRY and DICKSON, *J.*, 1913, **103**, 1069.

limits of complete transparency of the medium, and (ii) compounds which do not conform to this equation, although their dispersion can usually be represented by using two of these terms. On theoretical grounds it has been suggested by Lowry and Walker in 1924 [1] and by Kuhn in 1929 [2] that the optical rotation of every optically-active molecule must include a long series of partial rotations, since all the electrons are influenced to a greater or smaller extent by the dissymmetry of the molecule. *Simple rotatory dispersion* is therefore an ideal which can only be realised when the partial rotations of the different electrons can be expressed to a sufficiently close approximation by a single dispersion-constant.

(*b*) It is convenient to describe as examples of PSEUDO-SIMPLE ROTATORY DISPERSION those cases in which the deviations from the simple equation are too minute to be detected readily, although the existence of a second term can be inferred or demonstrated in other ways. Thus two cases (*sodium tartrate* in aqueous solution,[3] and liquid *octyl oxalate* [4]) have been discovered in which the rotatory dispersion can be expressed to a close approximation by Biot's Law of Inverse Squares. This close agreement must, however, be fortuitous, since it would imply that the rotatory dispersion was controlled by an absorption band at zero wave-length. The complexity of the rotatory dispersion was first demonstrated by the discovery of small systematic deviations between the observed and calculated rotations, although these could not have been detected in measurements of a lower order of accuracy; but, in the case of sodium tartrate, the conclusion that the deviations were systematic and not merely fortuitous was confirmed [5] by measurements in the ultra-violet of more dilute solutions, which disclosed the existence of anomalies, corresponding closely with those which had been predicted by using a two-term equation to represent the data.

(*c*) The rotatory dispersion of compounds, such as tartaric acid, camphor, or the menthyl and bornyl xanthates, which exhibit selective absorption outside the Schumann region, is nearly always complex and often anomalous, since the equations include (i) a low-frequency term (with a characteristic frequency approximating to that of the absorption band) which may be associated with the unsaturated or chromophoric centres in the molecule, and (ii) a high-frequency term (with a frequency in the Schumann region) which may be associated with the fixed centres of asymmetry in the molecule. A remarkable example of simple rotatory dispersion is provided, however, by the *tetra-acetyl-μ-arabinose*.

$$H . [CHOAc]_4 . CHO,$$

[1] LOWRY and WALKER, *Nature*, 1924, **113**, 566.
[2] W. KUHN, *Ber.*, 1929, **62**, 1727; *Z. ph. C.*, 1929, B. 4, 14; *T.F.S.*, 1930, **26**, 293.
[3] LOWRY and AUSTIN, *Phil. Trans.*, 1922, A. **222**, 295.
[4] LOWRY and RICHARDS, *J.*, 1924, **125**, 1596.
[5] LOWRY and VERNON, *P.R.S.*, 1928, A. **119**, 706–709; compare DESCAMPS, *Thesis*, Brussels, 1928; *C.R.*, 1927, **185**, 116.

where the partial rotations of high frequency, which are associated with the three asymmetric carbon atoms of the $[CHOAc]_4$ chain, happen to cancel out, so that the rotatory power of the whole molecule can be represented by a single low-frequency term with the characteristic frequency of the aldehydic group.[1]

This phenomenon may be imitated in many cases in which the rotatory dispersion is dominated by an adjacent absorption band, thus giving rise to a form of "pseudo-simple" rotatory dispersion with a characteristic frequency of long wave-length. An example of this kind is provided by *camphorquinone*, $C_8H_{14}\langle {CO \atop CO} \rangle$ (the two CO groups joined by a bond), a derivative of camphor, which has a light yellow colour, and exhibits a maximum of selective absorption in the blue at 4675 A.U.[2] The rotatory dispersion in the narrow region of transparency in the red, yellow and green (Table 10 [3]) can be represented by the "simple" equation

$$[\alpha] = -13{\cdot}170/(\lambda^2 - 0{\cdot}2235).$$

This equation has a characteristic frequency at 4730 A.U., which evidently corresponds with that of the absorption band at 4675 A.U. In subsequent experiments [4] the rotatory dispersion has been measured right through the absorption band, as far as 4219 A.U. for a solution

TABLE 10.—ROTATIONS OF CAMPHORQUINONE IN BENZENE AT 20°.

15·069 gms. of camphorquinone in 100 gms. of solution.
Density 0·90290. Length of tube 6 dm.

$$[\alpha] = 1{\cdot}225\ \alpha. \quad [M] = 1{\cdot}6611\ [\alpha].$$

$$[\alpha] = -\frac{13{\cdot}170}{\lambda^2 - 0{\cdot}22352}.$$

Wave-length.	Specific Rotations.		
	Obs.	Calc.	Diff.
	Degrees.	Degrees.	Degrees.
Li 6707·8	− 58·26	− 58·16	− 0·10
Cd 6438·5	− 68·60	− 68·93	+ 0·33
Zn 6362·3	− 72·66	− 72·64	− 0·02
Na 5893·0	− 105·98	− 106·42	+ 0·44
Cu 5782·2	− 117·96	− 118·93	+ 0·97 }
Hg 5780·1	− 118·42	− 119·10	+ 0·68 }
Cu 5700·2	− 128·21	− 127·65	− 0·56 }
Ag 5468·6	− 174·83	− 174·36	− 0·47 }
Hg 5460·7	− 176·40	− 176·34	− 0·06 }

[1] HUDSON, WOLFROM and LOWRY, *J.*, 1933, 1179.
[2] LOWRY and FRENCH, *J.*, 1924, **125**, 1921.
[3] LOWRY and CUTTER, *J.*, 1925, **127**, 614.
[4] LOWRY and GORE, *P.R.S.*, 1932, A. **135**, 13; compare LIFSCHITZ, *Z. ph. C.*, 1923, **105**, 49 and 51.

in *cyclo*hexane at 20° and 3659 A.U. for the vapour at 200° (Fig. 133, p. 309). These measurements have not disclosed with certainty the existence of a high-frequency partial rotation on the further side of the absorption band; but, since the *negative* maximum $[\alpha] = -450°$ at 4940 A.U. in cyclohexane is substantially larger than the *positive* maximum, $[\alpha] = +300°$ at 4440 A.U., it may be inferred that the high-frequency term is negative in sign, and that the rotatory dispersion in the narrow region of transparency in the visible spectrum is not really simple but should be classified as pseudo-simple, i.e. as involving two terms of Drude's equation, of which one is insignificant in comparison with the other.

The general problem of rotatory dispersion in unsaturated compounds is discussed in the following chapter.

CHAPTER X.

COMPLEX AND ANOMALOUS ROTATORY DISPERSION.

Application of Drude's Equation to Ethyl Tartrate.—In view of the fact that no accurate data were available, by means of which Drude could test the application of his equation to organic compounds exhibiting *normal* rotatory dispersion, it is not surprising that data in reference to *anomalous* rotatory dispersion were at least equally scanty. In 1914, however, Lowry and Dickson,[1] in addition to establishing the validity of Drude's one-term equation to express the normal rotatory dispersion of a number of simple organic compounds, were able to show that the anomalous rotatory dispersion of a 6-dm. column of *ethyl tartrate* at 20°, for 13 wave-lengths in the visual region from Li 6708 to Hg 4358 and for 5 wave-lengths in the photographic region (including negative rotations of — 20° and —40°) could be expressed by means of *two* terms of Drude's equation with opposite signs, thus :

$$\alpha = \frac{132\cdot78}{\lambda^2 - 0\cdot026} - \frac{103\cdot05}{\lambda^2 - 0\cdot061}.$$

This preliminary evidence was afterwards confirmed by a more rigid proof depending on measurements of the rotatory power for 18 wave-lengths in the visual region from Li 6708 to Hg 4358 and for 18 wave-lengths in the photographic region (including negative rotations up to — 88°) of a 6-dm. column of a sample of the ester which had been purified, not only by fractional distillation, but also by crystallisation.[2] The specific rotations in Table 11, which were about seven times smaller than the observed rotations, can be expressed by the formula

$$[\alpha] = \frac{25\cdot005}{\lambda^2 - 0\cdot03} - \frac{20\cdot678}{\lambda^2 - 0\cdot056}.$$

The maximum deviation between the observed and calculated values in the visual region is 0·03°, and the average error 0·0055° ; the maximum deviation in the photographic region is 0·3° and the average error was 0·055°, the majority of the deviations in each case being registered as ±.

[1] Lowry and Dickson, *T.F.S.*, 1914, **10**, 99.
[2] Lowry and Cutter, *J.*, 1922, **121**, 532–544.

Application of Drude's Equation to Anomalous Rotatory Dispersion.—The observations described above established the

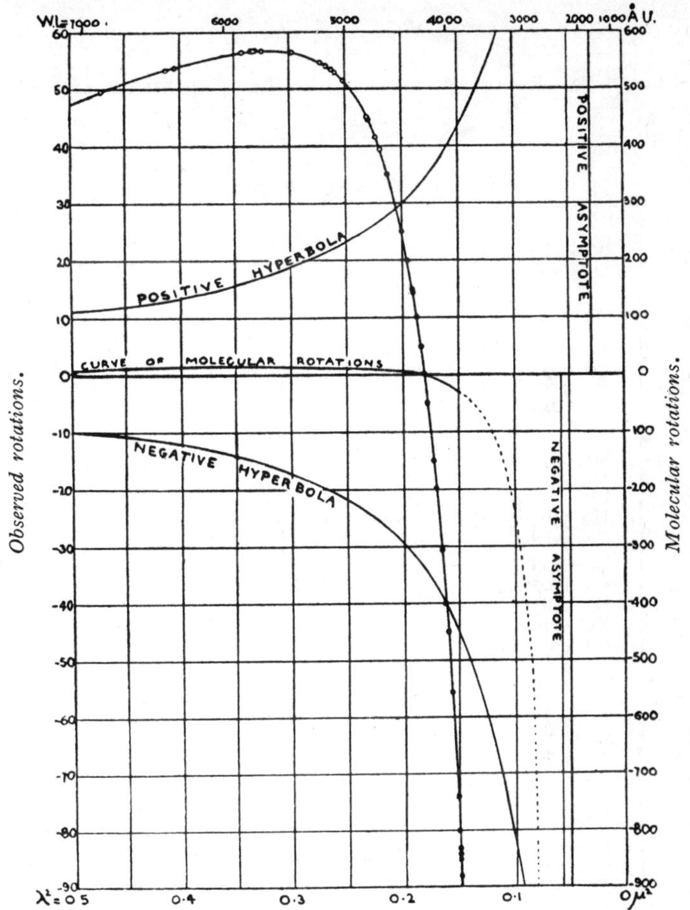

FIG. 43.—ANOMALOUS ROTATORY DISPERSION OF ETHYL TARTRATE.

In this figure the *observed rotations* of a 6-dm. column of the ester, ranging from + 57° to − 88°, are plotted against the squares of the wave-lengths in microns. The *molecular rotations* are plotted on a much smaller scale, together with positive and negative hyperbolæ representing the normal *partial rotations*, which when added give rise to the anomalous curve of *total molecular rotations.*

fact that the anomalous rotatory dispersion of ethyl tartrate can be expressed by two terms of Drude's equation, under the limitations set out by Arndtsen in 1858, namely, that the partial rotations must be *of opposite sign and of unequal dispersion.* These limitations can be expressed by writing the equation in the form

$$\alpha = \frac{k_1}{\lambda^2 - \lambda_1^2} - \frac{k_2}{\lambda^2 - \lambda_2^2}.$$

TABLE 11.—ROTATORY DISPERSION OF ETHYL TARTRATE AT 20°.

$$[\alpha] = \frac{25{\cdot}005}{\lambda^2 - 0{\cdot}03} - \frac{20{\cdot}678}{\lambda^2 - 0{\cdot}056}.$$

(*a*) *Visual Readings.*

Wave-length $\lambda \times 10000$.	Specific Rotations.		
	Obs.	Calc.	Diff.
	Degrees.	Degrees.	Degrees.
Li 6707·8	6·87	6·87	±
Cd 6438·5	7·38	7·38	±
Zn 6362·3	7·43	7·43	±
Na 5893·0	7·82	7·82	±
Hg 5790·7	7·87	7·87	±
Cu 5782·2	7·87	7·87	±
Hg 5769·6	7·88	7·88	±
Cu 5700·2	7·88	7·90	−0·02
Hg 5460·7	7·87	7·86	+0·01
Cu 5218·2	7·61	7·61	±
Cu 5153·3	7·48	7·48	±
Cu 5105·5	7·36	7·38	− 0·02
Cd 5085·8	7·34	7·33	+ 0·01
Zn 4810·5	6·25	6·28	− 0·03
Cd 4799·9	6·22	6·22	±
Zn 4722·2	5·75	5·75	±
Cd 4678·2	5·45	5·46	− 0·01
Hg 4358·3	1·98	1·98	±
(*b*) *Photographic Readings.*			
Fe 4603	4·9	4·9	±
Fe 4475	3·5	3·5	±
Fe 4415	2·8	2·8	±
Fe 4359	2·1	2·0	+ 0·1
Fe 4315	1·4	1·3	+ 0·1
Fe 4271	0·7	0·5	+ 0·2
Fe 4240	0	0·1	+ 0·1
Fe 4210	− 0·7	− 0·7	±
Fe 4148	− 2·1	− 2·1	±
Fe 4119	− 2·8	− 2·8	±
Fe 4071	− 4·2	− 4·2	±
Fe 4005	− 6·2	− 6·2	±
Fe 3967	− 7·7	− 7·7	±
Fe 3900	−10·3	−10·4	− 0·1
Fe 3889	−11·1	−10·8	+ 0·3
Fe 3872	−11·5	−11·6	− 0·1
Fe 3868	−11·8	−11·8	±
Fe 3860	−12·2	−12·2	±

A simple mathematical analysis then showed that all the anomalies cited in Chapter VIII could be predicted, and their position in the spectrum calculated, from the constants of the two-term equation which serves to express the rotatory dispersion of the medium. Thus the condition for a REVERSAL OF SIGN is that for some value of λ

$$\frac{k_1}{\lambda^2 - \lambda_1^2} = \frac{k_2}{\lambda^2 - \lambda_2^2} \quad \text{or} \quad \frac{\lambda^2 - \lambda_1^2}{\lambda^2 - \lambda_2^2} = \frac{k_1}{k_2}.$$

The condition for a MAXIMUM, namely, that $d\alpha/d\lambda = 0$, leads to the relation

$$\frac{\lambda^2 - \lambda_1^2}{\lambda^2 - \lambda_2^2} = \sqrt{\frac{k_1}{k_2}}.$$

Finally, the condition for a POINT OF INFLECTION, namely, $d^2\alpha/d\lambda^2 = 0$, leads to the relation

$$\frac{\lambda^2 - \lambda_1^2}{\lambda^2 - \lambda_2^2} = \sqrt[3]{\frac{k_1}{k_2}}.$$

If these equations are applied to the data for the rotatory dispersion of ethyl tartrate which are set out in Table 11, we can at once deduce the exact position of the various anomalies as follows:

Inflection	6735 A.U.
Maximum	5628 A.U.
Reversal of sign . . .	4245 A.U.

Quasi-anomalous Rotatory Dispersion.—The preceding analysis enables us to recognise an additional factor, not mentioned by Arndtsen, which is essential for the development of anomalous rotatory dispersion, namely, that *the partial rotation with the smaller dispersion-constant must have the larger rotation-constant*, and conversely, e.g. if $k_1 > k_2$, $\lambda_1 < \lambda_2$. If this condition is fulfilled, the high-frequency term has a larger numerator than the low-frequency term, and therefore gives rise to a larger partial rotation when the wave-lengths are long; but, as the wave-lengths approach towards the asymptote of the low-frequency term (corresponding perhaps with an absorption-band in the middle of the accessible ultra-violet region) the magnitude of this term grows rapidly and very soon swamps the high-frequency term. The ascending curve of total rotations is therefore dragged down and made to develop successively an inflection, maximum and reversal of sign, with the prospect of ultimate assimilation to an asymptote of opposite sign, if observations could be continued and the equation remained valid up to this point. On the other hand, if the fast-growing low-frequency term has the larger numerator, and is therefore already greater than the high-frequency term when the wave-lengths are long, there will be no possibility of reversing the sign of the rotation, and the equations given in the preceding paragraphs for calculating the positions of the three chief anomalies will all yield imaginary values for λ. The dispersion will therefore be normal, by reason of the relative magnitudes of the two terms. This type of dispersion must be classified as COMPLEX BUT NORMAL. On the other hand, since the relative magnitude of the two partial rotations is often altered by changes of temperature, solvent and concentration, without producing any important simultaneous alteration in the dispersion-constants, it is convenient to apply the term QUASI-ANOMALOUS [1] to those cases in which all the other conditions for anomalous dispersion are fulfilled, i.e. those cases in which two partial rotations of opposite sign and unequal dispersion fail to produce anomalous dispersion only because the relative magnitude of the two rotation-constants of the Drude equation is reversed.

Quartz is perhaps the commonest example of this type of rotatory dispersion, since the two ultra-violet terms are opposite in sign and we can ignore the small constant term in equation (i) (p. 258). Examples are also found amongst those derivatives of tartaric acid which are lævorotatory in the visible spectrum. Some of these would exhibit all the usual anomalies if the rotations could be observed in the infra-red, e.g. *ethyl tartrate in carbon tetrachloride,*[2] for which the numerical data are as follows:

$$[\alpha]_{6708} = +0{\cdot}20^\circ,\quad [\alpha]_{5893} = -1{\cdot}34^\circ,$$
$$[\alpha]_{5461} = -3{\cdot}11^\circ,\quad [\alpha]_{4358} = -17{\cdot}17^\circ,$$

$$\alpha = \frac{21{\cdot}65}{\lambda^2 - 0{\cdot}03} - \frac{20{\cdot}13}{\lambda^2 - 0{\cdot}058}.$$

Reversal of sign	6550 A.U. (observed).
Maximum	9020 A.U. (calculated).
Inflection	10940 A.U. (calculated).

In this case the narrow margin by which $21{\cdot}65 > 20{\cdot}13$ suffices to make the dispersion *anomalous;* but this margin vanishes and gives rise to a *quasi-anomalous* dispersion in solutions in *ethylene dibromide,*[3] where

$$[\alpha]_{6708} = -6{\cdot}66^\circ,\quad [\alpha]_{5893} = -10{\cdot}56^\circ,$$
$$[\alpha]_{5461} = -14{\cdot}11^\circ,\quad [\alpha]_{4358} = -36{\cdot}77^\circ,$$

$$\alpha = \frac{18{\cdot}08}{\lambda^2 - 0{\cdot}03} - \frac{19{\cdot}335}{\lambda^2 - 0{\cdot}061}.$$

In the latter case, the form of the equation indicates that the solution would be lævorotatory at all wave-lengths on the less-refrangible side of the nearer asymptote, and that no reversal of sign would be found even if the observations were extended indefinitely into the infra-red region of the spectrum.

"Complex but Normal" Dispersion.—Another type of dispersion which is "normal," but nevertheless "complex," can be

[1] LOWRY and CUTTER, *J.*, 1925, p. 608, footnote.
[2] LOWRY and DICKSON, *J.*, 1915, **107**, 1173; LOWRY, *J.*, 1929, 2858–2863.
[3] LOWRY and DICKSON, *J.*, 1915, **107**, 1182.

developed by superposing *two partial rotations of similar sign* and unequal dispersion. It was pointed out as long ago as 1913 that

> "The accuracy of the observations would not usually be sufficient to detect the presence of a second term of similar sign unless there were a considerable difference in the magnitude of the two dispersion-constants." [1]

This warning is still valid, since two partial rotations with frequencies in the Schumann region near 1500 A.U. could only produce a simple or pseudo-simple rotatory dispersion in the visible or near ultra-violet region, even if they were of opposite sign; and a much wider separation would be needed to develop an obviously complex dispersion if the signs were the same. Nevertheless, a clear example of this type has been found in *α-bromocamphor*,[2] where the introduction of the halogen gives rise to an additional asymmetric carbon atom, which makes the high-frequency term positive in sign, like the low-frequency term. The specific rotations of this compound can be expressed up to the limits of experimental error by the equation

$$[\alpha] = \frac{13{\cdot}276}{\lambda^2 - 0{\cdot}10855} + \frac{19{\cdot}53}{\lambda^2 - 0{\cdot}0562}.$$

Graphical Representation of Drude's Equation.—When α is plotted against λ^2, Biot's equation is represented by a rectangular hyperbola, with the axes of rotatory power and wave-length as asymptotes. In the same way, each term of Drude's equation is represented by a rectangular hyperbola, but the vertical asymptote, which formerly passed through $\lambda^2 = 0$, must now be drawn through $\lambda^2 = \lambda_n^2$. The general result of adding (or subtracting) *two* terms of Drude's equation is shown in Fig. 44,[3] for the special case in which the two terms are

$$A = \frac{1}{\lambda^2 - 0{\cdot}03} \quad \text{and} \quad B = \pm\frac{1}{\lambda^2 - 0{\cdot}06}.$$

The total rotations of a mixture containing k parts of A and $1 - k$ parts of B are then given by the equation

$$\alpha = \frac{k}{\lambda^2 - 0{\cdot}03} \pm \frac{1 - k}{\lambda^2 - 0{\cdot}06}.$$

(i) When the two terms are of similar sign, the hyperbolas $+A$ and $+B$ lie on the same side of the axis. They approach two different asymptotes, but the curve of total rotation for any given value of k must always lie between the two limiting hyperbolas, $+A$ and $+B$. If we select the case in which $k = 0{\cdot}5$, so that

$$\alpha = \frac{0{\cdot}5}{\lambda^2 - 0{\cdot}03} + \frac{0{\cdot}5}{\lambda^2 - 0{\cdot}06},$$

[1] LOWRY and DICKSON, *J.*, 1913, **103**, 1075.
[2] CUTTER, BURGESS and LOWRY, *J.*, 1925, **127**, 1269.
[3] LOWRY, *J.*, 1915, **107**, 1200.

the resultant curve of total rotations lies midway between the limiting hyperbolas $+A$ and $+B$, and can be plotted by taking the average of the *ordinates* of $+A$ and $+B$. This curve has the general appearance of a rectangular hyperbola, but the asymptote towards which it tends shifts progressively with decreasing wave-length from a mean position between the asymptotes of A and B, towards

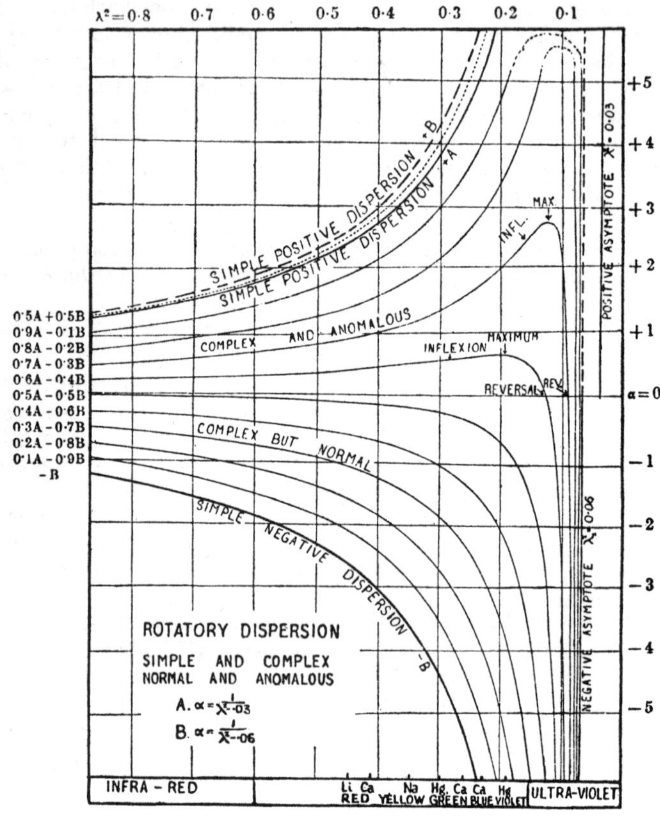

FIG. 44.—SIMPLE, NORMAL, COMPLEX AND ANOMALOUS ROTATORY DISPERSION.

Two of the anomalous curves, 0·9A − 0·1B and 0·8A − 0·2B, rise to maxima beyond the top of the diagram; the fragments of these curves have been joined by broken lines.

the asymptote of B, which finally controls it exclusively. If, on the other hand, we were to take the average of the *abscissæ* of $+A$ and $+B$, we should obtain a true hyperbola represented by the equation

$$\alpha = \frac{1}{\lambda^2 - 0{\cdot}045}.$$

In the former case the dispersion is *complex but normal ;* in the latter case it is *simple ;* but the only difference between the two curves depends on plotting the mean values of a series of vertical instead of horizontal co-ordinates. In order to detect the complexity of the former curve, it is obviously necessary to study the rotations where the hyperbolas A and B are most widely separated, i.e. in the range of wave-lengths contiguous to the nearer asymptote, the approach to which is usually blocked by the growing absorptive power of the medium.

(ii) If the two partial-rotation terms are opposite in sign, e.g. $+A$ and $-B$, the curves of total rotation will form a complete family, covering the whole field between the positive hyperbola $+A$ and the negative hyperbola $-B$. If k is less than 0·5, the curves will remain on the negative side of the axis and will represent the type of *complex but normal* dispersion which has been described above as *quasi-anomalous.* If k is greater than 0·5, the curves will include an *inflection*, *maximum* and *reversal of sign*, and will obviously represent a type of rotatory dispersion which is both *complex* and *anomalous.* The anomalies will at first be in the infra-red, when k is only a little greater than 0·5, and the dispersion would probably be regarded as "normal" until the correct equation had been worked out. As k increases the anomalies move across the visible spectrum, and then disappear into the ultra-violet, where they are finally crowded together in the neighbourhood of the asymptote of B. The curves in which the anomalies have disappeared into the ultra-violet, like those in which the anomalies are all in the infra-red, will be approximately hyperbolic in the visible region of the spectrum; and the dispersion may come very near to being not merely "normal" but "simple," as the weighted mean of the two partial rotations approaches more and more nearly to the positive component $+A$. These "pseudo-simple" curves may often be unmasked, however, by the exceptional smallness of the dispersion, since the asymptote of a true hyperbola is unlikely to retreat beyond $\lambda_0^2 = 0{\cdot}02$. Smaller values of λ_0^2 may then be attributed to the disturbance caused by a term of opposite sign, the real existence of which may be demonstrated by extending the observations further into the ultra-violet.

Rotatory Dispersion in Relation to Absorption: (*a*) Saturated Compounds.—We have seen that the asymptote, towards which the rotations of the secondary alcohols tend, lies in the Schumann region at about 1500 A.U. Until measurements of absorption in this region become available, we have no means of knowing whether the absorption bands which we have postulated actually exist; and this state of ignorance is characteristic of all cases of simple rotatory dispersion in saturated organic compounds.

When we pass on to consider the rotatory dispersion of tartaric acid and its esters we are confronted by a very similar situation. Thus, in the case of ethyl tartrate we have postulated the existence of two characteristic wave-lengths:

$$\lambda_1^2 = 0{\cdot}03 \quad \lambda_1 = 1730 \text{ A.U.}$$
$$\lambda_2^2 = 0{\cdot}056 \quad \lambda_2 = 2370 \text{ A.U.}$$

The first of these wave-lengths is once more in the Schumann region, and its relation to the absorption-spectrum of the ester has not yet been investigated experimentally. The second wave-length also is very near to the limit of observation by the methods generally in use, and no direct observations of the head of the absorption-band are yet available. Measurements of absorption in aqueous solutions of tartaric acid have, however, been made recently by Legris, working under the direction of Bruhat.[1] These measurements, which were extended as far as 2378 A.U., did not reach the maximum, but a graphical method of extrapolation, based upon the Ketteler-Helmholtz formula (p. 395) showed that the head of the band could be located, with an error not exceeding 20 A.U., at

$$\lambda^2_0 = 0{\cdot}0543, \quad \lambda_0 = 2330 \text{ A.U.}$$

This value is sufficiently close to the characteristic wave-length 2370 A.U., deduced from the rotatory dispersion of ethyl tartrate, to justify the conclusion that the dispersion of the ester is controlled by a real absorption band in this region. Bruhat was also able to show that the rotatory dispersion of aqueous solutions of tartaric acid can be expressed more accurately by the formula

$$[\alpha] = \frac{A}{\lambda^2 - 0{\cdot}030584} - \frac{B}{\lambda^2 - 0{\cdot}054289}$$

with characteristic wave-lengths

$$\lambda_1 = 1750 \text{ A.U.} \quad \lambda_2 = 2330 \text{ A.U.}$$

than by the formula

$$[\alpha] = \frac{k_1}{\lambda^2 - 0{\cdot}03} - \frac{k_2}{\lambda^2 - 0{\cdot}074}$$

which was used by Lowry and Austin [2] to express the rotatory dispersion of the aqueous acid for 9 wave-lengths in the visible spectrum. The dispersion-constant of the negative (low-frequency) term, as deduced in 1922, must be modified in any case in view of the fact that Descamps,[3] by working with dilute solutions, was able to extend his observations of optical rotatory power to 2537 A.U., i.e. nearly 200 A.U. beyond the wave-length $\lambda_2 = 2720$, used to calculate the rotations in the visual region. The modified formula proposed by Bruhat has the merit of covering Descamps' observations in the ultra-violet, as well as the visual observations of Lowry and Austin.

[1] BRUHAT, *T.F.S.*, 1930, **26**, 400–411.
[2] LOWRY and AUSTIN, *Phil. Trans.*, 1922, A. **222**, 293.
[3] DESCAMPS, *Thesis*, Brussels, 1928; *C.R.*, 1927, **184**, 1543.

Rotatory Dispersion in Relation to Absorption (*b*) **Unsaturated Compounds.**—In order to be able to make a direct experimental comparison between (i) the wave-lengths of maximum absorption, and (ii) the wave-lengths which appear in the rotation-constants of equations of rotatory dispersion, it is necessary to make use of unsaturated compounds, such as the ketones, which have bands in a readily-accessible region of the visible or ultra-violet spectrum, instead of using compounds like the alcohols and carboxylic acids, which have absorption bands in the Schumann region or near the extreme limit of easy observation in the ultra-violet. Thus Pickard and Hunter in 1923 [1] found that the rotatory dispersion of *d-γ-nonyl nitrite*, $C_6H_{13}\cdot CH(C_2H_5)\cdot O\cdot N{:}O$, for a column of 100 mm. could be expressed by the equation

$$\alpha = \frac{0\cdot 76}{\lambda^2 - 0\cdot 135} + \frac{0\cdot 43}{\lambda^2}.$$

The dispersion-constant $\lambda_0^2 = 0\cdot 135$ corresponds with a characteristic wave-length $\lambda_0 = 3680$ A.U. Measurements of refraction gave rise to a dispersion-constant with characteristic wave-length $\lambda_0 = 3730$ A.U. Direct measurements of absorption revealed a maximum at 3670 to 3720 A.U., in good agreement with the wave-lengths deduced by extrapolation from measurements of refractive and rotatory dispersion.

An even simpler case on which to test the relation between rotatory dispersion and absorption is provided by the pseudo-simple rotatory dispersion of *camphorquinone*. In the range of wave-lengths to which the quinone is transparent, the rotatory dispersion tends towards an asymptote at 4730 A.U.,[2] although on entering the region of absorption it actually passes through maxima of opposite sign at 4945 and 4445 A.U.[3] The wave-length of maximum absorption at 4675 A.U.[4] is near enough to that deduced by extrapolation from the rotatory dispersion in the region of transparency (viz. 4730 A.U.) to prove the correlation of the two phenomena; but it shows a discrepancy of about 55 A.U.,* in the same direction as the difference of 40 A.U. between Bruhat's estimate of the position of the first absorption band in tartaric acid and the wave-length of the first dispersion-constant of ethyl tartrate (p. 144). This discrepancy can be accounted for if we remember that the *selective* absorption is superposed on a curve of *general* absorption tending towards a very high maximum in the inaccessible region of shorter wave-lengths. If these two absorptions are added together the maximum of *total* absorption, which is recorded as the "head" of the absorption band, will necessarily occur at a shorter wave-length than the maximum

1 Pickard and Hunter, *J.*, 1923, **123**, 434–444.
2 Lowry and Cutter, *J.*, 1925, **127**, 604–615.
3 Lowry and Gore, *P.R.S.*, 1932, A. **135**, 13.
4 Lowry and French, *J.*, 1924, **125**, 1921.
* The average wave-length of the two rotation-maxima, viz. 4695 A.U., is 20 A.U. greater than that of the head of the band at 4675 A.U.

of selective absorption, which determines the dispersion-constant of the Drude equation, so that some discrepancy in this direction is to be expected. An alternative and more general interpretation has, however, been suggested by Kuhn.[1]

A fortunate accident made it possible to detect the existence of a similar discrepancy in a series of camphor derivatives. The rotatory dispersion of a solution of camphor in benzene can be expressed by the equation [2]

$$[\alpha] = \frac{29{\cdot}384}{\lambda^2 - 0{\cdot}0872} - \frac{20{\cdot}138}{\lambda^2 - 0{\cdot}05428}.$$

The low-frequency dispersion-constant, $\lambda_1^2 = 0{\cdot}0872$, corresponds to a wave-length $\lambda_1 = 2950$ A.U., which is greater by 70 A.U. than the observed "head of the band" at 2880 A.U. of a film of camphor in benzene. In the series of halogen derivatives of camphor [3] the following values were recorded :

TABLE 12.—ROTATORY DISPERSION OF HALOGEN DERIVATIVES OF CAMPHOR.

	k_1.	k_2.	λ_1^2.	λ_2^2.	λ_1 (A.U.).	λ_α * (A.U.).
α-Chlorocamphor .	+12·938	+ 7·5189	0·09882	0	3140	3050
α′- ,, . .	+13·733	− 9·401	0·10881	0	3300	3100
α-Bromocamphor .	+13·276	+19·530	0·10855	0·0562	3290	3120
α′- ,, .	+10·232	−26·393	0·10630	0·03525	3260	3140
β- ,, .	+ 9·7419	− 8·1869	0·09915	0	3150	2930
αα′-Dibromocamphor	+14·666	− 8·1713	0·11296	0	3770	3230
αβ- ,, ,,	+16·134	+ 5·8907	0·09863	0	3140	3080
α′β- ,, ,,	Not	expressed	by a two-	term equ	ation.	
απ- ,, ,,	+25·163	− 5·741	0·10476	0	3240	3130
α′π- ,, ,,	+34·648	−15·659	0·10470	0	3235	3090

* This column shows the wave-length of maximum absorption of light.

In Table 12 the average value of the characteristic wave-length λ_1 of the low-frequency partial rotation is 3220 A.U., whilst the average wave-length λ_α of the head of the ketonic band is 3080 A.U. There is, therefore, a difference of about 140 A.U. between the wave-lengths deduced by the two methods.

Induced Dissymmetry as a Cause of Anomalous Rotatory Dispersion.—In a note on "Induced Asymmetry of Unsaturated Radicals in Optically-active Compounds" Lowry and Walker in 1924 [4] suggested that the powerful influence of unsaturated and chromophoric groups on optical rotatory power might be due to an INDUCED ASYMMETRY of the chromophoric group, whereby, for instance,

[1] KUHN and BRAUN, *Z. ph. C.*, 1930, B. **8**, 293.
[2] LOWRY and CUTTER, *J.*, 1924, **125**, 612.
[3] CUTTER, BURGESS and LOWRY, *J.*, 1925, **127**, 1260–1274.
[4] LOWRY and WALKER, *Nature*, April 19, 1924, **113**, 565–566.

the carbon atom of the >C=O group of camphor was rendered asymmetric in the unsymmetrical environment of the optically-active molecule. In a subsequent note,[1] following Pasteur's nomenclature, this phenomenon was described more correctly as INDUCED DISSYMMETRY, since the molecules in which it occurs may possess a high degree of axial symmetry.

This theory is discussed more fully in Chapter XXXI, but it may be referred to here as affording an alternative interpretation of the origin of anomalous rotatory dispersion in unsaturated molecules. The essential condition for this anomaly (p. 109) is the superposition of two partial rotations of opposite sign and unequal dispersion. Since a chromophoric group has necessarily a lower characteristic frequency than the saturated radicals of an asymmetric carbon atom, this anomalous rotatory dispersion can readily be produced if the "induced dissymmetry" of the chromophoric group is opposite in sign to that of the "fixed asymmetry" of the asymmetric carbon atoms. This possibility was therefore indicated in the following terms:

> "Induced asymmetry need not lead to an optical rotation of the same sign as that of the fixed asymmetric groups by which it is controlled. Since these will usually have a lower dispersion than the chromophoric groups, anomalous rotatory dispersion may be developed, if the rotatory power of the fixed groups is sufficiently great. In that case, the anomalous dispersion of the bornyl dixanthide examined by Tschugaeff may be due to the same cause as in the case of his menthyl camphorsulphonate, the principal difference being that the second radical has only an induced asymmetry in the former case, as compared with a fixed asymmetry in the latter case."[2]

Subsequent experience has justified the formulation of a general rule according to which the FIXED ASYMMETRY of the saturated groups and the INDUCED DISSYMMETRY of the unsaturated groups usually give rise to partial rotations of opposite sign. To this general rule there are many apparent exceptions, e.g. α-chloro-, α-bromo-, and αβ-dibromo-camphor,[3] potassium α-chloro- and α-bromo-camphor-sulphonates,[4] and nicotine;[5] but in the camphor series at any rate these are all associated with the development of an additional asymmetric carbon atom, e.g.

```
              ,H
          /C<
         /  | \Br
$C_8H_{14}$<    |
         \  |
          \C=O
```

[1] LOWRY, *Nature*, 1933, **131**, 566. [2] *Ibid.*, 1924, **113**, 566.
[3] CUTTER, BURGESS and LOWRY, *J.*, 1925, **127**, 1260.
[4] RICHARDS and LOWRY, *J.*, 1925, **127**, 1503.
[5] LOWRY and LLOYD, *J.*, 1929, 1771.

in α-bromocamphor. It therefore seems likely that this additional asymmetric carbon atom contributes a large additional high-frequency partial rotation, of opposite sign to that of the two asymmetric carbon atoms of the parent compound, and so reverses the sign of the high-frequency term, but without destroying the opposition in sign between the unsaturated carbonyl-radical and the camphor-nucleus which continues to dominate it.

CHAPTER XI.

CIRCULAR DICHROISM.

Dichroism and Circular Dichroism.—The phenomenon of DICHROISM, i.e. the unequal absorption of the ordinary and extraordinary rays by doubly refracting crystals, was observed by Biot [1] in 1815 in crystals of tourmaline. Crystals of appropriate colour absorb the ordinary ray throughout the spectrum, whilst the extraordinary ray is transmitted as a plane-polarised beam over a limited range of wave-lengths. Billet [2] in 1859 suggested that it should be possible to discover media, analogous to tourmaline, which would transmit only one type of *circular* vibration. The unequal absorption of oppositely polarised circular vibrations was first observed by Haidinger [3] and subsequently by Dove [4] in amethyst quartz, and is now known as CIRCULAR DICHROISM.

Circular dichroism in optically-active absorbing solutions was described by Cotton in 1895 in a paper [5] entitled *Recherches sur l'Absorption et la Dispersion de la Lumière par les Milieux Doués du Pouvoir Rotatoire.* After referring to Biot's discovery of dichroism, Cotton continues as follows :

> " In an optically-active medium, there are also two kinds of rays, a right circularly polarised ray and a left circularly polarised ray, which are propagated with different velocities, and it is seen, in some cases, that a ray is sharply separated, in traversing the medium, into two circularly polarised rays of opposite sign."
>
> " If this circular double refraction is compared with crystalline double refraction, a very close analogy is noticed between the two types of phenomenon. It is therefore natural to ask if there are not active substances *which absorb unequally a left and a right ray.*"

Cotton's Demonstration of Circular Dichroism in Solutions of Potassium Chromium Tartrate.—Cotton's observations were made with a double circular polariser, consisting of a Nicol prism and two quarter-wave plates placed side by side so that their principal sections were inclined at 90° to one another as in Fig. 45.

[1] BIOT, *Bull. Soc. Philom.*, 1815, 26–27; BREWSTER, *Phil. Trans.*, 1819, 11–28.
[2] BILLET, *Traité d'Optique Physique* (Paris, 1858/9), **2**, 7.
[3] HAIDINGER, *Pogg. Ann.*, 1847, **70**, 531–544.
[4] DOVE, *ibid.*, 1860, **110**, 279.
[5] COTTON, *A.C.P.*, 1896, [vii], **8**, 347; compare *C.R.*, 1895, **120**, 989 and 1044.

In this figure, OP represents the principal section of the Nicol polariser placed behind the double quarter-wave plate and OS and OS′ the principal sections of the two halves of the latter. When OP makes an angle of 45° with OS the two halves of the plate transmit rays which are of equal intensity but are circularly polarised in opposite directions. When light from a sodium flame was used, and the double circular polariser was viewed through a cell containing a solution of *potassium chromium tartrate*, the two halves of the plate appeared unequally illuminated. One of the circularly polarised rays was thus absorbed more strongly than the other. When a 10 per cent. solution was used in a 1 cm. cell, the difference between the intensities was of the order of one-quarter of their mean value.

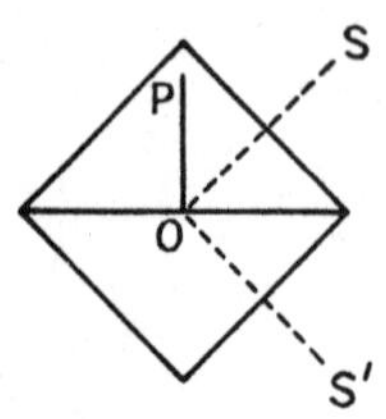

FIG. 45.—DOUBLE CIRCULAR POLARISER (Cotton).

Cotton showed, by a variety of methods, that this difference was due entirely to the difference in sign of the circular vibrations. Thus :

(*a*) When the double circular polariser was turned through an angle of 45° so that both halves transmitted plane polarised light, the two halves appeared equally illuminated.

(*b*) With the double circular polariser in its former position, the directions of the circular vibrations were reversed by interposing a half-wave plate. This reversed the appearance of the two halves of the quarter-wave plate, that half which was originally brighter being now darker than the other.

(*c*) An identical result was obtained by reversing the sign of the circular vibrations by reflection.

Similar results were obtained with solutions of *potassium copper* d-*tartrate*, but with the *l*-tartrate the effects were reversed ; and no trace of unequal absorption could be detected with the inactive racemate. When white light was used in place of monochromatic radiation the two halves of the double circular polariser appeared differently coloured.

By analogy, Cotton proposed the term CIRCULAR DICHROISM to describe the unequal absorption of left and right circularly polarised light. This phenomenon, and the anomalous rotatory dispersion in the region of absorption which accompanies it, are now known as the COTTON EFFECT.

Unpolarised Light is, in part, Circularly Polarised by an Absorbing Optically-active Medium.—Ordinary light can be resolved into (i) two *plane polarised* rays of equal intensity and polarised in planes at right angles to one another, or (ii) two *circularly polarised* rays of equal intensity and opposite sign. In passing through a circularly dichroic medium one of the circular components will be selectively absorbed, so that the other will predominate in the transmitted radiation. Cotton verified this by reversing the apparatus just described. Light was thus made to pass successively through the solution, the "split" quarter-wave plate and the Nicol prism. The double circular polariser now became a double circular analyser, of which one-half transmitted a left but not a right circular vibration and the other half a right but not a left circular vibration. The two halves of the quarter-wave plate again appeared unequally bright. The natural light falling on the active solution had therefore become to some extent circularly polarised, by the unequal absorption of its two circularly polarised components.

This result led Cotton to predict that almost pure circularly polarised light might be produced by selective absorption of ordinary light in a medium exhibiting circular dichroism in a very high degree, e.g. amethyst quartz, or cinnabar, the rotatory power of which is very much greater than that of quartz. This conversion of ordinary light into circularly polarised light was realised by Giesel [1] and by Stumpf [2] in certain "liquid crystals" (p. 346), but the non-transmission of one of the circular components was due to selective *reflection* and not to selective *absorption* (p. 347).

Direct Comparison of the Absorption of Left and Right Circularly Polarised Light.—Cotton's qualitative study of circular dichroism was followed by a quantitative investigation. The intensity of a beam of monochromatic light, which had been partially absorbed as a left or right circular vibration by the active solution, was compared with that of a beam from the same source which had not traversed the solution. The sign of the circular vibration traversing the solution could be reversed at will merely by rotating the polarising Nicol through 90°. By methods which are described in Part II (p. 240), Cotton showed that a 1 cm. column of a 10 per cent. solution of potassium chromium tartrate reduced the amplitude of a left circularly polarised ray to the extent of 91·2 per cent., and of a right circularly polarised ray by 92·3 per cent.

Production of Elliptical Polarisation by the Unequal Absorption of Circularly Polarised Rays.—When a plane polarised ray passes through an optically-active medium, which absorbs the two circular components to unequal extents, not only is the plane of polarisation rotated, but the ray becomes elliptically polarised as shown in Fig. 46.

A simple geometrical construction shows that the major and minor axes of the resultant elliptical vibration are equal to the sum and difference respectively of the amplitudes of the two circular components. The angle through which the major axis has been rotated is equal to half the difference of phase between the circular components; and the sign of the ellipse is the same as that of the circular component with the greater amplitude. A complete polarimetric study of circularly dichroic media includes the measurement of (*a*) the rotation (the angle between the incident vibration and the major axis of the ellipse) and (*b*) the circular dichroism, which can

[1] GIESEL, *Phys. Zeit.*, 1910, **11**, 192.
[2] STUMPF, *ibid.*, 780; *Ann. der Physik*, 1912, **37**, 351.

be expressed either as the difference between the absorption coefficients for the two circular components or as the ellipticity of the emergent vibration.

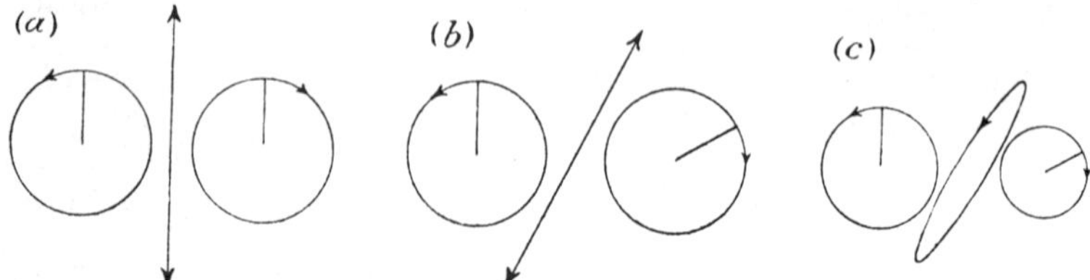

FIG. 46.—ELLIPTICAL POLARISATION.

(*a*) Equal velocities of transmission give *no rotation*. (*b*) Unequal velocities of transmission give *rotation*. (*c*) Unequal velocities and unequal absorptions give *rotation* and *elliptical polarisation*.

Experimental Demonstration of the Ellipticity of the Emergent Light.—The conversion of a plane polarised ray into an elliptically polarised ray by its passage through a circularly dichroic medium was demonstrated by Cotton with the help of a BRAVAIS DOUBLE PLATE. This consists of a thin plate of quartz, cut parallel to the optic axis, of such a thickness as to produce a phase-retardation of 2π for light of a particular wave-length in the yellow-green part of the spectrum. This plate is cut in two along a direction inclined at 45° to the axis of the crystal and cemented together again after turning one half completely over.[1]

(*a*) The Bravais double plate was placed, with its dividing line horizontal, between a polariser and an analyser set mutually at extinction and so that the principal section of the former made an angle of 45° with the axes of the Bravais double plate. The polariser was illuminated with sunlight reflected from a heliostat. An image of the Bravais double plate was focussed by an achromatic lens on the slit of a spectroscope. Two spectra, corresponding to the two halves of the Bravais double plate, can then be seen, one above the other, in the eyepiece of the spectroscope. These spectra are crossed by vertical dark bands, each corresponding to a wave-length for which the double plate produces a phase retardation of a multiple of 2π. Light of this particular wave-length is unaltered in traversing the Bravais double plate and is therefore cut off by the analyser, thus producing the dark bands in the spectrum. On either side of a dark band the Bravais plate polarises the incident light elliptically, the ellipticity being greater the further the wave-length is removed from that corresponding to the dark band. On one side of a dark band the vibrations are right elliptical and on the other side left elliptical. Both halves of the Bravais double plate give elliptical vibrations of the same amplitude but opposite in sign. This is represented in Fig. 47 (*a*).

[1] Fig. 45, p. 150, may be used to illustrate Bravais' "double plate" as well as Cotton's "Double circular polariser."

(*b*) A circularly dichroic solution is now placed between the polariser and the Bravais double plate. If the solution polarises the light elliptically the dark bands in each field will move to the particular wave-length for which the ellipticity produced by the solution exactly counterbalances that due to the Bravais plate. For example, if the solution produces an elliptical vibration identical with the ellipses marked X in Fig. 47 (*a*), but left instead of right, the dark bands will move to the positions formerly occupied by these ellipses as shown in Fig. 47 (*b*). If the solution produces an

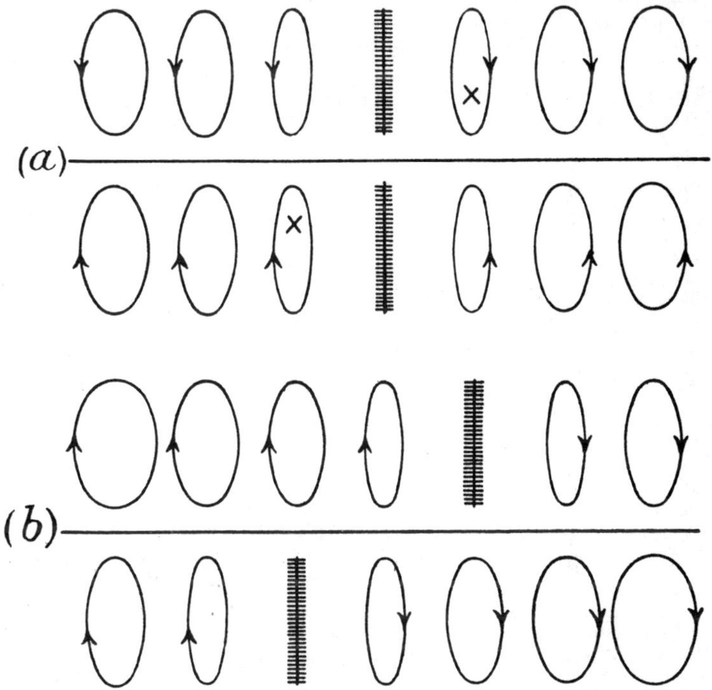

FIG. 47.—DETECTION OF ELLIPTICAL POLARISATION BY MEANS OF BRAVAIS' DOUBLE PLATE.

elliptical vibration of opposite sign, the dark bands will move in the opposite direction. Since the solution also causes the major axis of the ellipse to rotate, the centres of the dark bands will no longer be completely black unless the analyser is rotated through a corresponding angle.

Cotton observed this displacement of the dark bands, showing that a plane polarised ray is converted into an elliptically polarised ray in traversing a medium which exhibits circular dichroism. It was, in fact, by this method that Cotton first observed the phenomenon of circular dichroism when examining Fehling's solution.

The number of dark bands can be increased by using a thicker Bravais double plate. By this means the elliptical character of the incident vibrations can be studied over the whole of the visible spectrum without changing the Bravais plate. A modification of this experimental method was used by Cotton to measure quantitatively the rotation and ellipticity produced by circularly dichroic media. An account of the method is given in Chapter XIX.

Cotton's Experimental Results.—By the methods described above, Cotton investigated the circular dichroism and rotatory dispersion of a number of coloured solutions in the region of an absorption band. In every case in which the two circular components of plane polarised light were absorbed to unequal extents the curve of rotatory dispersion was also found to follow an anomalous course. Thus Fig. 48 shows the behaviour of a solution of *potassium chromium tartrate*, which exhibits the phenomena of circular dichroism and anomalous rotatory dispersion in a very marked degree.

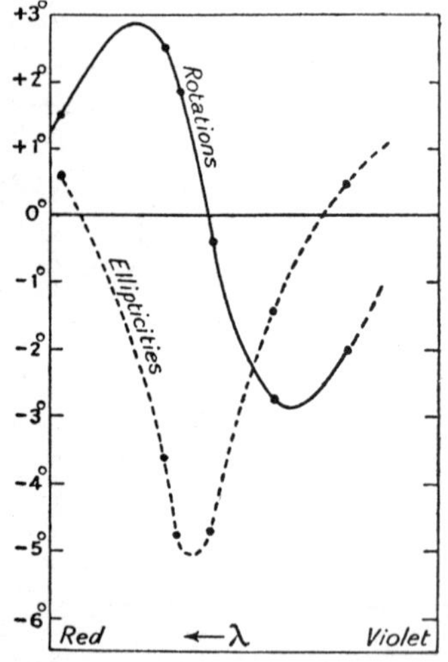

FIG. 48.—ELLIPTICAL POLARISATION AND ANOMALOUS ROTATORY DISPERSION IN POTASSIUM CHROMIUM TARTRATE (Cotton).

These and other similar observations led Cotton to the following conclusions:—

(*a*) The curve of ellipticities shows a maximum in a region of the spectrum where the absorption is most intense.

(*b*) The rotatory power changes sign in the region of absorption and the curve of rotatory dispersion is very markedly anomalous.

(*c*) The maximum of circular dichroism corresponds to a point of inflection in the curve of rotatory dispersion.

(*d*) The relationship between the rotatory power and the ellipticity (i.e. between the difference in velocity of propagation and the difference in absorption of the two circular components) does not appear to be a simple one.

Babinet[1] had suggested that in dichroic crystals the ray which is transmitted with the greater velocity is less absorbed, but numerous exceptions to this rule were discovered. Cotton's observations showed that it is equally invalid when applied to *circularly* dichroic media. Valid rules connecting the sign and magnitude of the rotatory power

[1] BABINET, *C.R.*, 1838, **7**, 832–833.

with the circular dichroism were, however, formulated subsequently by Natanson [1] and by Bruhat [2] (p. 427).

Chemical Combination is Necessary for the Production of Circular Dichroism.—Cotton detected circular dichroism and anomalous rotatory dispersion in coloured solutions of *potassium, sodium* and *ammonium copper d-* and *l-tartrates*, of *sodium, potassium* and *ammonium chromium tartrates*, of *calcium copper saccharate* and of normal and acid *copper malates*. The circular dichroism of the three last compounds was too small to be established with certainty, but the curves of rotatory dispersion were distinctly anomalous.

In all of these solutions the optical activity and the absorption were produced by the same molecules. Cotton showed that circular dichroism did not occur in mixtures in which the activity and absorption were due to different molecules. For example, no change of rotatory power was observed when a solution of sugar was coloured with magenta, nor were the transmitted vibrations rendered elliptical. A similar negative result was observed with a solution of sodium bitartrate containing aniline-violet, although, when this solution was evaporated, coloured crystals were obtained, which showed traces of crystalline dichroism.

Circular Dichroism and Anomalous Rotatory Dispersion in Coloured Compounds.—(*a*) *Coloured Tartrates, Lactates and Malates.*—Cotton's measurements of the circular dichroism of *copper* and *chromium tartrates* were repeated by a different experimental method in 1905 by McDowell,[3] who also investigated *cobalt* and *nickel tartrates* and *copper malate*. Measurements of the rotatory dispersion of a number of coloured tartrates and of solutions containing *mannitol* and chromium salts, were made in 1910 by Grossman and Loeb; [4] similar measurements of the rotatory dispersion of coloured tartrates were made in 1912 by Olmstead [5] who also added *manganese tartrate, chromium lactate* and *chromium malate* to the list. Three years later Bruhat [6] observed the circular dichroism of a number of these coloured tartrates, together with that of *uranyl tartrate*. More recently, measurements of the circular dichroism of alkaline solutions of *copper tartrate* have been made by de Mallemann and Gabiano.[7]

Experiments on the conditions under which the "Cotton effect" is developed in solutions of the coloured tartrates were made by Mathieu [8] in 1931. Freshly precipitated chromium hydroxide, dissolved in tartaric acid at 0° C., showed no trace of circular dichroism, and measurements of conductivity proved that the metal was ionised;

[1] NATANSON, *Journ. de Phys.*, 1909, [iv], **8**, 321.
[2] BRUHAT, *Ann. de Phys.*, 1915, [ix], **3**, 232 and 417.
[3] McDOWELL, *Phys. Rev.*, 1905, **20**, 163.
[4] GROSSMANN and LOEB, *Z. ph. C.*, 1910, **72**, 93.
[5] OLMSTEAD, *Phys. Rev.*, 1912, **35**, 31.
[6] BRUHAT, *Ann. de Phys.*, 1915, [ix], **3**, 232 and 417.
[7] DE MALLEMANN and GABIANO, *C.R.*, 1927, **185**, 350.
[8] MATHIEU, *C.R.*, 1931, **193**, 1079; 1932, **194**, 1727.

the subsequent formation of a complex anion was accompanied, however, by a change in the colour of the solution and by the development of circular dichroism. *Sodium cobaltous tartrate*, which exists as the double salt $Na_2Co(C_4H_4O_6)_2$, nH_2O, and as the complex salt $Na_2[CoC_4H_4O_6]$, nH_2O, gave similar absorption spectra in both forms; but the rotatory dispersion of the double salt was only very slightly anomalous, whereas that of the complex salt was markedly anomalous with strong circular dichroism. Mathieu therefore concluded that a chromophoric group gives rise to circular dichroism only when attached to a dissymmetric radical by a homopolar linkage, and not when joined to it through an electrovalency or heteropolar linkage.

(*b*) *Coloured Derivatives of Camphor.*—The effect of the types of valency bond on the production of circular dichroism and optical activity had already been studied by Lifschitz.[1] For this purpose he measured the rotatory dispersion of a number of coloured metallic derivatives of camphor, including the *ferric*, *acid copper*, *normal copper*, *chromic*, *cobaltic*, *cobaltous*, *nickel*, *uranyl* and *aluminium* derivatives of *hydroxymethylenecamphor*, and the *ferric*, *chromic* and *cobaltous* derivatives of *nitrocamphor* in various organic solvents. All but one of these compounds are brightly coloured on account of selective absorption in the visible spectrum. Solutions in benzene and in methyl alcohol of the chromic derivative of hydroxymethylenecamphor showed strong circular dichroism in the visible spectrum, accompanied by a maximum and minimum in the curve of rotatory dispersion, but in other cases the rotatory dispersion was normal. Lifschitz concluded that circular dichroism and anomalous rotatory dispersion were only produced when the metallic atom was united to the dissymmetric organic radical by primary, and not by secondary valencies.

Lifschitz [1] also measured the rotatory dispersion of *camphorquinone* and its *phenylhydrazone* in the neighbourhood of their absorption bands, and established the existence of anomalous rotatory dispersion in the region of absorption. The rotatory dispersion of camphorquinone, over a narrow range of complete transparency, can be represented approximately by a single term of Drude's equation,[2] but in the region of absorption a complete range of anomalies is developed, including positive and negative maxima on either side of a reversal of sign, as may be seen in Fig. 133 (p. 309).[3]

The circular dichroism of camphorquinone was detected by Wedeneewa [4] in 1923. Her observations (Fig. 168, p. 405) disclosed a marked discrepancy between the position of the maximum of circular dichroism (4800 A.U.) and the maximum of absorption of unpolarised light (4700 A.U.). In all of the coloured compounds investigated by Cotton, McDowell, Olmstead and Bruhat the two

[1] LIFSCHITZ, *Z. ph. C.*, 1923, **105**, 27; 1925, **114**, 485.
[2] LOWRY and CUTTER, *J.*, 1925, **127**, 614.
[3] LOWRY and GORE, *P.R.S.*, 1932, A. **135**, 22.
[4] WEDENEEWA, *Ann. der Physik*, 1923, **72**, 122.

maxima had occurred at approximately the same wave-length, but similar divergencies have recently been observed in the ultra-violet absorption bands of *camphor* (Fig. 169, p. 406) and of *camphor-β-sulphonic acid* (Fig. 170, p. 407). Their significance is discussed in Chapter XXXI.

Pfleiderer[1] in 1926 repeated the measurements of Lifschitz and of Wedeneewa on the rotatory dispersion of camphorquinone, and showed that the rotatory dispersion of the phenyl derivatives of methylene camphor is also anomalous.

(*c*) *Caryophyllene Nitrosite.*—Mitchell[2] measured the circular dichroism of *caryophyllene nitrosite*, $C_{15}H_{24}N_2O_3$, a blue compound with a narrow absorption band in the red region of the spectrum. The method used was a modification of Cotton's original method, since direct measurements were made of the absorption of left and right circularly polarised light. In studying *bornylene nitrosite*, $C_{10}H_{16}N_2O_3$, however, Mitchell and Cormack[3] used Bruhat's method of measuring ellipticities. This compound is of interest since the circular dichroism, and the associated partial rotation in the visible spectrum, are governed by changes of temperature. Thus the dimeric form of the compound, which is stable at low temperatures, is colourless and shows a normal rotatory dispersion; but the monomeric form, which is produced above 63°, is blue, with an optically-active absorption band in the red, which gives rise to circular dichroism and to anomalous rotatory dispersion in the region of absorption. The data of Mitchell and Cormack for this compound are set out in Fig. 49.

(*d*) *Xanthates and Thiourethanes.*—Between the years 1903 and 1914 Tschugaeff[4] and his pupils prepared a number of optically-active xanthates and thiourethanes and studied their absorption spectra and rotatory dispersion in the visible spectrum. The compounds thus investigated included the methyl, ethyl, propyl, benzyl, di- and tri-phenyl methyl, methylene, ethylene and tri-methylene xanthates of *d*- and *l*-borneol, *l*-menthol, fenchol, thujyl alcohol and cholesterol, and the dixanthides, xanthamides, xanthic thioanhydrides, mono- and di-thiourethanes of these secondary alcohols. (i) The mono- and di-thiourethanes are bright red in colour by reason of a weak absorption band in the green-blue region of the spectrum. Tschugaeff showed that the rotatory dispersion was markedly anomalous in the region of absorption, and Bruhat[5] in 1923 observed the circular dichroism of the red diphenyl *d*- and *l*-bornyl dithiourethanes, both in solution and in the fused state.

[1] PFLEIDERER, *Z. Physik.*, 1926, **39**, 663.
[2] MITCHELL, *J.*, 1928, 3258.
[3] MITCHELL and CORMACK, *J.*, 1932, 415.
[4] TSCHUGAEFF, *J. Russ. Phys. Chem. Soc.*, 1903, **35**, 1116; 1904, **36**, 988 and 1253; *Ber.*, 1909, **42**, 2244; TSCHUGAEFF and GASTEFF, *ibid.*, 4631; TSCHUGAEFF and OGORODNIKOFF, *Z. ph. C.*, 1910, **74**, 503; 1912, **79**, 471; 1913, **85**, 481; TSCHUGAEFF and PASTANOGOFF, *Z. ph. C.*, 1913, **85**, 553.
[5] BRUHAT, *Ann. de Phys.*, 1915, **3**, 232 and 417.

(ii) The other compounds, which have an absorption band in the near ultra-violet, are either colourless, or are pale yellow or orange when the foot of the band extends into the visible spectrum. Since Tschugaeff's measurements of rotatory dispersion were not extended into the ultra-violet, the Cotton effect in the region of absorption was not observed; but the rotatory dispersion in the visible spectrum was anomalous (like that of the colourless tartrates), owing to the superposition of partial rotations of opposite sign, with characteristic

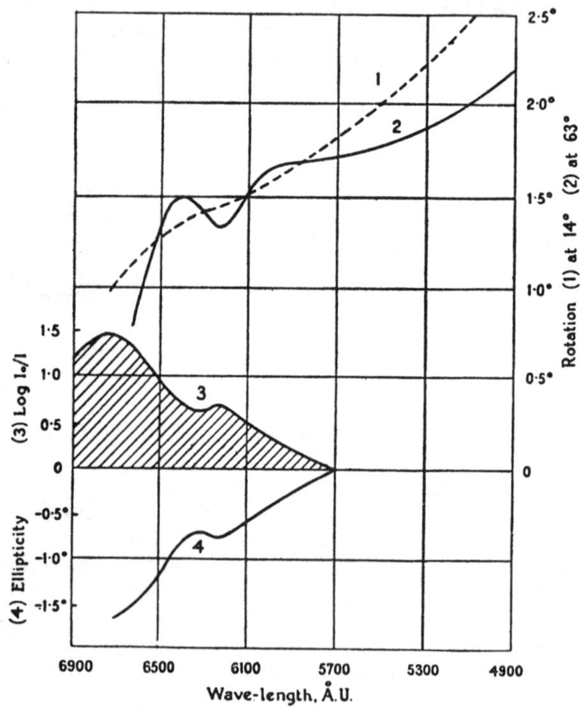

FIG. 49.—ROTATORY DISPERSION, ABSORPTION AND CIRCULAR DICHROISM OF BORNYLENE NITROSITE (Mitchell and Cormack).
(*From Mitchell's "Cotton Effect," by courtesy of Messrs. G. Bell & Sons.*)

frequencies in the ultra-violet. More recent measurements in the ultra-violet have, however, disclosed the Cotton effect in many of these compounds.[1]

(*e*) *Co-ordination Compounds.*—A fresh impetus to the study of the Cotton effect was given by Werner's [2] discovery in 1911, of optical activity in coloured co-ordination compounds containing

[1] LOWRY and HUDSON, *Phil. Trans.*, 1933, A. **232,** 117.

[2] WERNER, *Ber.*, 1911, **44,** 1887, 2445, 3132, 3272, 3279; 1912, **45,** 121, 433, 865, 1228, 3061, 3281, 3287, 3294; 1913, **46,** 3674; 1914, **47,** 1954, 1961, 2171, 3087, etc.

no asymmetric carbon atom. In recent years a very large number of these compounds have been prepared in optically-active forms.[1] They are very suitable for measurements of circular dichroism and of rotatory dispersion in the region of absorption, since they possess weak absorption bands in the visible spectrum, which are often associated with very great rotatory power, but the mathematical analysis of the experimental results is made very difficult by the complexity of the absorption and rotation curves. From amongst a large number of measurements of rotatory dispersion in the region of absorption,[2] the following may be selected by reason of the fact that the circular dichroism has also been measured :

Potassium cobaltic oxalate.[3]
,, iridium oxalate [4]
cyclo-Pentylenediaminecobaltic chloride.[4]

(*f*) Experiments on circular dichroism and anomalous rotatory dispersion in *ultra-violet absorption bands* are described in Part IV, pp. 405, 433, 439, 445, etc.

[1] JAEGER, *Spacial Arrangements of Atomic Systems and Optical Activity*, Cornell University Press, 1930.

[2] See KUHN and BEIN, *Zeitschr. anorg. Chem.*, 1934, **216**, 331.

[3] KUHN and SZABO, *Z. ph. C.*, 1931, B. **15**, 59.

[4] BRUHAT, *Bull. Soc. Chim.*, 1915, [iv], **17**, 223.

CHAPTER XII.

MAGNETIC ROTATORY POWER.

Faraday's Experiments.—In 1846 FARADAY wrote, "I have long held an opinion, almost amounting to conviction . . . that the various forms under which the forces of matter are made manifest have one common origin. . . ."[1] One of the direct consequences of this belief was his discovery, in 1845, of the phenomenon of the rotation of the plane of polarisation of a beam of light traversing an isotropic medium along the lines of force of a magnetic field, i.e. of MAGNETIC ROTATORY POWER.

As early as 1822 Faraday's *Diary* shows that he was engaged on experiments to discover an effect on a beam of polarised light when it was passed through water under electrolysis. Negative results led him in 1833 to abandon these attempts in favour of experiments with dielectrics. Experiments on these lines, made with his own heavy lead borate glass, again yielded no results, but time has shown how near he then was to the discovery of ELECTRICAL DOUBLE REFRACTION.

On September 13, 1845, Faraday set out to discover whether any effect on a polarised ray was produced by magnetic, rather than electric fields. The ray was passed across the pole of his electromagnet, or reflected from its polished surface, under all the various conditions which Faraday's genius could contrive, but still he had to record "no effect." At last he tried an arrangement in which a ray of light issuing from an Argand lamp, and polarised in a horizontal plane by reflection from a surface of glass, was passed through a Nicol's eyepiece revolving on a horizontal axis. Between the polarising mirror and the eyepiece were two powerful electromagnetic poles, separated from each other by about 2 inches in the direction of the line of the ray. The ray passed either by the side of them, or between them and the substance to be examined was laid across the poles of the magnet. Using a plate of lead borate glass 2 inches in thickness, he found that:

> "It gave no effects when the *same magnetic poles* or the *contrary* poles were on opposite sides (as respects the course of the polarised ray)—nor when the same poles were on the same

[1] FARADAY, *Phil. Mag.*, 1846, [iii], **28**, 294.

side either with the constant or intermitting current—BUT when contrary magnetic poles were on the same side there *was an effect produced on the polarised ray*, and thus magnetic force and light were proved to have relation to each other."[1]

Under these conditions, when the analyser was set to extinction, the polarised ray was transmitted again on passing the current through the electromagnet; but the image of the flame could be extinguished by rotating the eyepiece to the right or to the left according to the direction of the current in the electromagnet. It was therefore clear that, under the influence of the magnetic field, the plane of polarisation of the light had been rotated about its direction of propagation. Similar effects were observed in water and turpentine, and in a large number of aqueous and alcoholic solutions, oils, resins, and fused solids, but not in gases.

Faraday also examined substances exhibiting natural rotatory power, and records that "bodies which have rotatory power *per se* do not seem by that to be capable of assuming more freely the magneto power."[2]

The following laws governing magnetic rotations were established in a qualitative way by Faraday:

(i) the sense of the rotation is determined by the direction of the magnetic field, and not by the direction in which the light passes through it;

(ii) the rotation is proportional to the length of the light path in the substance;

(iii) it is also affected by the inclination of the field to the direction of the light rays.

In accordance with (i) Faraday points out[3] that magnetic rotatory power differs from natural rotatory power in that the observed rotations can be multiplied, instead of being neutralised, by sending the light backwards and forwards through the medium.

Other Early Experiments.—The announcement of Faraday's discovery led other workers to repeat his experiments.[4, 5] Thus BECQUEREL[6] in 1846 used a modification of Faraday's apparatus, in which the substance under investigation was placed *between* the poles of the electromagnet, and the light was passed into it through holes pierced in the pole pieces. He observed that the rotation varied with the wave-length of the light, but could be compensated by means of a sugar solution placed in the path of the light, outside the magnetic field. He concluded that the magnetic rotation varied with wave-length according to the law of inverse squares which Biot had proposed for natural rotatory power.

[1] FARADAY, *Diary*, Folio, Bell's edition (1932), vol. iv, entry No. 7504, p. 264.
[2] *Ibid.*, entry No. 7687, p. 286.
[3] FARADAY, *Phil. Mag.*, 1846, **29**, 153.
[4] POUILLET, *C.R.*, 1846, **22**, 135.
[5] RUHMKORFF, *A.C.P.*, 1846, [iii], **18**, 318.
[6] E. BECQUEREL, *ibid.*, **17**, 437.

Bertin[1] discovered that CS_2 and $SnCl_4$ produce rotations which are much larger than any previously observed in other liquids. Wiedemann[2] placed the substances which he wished to examine in the magnetic field produced by a long solenoid, as suggested by Faraday,[3] and showed that the rotation was proportional to the current. He found that Biot's law did not represent the magnetic rotatory dispersion of limonene and turpentine, but asserted that the magnetic rotation was nevertheless proportional to the natural optical rotation for the same wave-length. This erroneous statement, to which further reference is made below (p. 167), has persisted in the literature under the title of Wiedemann's law.

Verdet's Experiments.—A precise quantitative investigation of the Faraday effect was described by Verdet in four papers published in the *Annales de Chimie et de Physique* between 1854 and 1863. Verdet[4] first developed arrangements for producing a uniform magnetic field, and used a filter of ammoniated copper sulphate to produce a beam of uniform colour with a maximum transmission at the *G* line of the solar spectrum. His experiments with heavy flint glass and with carbon bisulphide demonstrated the proportionality of rotation to the length of the light path in the medium, and established a strict proportionality between the strength of the magnetic field and the rotation of the plane of polarisation for a given wave-length; but when the beam of light was not parallel to the magnetic field, the rotation was proportional to the component in the direction of the beam and therefore to the cosine of the angle of inclination. These three results are summarised in Verdet's law,

$$\omega = \delta l \mathrm{H} \cos \theta$$

in which the constant of proportionality δ is known as Verdet's constant.

The name magnetic rotatory power was used for the first time by Verdet in 1858.[5] He also introduced the practice of comparing the magnetic rotatory power of a substance with that of water under identical conditions.

A direct relationship between the refractive index of a substance and its magnetic rotatory power had been suggested by de la Rive in 1853.[6] After examining a large number of substances, Verdet concluded that no simple relationship exists; but he was able to show that the rotation produced by a solution is the sum of the rotations produced by a solvent and solute alone, and that the rotation thus produced by the solute is proportional to its concentration. The constant of proportionality, rotation divided by proportion of solute, he called the molecular magnetic rotatory power of the solute.

[1] Bertin, *C.R.*, 1848, **26,** 216; *A.C.P.*, 1848, [iii], **23,** 5.
[2] Wiedemann, *Pogg. Ann.*, 1851, **82,** 215.
[3] Faraday, *Phil. Mag.*, 1846, [iii], **29,** 153.
[4] Verdet, *A.C.P.*, 1854, [iii], **41,** 370; 1858, [iii], **43,** 37.
[5] Verdet, *ibid.*, **52,** 129.
[6] de la Rive, *Traité d'Electricité,* 1853, **1,** 555.

Verdet's work on solutions led him to a discovery of fundamental importance, namely, that iron salts make a " negative " contribution to the rotatory power of an aqueous solution, i.e. they produce a partial rotation which is in the opposite sense to that of water. This effect had been suspected, but not demonstrated, by Becquerel. It was observed by Verdet in solutions of other inorganic salts, and has formed the subject of a study of PARAMAGNETIC ROTATORY POWER, which has proved to be of value in atomic and molecular physics.[1]

In his fourth paper [2] Verdet discusses Wiedemann's observations on magnetic rotatory dispersion. He showed that the rotation ω is only *approximately* proportional to $1/\lambda^2$, since the product $\omega\lambda^2$ increases as λ^2 decreases. Deviations from the Law of Inverse Squares were most pronounced in substances which exhibit high dispersive power, coupled with a high magnetic rotatory power. Tartaric acid, which exhibits an " anomalous " natural rotatory dispersion, nevertheless showed a " normal " magnetic rotatory dispersion. Wiedemann's " law," if valid at all, could therefore only be applied to a limited range of compounds.

Magnetic Rotatory Power and Chemical Constitution.—(*a*) *de la Rive*.—The relationship between magnetic rotatory power and chemical constitution was studied in 1870 by DE LA RIVE,[3] who used for this purpose Verdet's method of comparing the observed rotations with those produced by water under identical conditions, in order to eliminate errors due to lack of uniformity in the magnetic field. The ratio of the two rotations he called the SPECIFIC MAGNETO-ROTATORY POWER of the substance.

He determined this quantity, and its variation with temperature, for a number of organic compounds. For some liquids, the rotation seemed to be strictly proportional to the density, but no general relationship of this type could be established. Important effects were, however, produced by the processes of addition and substitution. Thus, addition of oxygen reduced the magnetic rotatory power whilst the introduction of nitrogen increased it; halogen compounds gave very large rotations. Since isomers did not give identical values, de la Rive concluded " that the mode of combination, and particularly the mode of grouping of the atoms, consequently the atomic volume, has, independently even of the nature of the atoms, a great influence upon the intensity of the magneto-rotatory power."

(*b*) *Perkin*.—The suggestion of de la Rive attracted the attention of W. H. PERKIN, senr.[4] In order to obtain comparable data, Perkin concluded that de la Rive's method of comparing the rotations produced by equal lengths of liquid must be replaced by a comparison of equal columns of *vapour*, containing equal numbers of molecules.

[1] LADENBURG, *Zeits. f. Physik*, 1925, **34**, 898; 1927, **46**, 168; see BECQUEREL, *Proc. Akad. Wet., Amsterdam*, 1930, **33**, 913, for a complete bibliography.

[2] VERDET, *A.C.P.*, 1863, [iii], **69**, 415.

[3] DE LA RIVE, *Phil. Mag.*, 1870, [iv], **40**, 393.

[4] PERKIN, *J.*, 1882, **41**, 330.

Since, however, a direct investigation of the magnetic rotatory power of vapours was difficult, he proposed to reduce his observations on liquids to values corresponding with equal lengths of vapour. For this purpose it was only necessary to multiply the observed rotations in de la Rive's formula for specific magneto-rotatory power by the molecular weights and to divide them by the densities of the compound and of water respectively. The MOLECULAR COEFFICIENT OF MAGNETIC ROTATION OR MOLECULAR ROTATORY POWER is thus defined by the relation

$$[\Omega] = \frac{W}{d}M \div \frac{\omega_W M_W}{d_W},$$

where ω, d, M are the observed rotation, density, and molecular weight of the substance, and ω_W, d_W, M_W are the corresponding quantities for water under identical conditions.

In 1884 the results for 142 compounds belonging to various homologous aliphatic series were published and analysed,[1] and in 1896[2] rotations were given for 180 aromatic compounds. It was shown that the addition to a compound of a CH_2 group is accompanied by an increase of 1·023 units in the molecular magnetic rotatory power, and that in any homologous series the molecular magnetic rotatory power of a molecule containing n carbon atoms can be expressed very accurately by the formula

$$\text{Molecular rotation} = s + 1{\cdot}023n,$$

where s is a "series constant."[1] Perkin's constants for 26 such series are reproduced in Table 13, but the constant increment of 1·023 units could not be analysed satisfactorily into "atomic rotations" for the carbon and hydrogen atoms, and thence for other atoms, in the way which is possible in the case of *molecular refractions*. Whilst, therefore, there is an analogy between these two properties, the influence of chemical constitution on magnetic rotatory power is much more pronounced. Comparative measurements of magnetic rotation in closely related compounds could therefore be used to throw light on the constitution of compounds of doubtful molecular structure, where ordinary chemical methods are not available.[3]

(*c*) *Lowry.*—In order to cover a large range of compounds, Perkin's measurements were limited to a single wave-length, namely, that of the sodium D lines, although the possibility of using lithium and thallium flames was also indicated.[4] In order to measure the magneto-rotatory dispersion of a number of simple compounds, Lowry[5] used a modification (Fig. 50) of the apparatus developed by Perkin.[6] The magneto-rotatory dispersion, expressed by the ratio of the rotations for the violet and green lines of mercury, was found to be

[1] PERKIN, *J.*, 1884, **45**, 421–580. [2] *Ibid.*, 1896, **69**, 1025–1257.
[3] *Ibid.*, 1902, **81**, 177 and 292; 1905, **87**, 1491; 1906, **89**, 33 and 849; 1907, **91**, 806. [4] *Ibid.*, 1884, **45**, 425; 1906, **89**, 617.
[5] LOWRY, *J.*, 1913, **103**, 1062 and 1322; LOWRY and DICKSON, *J.*, 1913, **103**, 1067; LOWRY, *J.*, 1914, **105**, 81.
[6] PERKIN, *J.*, 1906, **89**, 608.

TABLE 13.—FORMULÆ FOR CALCULATING THE MOLECULAR MAGNETIC ROTATIONS OF HOMOLOGOUS SERIES (Perkin).

Substance.	Formula.	Series Constant.
Paraffins	C_nH_{2n+2}	0·508 + n(1·023)
,, *iso*	,,	0·621 ,,
Alcohols	$C_nH_{2n+2}O$	0·699 ,,
,, *iso* and *sec.*	,,	0·844 ,,
Oxides	,,	0·642 ,,
,, *iso*	,,	0·932 ,,
Aldehydes	$C_nH_{2n}O$	0·261 ,,
,, *iso* and ketones	,,	0·375 ,,
Acids	$C_nH_{2n}O_2$	0·393 ,,
,, *iso*	,,	0·509 ,,
Formic esters (ethyl and above)	,,	0·495 ,,
Acetic esters (ethyl and above)	,,	0·370 ,,
Ditto *iso*	,,	0·485 ,,
Methyl esters	,,	0·273 ,,
Ethyl esters and above	,,	0·337 ,,
Ditto *iso*	,,	0·449 ,,
Methyl esters, succinic series	$C_nH_{2n-2}O_4$	0·093 ,,
Ethyl ,, ,, ,,	,,	0·196 ,,
Ditto *iso*	,,	0·422 ,,
Chlorides	$C_nH_{2n+1}Cl$	1·988 ,,
,, *iso* and *sec.*	,,	2·068 ,,
Bromides	$C_nH_{2n+1}Br$	3·816 ,,
,, *iso* and *sec.*	,,	3·924 ,,
Iodides	$C_nH_{2n+1}I$	8·011 ,,
,, *iso* and *sec.*	,,	8·099 ,,
Ethyl esters, unsaturated	$C_nH_{2n-2}O_2$	1·451 ,,

remarkably constant in a large number of simple aliphatic compounds. Thus the values for 44 aliphatic compounds as set out in Table 14 varied only from 1·631 to 1·639. More complex aliphatic compounds, such as glycol, glycerol, ethyl tartrate and methyl camphorcarboxylate, gave similar values; but the double bond in allyl alcohol raised the

TABLE 14.—MAGNETIC ROTATORY DISPERSION OF ALIPHATIC COMPOUNDS (Lowry and Dickson).

		Dispersion-ratio $\alpha_{4359}/\alpha_{5461}$.
Two paraffins	C_6 to C_8	1·635
Three primary alcohols	C_4 to C_8	1·636
Nine methyl-alkyl-carbinols	C_4 to C_{12}	1·637
Five ethyl ,, ,,	C_6 to C_{11}	1·632
Six *iso*propyl ,, ,,	C_6 to C_{12}	1·635
Four methyl alkyl ketones	C_3 to C_8	1·639
Four *iso*propyl ,, ,,	C_7 to C_{12}	1·635½
Six fatty acids	C_1 to C_5	1·634
Seven esters	C_3 to C_5	1·631
Average for 44 substances	—	1·635

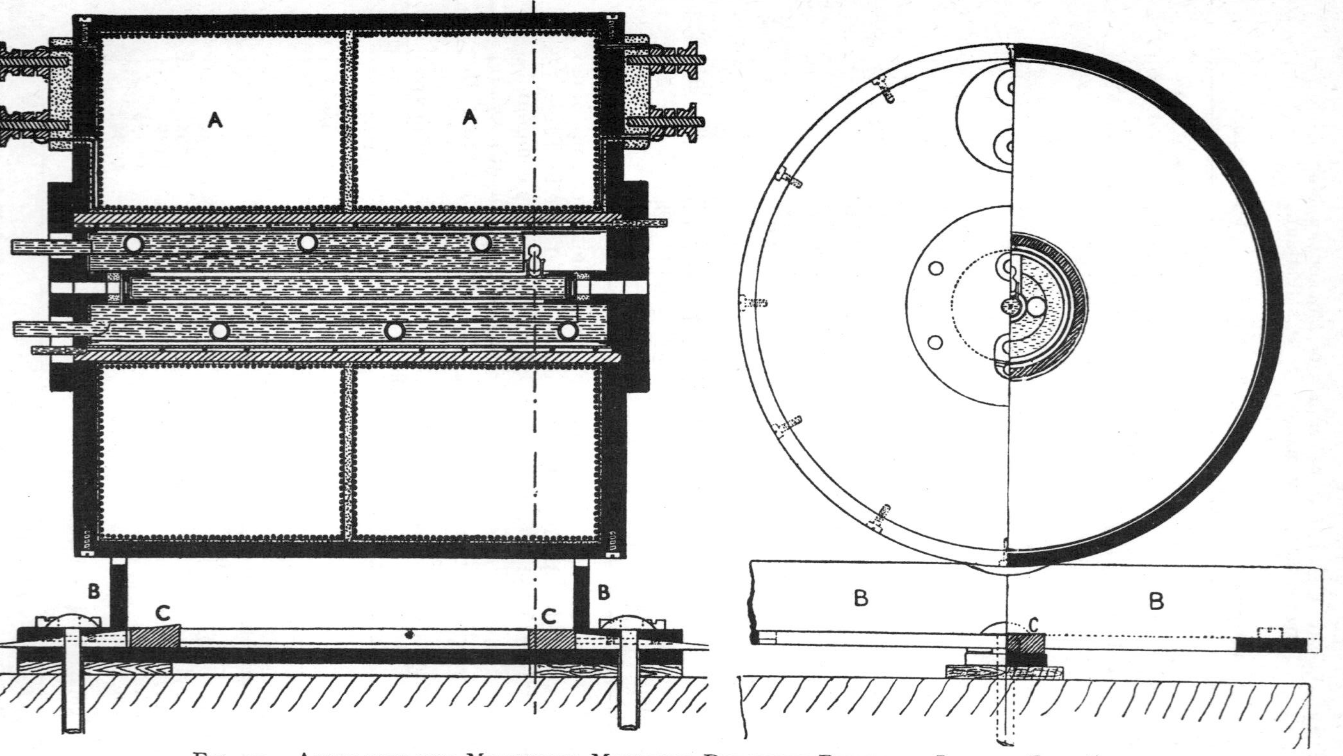

Fig. 50.—Apparatus for Measuring Magnetic Rotatory Power of Liquids (Lowry).

The liquid is contained in a central polarimeter tube with a side-tube for filling and emptying. It is surrounded by a double water-jacket, fed from a thermostat, to protect it from the heat of the magnet and to keep it at constant temperature. The electromagnet is wound on a brass tube with a water-lining to conduct away the heat of the magnet. The lines of force are concentrated through the liquid by internal perforated pole-pieces in circuit with a soft-iron casing.

ratio to 1·672, and larger values were also observed in compounds containing an aromatic nucleus.

The Relationship between Natural and Magnetic Rotatory Power: Wiedemann's Law.—Wiedemann's Law (p. 162) was based on measurements of the ratio of the magnetic to the optical rotations of turpentine for five wave-lengths in the solar spectrum. From these measurements he concluded that:

> "These numbers agree so well together that one may assume that the law of proportionality of the rotation of the plane of polarisation produced by the current in light of different wave-lengths, with the rotation already existing in turpentine-oil may be regarded as correct."

This statement was shown almost immediately to be incorrect;[1] but nevertheless remained almost unchallenged until Disch[2] in 1903 again found marked deviations, particularly with substances exhibiting anomalous rotatory dispersion. Similarly Darmois[3] in 1911 concluded that "the law of proportionality was quite inexact, and that Wiedemann's result was the result of pure chance." On the other hand, Lowry's early experiments on quartz[4] established an exact proportionality, within the limits of experimental error, between the two rotatory powers of this substance over a wide range of the visible spectrum. Since a similar result was obtained by Dahlen with sodium chlorate[5] the law may perhaps be true for optically-active *crystals*.

An extensive series of measurements by Lowry, Pickard and Kenyon[6] on optically-active *liquids*, showed that the law fails, not only for the more complex compounds, in which anomalous rotatory dispersion or structural influences might be expected to exert a disturbing effect, but also for quite simple molecules, such as the optically-active alcohols. Thus the following characteristic dispersion ratios were found for the series of carbinols:

Methyl series	$CH_3 . \mathbf{C}H(OH) . R$	(opt.)	1·651
		(mag.)	1·637
Ethyl series	$C_2H_5 . \mathbf{C}H(OH) . R$	(opt.)	1·639
		(mag.)	1·632
iso*Propyl series*	$(CH_3)_2CH . \mathbf{C}H(OH) . R$	(opt.)	1·663
		(mag.)	1·635

In more complicated substances, the discrepancy is still larger. Thus for *methyl camphorcarboxylate*, the two dispersion-ratios are 2·048 (opt.), 1·632 (mag.); whilst for *nicotine* they are 1·831 (opt.) and 1·664 (mag.).[7]

[1] VERDET, *A.C.P.*, 1863, [iv], **69**, 415.
[2] DISCH, *Ann. der Physik*, 1903, **12**, 1153.
[3] DARMOIS, *A.C.P.*, 1911, [viii], **22**, 247–281, 495–590, p. 589.
[4] LOWRY, *Phil. Trans.*, 1912, A. **212**, 295.
[5] DAHLEN, *Zeitschr. wiss. Phot.*, 1915, **14**, 315.
[6] LOWRY, PICKARD and KENYON, *J.*, 1914, **105**, 94.
[7] LOWRY and ALLSOPP, *J.*, 1932, 1613.

The magnetic rotatory dispersions of many simple optically-active liquids can, however, be represented very accurately by one term of Drude's equation, $\alpha = k/(\lambda^2 - \lambda_0^2)$, although its validity cannot be tested so drastically as in the case of simple optical rotatory dispersion on account of the lower accuracy obtainable in the magnetic measurements. This fact might appear at first sight to predict proportionality between the two rotatory powers, but this would depend on identity of the values of λ^2_0 in the respective equations, which is not realised in practice. Thus, for *nicotine*, λ^2_0 (mag.) = 0·027, λ^2_0 (opt.) = 0·060.

It is interesting to note that Salceanu[1] has recently observed a proportionality between the optical and magnetic rotatory powers of molten *menthol*, but did not find it in molten *camphor* nor in molten *carvone*. Here again the agreement seems to have been a pure coincidence.

Molecular Rotatory Power of Gases.—Faraday's failure to detect any rotatory power in gases was obviously due to the insensitiveness of his apparatus, and not to the complete absence of the effect. The existence of this effect was established about 1880 by H. Becquerel[2] and by Kundt and Röntgen.[3] Becquerel used a tube 3·27 metres long, with a diameter of 12·2 cm., in a solenoid capable of carrying a steady current of 23 amperes, and obtained rotations of about half a degree when the light was reflected to and fro nine times along a tube containing O_2, N_2, air, CO_2, N_2O, SO_2, or C_2H_4 at atmospheric pressure. On the other hand, Kundt and Röntgen increased the rotations by compressing the gases to a pressure of several atmospheres. This method was subsequently used in an improved form by Siertsema.[4]

Bichat[5] in 1878 investigated the influence of change of state from liquid to vapour on the rotatory powers of CS_2 and SO_2. This work has recently been extended systematically by Gabiano,[6] who has measured the rotations given by a large number of gases and vapours, and in some cases, by their liquids, in connection with the molecular theory of magnetic rotatory power of which an account is given in Chapter XXXIV.

[1] Salceanu, *C.R.*, 1931, **192**, 1218.
[2] H. Becquerel, *A.C.P.*, 1880, [v], **21**, 289.
[3] Kundt and Röntgen, *Ann. der Physik*, 1879, **8**, 278; 1880, **10**, 257.
[4] Siertsema, *Arch. Néerlandaises*, 1899, **2**, 291.
[5] Bichat, *Journ. de Phys.*, 1879, [i], **8**, 204; 1880, [i], **9**, 275.
[6] Gabiano, *Thesis*, Nancy, 1933.

PART II.—*POLARIMETRY.*

CHAPTER XIII.

POLARISING PRISMS.

General Principles.—Measurements of rotatory power and rotatory dispersion are made with a POLARIMETER, the essential optical parts of which are a POLARISER and an ANALYSER. The incident light is rendered plane-polarised by the polariser and is observed through the analyser. If the PRINCIPAL PLANES of the polariser and analyser (see below) are parallel, the emergent light has its maximum intensity. On rotating the analyser the intensity decreases and becomes zero when the principal planes are at right angles. If the analyser is rotated through a further 90°, maximum brightness is restored. In practice it is found that the position of minimum brightness can be adjusted much more accurately than that of maximum brightness. In this position the polariser and analyser are said to be set to extinction. If a substance which rotates the plane of polarised light is now interposed between the polariser and the analyser the emergent light will, in general, no longer have minimum brightness, and the analyser must be rotated to restore it to the extinction position. The angle through which the analyser has to be rotated is a measure of the rotation produced by the optically-active substance. Biot adopted a convention which described as dextrorotatory those media which required a clockwise movement of the analyser to restore the extinction. An opposite convention (p. 26), introduced by Herschel, has now been abandoned.

In the earliest experiments, polarisation was produced by reflection (p. 3). A crystal or plate of calcite was used as a polariser or analyser by Biot (p. 179), Joubin (p. 214), Dongier[1] and Darmois.[2] At normal incidence the ordinary ray passes undeviated through two parallel faces of the crystal, whilst the extraordinary ray is deflected and can be cut off by a screen. DOUBLE IMAGE PRISMS, in which a similar separation is effected in prisms of quartz and calcite are still in use (pp. 176–178); but in other polarising prisms the two rays are separated completely by selective total reflection of one ray.

Nicol Prism.—The NICOL PRISM[3] is constructed from a natural crystal of Iceland spar. The cleavage surfaces of the crystal

[1] DONGIER, *A.C.P.*, 1898, [vii], **14,** 331.

[2] DARMOIS, *A.C.P.*, 1911, [viii], **22,** 267.

[3] NICOL, *New Phil. Journ., Edinb.*, 1828, **6,** 83; 1839, **27,** 332. See also POGGENDORF (?), *ibid.*, 1834, **16,** 372; *Pogg. Ann.*, 1833, **29,** 182.

form a rhombohedron bounded by six similar parallelograms whose angles are 101° 55′ and 78° 5′. Two opposite solid angles are contained by three obtuse angles, and the other four solid angles are contained by one obtuse and two acute angles. The OPTIC AXIS of the crystal is a direction parallel to a line passing through one of the blunt corners and making equal angles with the three edges which meet there. The crystal has three PRINCIPAL SECTIONS each of which contains the optic axis and is perpendicular to a pair of opposite faces of the rhomb.

ABCD in Fig. 51 represents a crystal of Iceland spar whose end faces AD and BC are bounded by edges of equal length. A and C are the two blunt corners and AB and DC are the two lateral blunt edges. A plane through ABCD is then a principal section of the crystal. The natural end-faces AD and BC are inclined at an angle of 71° to the blunt edges AB and DC.

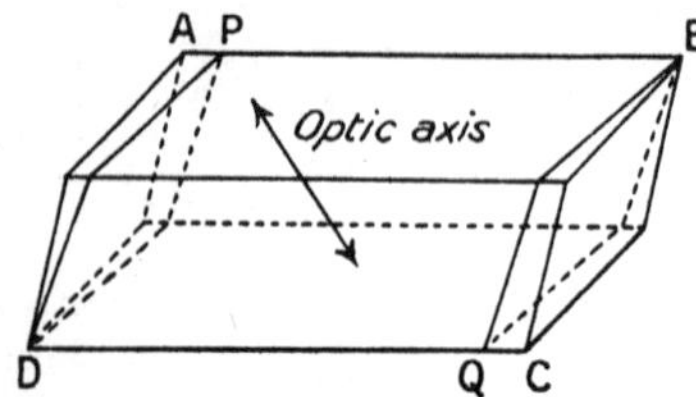

FIG. 51.—CRYSTAL OF ICELAND SPAR.

These natural end faces are first cut off along DP and BQ in a direction at right angles to the principal section through ABCD so that the new end faces, when polished, are inclined at 68° to the lateral blunt edges AB and DC.

The crystal is then cut in two along a plane at right angles to the principal section through PBQD and at right angles to the new end faces. The cut surfaces are polished and cemented together again with a thin film of Canada balsam whose refractive index is intermediate in value between those of Iceland spar for the ordinary and extraordinary rays.

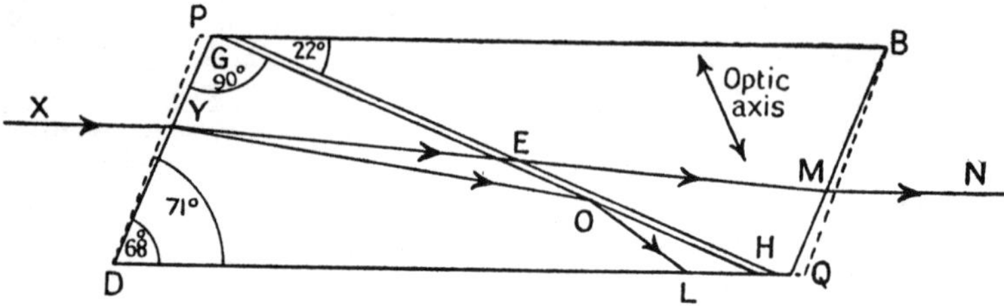

FIG. 52.—SECTION THROUGH THE PRINCIPAL PLANE OF A NICOL PRISM.

Fig. 52 is a section through the principal plane PBQD of the Nicol prism. The dotted lines show the original positions of the natural end faces of the crystal. GH is the thin film of Canada balsam, the plane of which is perpendicular to the new end faces and makes an angle of 22° with the lateral edges. A beam of unpolarised light XY incident on one of the end faces of the Nicol prism is divided into an ordinary ray YO and an extraordinary ray YE, the latter travelling with the greater velocity. These two rays are plane polarised in directions mutually at right angles. The inclination of the film of Canada balsam is such that, when the incident ray XY is parallel to the lateral faces of the prism, the ordinary ray YO strikes the film of Canada balsam at an angle greater than the critical angle for the interface, and is totally reflected along OL, whilst the extraordinary ray YE strikes the Canada balsam at an angle less than the critical angle for the interface and is therefore transmitted along EM and emerges along MN. Since the vibrations of the extraordinary ray take place in the principal plane, the vibrations of the emergent ray EMN are in the plane of the paper. The lateral faces of the prism are blackened so as to absorb the ordinary ray at L.

If the ray XY is slowly tilted in the direction shown in Fig. 53 (*a*) the ordinary refracted ray YO will strike the film of Canada balsam at a smaller and smaller angle of incidence. When this angle becomes smaller than the critical angle for the ordinary ray at the calcite-balsam interface, this ray will cease to be totally reflected; it will therefore penetrate the Canada balsam and be transmitted through the prism along the path YOJK. The light emerging from the

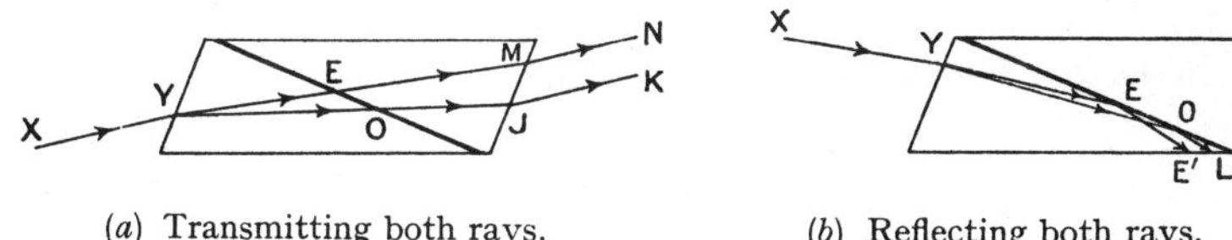

(*a*) Transmitting both rays. (*b*) Reflecting both rays.

FIG. 53.—NICOL PRISMS.

Nicol prism will therefore no longer be plane polarised. If, on the other hand, the incident ray XY is slowly tilted in the opposite direction, as shown in Fig. 53 (*b*), the extraordinary refracted ray YE will not only strike the Canada balsam at a greater angle of incidence, but will be inclined at a smaller angle to the optic axis of the Iceland spar, thus producing an increase in the refractive index of the extraordinary ray. When this index exceeds the refractive index of Canada balsam the extraordinary ray will be totally reflected along EE′. No light will then emerge from the end face of the Nicol prism.

A consideration of these two extreme cases shows that, for plane polarised light to be produced by a Nicol prism, the incident beam must be inclined at an angle of less than 14° to 15° to the long sides of the prism. If, therefore, a convergent or divergent beam of light is to be completely plane polarised by a Nicol prism, the degree of convergence or divergence must not exceed 29°. This angle is termed the FIELD of the Nicol prism.

Since the end faces of an ordinary Nicol prism are obliquely inclined to the length of the prism, and therefore to the direction of the incident light, the ray suffers a lateral displacement in passing through the prism. The emergent plane polarised ray therefore moves in a circle when the Nicol prism is rotated. Several text-books on optics state that it is customary to cut the end faces of the Nicol prism perpendicular to the long sides in order to prevent this lateral displacement of the emergent ray. This is not the case, because the extraordinary ray will always be deviated unless the incident ray is parallel or perpendicular to the optic axis of the crystal.

A further disadvantage of the Nicol prism is that complete extinction of the *entire* field of vision is not produced when a ray of light passes successively through two Nicols whose principal sections are at right angles to one another. The field, instead of appearing uniformly black, is crossed by a dark band inclined at 45° to the principal sections of the two Nicols. This band was described in 1879 by Landolt,[1] and its existence was subsequently explained by Lippich.[2] It has been shown that it is necessary to rotate one of the Nicols through about ½° in order to make this dark band move completely across the field. Since the dark band has very diffuse edges it is not possible to get the band in the centre of the field with any degree of accuracy. Bruhat and Mlle. Hanot [3] have shown that the error involved is several minutes of arc. For this reason, very accurate polarimetric measurements are impossible with ordinary Nicol prisms.

Modified Forms of Nicol Prism.—(i) In order to ensure the transmission of all the extraordinary refracted rays Nicol [4] and Radicke [5] devised a form of Nicol prism in which the film of Canada balsam was perpendicular to the optic axis of the Iceland spar, and the end faces of the prism were cut normal to the lateral faces.

(ii) This prism was further modified by Hartnack and Prazmowski [6] who introduced a film of linseed oil in place of the film of Canada balsam. Its refractive index is much smaller than that of Canada balsam, and is almost equal to the principal extraordinary refractive index for Iceland spar. This alteration reduced the critical angle for the ordinary ray and thus ensured more complete total reflection. The field of this modified form of Nicol prism was 35° (as compared with 29°) when the film of linseed oil was inclined at 17° to the lateral blunt edges.

(iii) Jamin [7] inserted a thin plate of Iceland spar, cut parallel to the optic axis, as a diagonal partition in a rectangular vessel of carbon bisulphide, the refractive index of which is slightly less than the ordinary refractive index of Iceland spar. Part of the incident beam, corresponding to the extraordinary ray in Iceland spar, is totally reflected at the surface of the latter. The emergent beam,

[1] LANDOLT, *Handbook of the Polariscope*, English translation, 1882, p. 100.
[2] LIPPICH, *Sitz. Akad. Wiss., Wien*, 1882, **85**, 268.
[3] BRUHAT and HANOT, *Journ. de Phys.*, 1922, **3**, 46.
[4] NICOL, *New Phil. Journ., Edinb.*, 1839, **27**, 332; *Pogg. Ann.*, 1840, **49**, 238.
[5] RADICKE, *ibid.*, **50**, 25–34.
[6] HARTNACK and PRAZMOWSKI, *A.C.P.*, 1866, [iv], **7**, 181.
[7] JAMIN, *C.R.*, 1869, **68**, 221.

corresponding to the ordinary ray in Iceland spar, is therefore plane polarised.

Prisms consisting of a thin plate of Iceland spar set obliquely in a vessel of liquid or between prisms of glass are termed "SENSITIVE STRIP" POLARISERS. Several forms have been proposed. Thus Brace[1] used monobromonaphthalene in place of carbon bisulphide. Bertrand[2] increased the field of the prism to 44° by placing the plate

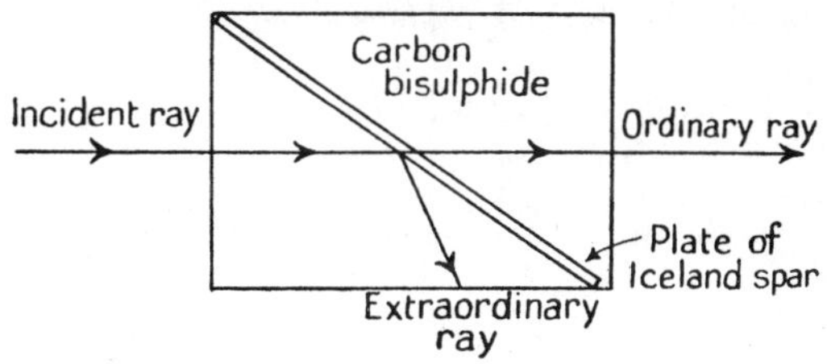

FIG. 54.—JAMIN'S POLARISING PRISM.

of Iceland spar between two right-angled prisms of flint glass of refractive index almost equal to that of Iceland spar for the ordinary ray. A prism with an even larger field of 54° was devised by Feussner,[3] who replaced the Iceland spar by a thin plate of sodium nitrate, the crystals of which are much more strongly doubly refracting than those of calcite.

Foucault Prism.—Nicol prisms of large aperture are costly, since long crystals of Iceland spar are very rare. In the FOUCAULT

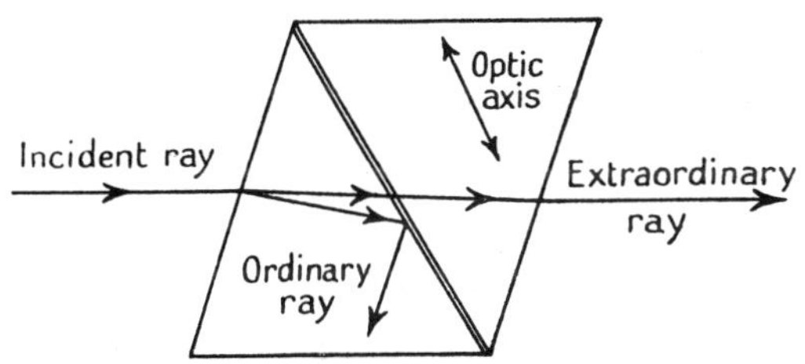

FIG. 55.—FOUCAULT PRISM.

PRISM[4] (FIG. 55) this difficulty is overcome by using a film of air instead of Canada balsam. A natural crystal of Iceland spar is cut along a plane at right angles to a principal section, as for an ordinary Nicol prism; but the natural end faces of the crystal are retained, and the plane of the cut, instead of being perpendicular to the end faces of the prism, makes an angle of 59° with the lateral blunt edges

[1] BRACE, *Phil. Mag.*, 1903, [vi], **5**, 161.
[2] BERTRAND, *C.R.*, 1884, **99**, 538.
[3] FEUSSNER, *Zeit. Instr.*, 1884, **4**, 41.
[4] FOUCAULT, *C.R.*, 1857, **45**, 238.

of the crystal. The cut faces are polished and placed together again with a thin film of air between them.

As in the Nicol prism, the ordinary ray is totally reflected while the extraordinary ray is transmitted. The limiting angles for the transmission of the extraordinary and the reflection of the ordinary rays are, however, very close together, so that all the rays incident on a Foucault prism must be very nearly parallel to one another and to the lateral faces of the prism in order that the emergent light may be completely plane polarised. The field of a Foucault prism is, in fact, only 8° as compared with 29° for a Nicol prism.

A further disadvantage of the Foucault prism is that the intensity of the extraordinary ray is appreciably reduced by partial reflection from both faces of the air film, and, since the amount of light thus lost varies with the angle of incidence, the field of vision is not uniformly illuminated. This defect is more pronounced if the two cut faces of the crystal are not exactly parallel.

Foucault prisms can be used with light of wave-lengths down to the limit of transmission by calcite at about 2300 A.U. Nicol prisms, on the other hand, can only be used in the visible and near ultra-violet, since the film of Canada balsam is only transparent to about 3500 A.U.

Glan Prism.—In the GLAN PRISM[1] (Fig. 56) the main disadvantages of the Nicol and Foucault prisms (that they do not give complete extinction of the whole field and that the emergent ray is laterally displaced) are avoided by cutting the prism in such a way that the optic axis of the crystal is perpendicular to the direction of the incident ray.

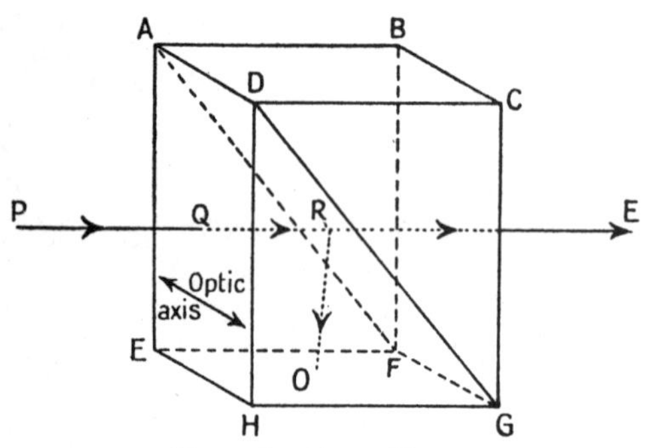

FIG. 56.—GLAN PRISM.

Each face of the prism is rectangular, the optic axis being parallel to the opposite edges AD, EH and BC, FG of the end faces. The prism is cut along the plane ADGF, which makes an angle of about 50° with the lateral faces. The two halves are separated by a film of air as in the Foucault prism. A ray PQ, incident normally on one of the end faces of the prism, is propagated along QR undeviated, the ordinary and extraordinary rays travelling in the same direction but with unequal velocities. The ordinary ray is totally reflected at the air film along the direction RO, while the extraordinary ray is transmitted undeviated along RE. The direction of vibration in

[1] GLAN, *Repert. Physik*, 1880, **16**, 570.

the emergent extraordinary ray is parallel to the optic axis of the Iceland spar. The field of the Glan prism is of the order of 8°.

Glazebrook Prism.—The GLAZEBROOK PRISM[1] (Fig. 57) differs from the Nicol prism in the same way that the Glan prism differs from the Foucault prism. The optic axis of the crystal is again made parallel to the opposite edges of an end face, but the prism is cut along a plane making an angle of about 20° with the lateral faces, and cemented together again with a film of Canada balsam.

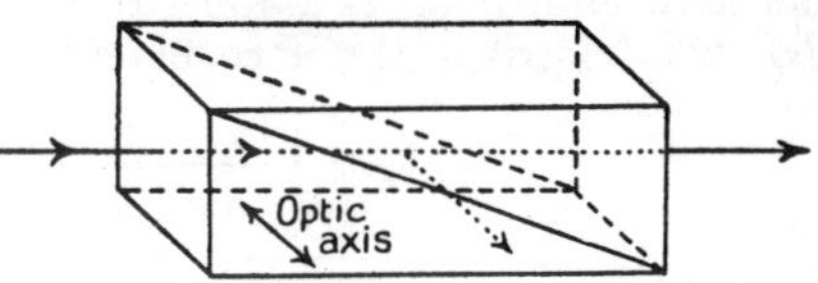

FIG. 57.—GLAZEBROOK PRISM.

As before, the extraordinary ray is transmitted, while the ordinary ray suffers total reflection at the film of Canada balsam. As in the Glan prism, when the incident ray is normal to the end face, the extraordinary ray is transmitted without deviation; but the field is only of the order of 4°.

Prisms[2] of the Glan and Glazebrook type have also been constructed, in which the optic axis is perpendicular to those edges of the end faces through which the film of air or Canada balsam passes, i.e. parallel to the edges AE and CG in Fig. 56. Such prisms have the disadvantage of a somewhat smaller field, since the refractive index for the extraordinary ray is no longer the same for all directions in the plane of symmetry of the prism. To increase the field of the Glazebrook prism, films of linseed oil[3] and poppy oil[4] have been used in place of Canada balsam, since these substances have refractive indices equal to the extraordinary refractive index of Iceland spar. These oils are, however, too fluid and difficult to free from tiny air bubbles.

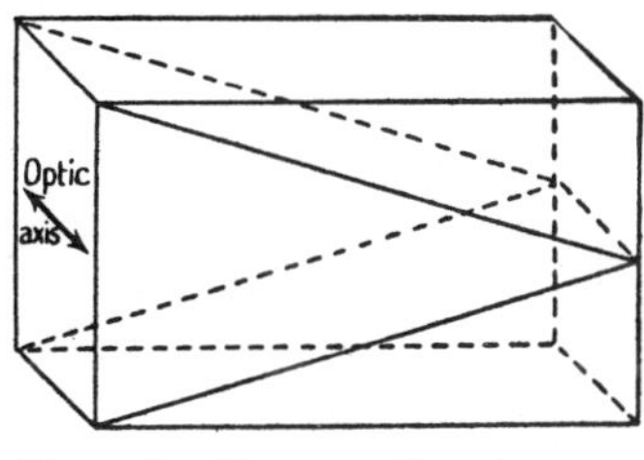

FIG. 58.—DOUBLE GLAZEBROOK PRISM.

A double Glazebrook prism,[5] as shown in Fig. 58, has also been devised.

In accurate polarimeters for use in the visible and near ultraviolet regions of the spectrum, Glazebrook prisms are almost always employed. For measurements beyond the limit of transmission by Canada balsam, Foucault and Glan prisms are used, but are subject to disadvantages already enumerated. Some workers have replaced the Canada balsam of the Glazebrook prisms by glycerine,[5] which is

[1] GLAZEBROOK, *Phil. Mag.*, 1883, [v], **15**, 352.
[2] THOMPSON, *ibid.*, 1886, [v], **21**, 476.
[3] LIPPICH, *Sitz. Akad. Wiss., Wien*, 1885, **91**, 1059.
[4] BÉNARD, *Bull. de la Direction des Recherches et Inventions*, 1920, **1**, 229.
[5] BRUHAT and CHATELAIN, *Rev. d'Opt.*, 1933, **12**, 1.

transparent down to about 2200 A.U.; but the most satisfactory prisms for ultra-violet polarimetry are double-image prisms, constructed from quartz instead of calcite, as described below.

Rochon Prism.—A ROCHON PRISM[1] (Fig. 59) consists of two right-angled prisms, cut differently with respect to the optic axis, and placed together in optical contact to form a single rectangular prism. In the prism ADC the direction of the optic axis (shown by the horizontal lines) is parallel to the direction of the incident beam PQ, which strikes the face of entry AD normally. In the prism

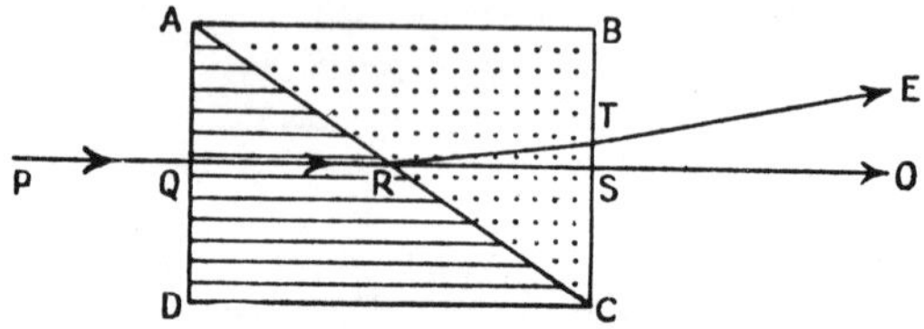

FIG. 59.—ROCHON PRISM.

ABC the optic axis (shown by dots) is parallel to the refracting edge C, i.e. perpendicular to the plane of the paper. Since the incident ray PQ is parallel to the optic axis in the first prism it is transmitted unchanged and undeviated along QR. At R it is incident obliquely on the second prism in which the optic axis is perpendicular to the incident light. It is therefore resolved into an ordinary ray (vibrations in the plane of the paper) which is transmitted undeviated along RSO, and an extraordinary ray (vibrations at right angles to the plane of the paper) which in quartz is refracted *towards* * the

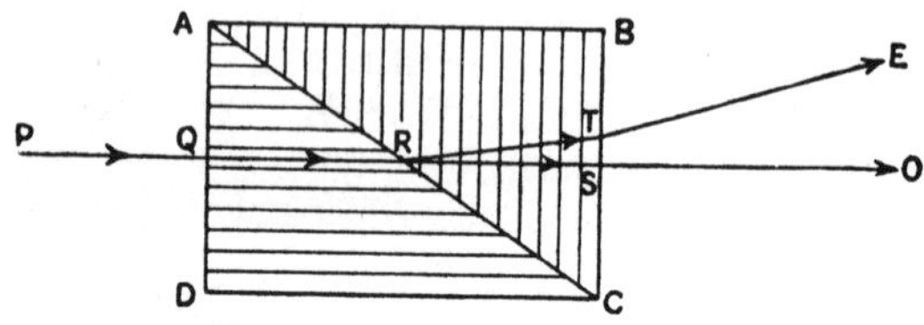

FIG. 60.—SENARMONT PRISM.

face AB along RTE, since the refractive index of quartz for the extraordinary ray is greater than for the ordinary ray.

The angular separation of the two plane polarised emergent rays is greater the more obliquely the ray PQR strikes the second prism. Thus, with a refracting angle ACB of 30°, calcite gives an angular separation of only 5°, but for an angle of 60° it is of the order of 14°. The deviated extraordinary ray can be cut off by an opaque screen, leaving only the plane polarised, undeviated, ordinary ray.

[1] ROCHON, *Journ. de Phys.*, 1811, **72**, 319–332.

* In calcite, the extraordinary ray would be refracted away from AB.

Senarmont Prism.—A SENARMONT PRISM [1] (Fig. 60) is identical with a Rochon prism, except that the optic axis of the second half of the prism is perpendicular to the refracting edge C (i.e. in the plane of the paper) and parallel to the vertical line BC. A ray PQ incident normally on the face AD is transmitted in exactly the same way as in a Rochon prism. In a crystal of calcite the optic axis is inclined at an angle of 45° to the natural faces of the crystal. Senarmont prisms of calcite are, therefore, usually made with refracting angles of 45°, since the natural faces of the crystal can then be used as the surfaces AC which are placed in contact.

Wollaston Prism.—The WOLLASTON PRISM [2] (Fig. 61) differs from a Rochon prism only in that the optic axis of the first half of

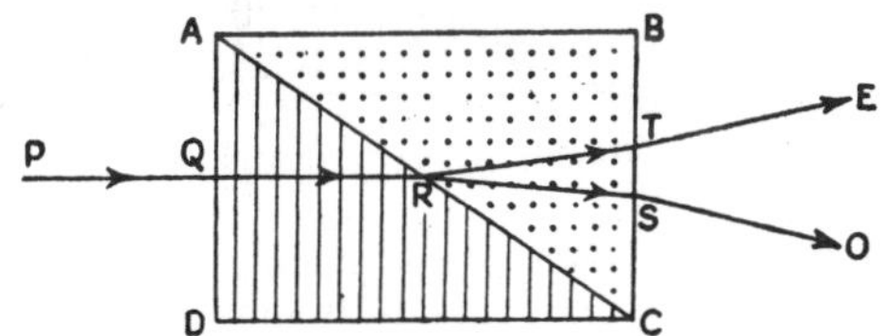

FIG. 61.—WOLLASTON PRISM.

the prism is parallel to the face of entry AD and perpendicular to the refracting edge A. In other words, it is the Rochon prism of Fig. 59 with the other face of the first prism as the face of entry. A ray PQ incident normally on the face AD is transmitted undeviated along QR, the ordinary and the extraordinary rays travelling in the same direction but with unequal velocities. The vibrations in the ordinary ray are perpendicular to the plane of the paper. This ray on entering the second prism at R is therefore transmitted as the extraordinary ray, since the vibrations are parallel to the optic axis in this prism.

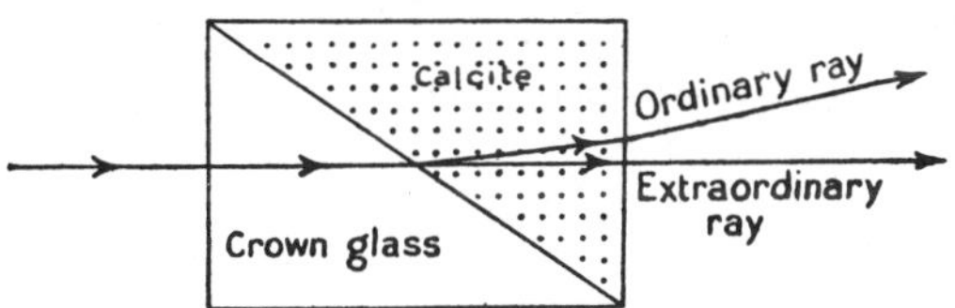

FIG. 62.—GLASS–CALCITE DOUBLE–IMAGE PRISM.

It is therefore refracted, in the case of quartz, towards the face AB along RTE. Similarly, the extraordinary ray in the first prism is transmitted as the ordinary ray in the second prism and is refracted away from the face AB along RSO. The angular separation of the two emergent rays is therefore approximately twice as great as in the Rochon and Senarmont prisms, but both rays are deviated and the image formed by the ordinary ray is no longer achromatic.

[1] DE SENARMONT, *A.C.P.*, 1857, [iii], **50**, 480.
[2] WOLLASTON, *Phil. Trans.*, 1820, **110**, 126.

Other Double-Image Prisms.—A prism which transmits the extraordinary ray undeviated has been made from two right-angled prisms, one cut from calcite so that the optic axis is parallel to the refracting edge and the other, of equal angle, cut from crown glass with a refractive index equal to the extraordinary refractive index of calcite. The path of a ray through such a prism is shown in Fig. 62.

Larger angular separation of the two rays can be obtained by

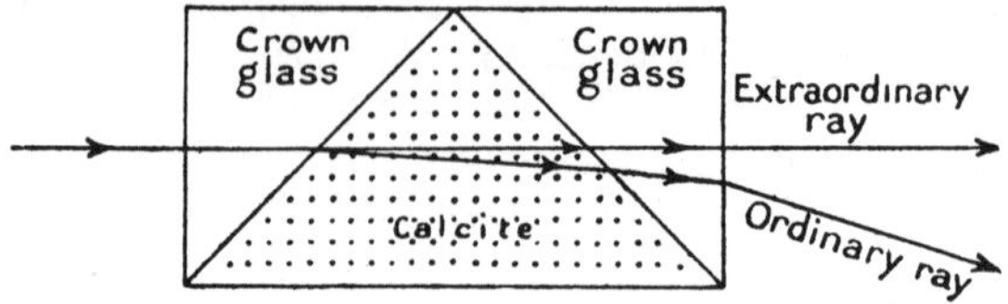

FIG. 63.—GLASS–CALCITE–GLASS DOUBLE–IMAGE PRISM.

increasing the refracting angle or by placing a calcite prism of large angle between two crown glass prisms as shown in Fig. 63.

Since these types of prism include a glass component, they cannot be used for measurements in the ultra-violet. Furthermore, the extraordinary ray can only be undeviated for one particular wavelength, since the refractive index of glass and the extraordinary refractive index of calcite are not identical throughout the spectrum. These prisms therefore have the same disadvantage as the Wollaston prism in not being achromatic.

Joubin (p. 214), Cotton and Descamps (p. 220), and Carvallo (p. 230) have used, as analyser and dispersive system combined, a refracting prism of calcite with the optic axis parallel to the refracting edge. Sodium light, incident normally on a 30° prism, gives an angular separation of 8° between the ordinary and extraordinary rays. A 60° prism at minimum deviation for $\lambda = 3100$ gives two spectra which do not overlap over the range 7000 to 2000 A.U. Furthermore, the dispersion of the ordinary spectrum is twice that of the extraordinary spectrum.

CHAPTER XIV.

POLARIMETERS.

Biot.—In Biot's original polarimeter,[1] sunlight from white clouds was plane polarised by reflection at the polarising angle from a sheet of glass. After passing through the optically-active medium, the plane polarised ray was analysed by means of a natural rhomb of calcite which could be rotated in the centre of a graduated circle. By placing a plate of red glass in the path of the incident light, Biot was able to measure the rotation for light of a uniform colour. The rotation of light of other colours was inferred from the sequence of tints observed with white light when the analyser was inclined at different angles to the polariser. In later experiments Biot used as analyser a double image prism (p. 176), devised by Rochon in 1801.

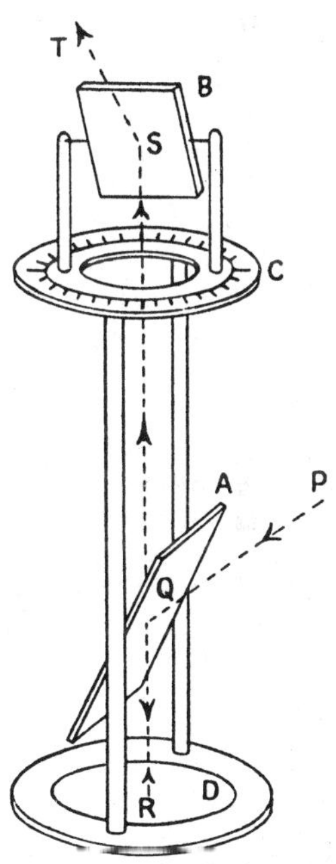

FIG. 64.—NÖRRENBERG'S POLARIMETER.

Nörrenberg.—The rotation of the plane of polarised light by an optically-active substance can be conveniently demonstrated and measured in an approximately quantitative manner by means of the apparatus devised by Nörrenberg [2] and shown in Fig. 64. A is a plate of transparent glass which can be tilted about a horizontal axis. B is a plate of black glass which can also be tilted in a similar manner. B is mounted on a ring which can rotate inside the horizontal graduated circle C. D is a plane horizontal silvered mirror. A ray of light PQ, incident on A at the polarising angle ($57\frac{1}{2}°$), is reflected vertically downwards along QR, and polarised in the plane of incidence. It is reflected normally from the silvered mirror at R, and travels vertically upwards along RQS. At S it is reflected by the black glass B which is also tilted so that the reflection takes place at the polarising angle. When the horizontal axis

[1] BIOT, *C.R.*, 1840, **11**, 413 ; *A.C.P.*, 1840, [ii], **74**, 401.
[2] See BERTIN, *A.C.P.*, 1863, [iii], **69**, 87.

of B is perpendicular to that of A, the reflected ray ST is extinguished. If, however, an optically-active medium is interposed between A and B in the path of the plane polarised ray QS, the reflected ray ST will, in general, no longer be extinguished. The rotation of the plane of polarisation by the medium is given by the horizontal angle through which B must be rotated in order to reproduce the extinction. This "new polarisation apparatus" was exhibited in Carlsruhe in 1858[1] as a means of recognising the directions of polarisation in a doubling-refracting crystal "of the fineness of a hair," and was described by Bertin (*loc. cit.*) in 1863.

Mitscherlich.—The use of Nicol prisms as polariser and analyser was introduced by Mitscherlich[2] in 1844.

The polarising Nicol P (Fig. 65) and a small convex lens L are mounted in a cylindrical brass case which can be rotated and clamped in any position by the screw S. The analysing Nicol A is enclosed in a cylindrical case which can be rotated by means of the handle H,

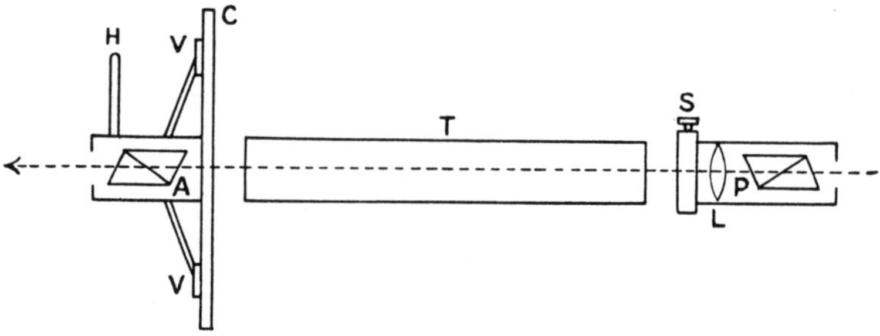

FIG. 65.—MITSCHERLICH'S POLARIMETER.

and carries two verniers V moving over a fixed circular scale C graduated in degrees. The substance to be examined is placed between the polariser and the analyser, e.g. in the tube T. After the invention of the Bunsen burner in 1866 a sodium flame was used as a source of monochromatic radiation.

The zero setting of the instrument is read off in degrees and tenths, after rotating the analyser until the field of vision is as dark as possible, in the absence of an optically-active medium. If a liquid is to be examined, the zero reading should be taken with the empty tube T in position, in order to allow for double refraction set up by strain in the end plates, which produce serious errors if they are screwed up too tightly. If a solution is to be examined the tube T should be filled with the pure solvent. Since the zero position depends on the sensitivity of the eye in estimating when the field of vision has minimum brightness, it is necessary to take the mean of a series of settings of the analyser. Any eccentricity of the Nicol prisms can be

[1] *Amtlicher Bericht über die 34 Versammlung Deutscher Naturforscher und Arzte in Carlsruhe im September, 1858*, p. 152.

[2] MITSCHERLICH, *Lehrb. der Chem.*, 4th edition, 1844, 361.

allowed for by taking a further series of readings after rotating the analyser through two right angles. Irregularities in the graduated scale can be eliminated by reading both verniers, since these move over opposite arcs of the scale.

On introducing the optically-active medium or tube of solution, the field of vision becomes brighter, and the analyser is rotated to restore maximum darkness. A series of readings is taken of the zero position of the analyser. If the analyser has to be rotated in a clockwise direction, the rotation is dextro and *vice versa*.

Unless the rotation is known to be relatively small, a single series of readings is insufficient to specify either its direction or its magnitude. Thus, the same effect would be produced by rotating the analyser through $\alpha°$ in a clockwise direction or through $180 - \alpha°$ in an anti-clockwise direction, or again, by rotating it through either of these angles together with a multiple of 180°. The true sign and magnitude of the rotation can be determined, however, by reducing the column of the optically-active medium to a known fraction of its former length, when the rotation will be reduced in the same proportion.

It has already been mentioned (p. 172) that when two Nicol prisms are " crossed " the whole of the field of vision is not uniformly dark, but is crossed by a dark band which has diffuse edges and which lies in the direction bisecting the angle between the principal planes of the two Nicols. In determining the zero of Mitscherlich's instrument the analyser is rotated until this dark band crosses the centre of the field. In measuring the rotatory power of a medium the analyser is rotated to produce a similar result; but, since the direction of the dark band varies with the orientation of the analyser, the aspect of the field of vision is not identical in the two cases. Furthermore, the dark band is only sharply defined when a very intense source of light is used. Owing to these defects the accuracy with which the analyser can be set to the extinction position is not very great; the error is of the order of several tenths of a degree.

The rotatory power for mean yellow rays can be measured, but with less accuracy, by employing a source of white light instead of a sodium flame. The zero of the instrument is found, as before, by rotating the analyser until the dark band occupies the centre of the field. On introducing the optically-active substance, however, a new phenomenon makes its appearance. If the substance has only a small rotatory power, the dark band, when brought back into the field by rotating the analyser, is seen to have a blue edge on one side and a red edge on the other, owing to the unequal rotation experienced by the various components of white light. The dark band corresponds to the extinction of the yellow components. With substances of high rotatory power the dark band is only hazily defined, and a succession of colours crosses the field of vision when the analyser is rotated. The extinction position for yellow light is obtained when red shading into blue occupies the centre of the field. With substances of very high rotatory power the red shades into blue through an intermediate greyish or reddish-violet tint. A slight rotation of the analyser in one direction transforms this into red, and in the other direction into blue. This change is very sensitive, and the reddish-violet shade is known as the SENSITIVE or TRANSITION TINT. Its appearance again corresponds to the extinction position of the analyser for yellow light.

More precise measurements of rotatory power were made by Mitscherlich with an improved form of the simple polarimeter described above. The plane polarised rays, before entering the solution tube, were passed through a glass cell containing bichromate solution to free the sodium light from green and blue rays. Since rotatory power varies considerably with temperature, the polarimeter tube was enclosed in a rectangular vessel of water which was heated to a uniform temperature. The exact setting of the analyser was facilitated by a fine adjustment consisting of a small cog-wheel working in the toothed edge of the circular scale which was thereby rotated past fixed verniers. A square aperture of 5 mm. side was inserted in the brass tube containing the polarising Nicol, and the tube containing the analysing Nicol was fitted with a small telescope, which was focussed on the edges of the square aperture of the polariser. With this instrument the range between the highest and lowest of a series of ten readings was of the order of two-tenths of a degree.

Soleil.—A DOUBLE-FIELD POLARISER was devised by Soleil.[1] A BI-QUARTZ PLATE was constructed by cementing together two semicircular discs of *d* and *l* quartz cut perpendicular to the optic axis (Fig. 66). The two halves are the same thickness, 3·75 mm., and produce a rotation of ± 82° in plane polarised sodium light.

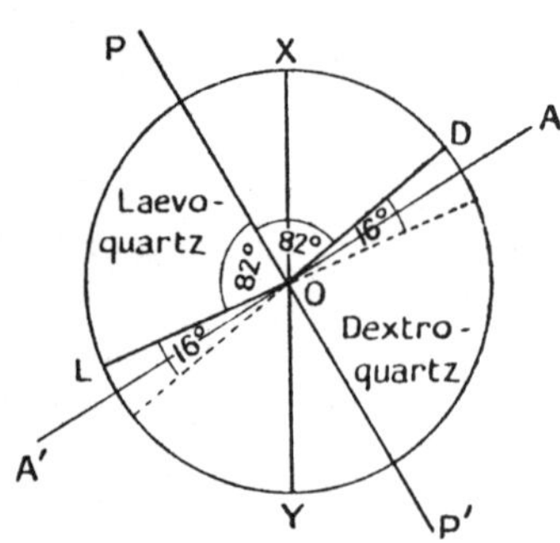

FIG. 66.—SOLEIL'S BI-QUARTZ PLATE.

The bi-quartz plate is placed immediately after the polarising Nicol, with the dividing line XY vertical. The plane of polarisation PP′ of a ray of sodium light incident on the bi-quartz plate is rotated through 82° to the left by one-half of the plate and through 82° to the right by the other half. Thus there emerge from the plate two rays, whose planes of polarisation OD and OL are inclined at 16° to one another. The analyser can be set, in two positions separated by 16°, to extinguish OD or OL, making either the right or the left half of the bi-quartz plate appear dark. If the analyser is set at AA′, midway between these two positions, the two halves will appear equally but weakly illuminated. This HALF-SHADOW SETTING depends

[1] SOLEIL, *C.R.*, 1845, **20**, 1805; 1847, **24**, 973; 1848, **26**, 162.

on the matching of two contiguous feebly illuminated fields, and gives much greater accuracy than the setting of a single-field instrument. If the analyser is rotated through 90° the two halves of the field of vision are again equally illuminated but with almost their maximum intensity. In this position the matching of the two fields cannot, however, be made with precision, because a slight rotation of the analyser produces no appreciable alteration in the relative intensities.

The Soleil bi-quartz plate was originally designed for use with white light in a special type of saccharimeter (p. 196) for comparing the rotatory power of specimens of cane sugar and is not often used in ordinary polarimeters. The particular thickness (3·75 mm.) of a bi-quartz plate was chosen because it produces, in the mean yellow radiations of white light, a rotation of exactly 90°. The red and blue components of white light are therefore rotated through angles which are smaller and greater than 90° respectively (Fig. 67). If the analyser is set so as to extinguish the yellow rays the two halves

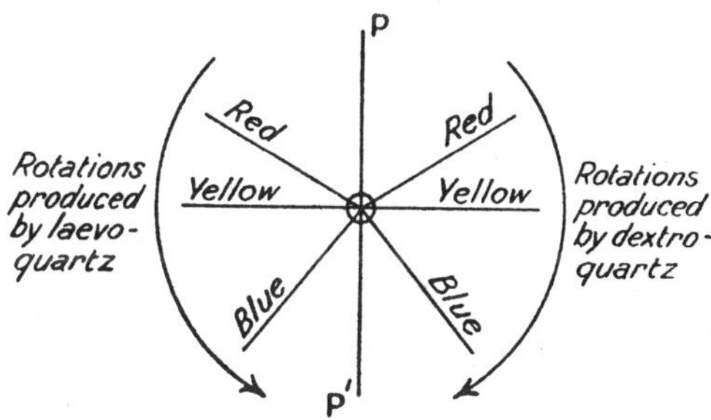

FIG. 67.—EFFECT PRODUCED BY USING A SOLEIL BI-QUARTZ PLATE WITH WHITE LIGHT.

of the field are equally illuminated by the reddish-violet, sensitive or transition tint, but a slight rotation in a clockwise direction makes the right-hand field appear red and the left-hand field appear blue and conversely. Rotations are measured by setting both halves of the field to give the same sensitive tint and refer to a wave-length (5500 A.U. approximately, for maximum visual intensity) which is slightly shorter than that of mono-chromatic sodium light (5893 A.U.).

Senarmont.—In Senarmont's [1] polarimeter one or more narrow vertical dark bands is produced in the field of the analyser by means of a pair of thin, right-angled, quartz wedges cut with their bases parallel to the optic axis. One wedge is of dextro- and the other of lævo-quartz and they are placed face to face to form a thin rectangular plate (Fig. 68).

A plane polarised ray incident at A, midway along the plate, traverses equal thickness of the two forms of quartz. Rays incident

[1] SENARMONT, *A.C.P.*, 1850, [iii], **28**, 279.

on the plate to the right of A traverse a greater thickness of dextro- than lævo-quartz and conversely. The plane of vibration is therefore rotated through an angle $\pm \alpha$ which is proportional to the difference between the thicknesses of the two types of quartz traversed. If the rays emerging from the plate are examined through an analyser, orientated to extinguish the original rays incident on the plate, then the field of the analyser will be crossed by a narrow vertical dark band corresponding to the unrotated ray passing through the plate at A. On either side of the dark band the field appears brighter and brighter, corresponding to the rays which are not extinguished, owing to the rotation of their planes of polarisation by the quartz plate. If the angles of the quartz wedges are large there will be a series of points on either side of A at which the difference in thickness of the wedges is sufficient to produce rotations of multiples of 180°. At these points also, vertical dark bands will appear in the field of the analyser. In practice the quartz wedges are made thin enough to give only the central dark band.

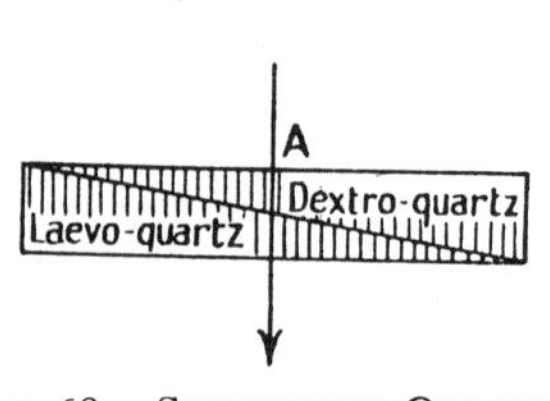

FIG. 68.—SENARMONT QUARTZ WEDGES.

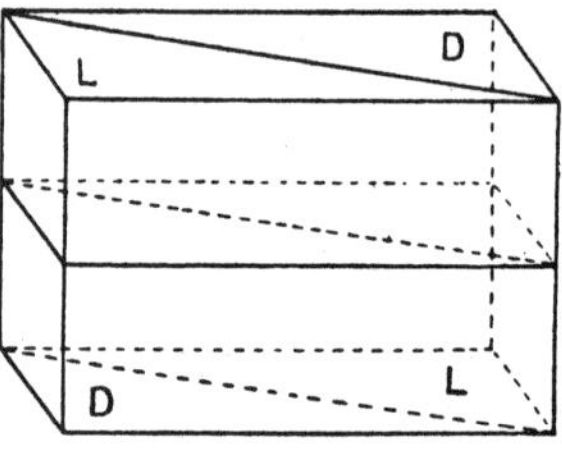

FIG. 69.—SENARMONT DOUBLE QUARTZ-WEDGE PLATE.

On introducing an optically-active substance between the polariser and the analyser, the dark band moves across the field of the analyser to the point which corresponds to an equal but opposite rotation in the quartz wedges. The analyser then has to be rotated through the required angle of rotation in order to restore the dark band to its original position.

In Senarmont's DOUBLE-FIELD POLARIMETER a double quartz-wedge plate (Fig. 69) is placed between the polariser and analyser. In the upper half of the plate the bases D and L of the dextro- and lævo-quartz wedges are on the right and left respectively. In the lower half of the plate the positions of the wedges are interchanged. In the absence of the optically-active medium the analyser is rotated until the vertical dark bands in the upper and lower fields are continuous; this is the zero position. On introducing an optically-active substance which is dextro-rotatory, the dark bands in the upper and lower fields move to the left and right respectively. The analyser is then rotated in a clockwise direction until the dark bands are again in the same vertical line.

Wild.—The most accurate single-field polarimeter is that of Wild,[1] whose method of measuring rotatory power depends on elimination of interference fringes produced by means of a Savart prism placed in front of the analyser.

A SAVART PRISM consists of two thin plates of quartz cut at an angle of 45° to the optic axis, and cemented together face to face so that their principal sections are at right angles as shown in Fig. 70. In Wild's polarimeter (Fig. 71) the analysing Nicol is fixed with its principal section horizontal, while the *polariser* P can be rotated by means of a toothed wheel G at the end of the shaft H. The rotation of the polariser is recorded by the movement of the circular scale C behind the fixed vernier V, and is read off with the help of the telescope T. The Savart prism S is fixed in the analyser mounting so that the bisector of the angle between the principal sections of its two halves is parallel to that of the analysing Nicol, i.e. horizontal. If the polariser is rotated until its principal section is perpendicular to that of the analyser, the field of the latter is crossed by a series of parallel horizontal interference fringes. The analyser mounting contains two fine cross-wires, so that when the polariser is in the above position the field of the analyser appears as shown in Fig. 72 (*a*). As the polariser is rotated from this position the interference fringes become fainter, and finally

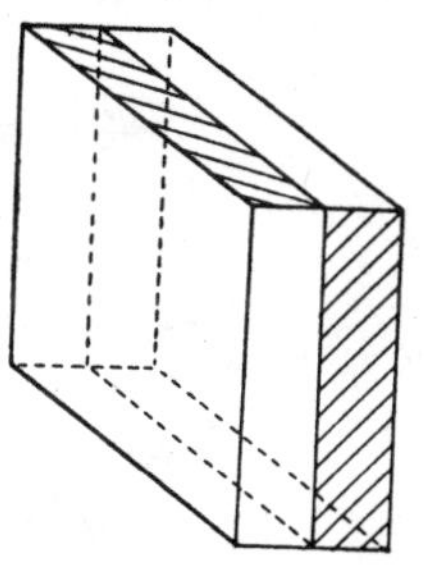

FIG. 70.—SAVART PRISM.

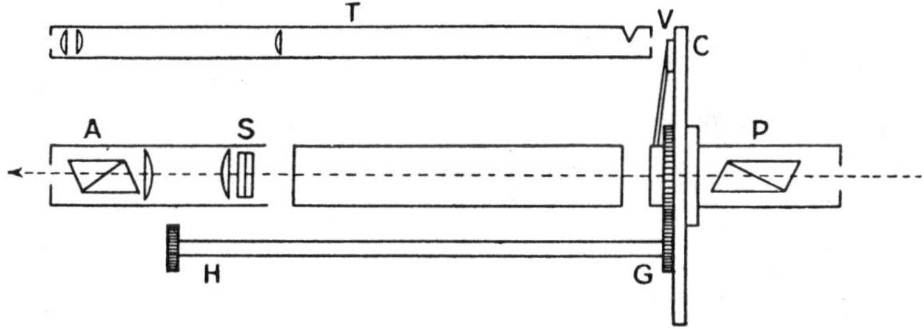

FIG. 71.—WILD'S POLARIMETER.

disappear in the centre of the field (Fig. 72 (*b*)), when the principal section of the polariser is parallel to the principal section of either half of the Savart prism. The polariser is adjusted until the ends of the remaining fringes are symmetrically placed with respect to the cross-wires. This is the zero position of the polariser.

On introducing an optically-active medium between the polariser and the Savart prism, the interference fringes once more cover the

[1] WILD, *A.C.P.*, 1864, [iv], **3**, 501 ; *Pogg. Ann.*, 1864, **122**, 626 ; *Ueber ein neues Polaristrobometer*, Berne, 1865.

field of the analyser. The polariser is rotated in the opposite direction to the rotation produced by the substance, until the appearance of the fringes is again as shown in Fig. 72 (*b*). The rotation of the medium is then equal but opposite in sign to the angle through which the polariser has been turned.

Rotations measured in different quadrants of the polariser circle will be found to differ appreciably. Van de Sande Bakhuyzen has

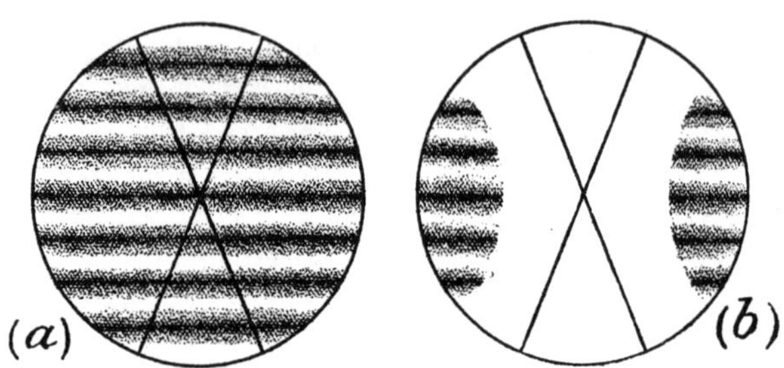

FIG. 72.—APPEARANCE OF THE FIELD OF VISION IN THE WILD POLARIMETER.

shown that these differences are due to defects in the construction of the Nicols and Savart prisms. These errors are corrected by taking the average of mean readings in each of the four quadrants. The experimental error involved in this method is usually of the order of less than 0·01°, i.e. the method compares favourably with the modern methods employing variable half-shadow polarisers.

Jellett.—In Jellett's double field polariser[1] two contiguous feebly illuminated fields are produced by means of a specially con-

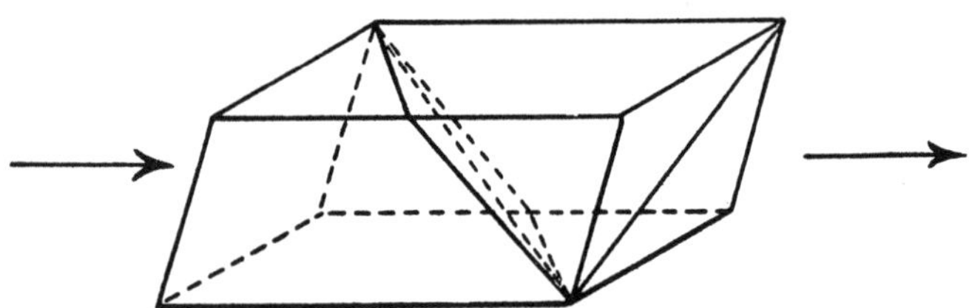

FIG. 73.—JELLETT DOUBLE-FIELD POLARISING PRISM.

structed Nicol prism. One-half of an ordinary Nicol prism is cut longitudinally as shown on the right in Fig. 73. A thin wedge is removed from one of the cut faces, and the three pieces are then cemented together with Canada balsam. The divided end is turned towards the analyser. The light emerging from the two halves of the polariser is then plane polarised in directions inclined at a small angle to one another. This is known as the SHADOW ANGLE. When

[1] JELLETT, *Brit. Assoc. Rep.*, 1860, **30**, 13.

the principal section of the analyser is perpendicular to the bisector of the shadow angle, the two fields of vision appear faintly but equally illuminated. The rotation is measured by setting the analyser so that the two fields are equal, with and without the optically-active medium.

Cornu.—In Cornu's double-field polarimeter [1] an ordinary Nicol prism is cut in half along a principal section through the shorter diagonals of the end faces. A thin wedge of angle $2\frac{1}{2}°$ is then removed from each of the new faces and the two halves are reunited. The principal sections of the two parts of the polariser are now inclined at an angle of 5°. If the principal section of the analyser is at right angles to the bisector of this shadow angle the two halves of the field appear equally bright as in Jellett's prism, and this is the setting used for taking measurements. A rotation of the analyser through the HALF-SHADOW ANGLE of $2\frac{1}{2}°$ on either side of this position extinguishes completely one or other of the fields.

Laurent.—If a plane polarised ray, in which the direction of vibration lies along ZP (Fig. 74) traverses a thin quartz plate, cut parallel to the optic axis ZX, that component ZO of the incident vibration which is parallel to the axis (and therefore corresponds to the extraordinary ray in quartz) suffers a phase retardation relative to the other component ZE corresponding to the ordinary ray in quartz. The magnitude of the retardation is proportional to the thickness of the plate. If the latter is of such a thickness as to produce a relative phase retardation of π the plate is termed a HALF-WAVE PLATE. In this case the retardation produces a component ZO′ opposite in direction to ZO. The component ZE being unchanged, the resultant effect of the plate is to change the direction of vibration of the incident ray from ZP to ZP′. If the direction of the incident vibration makes an angle θ with the optic axis of the plate, the plane of the vibration emerging from the plate is inclined at an angle of 2θ to its original direction.

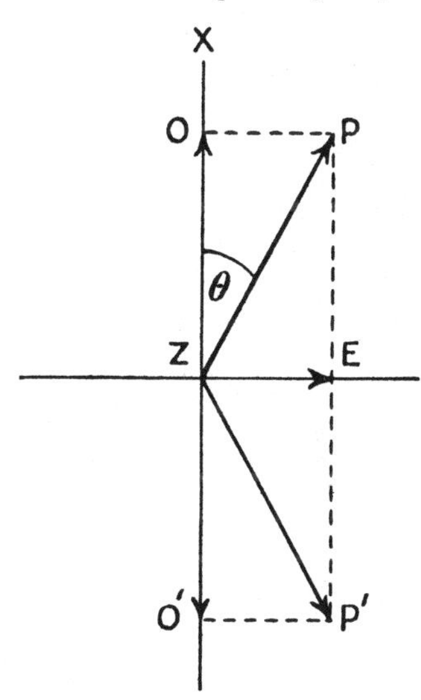

FIG. 74.—ILLUSTRATING THE ACTION OF THE HALF-WAVE PLATE IN THE LAURENT POLARIMETER.

This principle was employed by Laurent [2] in the first polarimeter to be constructed having a *variable half-shadow angle.*—A horizontal

[1] CORNU, *Bull. Soc. Chim.*, 1870, [ii], **14**, 140.
[2] LAURENT, *Journ. de Phys.*, 1874, [i], **3**, 183–186; *C.R.*, 1874, **78**, 349.

diagrammatic section of the optical parts of LAURENT'S POLARIMETER is shown in Fig. 75.

Immediately after the polarising prism P is placed a thin glass plate G to which is attached a quartz half-wave plate Q which covers only one-half of the field of the polariser. The diaphragm D serves to produce a circular field of vision which is divided vertically in two by the edge of the half-wave plate, on which the telescopic eyepiece T is focussed. If the principal section of the polariser is parallel to the optic axis in the half-wave plate the incident ray traverses the plate unchanged, and the two halves of the field of vision will appear equally bright or dark no matter what the position of the analyser A. If, however, they are inclined at a small angle θ to one another, the plane of polarisation of the light emerging from the half-wave plate is inclined at an angle of 2θ to that in the other half of the field. The half-shadow angle θ can be varied at will by altering the orientation of the polarising prism. The setting of the analyser with its graduated scale C is carried out as in the polarimeters of Jellett and Cornu

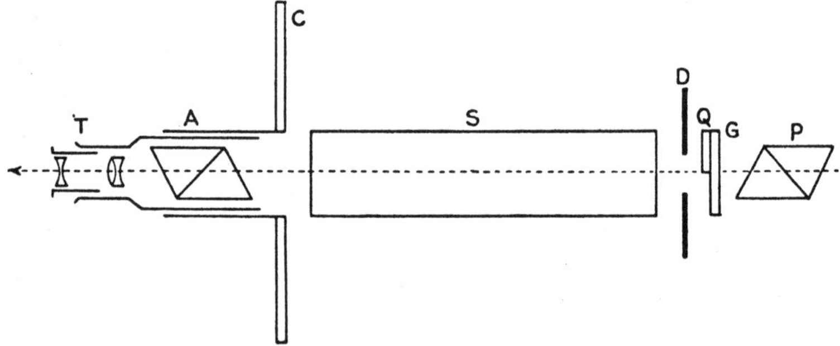

FIG. 75.—LAURENT POLARIMETER.

by matching the two fields, with and without the optically-active medium S.

The principal advantage of the Laurent and other similar instruments is that the magnitude of the half-shadow angle can be varied. The smaller this angle, the more accurately can the analyser be set for equal illumination of the two fields, since the difference between the two positions of the analyser which completely extinguish each field separately is equal to the shadow angle; i.e. with a small angle a very small rotation of the analyser produces a very marked inequality in the two fields. On the other hand, if the half-shadow angle is very small, both fields are very nearly completely extinguished together and the intensity is then insufficient for accurate matching. Thus with completely transparent media and a bright source of light a half-shadow angle of $2\frac{1}{2}°$ may give the most accurate results; but if the media are not completely transparent, it may be necessary to increase the half-shadow angle to 5° or even 10°, in order to get enough light through at the "extinction" position.

The principal disadvantage of the Laurent variable half-shadow is that it can only be used with monochromatic light of the particular wave-length for which the quartz plate produces an exact retardation of $\lambda/2$. For light of other wave-lengths other plates of different thicknesses are required.

Lippich.—The most satisfactory variable half-shadow polariser, from the point of view of convenience, simplicity and accuracy, is that devised by Lippich,[1] who placed a single large polarising prism immediately behind a second, smaller polarising prism covering only half the field (Fig. 76).

The principal section of the smaller prism is inclined at a small

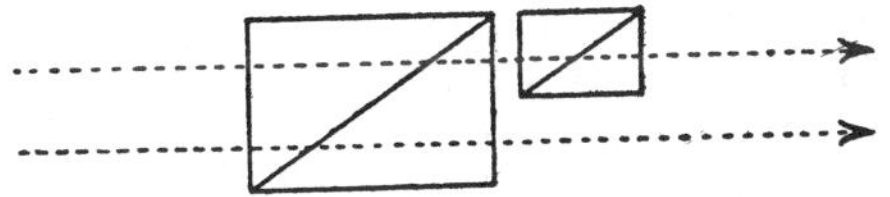

FIG. 76.—LIPPICH DOUBLE-FIELD POLARISER.

angle θ to that of the larger prism. The plane of polarisation of the light which traverses *both* prisms is therefore inclined at an angle θ, the shadow angle, to that of the light which traverses only the *one* large prism. The analyser is again set by matching the two halves of the field of vision ; but the principal section of the analyser is not then exactly perpendicular to the bisector of the shadow angle, since in one-half of the field the intensity of the light is reduced by absorption and reflection in the small prism and in the ratio $\cos\theta$ by the partial crossing of the large and small prisms. This effect is important for large values of the shadow angle, and also when measurements of rotation are made for wave-lengths at which the

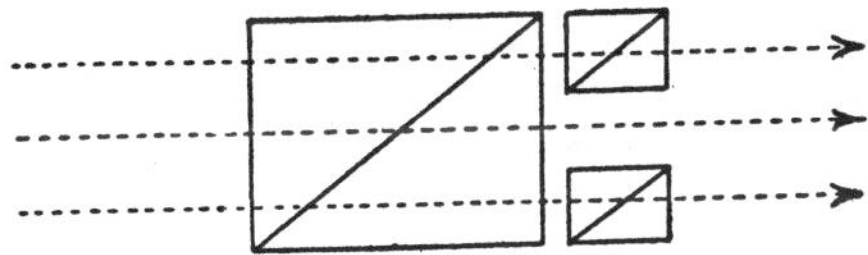

FIG. 77.—LIPPICH TRIPLE-FIELD POLARISER.

polarising prisms themselves begin to absorb an appreciable proportion of the light.

In order to secure a narrow dividing line between the two fields, care must be taken in setting the smaller prism, since otherwise the dividing line may be wide and black or the two fields may overlap and give rise to a bright streak across the centre.

Lippich also devised a TRIPLE-FIELD POLARISER (Fig. 77) in which the upper and lower thirds of the field of the main polarising prism are covered by two smaller prisms. These are set with their principal planes exactly parallel to one another, but the main prism is capable

[1] LIPPICH, *Sitz. Akad. Wiss., Wien*, 1885, **91**, 1059.

of rotation behind them. The telescopic eyepiece is focussed on the edges of the small prisms, so that the dividing lines between the three fields are sharply defined. The analyser is set by matching all three fields. This can only be done if the source of light and the eye are lined up symmetrically with reference to the two small prisms, whereas with a two-field polariser the two fields can be matched by balancing errors of alignment by an altered (and therefore erroneous) setting of the analyser.

Brace.—Brace[1] has constructed a very sensitive triple-field polariser by enclosing two " sensitive strips " of calcite, about 0·1 mm. thick, in the same cell of liquid (compare Fig. 54, p. 173) in such a way as to produce a half-shadow effect, with almost invisible lines of separation ; but the system is too fragile to be often used, except for measurements of high precision.

Other Double-Field Polarimeters.—Many authors have described double-field polarimeters in which half of the field of the polarising prism or analyser is covered by an optically-active substance producing a rotation of from 2° to 5°. Thus Glan[2] used a thin plate of quartz cut perpendicular to the optic axis, of a thickness of 0·1 or 0·2 mm., giving a shadow angle of 2° to 4°, whilst Poynting[3] suggested that the two halves of the field might be covered with robust quartz plates of similar sign but differing in thickness by a few tenths of a millimeter. A similar effect may be obtained by introducing a thin glass plate, covering half the field, into a cell filled with an optically-active liquid, e.g. limonene[4] or a sugar solution.[5] The shadow angle is then equal to the rotation produced by a column of the solution or liquid equal in length to the thickness of the glass plate. All these methods, however, have the disadvantage that the shadow angle, and therefore the zero-setting of the analyser, varies with the wave-length.

Polarimeter Tubes.—Liquids and solutions are enclosed in POLARIMETER TUBES with transparent end plates about 2 mm. thick. The tubes and end plates are usually made of glass, but for wave-lengths less than about 3500 A.U. the end plates must be made of carefully selected silica glass ; alternatively BALANCED QUARTZ PLATES of equal thickness but opposite sign may be used. Silica tubes have been used in photochemical experiments, and in order to avoid catalysis by the alkali of the glass, e.g. in experiments on mutarotation (p. 271). Brass tubes, sometimes lined with silver or gold, have also been used.

Polarimeter tubes are usually made in standard lengths of 1, 2, 4 and 6 dcm. The ends of the tubes are ground until the length is correct to ± 0·05 mm. If the rotatory power of a substance is very

[1] BRACE, *Phil. Mag.*, 1903, [vi], **5**, 161.
[2] GLAN, *Wied. Ann.*, 1891, **43**, 441.
[3] POYNTING, *Phil. Mag.*, 1880, [v], **10**, 18.
[4] DARMOIS, *A.C.P.*, 1911, [viii], **22**, 247–281 and 495–590.
[5] RAYLEIGH, *Phil. Trans.*, 1885, **176**, 343.

small and sufficient of it is available, long tubes should be used in order to increase the accuracy of the readings.

In measuring the rotatory dispersion of a medium in a region of absorption, it is necessary to reduce progressively either the concentration of the solution, or the length of column traversed by the light, as the absorption becomes more intense. The latter is the wiser course, because the specific rotation of many substances varies considerably with the dilution. For this purpose, tubes 1 cm. and even 1 mm. in length have been used.

The diameter of the tubes should be larger than the diameter of the beam of light traversing the polarimeter, in order to prevent the possibility of some light being reflected from the inside walls of the tube, thereby causing unequal illumination of the field of the analyser. Tubes with an internal diameter of 1 or 2 mm. only have, however, been constructed for use with substances of which very limited amounts are available.

The end plates are kept in position by metal caps with circular apertures for the passage of the light. These caps are screwed to threaded collars attached to the outside of the tube. Between the caps and the end plates is a washer, of rubber or cork, to hold the end plate firmly but gently in position. Care must be exercised in screwing on the metal caps, otherwise the end plates become strained and doubly refracting. Even slight strain is sufficient to produce an appreciable alteration in the zero of the analyser when observed through the empty tube. For the same reason cork is preferable to rubber as a material for the washers when solvents such as benzene are being used, since a slight leakage of the solvent may cause a rubber washer to swell and tighten up the end plate, thereby introducing an error into the readings.

The possibility of setting up strain in balanced quartz end plates is even greater. For this reason the most satisfactory method of keeping them in position is by means of cement. If a suitable cement is chosen the plates can be easily removed again. The end caps need then only be put on loosely for protection. Polarimeter tubes with the end plates permanently fused on have been constructed, but the plates, when so treated, are usually inferior from an optical point of view. Moreover, the tubes are difficult to clean and dry.

The earlier type of polarimeter tube had to be filled from one end, the end plate then being slid into position. The more modern forms have a side tube for filling (Fig. 50, p. 166). This not only allows the liquid to expand without straining the end plates, but it also admits of determining the temperature of the liquid inside the tube. The side tube may be closed with a cork or ground glass stopper or by a small disc of glass held in position by means of a clamp attached to the side tube.

Temperature Control.—Since rotatory power is influenced by changes of temperature, polarimeter tubes for accurate work are enclosed in a metal or glass jacket through which water at a constant temperature, e.g. 20°, may be circulated from a thermostat (Fig. 50, p. 166). The liquid inside the tube can be stirred most conveniently by allowing a bubble of air from the side tube to pass to and fro along the tube. A simple electrically heated and regulated thermostat will maintain a constant temperature within 0·01° C. By means of a suitably arranged pump, water at this temperature can be drawn straight from the main body of the thermostat through the jacketed polarimeter tube. In measuring the rotatory power of quartz,[1] water from a gas-heated and regulated thermostat was drawn by means of a pump through the jacket of the quartz column and returned to the thermostat; the temperature of the return flow showed a fall of 0·01° of which only 10 per cent. occurred before the water reached the polarimeter tube.

In order to maintain a working temperature considerably above or below the atmospheric temperature, Landolt,[2] Bruhat[3] and Patterson[4] have enclosed the whole polarimeter tube in a small thermostat between the polariser and analyser.

When working at low temperatures care must be taken to prevent moisture from condensing on the outer surfaces of the end plates. Conversely, when working with vapours, a high temperature is conveniently maintained by surrounding the polarimeter tube with a metal cylinder electrically heated by means of an insulated coil of wire wrapped on the outside; but in this case the vapour tends to condense on the inner surfaces of the end plates. This may be prevented by allowing the heating jacket to overlap the ends of the tube, or by using double end plates fused on to the tube, the space between the plates being evacuated.[5]

[1] LOWRY, *Phil. Trans.*, 1912, A. **212**, 261.
[2] LANDOLT, *Ber.*, 1895, **28**, 3102.
[3] BRUHAT, *Ann. de Phys.*, 1915, **3**, 232 and 417.
[4] PATTERSON, *J.*, 1927, 1717.
[5] LOWRY and GORE, *P.R.S.*, 1932, A. **135**, 13–22.

CHAPTER XV.

SACCHARIMETERS.

General Principle.—Polarimeters specially constructed for determining the quantity of sugar in a solution by measuring its rotatory power, are termed saccharimeters. The working of these instruments depends on the validity of two laws established by Biot,[1] that the rotatory power of a solution of cane sugar is proportional to (*a*) the length of the column, and (*b*) the concentration of the solution. The latter proportionality is not exactly correct, the specific rotation of cane sugar solution decreasing slightly as the concentration increases ; this necessitates the introduction of a small correction in making precise measurements.

It would obviously be possible to determine the percentage of sucrose in a solution by measuring its rotatory power on any of the polarimeters previously described, and comparing the rotation with that of a standard solution. To facilitate the measurement, however, special saccharimeters are used. These differ from ordinary polarimeters in only two respects : (i) Instead of rotating either the polariser or the analyser, the rotation of the sugar solution is measured with the help of a QUARTZ WEDGE COMPENSATOR of opposite sign. (ii) The compensator carries a longitudinal scale on which the concentration of the sugar solution can be read off directly, when the compensator has been set to give a zero total rotation.

The Compensator.—The simplest form of quartz wedge compensator, devised in 1845 by Soleil,[2] consists of a thin plate of quartz cut perpendicular to the optic axis and divided along a diagonal into two thin right-angled quartz wedges as shown in Fig. 78 (*a*). The two wedges can be slid one over the other as shown in Fig. 78 (*b*) so that the rotation produced in a plane polarised ray XX′ passing normally through the compensator may be varied by altering the thickness of quartz traversed. In the absence of a sugar solution the compensator must be capable of giving a zero rotation. This can be effected in two ways : (i) The compensator includes a rectangular quartz plate of opposite rotation to that of the wedges and of such a thickness that, when the wedges are in an intermediate position (as in Fig. 79 *a*), it counterbalances their rotation ; by moving either

[1] BIOT, *Mém. Acad. Sci.*, 1835, **13**, 39–175.

[2] SOLEIL, *C.R.*, 1845, **21**, 426 ; 1847, **24**, 973 ; 1848, **26**, 162.

of the wedges one way or the other the compensator may be made to produce dextro- or lævo-rotation. (ii) An alternative method is to use two exactly equal wedges, one of dextro- and the other of lævo-quartz, arranged as in Fig. 79 (*b*).

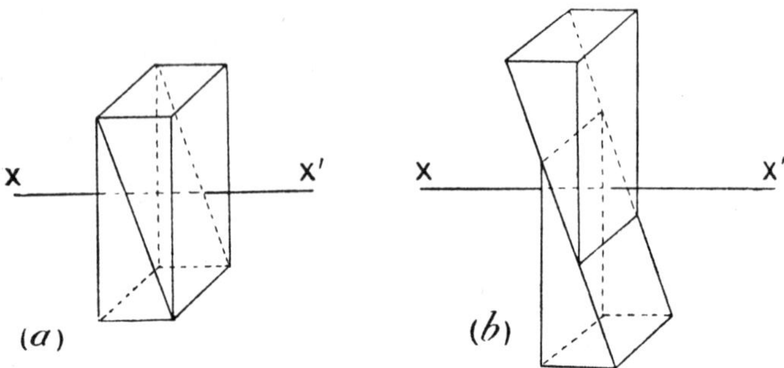

FIG. 78.—QUARTZ WEDGE COMPENSATOR.

The angle of the wedges is usually of the order of 3° and one or both of them may be movable. Very great care must be exercised in selecting suitable quartz, and great precision is needed in constructing, mounting and aligning the wedges, and in ensuring that they move in a direction exactly perpendicular to the optical axis of the instrument.

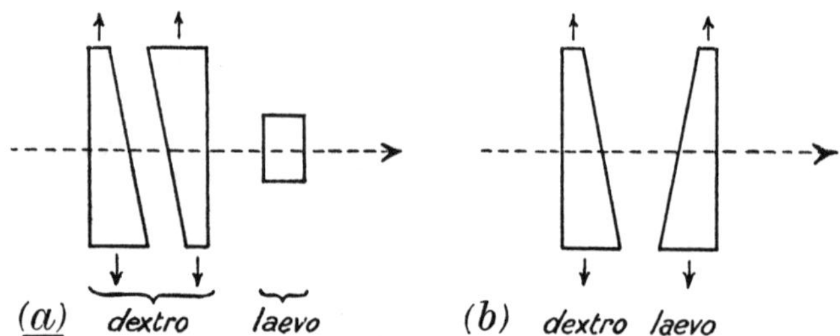

FIG. 79.—QUARTZ WEDGE COMPENSATORS.

(The short arrows indicate the directions in which the wedges may be moved.)

The Light Source.—By a fortunate accident the rotatory dispersion of quartz in the visible spectrum happens to be almost identical with that of cane sugar. For this reason, if the rotation of a sugar solution for one wave-length is exactly compensated by the quartz wedges, it is also compensated for other wave-lengths. It is therefore possible to work with a source of white light, which is usually steadier and more intense than a monochromatic source, and to avoid the use of a monochromator.

When strong solutions ($>$ 40 per cent.) are used, the two halves of the field of a double field polariser do not show quite the same tint, one appearing very slightly reddish and the other very slightly bluish. This effect is due to inequality in the rotatory dispersions of sugar and quartz, especially in the blue region of the spectrum and can be mitigated by inserting a bichromate filter to cut off the blue radiations. A bichromate filter also serves to eliminate the increased proportion of blue radiations in modern high-temperature electric lamps, as compared with the older gas and oil lamps, and thus reduces all sources of white light to a rough equality by absorbing all the components of shorter wave-length than the green. The standard bichromate filter for use with saccharimeters is a cell 1·5 cm. thick filled with a 6 per cent. solution of potassium bichromate.

The Saccharimeter Scale.—The movement of the quartz wedge compensator necessary to counterbalance exactly the rotation produced by a sugar solution is recorded by means of a vernier which slides over a linear scale. In order to be able to read the concentration of the solution directly, this saccharimeter scale is graduated in terms of a standard sugar solution. The zero point on the scale is the position of the vernier in the absence of a sugar solution when the rotation produced by the compensator is nil. The 100 point on the scale is the vernier reading when the standard sugar solution is placed in position.

At the meeting of the International Sugar Commission in Paris in 1900 the standard sugar solution was defined as 26 grams of sucrose dissolved in water and made up to a volume of 100 c.c. at 20° C. This solution was to be contained in a 2 dcm. tube and the graduation of the scale was to be made at 20° C. The space between the zero and the 100 point is divided into 100 equal parts, each of which is termed a degree sugar, 1° sugar or 1° S. The vernier is constructed to read to 0·1° sugar, which corresponds to a rotation for sodium light of 0·034657°. The accuracy of a saccharimeter is therefore comparable with that of the most up-to-date polarimeter. To determine the concentration of an unknown sugar solution the vernier reading is multiplied by 0·26. In accurate work it is necessary, of course, to make corrections for variations of specific rotation with temperature, concentration, etc.

Until the introduction of the International Sugar Scale in 1900 there was some confusion because French and German workers employed different methods for graduating the saccharimeter scale. Thus, on the French scale the 100° point was fixed by the rotation, for sodium light, of a control plate of dextro-quartz 1 mm. thick. The 100° point on the German scale,[1] on the other hand, was fixed by the rotation of a standard sugar solution having a density of 1·100 at 17·5° C.; this solution contained 26·048 grams of sugar in 100 c.c. at 17·5° C.

Saccharimeter scales are usually graduated to read from ± 5° S. to ∓ 105° S., the actual length of the scale being of the order of 3 cm.

[1] VENTZKE, *Journ. prakt. Chem.*, 1842, **25**, 65–84; 1843, **28**, 101–116.

The present-day method of calibrating saccharimeter scales is by the use of quartz control-plates previously standardised against a standard sugar solution.

Soleil - Duboscq - Ventzke - Scheibler Saccharimeter.—The first saccharimeter was devised and constructed by Soleil [1] in 1848. It was later modified and improved by Duboscq,[2] Ventzke [3] and Scheibler.[4] It consists essentially of a Soleil polarimeter with a bi-quartz plate for the production of a double field, in which is incorporated a quartz wedge compensator with saccharimeter scale. The principal optical parts of the instrument are shown in Fig. 80.

White light from a source L traverses, in turn, the regulator R consisting of a Nicol prism and a thin quartz plate cut perpendicular to the optic axis, the polariser P, the Soleil bi-quartz plate Q, the solution tube S, the quartz wedge compensator C, the analyser A, and the telescope T. The polariser and analyser are fixed with their principal sections parallel. Ignoring for a moment the regulator R, the compensator is adjusted, in the absence of the sugar solution, until the two halves of the bi-quartz plate are equally illuminated with the sensitive transition tint already described (p. 182). In this

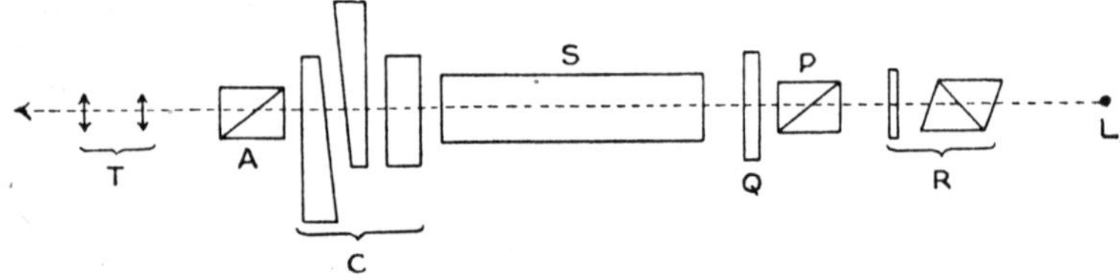

FIG. 80.—SOLEIL–DUBOSCQ–VENTZKE–SCHEIBLER SACCHARIMETER.

position the vernier of the compensator should read zero on the saccharimeter scale; if not, the scale itself can be adjusted to zero. The sugar solution contained in a 2 dcm. tube is then introduced, causing the two fields of the bi-quartz to appear unequal. The compensator is again adjusted until the two halves match; from the reading on the saccharimeter scale the concentration of sucrose is deduced.

If the sugar solution is at all coloured, or if a different source of white light is used, which gives the coloured rays in a somewhat different proportion, the two fields when matched may not have the sensitive transition tint. The object of the regulator R is to restore this sensitive tint. By rotating the Nicol prism of the regulator relative to the polariser, some of the coloured rays produced by rotatory dispersion in the quartz plate are extinguished or reduced in intensity in traversing the polariser. Thus the sensitive tint may be restored simply by adjusting the Nicol prism of the regulator.

[1] SOLEIL, *C.R.*, 1848, **26**, 162.
[2] DUBOSCQ and SOLEIL, *C.R.*, 1850, **31**, 248.
[3] VENTZKE, *Journ. prakt. Chem.*, 1842, **25**, 65; 1843, **28**, 101.
[4] SCHEIBLER, *Zeit. Ver. Rübenzuckerind.*, 1870, 609.

The correction for the variation of rotatory power with concentration has been calculated, and tables drawn up in which the true concentration of the sugar solution can be looked up for each reading of the saccharimeter scale.

Soleil-Duboscq Saccharimeter.—The original saccharimeter[1] devised by Soleil and improved by Duboscq differed from the one described above only in the following particulars. The regulator for restoring the transition tint was mounted in the telescope of the analyser instead of in front of the polariser. Both quartz wedges were also capable of motion, whereas in the latter form of instrument only one wedge was movable. The scale of the Soleil and Duboscq instrument was also graduated in the French way using a quartz control plate, whereas in the other instrument the German scale was used.

Saccharimeters with a Laurent Variable Half-Shadow Angle.—The Soleil saccharimeter is equivalent in principle to a polarimeter with a fixed half-shadow angle and therefore has the same disadvantages as the latter instrument in that its sensitivity cannot be varied. Many raw sugar solutions which have to be tested are highly coloured, so that, unless the intensity of the light can be increased by enlarging the half-shadow angle, it is impossible to obtain readings. For this reason saccharimeters were constructed on the same principle as the Laurent variable half-shadow polarimeter.

The double field is then produced by a quartz half-wave plate (p. 187). In saccharimeters of the Laurent type, as in all saccharimeters employing a true half-shadow polariser, the principal section of the analyser is set perpendicular to the bisector of the half-shadow angle, so that the two fields match in the absence of a sugar solution and with the compensator set to zero. If the half-shadow angle is altered the analyser must still be in the same position relative to the polariser. In the Laurent instrument this needs no special adjustment, for if the half-shadow is varied by rotating the polariser the direction of the bisector of the shadow angle remains fixed and no movement of the analyser is necessary.

This is an advantage over the Lippich type of saccharimeter described below. The Laurent instrument, however, has the drawback that the part of the field of the polariser which is covered by the half-wave plate always appears coloured when white light is used, owing to the difference in phase retardation produced for different wave-lengths. In no position, therefore, can the two fields be matched exactly.

Saccharimeters with a Lippich Variable Half-Shadow Angle.—All modern saccharimeters for accurate work are constructed on the same principle as the Lippich two-field polarimeter. The double field is produced, as in the latter instrument, by covering one-half of the field of the main polarising prism with a similar but smaller prism. In this type of saccharimeter the half-shadow angle

[1] Duboscq and Soleil, *C.R.*, 1850, **31**, 248.

is usually altered by rotating the larger polarising prism; this causes a rotation of the bisector of the shadow angle by an amount equal to half the angle turned through by the larger prism. It is therefore necessary to be able to rotate the analyser also, in order to keep its principal section perpendicular to this bisector.

Saccharimeters have been constructed in which this adjustment is made automatically by having the polariser and analyser geared together in such a way that, when the former is turned through an angle α, the latter also turns through an angle $\alpha/2$. It has, however already been pointed out on page 189 that, for equal matching of the two fields, the analyser must be set slightly inclined to the perpendicular position referred to above, on account of the decrease in intensity of the light which traverses both the larger and smaller prisms of the Lippich arrangement. This small deviation from the exact perpendicular orientation varies with the value of the half-shadow angle, and Bates[1] has shown that it is very nearly a linear

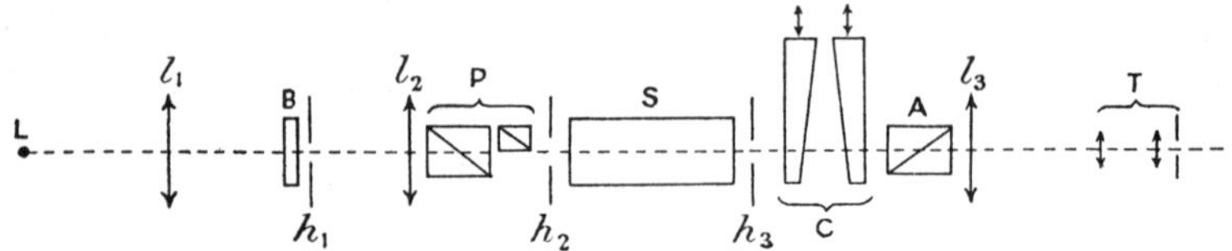

FIG. 81.—SACCHARIMETER OF THE LIPPICH DOUBLE-FIELD TYPE.

L, source of white light.

l_1, l_2 and l_3, lenses. (l_1 focusses an image of the light source on the diaphragm h_1, which is placed at the focus of l_2 so that parallel light traverses the instrument.)

h_1, h_2 and h_3, diaphragms to limit the diameter of the light beam.

B, bichromate cell.

P, Lippich two-field polariser.

S, polarimeter tube containing sugar solution.

C, quartz wedge compensator.

A, analyser.

T, telescopic eyepiece focussed on the polariser.

variation. It may therefore be taken into account by slightly altering the ratio of the gears operating the polariser and analyser. Saccharimeters conforming to these requirements have been constructed and perfected to such a degree that no alteration in the matching of the two fields can be detected even when the half-shadow angle is increased from 2·5° to 15°.

The accuracy of instruments of this type is such that, with numerous other refinements in design, readings may be obtained which are correct to 0·01° sugar, corresponding to an angular rotation of 0·003° approximately.

The complete optical system of a saccharimeter of the Lippich type is represented diagrammatically in Fig. 81.

Saccharimetric problems, for example the determination of cane sugar in the presence of reducing sugars, are beyond the scope of this book.

[1] BATES, *Bull. Bur. Stand.*, 1908, **4**, 461; 1908, **5**, 193.

CHAPTER XVI.

MEASUREMENT OF ROTATORY DISPERSION.

(*a*) In the Visible Spectrum.

Fizeau and Foucault.—Although Biot's researches on optical rotatory power always included measurements of rotatory dispersion, his measurements were liable to large errors, since the only wave-lengths which were then available were Newton's mean values for the various components of white light. The first measurements of rotatory power for a large series of accurately known wave-lengths were made by a method which was described by Fizeau and Foucault in 1845.[1] In this method (which was used by Broch [2] in 1846, and was used subsequently with minor refinements by Wiedemann [3] and Arndtsen [4]) sunlight is reflected horizontally by means of a heliostat, and passes successively through a Nicol polariser, the optically-active medium, a Nicol analyser and a spectroscope. The zero reading, in the absence of the substance under investigation, is obtained by rotating the analyser until the spectrum in the field of the spectroscope is as dark as possible. On introducing the substance, the solar spectrum (including the Fraunhofer lines) once more makes its appearance. On rotating the analyser a vertical dark band, corresponding to the particular ray which is extinguished by the analyser, appears and moves across the spectrum. A vertical cross-wire in the eyepiece of the spectroscope is made to coincide with one of the Fraunhofer lines, and the analyser is rotated until the dark band moves into alignment with the cross-wire. The angle through which the analyser has been rotated is the rotation of the substance for the Fraunhofer line, whose wave-length is known. This is repeated for other Fraunhofer lines throughout the visible solar spectrum.

If the rotatory dispersion of the substance is large, more than one dark band may be visible at the same time. In this case the rotations for the rays whose positions in the spectrum are occupied by two adjacent dark bands differ by 180°.

[1] Fizeau and Foucault, *C.R.*, 1845, **21**, 1155.

[2] Broch, Dove's *Repert. der Physik*, 1846, **7**, 113; *A.C.P.*, 1852, [iii], **34**, 119.

[3] Wiedemann, *Pogg. Ann.*, 1851, **82**, 215–232.

[4] Arndtsen, *A.C.P.*, 1858, [iii], **54**, 403.

Fig. 82 shows the arrangement of the optical parts in a polarimeter of this type as used by Stefan[1] and by von Lang.[2] The light source L is placed at the focus of the lens l_1 so that a parallel beam of light, limited by the diaphragms h_1 and h_2, traverses the polariser P, the optically-active medium S and the analyser A which is capable of rotation. The lens l_2 focusses an image of the diaphragm h_1 on the slit h_3 of the spectroscope. The lenses l_3 and l_4 focus the spectrum produced by the prism p on the plane R. The spectrum is examined by means of the eyepiece l_5.

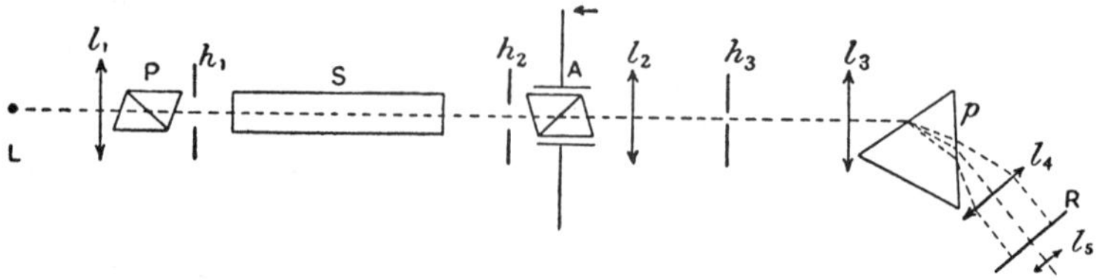

FIG. 82.—POLARIMETER OF THE TYPE USED BY STEFAN AND BY VON LANG.

Stefan and von Lang.—Stefan[3] measured the rotatory dispersion of a column of quartz by a modification of the method of Fizeau and Foucault, in which white light from an artificial source was used instead of sunlight. The goniometer scale of the spectroscope was calibrated independently with the help of the Fraunhofer lines of the solar spectrum. It was then possible to deduce the wave-lengths corresponding with the centres of a series of dark bands which appeared in the continuous spectrum of the artificial source when a column of quartz was introduced between the polariser and analyser. These dark bands corresponded with differences of 180° in the observed rotation of the column for the wave-lengths in question.

von Lang[4] used a similar method, but calibrated the spectroscope by introducing a salt of lithium, sodium or thallium into a Bunsen flame and bringing the vertical cross-wire of the eyepiece into coincidence with the red lithium, yellow sodium or green thallium line. The Bunsen burner was then replaced by a source of white light and the analyser was rotated until the dark band in the continuous spectrum coincided with the cross-wire in one of these positions. In this way the rotation was determined for each of the preceding lines, as well as for the red, green-blue and blue-violet lines from a hydrogen tube.

Neither of these methods gives very accurate results, since the dark extinction bands are rather broad and diffuse at the edges; they cannot therefore be set to coincide precisely with the cross-wire,

[1] STEFAN, *Sitz. Akad. Wiss., Wien,* 1864, **50,** 88.
[2] VON LANG, *Pogg. Ann.*, 1875, **156,** 422.
[3] STEFAN, *Sitz. Akad. Wiss., Wien,* 1864, **50,** 88.
[4] VON LANG, *Pogg. Ann.*, 1875, **156,** 422.

especially when the rotation is small. With moderate rotations the experimental error is of the order of 1°, but with quartz von Lang obtained results which were only liable to an error of 0·2°.

Wiedemann.—The first double field polariser for measuring rotatory dispersion was constructed by Wiedemann [1] and used later by Fleischl.[2] In this instrument a Soleil bi-quartz plate of variable thickness was placed, with its dividing line horizontal, immediately after the polarising prism. The thickness of the bi-quartz plate could be adjusted by constructing it of thin quartz wedges placed face to face, so that they could slide one over the other in the direction of the diameter of the plate. Fig. 83 shows a diagrammatic representation of such a bi-quartz plate of variable thickness. This produces two spectra, one above the other, in the field of the spectroscope. On rotating the analyser a dark band appears in each of the two spectra ; these bands move together until they become continuous in that part of the spectrum for which the rays have been rotated through ± 90° by the two halves of the bi-quartz plate. On introducing an optically-active medium, the two bands separate and the analyser is rotated until they are again continuous at exactly the same point in the spectrum as before. The movement of the analyser represents the rotation for the point in the spectrum occupied by the centre of the dark bands when in alignment. Similar readings can be obtained for other wave-lengths by varying the thickness of the bi-quartz plate, thereby shifting the dark bands along the spectrum.

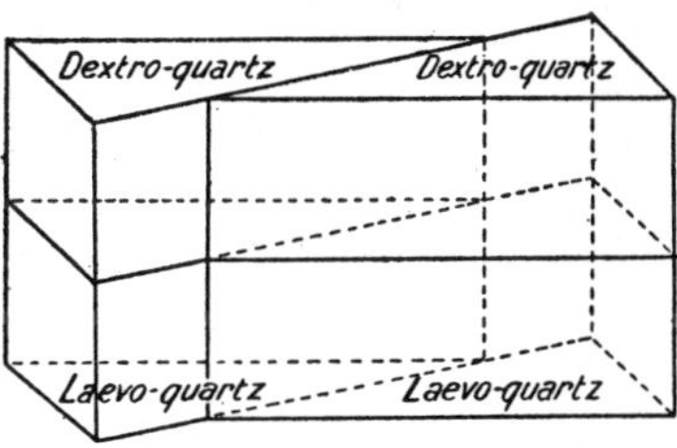

FIG. 83.—BI-QUARTZ PLATE OF VARIABLE THICKNESS.

Lippich.[3]—When a Lippich double-field polariser is used with a continuous spectrum there is, in the absence of an optically-active medium, a certain position of the analyser for which the two spectra are equally but feebly illuminated throughout their length. This is the zero position. On introducing the optically-active medium and turning the analyser, a dark extinction band appears in each spectrum, but the two bands are slightly displaced relatively to one another by an amount depending on the value of the shadow angle. At a point midway between the two bands the two spectra are equally illuminated. As the analyser is rotated the two bands move across the spectrum together, in the same relative positions. The reading of the analyser circle gives the rotation for the wave-length at which the two slightly-separated extinction bands are of the same intensity.

[1] WIEDEMANN, *Lehre von der Elektrizität*, 1883, **3**, 914.
[2] FLEISCHL, *Repert. Physik*, 1885, **21**, 323.
[3] LIPPICH, *Sitz. Akad. Wiss., Wien*, 1885, **91**, 1070.

If the rotation is small the dark bands are very diffuse. In such cases Lippich passed the light first through the spectroscope, the eyepiece of which was replaced by a second slit, which cut off a narrow strip of the spectrum to serve as a source of light for the polarimeter. The analyser was rotated until the two halves of the strip were equally bright. In this arrangement the darkest parts of the extinction-bands in the upper and lower spectra are hidden by the sides of the slit and only the equally illuminated strip is seen as in Fig. 84.

Glan [1] used the same method as Lippich, except that the spectroscope remained in its original position beyond the analyser. The field of vision was narrowed down by means of a vertical slit placed in front of the eyepiece. This method has the advantage that only the eyepiece of the spectroscope and the slit attached to it need be moved in order to bring a new part of the spectrum into view, whereas in Lippich's method a fresh alignment of the light source and spectroscope was necessary for each new wave-length.

In both methods the mean wave-length of the light may be found

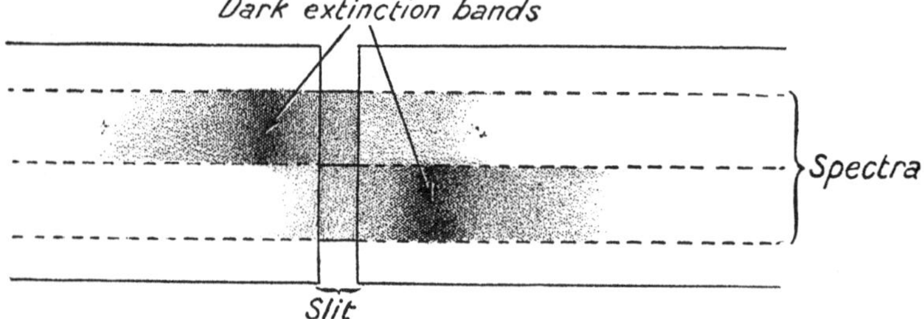

FIG. 84.—ILLUSTRATING LIPPICH'S METHOD OF MEASURING ROTATORY DISPERSION.

by calibrating the spectroscope in terms of the Fraunhofer lines of the solar spectrum, or of the monochromatic radiations emitted by the flames of lithium, sodium, thallium, etc. Alternatively Bencowitz [2] in 1928 proposed to calibrate the monochromator-spectroscope of Lippich's method by setting it so that the polarimeter gave a correct reading for the rotatory power of quartz for some well-known wave-length, e.g. Hg 5461. The wave-length thus recorded was not the true OPTICAL MEAN WAVE–LENGTH or OPTICAL MASS CENTRE of the patch of continuous spectrum transmitted by the monochromator, but the POLARIMETRIC OPTICAL MEAN, i.e. a weighted mean depending on the rotatory dispersion of the medium. It was claimed that by this method the convenience of a continuous source of light, as compared with a flickering arc, could be combined with the precision of readings with monochromatic light. In particular, it was found that twenty-four rotations determined in this way differed from those determined with the green line of a mercury arc by an average of

[1] GLAN, *Wied. Ann.*, 1891, **43**, 441.
[2] BENCOWITZ, *Journ. Phys. Chem.*, 1928, **32**, 1163.

0·01° and a maximum of 0·03°. The principle of the method is sound when applied to substances whose rotatory dispersion is similar to that of quartz, but it is obviously invalid when applied to media which exhibit anomalous rotatory dispersion, since in such cases the "polarimetric optical mean" might lie on the opposite side of the ordinary optical mass centre of the transmitted light.

Landolt.—In order to avoid the use of a monochromator Landolt[1] employed coloured filters placed between the source of light and the polarimeter in order to produce approximate homogeneity in the red, green, light blue and dark blue parts of the spectrum. It is, however, impossible to find filters which will transmit only narrow strips of a continuous spectrum, and the method breaks down entirely when the rotation of the medium exceeds a few degrees, since it is then impossible to extinguish the large sections of spectrum transmitted by the filters.

Perkin.—W. H. Perkin, senr.,[2] purified the light from a sodium, lithium or thallium flame by mounting a direct-vision spectroscope in front of the eyepiece of the polarimeter. This simple device is also sufficient to resolve the lines of the mercury arc (p. 210), using by preference a *dense* prism to separate the yellow and green lines as widely as possible, and a *light* prism in order to separate the violet line from the green and yellow lines without undue absorption by the glass.

Lowry.—The SPECTRO-POLARIMETER[3] used to measure the rotatory dispersion of quartz and of many organic compounds is shown in Fig. 85. Its principal features are (i) a CONSTANT-DEVIATION SPECTROSCOPE, specially adapted for use as a monochromator, and supplemented by a DIRECT-VISION SPECTROSCOPE to eliminate stray light, (ii) the invariable use of MONOCHROMATIC OR MULTICHROMATIC LIGHT instead of a continuous spectrum.

Multichromatic light from the source L is condensed by the lens l_1 on the slit h_1 of the collimator of a *monochromator* with a constant deviation prism p. The telescope of the monochromator is not furnished with a second slit as in earlier instruments, but, since the focal length of l_3 (22″) is twice as great as that of l_2 (11″), a series of magnified images of the slit h_1 is formed in the plane of a slit h_2, in close proximity to the triple field of the polariser P. By turning the drum of the constant-deviation spectroscope, the image of any selected spectral line can be projected through the slit h_2, and thus brought into the field of view of the polarimeter.

The *polariser* P is formed from one large and two small Glazebrook prisms, arranged so as to give a variable half-shadow angle. The *analyser* A is a single Glazebrook prism mounted in a circle graduated to 0·01° and reading to 0·002° with the help of two illuminated verniers. The setting of the analyser is made with a coarse and fine

[1] LANDOLT, *Ber.*, 1894, **27**, 2872.
[2] PERKIN, J., 1906, **89**, 616.
[3] LOWRY, *P.R.S.*, 1908, A. **81**, 472 ; *Phil. Trans.*, 1912, A. **212**, 261.

adjustment. The telescope l_4 and l_5 of the analyser is fitted with a Perkin eyepiece to throw out stray light, and the dispersive power of the optical system can be increased by the addition of a Rutherford prism.

The width of the slit h_1 is adjusted so that the line selected from the multichromatic spectrum, e.g. of a metallic arc, is not overlapped by adjacent lines; the slit h_2 is then adjusted so that the adjacent lines are excluded from the field of view. In many cases the aperture of the triple field can be filled almost completely by a single spectral line, but narrower slits must be used if it is desired to separate a pair of contiguous lines, such as the yellow mercury doublet (5769 and 5790 A.U.). As a limiting case, the rotatory power of a long column of quartz for the two lines of the yellow sodium doublet (5890 and 5896) was read by using a grating and two extra dense 30° prisms (refractive index 1·92) to form the constant-deviation system,

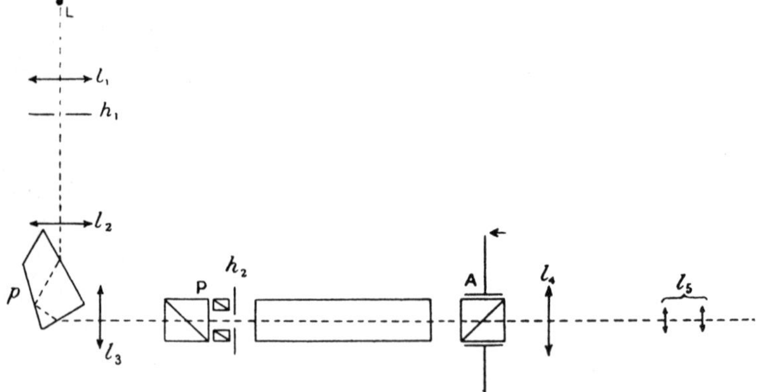

FIG. 85.—SPECTRO-POLARIMETER (Lowry, 1908).

and supplementing the direct-vision prism on the eyepiece by a Rutherford prism.

Lowry and Coode-Adams.—The number of lines in the red region of the visible spectrum is very limited, and they are not easy to read visually on account of the low sensitivity of the eye to red light, and the continuous radiation which always accompanies them. For this reason Lowry and Coode-Adams [1] devised a photographic method for measuring the rotatory dispersion of a column of quartz in the red, yellow and green regions. The quartz was mounted in a triple-field polarimeter set to extinction in series with a spectrograph (Fig. 95, p. 218). The continuous spectrum from a carbon arc was then photographed on a panchromatic plate specially sensitised for red radiations. The plate was crossed by vertical dark extinction bands, each corresponding to a rotation of 180° more than the adjacent band on the side of longer wave-length. The dark

[1] LOWRY and COODE-ADAMS, *Phil. Trans.*, 1927, A. **226**, 391.

bands in the central field (Fig. 86) were slightly displaced, relatively to those in the two outer fields, on account of the half-shadow angle of the triple-field polariser; the dotted lines show the "extinction-positions" at which the three fields are equally but feebly illuminated. Since the polariser and analyser were crossed, each extinction represents a rotation of $n \times 180°$, the value of n being deduced (as described on p. 181) from observations with a short column of quartz. A uniform scale of frequencies was provided by photographing (with the same spectrograph) the interference bands produced by the air-film of an ETALON. The wave-length of the extinctions were then deduced by reading off with a micrometer the positions of the dark bands on both plates relatively to a line spectrum, which was superposed on the continuous background in order to provide a series of reference lines of known wave-length.

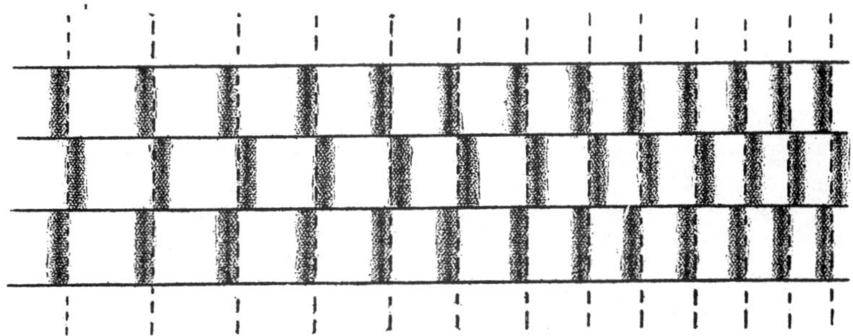

FIG. 86.—EXTINCTION BANDS IN A SPECTRO-POLARIMETER WITH A TRIPLE-FIELD POLARISER (Lowry and Coode-Adams).

Rotatory Power of Crystals.—The rotatory power of isotropic and of uniaxial crystals (e.g. sodium chlorate and quartz, pp. 337 and 339) can be measured without difficulty by the methods described above. In order to measure the rotatory dispersion of biaxial crystals (p. 339) Longchambon[1] developed an improved form (Fig. 87) of an apparatus already used by Dufet[2] and by Wallerant.[3]

Light from the slit h_1 of a high dispersion monochromator is rendered parallel by the lens l_1 before traversing the polariser P and the crystal C. A diaphragm h_2, with an aperture of only 0·5 mm. diameter, is placed at the focal plane of a lens l_2 with a focal length of about 6 metres. This ensures that the beam which enters the analyser A cannot converge or diverge to the extent of more than a few minutes of arc while traversing the crystal C. A half-shadow system was introduced by covering one-half of the aperture in h_2 with a half-wave plate H. With the help of the lens l_3 the eye is focussed,

[1] LONGCHAMBON, *Bull. Soc. fr. Minér.*, 1922, **45**, 161–252; *Thesis*, Paris, 1923.

[2] DUFET, *Journ. de Phys.*, 1904, [iv], **3**, 757.

[3] WALLERANT, *C.R.*, 1914, **158**, 91.

through the aperture h_3, on the dividing line of the half-shadow analyser.

The crystal C is mounted on an adjustable platform and is aligned with the help of cross-wires mounted in front of the centre of the aperture in the diaphragm h_2. The latter is removed and the crystal adjusted until the centre of the interference pattern coincides with the cross-wires. The diaphragm h_2 is then replaced.

FIG. 87.—POLARIMETER FOR MEASURING THE ROTATORY DISPERSION OF BIAXIAL CRYSTALS (Longchambon).

The apparatus is used in the same way as other double-field instruments, but the crystal must be oriented afresh for each monochromatic radiation, in order to allow for the different directions of the optic axes.

Rotatory Power of "Liquid Crystals."—The apparatus used by Lehmann[1] to measure the very large rotations produced by

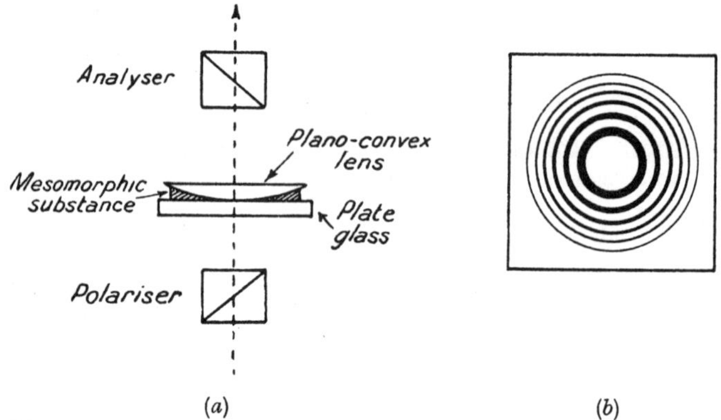

FIG. 88.—(*a*) LEHMANN'S APPARATUS FOR MEASURING THE ROTATORY POWER OF MESOMORPHIC SUBSTANCES.
(*b*) APPEARANCE OF THE FIELD OF VISION.

thin films of "liquid crystals" (p. 346) is shown in Fig. 88 (*a*). The mesomorphic substance is placed underneath a plano-convex lens resting on a piece of plate glass, so that its thickness increases towards the periphery of the lens. It is then inserted between the polariser

[1] LEHMANN, *Ann. der Phys.*, 1900, **2**, 649.

and analyser of a polarising microscope, which is set to extinction and illuminated by monochromatic light. Concentric dark extinction rings are seen (Fig. 88 *b*) which resemble "Newton's rings."

Each dark ring corresponds to light which has been rotated through $n \times 180°$ where n is an integer. The analyser is capable of lateral motion (recorded on a vernier), by which means the radius of each successive dark ring is measured. If r is the radius of the nth dark ring and R is the radius of curvature of the surface of the lens in contact with the mesomorphic substance, then the thickness l of substance which has caused a rotation of $n \times 180°$ is approximately $\frac{r^2}{2R}$. In practice the lens will not make contact with the plane glass surface but will rest on the film of mesomorphic substance, so that

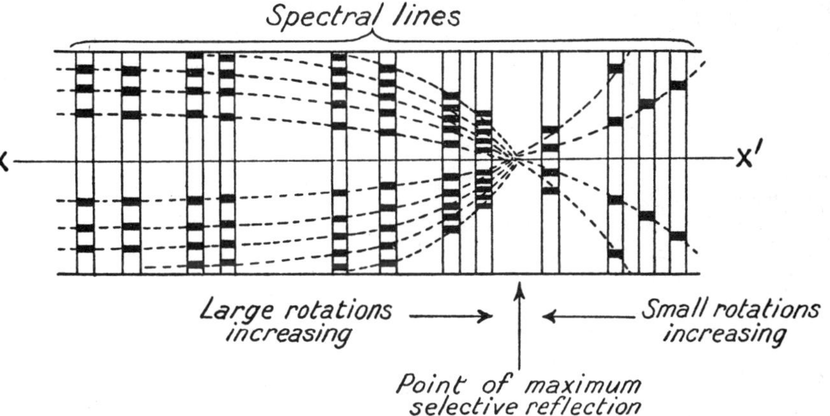

FIG. 89.—APPEARANCE OF THE FIELD OF VISION IN STUMPF'S METHOD OF MEASURING THE ROTATORY DISPERSION OF MESOMORPHIC SUBSTANCES.

the thickness producing a rotation of 180° will be equal to the difference in thickness corresponding to any two consecutive dark rings, i.e. $\frac{r_1^2 - r_2^2}{2R}$. The rotation ρ per mm. will then be given by

$$\rho = \frac{2R \times 180°}{r_1^2 - r_2^2}$$

where r_1 and r_2 are the radii of two adjacent dark rings. If a clockwise rotation of the analyser increases the size of the dark rings the rotation of the mesomorphic substance is dextro and vice versa.

The rotatory dispersion of "liquid crystals" may be determined by this method by using monochromatic radiations of different wave-lengths produced by means of a monochromator.

An alternative method due to Stumpf[1] consists in focussing on

[1] STUMPF, *Phys. Zeit.*, 1910, **11**, 780; *Ann. der Phys.*, 1912, **37**, 351.

the slit of a spectroscope an image of a diameter of the lens system described above, the latter being illuminated either with white light or preferably with a source of illumination giving a number of intense lines. Each dark ring produces a dark patch on either side of the centre of each spectral line. The greater the rotation produced for any particular wave-length [1] the closer together will be the dark patches on the corresponding spectral line. The type of effect which may be observed in the region of a point of maximum selective reflection (p. 347) is shown in Fig. 89. The distances of two consecutive dark patches on any particular spectral line from the line XX′ correspond to the radii r_1 and r_2. Stumpf's method is applicable to the ultra-violet as well as to the visible spectrum.

Sources of Monochromatic Light for Measurements of Rotatory Dispersion.—(*a*) *Sodium.*—The sodium D line, with which nearly all polarimetric measurements were formerly made, may be produced by introducing a sodium salt (usually the chloride) into the flame of a Bunsen burner, e.g. of the Meker type. The salt may be held in the flame in a platinum spoon or on a platinum wire, or the illuminating gas may be charged with dust by passing it through a mixture of fine sand and sodium carbonate before it enters the burner. A slightly greater intensity can be produced by using the bromide instead of the chloride. The intensity may also be increased by using an oxygen coal-gas flame, a globule of salt being introduced into the edge of the flame surrounding the oxygen inlet tube.[1]

As a rule no attempt is made to separate the yellow sodium doublet line ($\lambda_{D_1} = 5896$ and $\lambda_{D_2} = 5890$), the optical mean wave-length of the unresolved doublet being taken as $\lambda_D = 5893$. For the separation of the sodium doublet Lowry [2] found a sodium oxy-coal-gas flame was not sufficiently intense. He therefore used a carbon arc with a thread of glass in the hollow core of the upper carbon. The disadvantage of this method was that the lines frequently became reversed, giving a patch of yellow light crossed by two very narrow black Fraunhofer lines.

A bright sodium spectrum without reversal can be produced [2] by introducing sodium salts into a silver or brass arc by tipping the lower electrode with a spiral of wire in which sodium chloride or carbonate is placed. The intensity produced by this arrangement is insufficient for resolution of the sodium doublet, but quite suitable for ordinary polarimetric work in which this resolution is not required.

The blue and red rays which accompany the yellow sodium lines can be eliminated by filtering the light through solutions of potassium bichromate and uranyl sulphate respectively, the concentrations of which can be adjusted so that the OPTICAL MASS-CENTRE of the transmitted light is the same as that of the yellow doublet; [3] but a direct vision spectroscope attached to the eyepiece (p. 203) provides a more satisfactory guarantee of the spectral purity of the light.

[1] LOWRY, *Phil. Trans.*, 1912, A. **212,** Fig. 2 A, p. 270. [2] *Ibid.*, p. 271.
[3] LIPPICH, *Sitz. Akad. Wiss., Wien*, 1890, **99,** 695; *Zeit. Instr.*, 1892, **12,** 333.

An enclosed sodium lamp devised by Reger[1] in 1931 is illustrated in Fig. 90. It burns steadily, with an intensity which can be regulated by means of a rheostat, and is a most convenient source of sodium light.

In the alternating current lamp, Fig. 90 (*a*), both electrodes FF are spiral filaments. A small bead of sodium is placed in the lower filament and is partially vaporised by closing the switches S_1 and S_2,

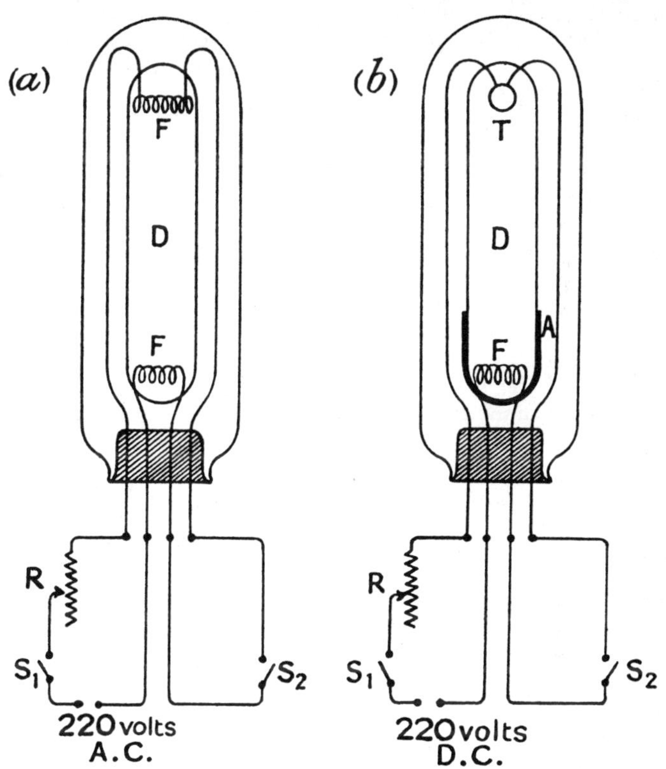

FIG. 90.—ENCLOSED SODIUM LAMPS.
(*a*) For use with alternating current.
(*b*) For use with direct current.

so that both filaments become heated. As soon as the lower filament is surrounded with a uniform glow the switch S_2 is thrown out and the current continues to pass through the discharge space D between the electrodes. The intensity of the illumination is then controlled by means of the rheostat R.

In the direct-current lamp, Fig. 90 (*b*), the lower hot filament F is the kathode, whilst a small bead of tungsten T takes the place of

[1] REGER, *Zeit. Instr.*, 1931, **51**, 472.

a filament at the anode. The lamp is started up in precisely the same manner as that described above. To prevent the heat developed at the kathode from disintegrating the packing at the base of the lamp the lower electrode is surrounded by an asbestos mantle A. Sodium lamps of both types are now constructed for use with either 110 or 220 volts.

(*b*) *Mercury.*—In recent years the mercury vapour lamp has displaced the sodium flame as the most convenient source of monochromatic light in polarimetry. It gives four very intense lines which are eminently suitable for polarimetric observations, namely, Hg 5791 and 5770 (yellow), Hg 5461 (green), Hg 4358 (violet).

(i) The two yellow lines, 21 A.U. apart, are usually separated from the green and violet lines with a direct-vision spectroscope, and are then read as a single line; but they can be resolved quite easily by a monochromator of average dispersive power.

(ii) The green line, Hg 5461, has now become the standard wave-length for polarimetric measurements on account of its brightness and spectral purity, and the ease with which it can be produced and read. The echelon spectroscope shows that it is accompanied by satellite lines, but these lie within a range of less than 1 A.U., and it has been proved that no detectable alteration in the mean optical wave-length of the green line is produced by modifications in the construction or running of the lamp.

(iii) The violet line, Hg 4358, is important, because it provides the only source of light in this region of the spectrum which can be read with ease and accuracy. The light blue line, Hg 4916, is not intense enough for polarimetric work, while the two extreme violet lines, Hg 4078 and 4046, are in a spectral region to which the eye is not very sensitive. The violet line 4358 is also important, because the dispersion of an optically-active medium is commonly expressed by recording the ratio of the rotations for the green and violet mercury lines.

Coloured filters have been proposed for the isolation of the various lines of the mercury arc, but they have been completely superseded by the direct-vision spectroscope. With substances of high rotatory power the glare from the yellow and green lines may be very great when readings are being made with the violet line. In this case the yellow and green rays may be excluded by using a violet filter attached to the Perkin eyepiece of the analyser, in order to avoid the loss of light which is incurred by using a monochromator.

(*c*) *Lithium.*—The lithium spectrum includes three lines which are suitable for polarimetric observations, namely, Li 6708 (red), Li 6104 (orange), Li 4602 (blue). The red line can be produced in adequate strength by introducing lithium chloride or carbonate into an oxy-coal-gas burner, but the orange and blue lines must be produced in an arc (compare sodium). Since the lines are very widely spaced only a Perkin eyepiece need be used to separate them. The orange line serves to bridge an awkward gap between the red and yellow, but the blue line (a very close doublet) is rather redundant in view of its proximity to the much more satisfactory mercury violet line.

(*d*) *Thallium.*—The flame spectrum of thallium gives an intense pure green line at $\lambda = 5350$ A.U., but is difficult to maintain owing to the rapidity with which thallium salts volatilise. A steady source of light has been obtained by vaporising thallium chloride in a small silica bulb, through which a stream of oxygen is led to a small jet in the centre of a Bunsen flame.[1]

(*e*) *Cadmium.*—The arc spectrum of cadmium furnishes the following intense lines: Cd 6438 (red), Cd 5086 (green), Cd 4800 (light blue), Cd 4678 (dark blue). The red and green cadmium lines are of such perfect spectral purity that the standard metre has been determined to eight significant figures in terms of their wave-lengths. Cadmium itself is too fusible to be burnt in an open arc, but a brilliant source of light is provided by an open arc burning between electrodes of silver-cadmium alloy.[2] It is difficult, however, to make the arc run smoothly on account of the ready volatility of cadmium as compared with silver; this sometimes causes the arc to splutter violently, even when the electrodes are kept rotating in opposite directions by means of an electric motor. On account of stray light the red line is not very easy to read, while the dark blue line often gives trouble owing to its poor visibility.

A steadier and purer source of cadmium light can be produced by means of an enclosed cadmium arc working on the same principle as a mercury or sodium vapour lamp. The first enclosed cadmium arc was used by Michelson and Benoit [3] for determining the standard metre in terms of the wave-length of the monochromatic red cadmium line. It was a glass discharge tube with aluminium electrodes in which a small quantity of cadmium was vaporised by heating the tube in an electric furnace; the discharge was then maintained by means of a transformer. This lamp was subsequently improved by Pérard [4] and again by Sears and Barrell [5] who used it for redetermining the standard metre in terms of the red and green cadmium lines.

An enclosed cadmium arc for polarimetric work was devised by Lowry and Abram [6] in 1914, but a more satisfactory type was described by Sand [7] in the following year and used by Lowry and Austin.[8] The arc was similar in principle to a mercury vapour lamp, the molten cadmium being prevented from sticking to the inside of the lamp by the addition of a small quantity of zirconia powder. This lamp proved to be admirably adapted for polarimetric work, but

[1] LOWRY, *Proc. Chem. Soc.*, 1912, 65; *Phil. Trans.*, 1912, A. **212**, Fig. 2B, p. 270.
[2] LOWRY, *Phil. Mag.*, 1909, [vi], **18**, 320–327.
[3] MICHELSON, *Trav. Bur. int. Poids. Mes.*, 1895, **11**, 35.
[4] PÉRARD, *Rev. d'Opt.*, 1928, **7**, 1.
[5] SEARS and BARRELL, *Phil. Trans.*, 1932, A. **231**, 75.
[6] LOWRY AND ABRAM, *T.F.S.*, 1914, **10**, 103.
[7] SAND, *Brit. Assoc. Rep.*, 1915, **85**, 386.
[8] LOWRY and AUSTIN, *Phil. Trans.*, 1922, A. **222**, 265.

could not be reproduced with an adequate life, and did not therefore come into general use. A cadmium lamp, working on the same principle as Reger's sodium lamp is, however, now available, and promises to provide an adequate solution of the important problem of making the cadmium spectrum easily available for polarimetric work.

(*f*) *Silver*.—The silver-cadmium arc also gives two intense silver lines in the green part of the spectrum, one of them being a close doublet like the D line of sodium: Ag 5472, 5466 (light green), Ag 5209 (dark green). Both lines are very intense and easy to read, especially when pure silver electrodes are used instead of silver-cadmium alloy, but the doublet is too near the green mercury line, Hg 5461, to be worth reading except in special cases.

(*g*) *Copper* and *Zinc*.—By using electrodes of pure copper and a zinc-silver or zinc-copper alloy the following series of lines are made available for measurements of rotatory dispersion:

Cu 5782 and Cu 5700 (yellow), Cu 5220, 5218, Cu 5153, Cu 5106 (green), Zn 6362 (red), Zn 4811, Zn 4722, Zn 4680 (blue).

If the arc does not flicker the two yellow and three green copper lines can be read easily and accurately. The zinc lines are not so easy to read on account of continuous radiation in the red and lack of sensitivity of the eye in the blue. These blue lines are indeed more difficult to read than the violet mercury line Hg 4358, although this is 320 A.U. nearer the ultra-violet than the most refrangible zinc blue line.

A table of wave-lengths of monochromatic radiations commonly used in visual spectropolarimetry is given on p. 422.

CHAPTER XVII.

MEASUREMENT OF ROTATORY DISPERSION.

(*b*) In the Ultra-violet.

Ultra-violet Polarimetry.—The principal features of ultra-violet spectro-polarimetry are (i) the production of suitable intense ultra-violet radiation, and (ii) the construction of optical parts which will transmit the rays in question.

(*a*) *Sources of Ultra-violet Light.*—The principal sources of light which have been employed in ultra-violet polarimetry, and the shortest useful wave-length of each are given below :—

Solar spectrum	→ 3000 A.U.	(approximately).
Carbon arc	→ 2500 ,,	,,
Mercury vapour lamp	→ 2480 ,,	,,
Iron arc	→ 2327 ,,	,,
Cadmium spark	→ 2100 ,,	,,
Copper spark	→ 2000 ,,	,,
Zinc-aluminium spark	→ 2500 ,,	,,
Intense aluminium spark	→ 1850 ,,	,,

Since a many-line spectrum is generally required, the commonest sources of light are the iron arc, which becomes too weak to be useful beyond 2327 A.U., and the tungsten-steel arc, which gives a more crowded and uniform array of lines of known wave-length.

(*b*) *Limits of Transmission.*—The transmissibility of the various materials used in the construction of the optical parts of polarimeters is given below :—

Glass (heavy flint)	→ 3800 A.U.	(approximately).
,, (ordinary flint)	→ 3500 ,,	,,
,, (crown)	→ 3300 ,,	,,
Uviol glass	→ 3100 ,,	,,
Glass (thin plate, $\frac{1}{20}$ mm.)	→ 2650 ,,	,,
Canada balsam	→ 3400 ,,	,,
Iceland spar (4 cm.)	→ 2400 ,,	,,
Glycerine (thin film)	→ 2300 ,,	,,
Quartz	→ 1800 ,,	,,
Fluorspar	→ 1000 ,,	,,

Thus, for polarimetric work in the middle ultra-violet, glass must be replaced by calcite or quartz ; and Canada balsam must be eliminated by using Foucault, Glan or double-image prisms in the place of Nicols.

Soret and Sarasin.—Soret and Sarasin [1] in 1882 determined the rotatory power of quartz down to a wave-length of 2143 A.U. by a modification of the methods of Fizeau and Foucault, Broch and von Lang, but with Foucault prisms in place of Nicols. The source of light was an electric spark between cadmium electrodes, giving a series of ten intense lines in the ultra-violet. These ultra-violet lines were rendered visible by focussing them on a fluorescent screen of uranium glass or on a thin-walled vessel containing a solution of æsculine. The position of the vertical dark bands in the field of the analyser was adjusted as in the visible spectrum to coincide with the ultra-violet lines of the cadmium spectrum.

Guye.—Guye [2] in 1889 investigated the rotatory dispersion of crystalline sodium chlorate in the ultra-violet region of the spectrum by a modification of the method of Wiedemann (p. 201). Foucault prisms were used in the place of Nicols, while the bi-quartz plate of variable thickness was placed immediately in front of the analyser. The glass prism of the spectroscope was replaced by one of Iceland spar, which is completely transparent down to a wave-length of 2500 A.U. approximately; the lenses were of quartz. The polariser was rotated instead of the analyser in taking observations. Sunlight and a cadmium spark were used as sources of light, the ultra-violet spectrum being made visible, as in Soret and Sarasin's method, by means of a fluorescent screen. The rotations were measured, as in Wiedemann's experiments, by bringing the dark bands in the upper and lower fields into alignment.

Joubin.—A photographic method of measuring rotatory dispersion in the ultra-violet, which differs somewhat from any previously described, was devised by Joubin [3] in 1889.

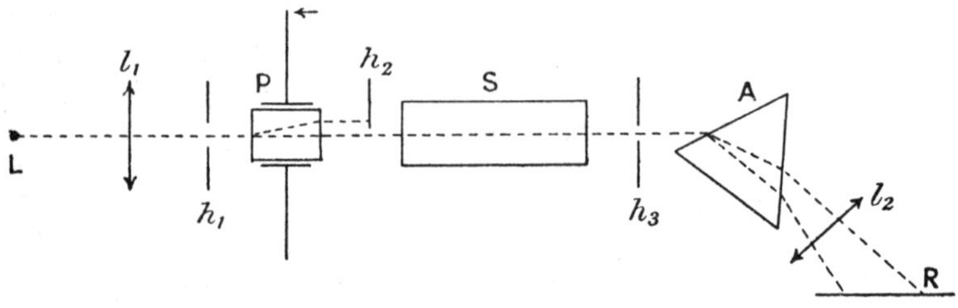

FIG. 91.—JOUBIN'S ULTRA-VIOLET POLARIMETER.

The polariser P (Fig. 91) consists of a single large crystal of Iceland spar. A 60° prism A of the same material, serves both as an analyser and as a dispersive system for the production of the spectrum. An image of the light source L is focussed by means of the lens l_1 on a small circular aperture in the screen h_1. The deviated extraordinary

[1] SORET and SARASIN, *C.R.*, 1882, **95**, 636; *Arch. Sc. Phys. Nat., Genève*, 1882, **8**, 5, 97, 201. See also *C.R.*, 1875, **81**, 610; 1876, **83**, 818; 1877, **84**, 1362.
[2] GUYE, *C.R.*, 1889, **108**, 348.
[3] JOUBIN, *A.C.P.*, 1889, [vi], **16**, 78.

ray from the polariser is cut off by the screen h_2, so that only the undeviated plane-polarised ordinary ray traverses the optically-active medium S. The lens l_2 serves to focus an image of the spectrum produced by A on the photographic plate R. The spectrum formed on R consists of a series of circular images of the aperture in the screen h_1, each circular patch corresponding to a particular monochromatic radiation emitted by the light source. Photographs are taken for various rotations recorded on the circular scale of the polariser, and the particular radiation extinguished by the analyser can be determined readily, since the ordinary and extraordinary rays transmitted by the prism A are separated so widely that the two spectra do not overlap.

Nutting.—Joubin's method is rather laborious, because complete extinction of one of the monochromatic radiations is only obtained for a particular setting of the polariser, which can only be determined by trial and error. This was avoided by Nutting,[1] who devised an ingenious method whereby the rotatory power of a medium was determined for a whole series of ultra-violet wave-lengths by taking two

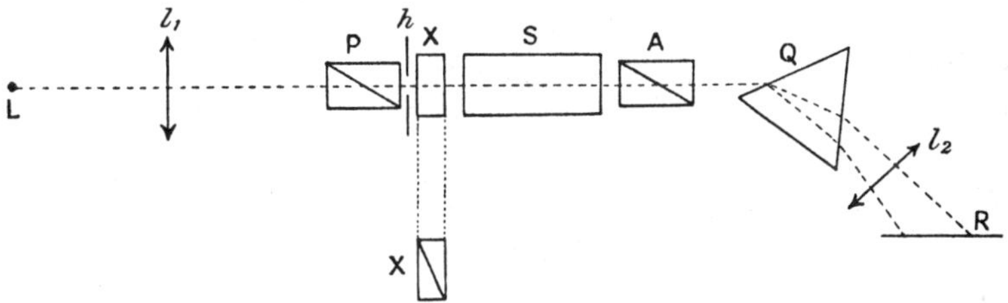

FIG. 92.—NUTTING'S ULTRA-VIOLET POLARIMETER.

photographs only, one with and one without the substance in position.

The peculiar optical feature of Nutting's apparatus (Fig. 92) is a pair of quartz wedges placed together to form a rectangular prism X, identical, except in dimensions, with that devised by Senarmont (Fig. 68, p. 184), but used in a different way.

The polariser P and analyser A are Glan prisms, with their principal sections parallel to one another. Immediately before the polariser is a screen h with a narrow vertical slit; this is followed by the prism X (2 × 2 cm. in cross-section and 1·6 cm. long) placed so that the refracting edges of the wedges are horizontal. The optically-active medium S is placed between this prism X and the analyser A. An image of the light source L (a spark between electrodes of zinc-aluminium alloy) is formed by the long focus lens l_1 on the slit h_1, so that the rays traversing the prism X are almost parallel. The line spectrum formed by the dispersive prism Q is focussed on the photographic plate R by the lens l_2.

[1] NUTTING, *Phys. Rev.*, 1903, **17**, 1.

A ray passing through the centre of the special prism traverses equal thicknesses of dextro- and lævo-quartz; it is completely transmitted by the analyser, and forms a narrow *dark* patch at the centre of each spectral line on the photographic *negative*. Rays passing through the upper and lower halves of the slit traverse unequal thicknesses of dextro- and lævo-quartz and have their planes of polarisation rotated through various angles. Each spectral line is therefore crossed by dark patches throughout its length (Fig. 93), each dark patch corresponding to a rotation of some multiple of 180°. These patches are closer together on the lines of shorter wave-length, owing to the increased rotatory power of the quartz. The slit *h* is furnished with a horizontal cross-wire which is adjusted, in the absence of the optically-active substance, to form an image I I′ across the central dark patch of each spectral line. This serves as a line of reference on the photographic plate; the distances of successive dark patches on either side of this reference line represent rotations of $\pm n\pi$.

A photograph of the banded line spectrum is first taken in the absence of the active medium, and the distance of the centres of the dark patches from the reference line is measured by means of a travelling microscope. A second photograph is then taken with the medium in position. On the second plate the dark patches will be displaced along a spectral line by an amount which is proportional to the rotation for that particular wave-length, and in a direction determined by the sign of the rotation. The displacement is measured and, by comparison with the distance representing 180°, the observed rotation is computed for each spectral line. The time of exposure, which depends on the transparency of the substance and on the intensity of the different lines, varies from 30 seconds to 1½ hours.

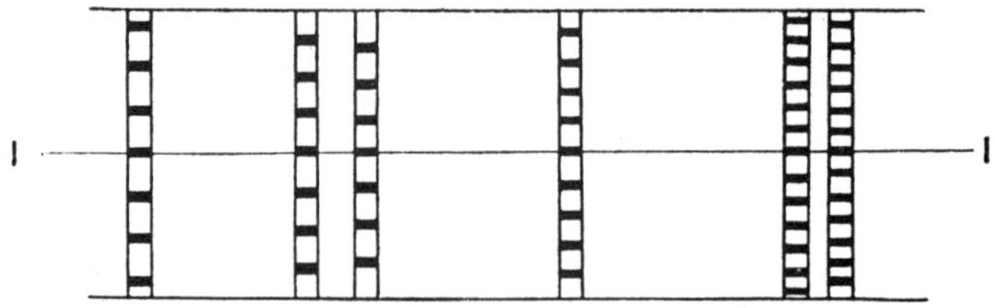

FIG. 93.—ILLUSTRATES THE TYPE OF PHOTOGRAPH OBTAINED WITH NUTTING'S ULTRA-VIOLET POLARIMETER.

The distance between successive dark patches (corresponding to 180° rotation) was 3 mm. and 0·6 mm. for wave-lengths 5893 and 2810 A.U. respectively. Since the dark patches are neither very narrow nor sharply defined the error in measuring the displacement is relatively large. Nutting gives no estimate of the error involved, but the observed rotations for quartz found by this method are given to 0·1°.

Landau.—The first double-field polarimeter for use in the ultra-violet was constructed by Landau,[1] and is shown in plan in Fig. 94.

The double field is produced by a Jellett-Cornu prism A, with a fixed shadow angle of 5°, in which the film of Canada balsam is replaced by glycerine. The analyser A is set with the dividing line between the two fields horizontal, and variations in intensity are produced by rotating the polarising Glan prism P in a graduated circle. An image of the light source L (an iron arc burning between rotating electrodes) is formed by the lens l_1 on the slit h of an ultra-violet spectroscope with camera attachment R. The spectroscope includes two quartz prisms p, p', and two achromatic quartz-fluorspar lenses l_2 and l_3. The slit h is placed close to the double-field analyser so that the spectrum focussed on the photographic plate is divided by a sharp horizontal line. In the zero position, the two halves of the spectrum are equally illuminated throughout. The setting of the

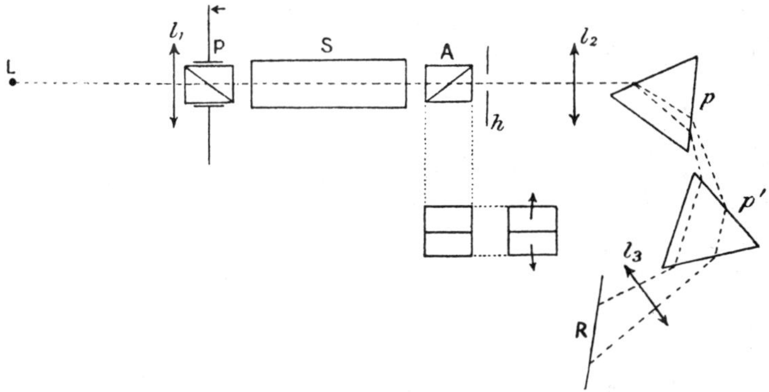

FIG. 94.—LANDAU'S ULTRA-VIOLET POLARIMETER.

polariser to produce this effect may be made either visually or photographically, the latter method being the more accurate on account of the unequal sensitivity of the eye to different colours.

On introducing the optically-active substance S the equality of the two spectra is destroyed. The polariser is then rotated through an angle, the approximate value of which can be deduced from measurements already made in the visible spectrum, and a series of photographs is taken for different positions. The wave-length of the lines which are equally illuminated in both strips of the spectrum are read off on the photographic plate with the help of charts of the iron spectrum. In this method, therefore, the usual procedure is reversed, since the wave-length which produces a given rotation is determined instead of conversely. For this purpose a many-lined spectrum is required in order that the position of equality of the two fields may coincide with a line of known wave-length. If this does

[1] ST. LANDAU, *Physik. Zeit.*, 1908, 9, 417.

not happen, the correct wave-length must be determined by interpolation, or by taking further photographs at small intervals of rotation, e.g. every 0·05°, until the "change-over" falls on a definite spectral line.

The time of exposure in Landau's measurements varied from 10 seconds to 6 minutes, and the experimental error was of the order of 0·02°. In Landau's apparatus as in W. H. Perkin's[1] the double field is produced by the analyser. The sharpness of the dividing line between the two fields is therefore not impaired when the medium is cloudy. Observations with a cloudy medium cannot, however, be very accurate because the light is partially depolarised by scattering.

Lowry.—Lowry's apparatus for measuring rotatory dispersion in the visible spectrum (Fig. 85, p. 204) can be adapted readily for use in the ultra-violet.[2] The monochromator is removed and an iron arc is placed in front of the polariser as a source of light. The

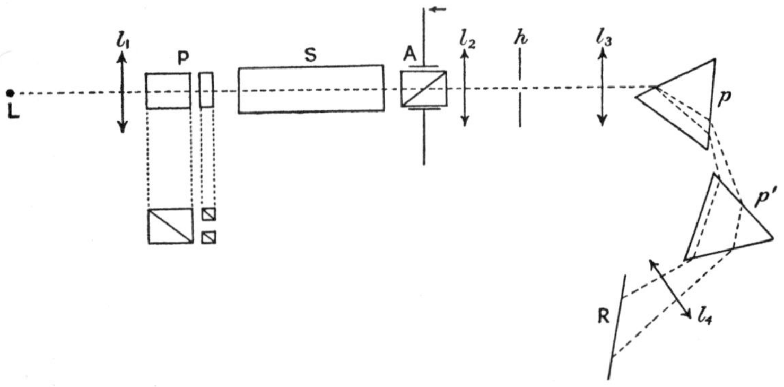

FIG. 95.—LOWRY'S ULTRA-VIOLET POLARIMETER.

Glazebrook prisms of the polariser are replaced by a triple-field polariser P, constructed from Foucault prisms, with a quartz lens l_1 set to produce a parallel beam of ultra-violet light. The Glazebrook prism of the analyser is also replaced by a Foucault prism A, and the telescope is replaced by a quartz-calcite achromatic doublet l_2. This lens focusses an image of the triple-field polariser on the vertical slit h of a quartz spectroscope similar to that used by Landau. A plan of the instrument is shown in Fig. 95.

This apparatus differs from Landau's in that a *triple-field* with a *variable half-shadow angle* is used instead of a *double-field* with a *fixed* half-shadow angle. The advantages of this system in the visual region have already been described (p. 190); its advantages in the photographic region are obviously still greater.

[1] W. H. PERKIN, Senr., *J.*, 1896, **69**, 1027, 1031.
[2] LOWRY, *P.R.S.*, 1908, **81**, 472; LOWRY and COODE-ADAMS, *Phil. Trans.*, 1927, A. **226**, 391.

W. Kuhn.—In order to measure rotatory dispersions down to a wave-length of 1850 A.U., W. Kuhn [1] constructed a polarimeter (Fig. 96) in which all the optical parts were of quartz or fluorspar.

A *half-shadow polariser* is constructed by cutting a Rochon prism (p. 176) horizontally as at P, removing a thin wedge of 5° angle from one of the cut faces and placing the two halves in optical contact. The two halves of the split prism then produce ordinary rays, whose planes of polarisation are inclined at a small fixed angle to one another.

The *analyser* A is a single Rochon (or Senarmont) prism mounted in a graduated circle, in such a way that the part in which the optic axis is parallel to the direction of the incident ray is furthest from the polariser. If the prism is used the wrong way round there is no extinction-position, since the ray entering the analyser is then rotated by a variable amount, proportional to the thickness of quartz traversed, before reaching the surface at which the ordinary and extraordinary rays are separated.

An image of the light-source L is focussed by the quartz lens l_1

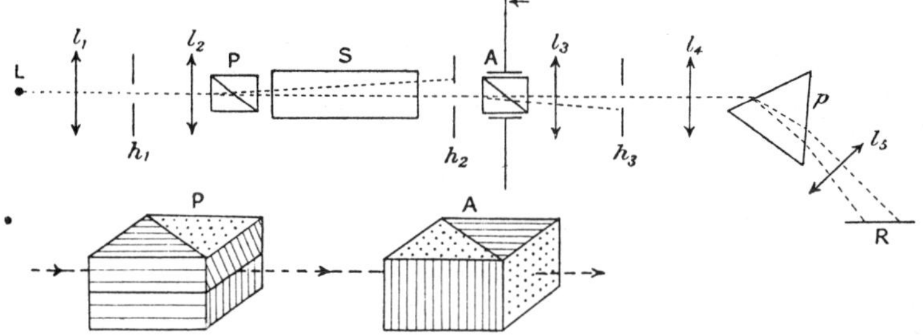

FIG. 96.—KUHN'S ULTRA-VIOLET POLARIMETER.

on a circular aperture in the screen h_1, which thus serves as a steady source of light. An image of this aperture is focussed by the quartz lens l_2 on a second screen h_2 placed in front of the analyser. This image is duplicated by the double-image polariser, but the extraordinary ray is cut off by the screen h_2 so that only the ordinary ray passes through. In an improved apparatus used by Lowry and Gore,[2] the deviation of the extraordinary ray was increased by increasing the length of the polarising prism to such an extent that this ray could be blocked out by a screen at a distance of about 14 cm., *before reaching the optically-active medium.*

On traversing the analyser, the ordinary ray is again divided into an ordinary undeviated ray and an extraordinary deviated ray. These rays form, by means of the achromatic quartz-fluorspar doublet l_3, an ordinary and an extraordinary image of the double-field polariser in the plane of the slit of the ultra-violet spectroscope. The latter is placed so that only the ordinary ray enters the slit and forms a

[1] KUHN, *Ber.*, 1929, **62**, 1727.
[2] LOWRY and GORE, *P.R.S.*, 1932, A. **135**, 13–22.

spectrum divided longitudinally into two. The apparatus is used in precisely the same way as those of Landau and Lowry.

Cotton and Descamps.—All the methods described hitherto for measuring rotatory dispersion in the ultra-violet are rather laborious; moreover, when working with a source of light which does not give a spectrum very rich in lines throughout its whole length, a considerable time may be spent in finding suitable settings of the analyser for particular spectral lines. With these considerations in view Cotton [1] devised a method whereby the "extinctions" for the various spectral lines were recorded automatically. The apparatus (Fig. 97) was constructed and perfected by Descamps,[2] who has used it successfully to measure the ultra-violet rotatory dispersion of a large number of optically-active compounds. The mechanical arrangement and working of the instrument are rather complicated, but the optical system and the principle of the method are relatively simple.

The light source L is a silica mercury vapour lamp, perfected

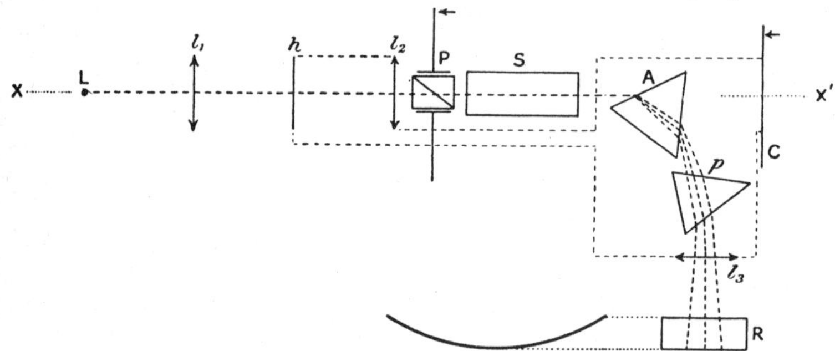

FIG. 97.—COTTON AND DESCAMPS' ULTRA-VIOLET POLARIMETER.

by Cotton, in which an arc burns between mercury surfaces about 1 cm. apart. The intrinsic brilliance of the illumination is very great; in addition the intensity of the light is very constant over a long period. An achromatic quartz-fluorspar triplet l_1, mounted with glycerine for transparency, throws an image of the light source on a minute square hole in the diaphragm *h*. This hole is formed by two slits at right angles to one another, one being parallel to the refracting edges of the prisms forming the dispersive system.

The light from the aperture is rendered parallel by the triplet lens l_2 and traverses the Glan prism polariser P and the optically-active medium S. It is then incident on a 60° prism of Iceland spar A which serves both as an analyser and as a dispersive system. The ordinary ray passes through a quartz prism *p* and is focussed by a lens l_3 on a photographic film at R, whilst the extraordinary ray, which is less refracted and not so widely dispersed, does not reach

[1] COTTON and DESCAMPS, *C.R.*, 1926, **182**, 22.

[2] DESCAMPS, *Rev. d'Opt.*, 1926, **5**, 481; *Réunions de l'Inst. d'Opt.*, 1934, 8–21.

the film. The two prisms A and p have their principal crystallographic axes parallel to their refracting edges, and both are placed in the position for minimum deviation of the ray $\lambda = 3025$ A.U. in the middle of the near ultra-violet; the angle of the quartz prism (47° $24'$) is such that this ray is deviated through exactly 90° in passing through the complete dispersing system.

The diaphragm h, the lenses l_2 and l_3, and the prisms A and p, are mounted rigidly on a platform which can turn on ball-bearings about the optical axis XX′ of the apparatus, the rotation being recorded on the circular scale C. The rest of the apparatus, including the light source L, the lens l_1, the polariser P, the optically-active medium S, and the large cylinder R, which carries the photographic film, is fixed in position. The moving parts do not execute a complete revolution, but oscillate slowly through an angle of 60°. This motion causes the mercury spectrum to move up and down the length of the photographic film, at the rate of about 1·1 cm. per angular degree. The dispersion (11·3 cm. from 4358 to 2536 A.U.) is so large that the lines Hg 3132 and Hg 3126 appear quite separate, but the doublet Hg 3655 and 3650 is not resolved although the line Hg 3663 is separated from it. Since scattered light would fog the photographic film, the light from the dispersive system is made to pass along a narrow blackened tube, which rotates with the moving part of the instrument. The end of the tube near the film is furnished with a very long horizontal slit through which the rays pass.

A photograph is first taken in the absence of the optically-active medium by allowing the moving part of the apparatus to oscillate to and fro several times until the record on the sensitised film is sufficiently dense. Reference lines are provided by stopping the moving system at each end of its oscillation (e.g. at 0° and 60°), and widening the slit of the diaphragm h to 2–3 mm., so that the mercury spectrum photographed at these two points consists of very wide but very short lines *aa*, *bb* (Fig. 98). This procedure is repeated with the active medium in position.

Each vertical streak in the photograph (Figs. 98 *a* and *b*) represents the trace of a line in the mercury spectrum. It is interrupted at a point X, corresponding to the extinction position of the analyser. The exact centre of the " extinction interruption " for each line is determined by means of a photoelectric microphotometer devised by Challonge and Lambert.[1] Its distance from the reference lines *a* and *b*, which are separated by 60°, then gives the extinction position of the analyser.

On the " zero " photograph (Fig. 98 *a*), the extinctions all occur at the same distance from the reference lines. Fig. 98 (*b*) shows at a glance that the medium exhibits anomalous rotatory dispersion, since the extinctions lie on both sides of the zero line, indicating the occurrence of a reversal of sign in the near ultra-violet. Rotations

[1] CHALLONGE and LAMBERT, *C.R.*, 1925, **180**, 924; *Rev. d'Opt.*, 1926, **5**, 404; **1931**, **10**, 405.

larger than 60° can be recorded on the film by rotating the polariser through an angle which can be recorded on its circular scale.

Descamps has estimated that the experimental error involved in measuring rotations by this method is of the order of 0·02°. It is therefore comparable with that of the best triple-field instruments.

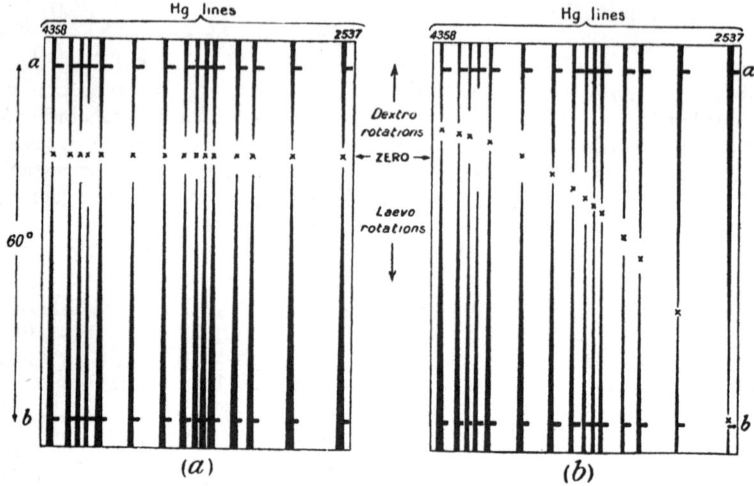

FIG. 98.—IMPRINTS OF SPECTRA IN COTTON AND DESCAMPS' POLARIMETER.
(*a*) In the absence of an optically-active medium.
(*b*) With an optically-active medium present.

Bruhat and Pauthenier.—Bruhat and Pauthenier[1] in 1926 described an alternative method for recording automatically the extinction position of the analyser, but only for one wave-length at a time. Their method consists in photographing the variations in intensity of pure monochromatic ultra-violet light, which has traversed a single-field polariser, for various orientations of the analyser. By means of a microphotometer, the exact extinction position is determined from the photographic record for the particular radiation. This record, however, is not continuous, but consists of a series of discrete impressions made while the analyser is stationary at successive positions. The actual extinction position is therefore deduced by the method of interpolation.

One of the essential requirements of Bruhat and Pauthenier's method is that the intensity of the light source shall remain very constant throughout a long series of measurements. For this reason a quartz mercury lamp is used as the source of light. The selected line is isolated in a state of spectroscopic purity by means of the apparatus shown in Fig. 99. The lenses are of quartz or are achromatic quartz-calcite doublets. The 60° dispersing prisms *pp'* are also

[1] BRUHAT, PAUTHENIER and COTTON, *C.R.*, 1926, **182**, 888; BRUHAT and PAUTHENIER, *Rev. d'Opt.*, 1927, **6**, 163.

of quartz with the optic axis perpendicular to the plane bisecting the refracting angle. The polariser and analyser are Glazebrook prisms with a film of glycerine. An opaque screen D carries three circular holes, h_1, h_2, h_3, each 1 mm. in diameter and 2·5 cm. apart. The vertical reflecting mirrors M and M′ each consist of a mercury surface behind a thin plate of quartz. The apparatus is aligned, and the lenses are focussed for a mercury line Hg 3660 in the middle ultra-violet region.

An image of the light source L is focussed by the lens l_1 on to the aperture h_1. Rays from this aperture are (i) rendered parallel by the quartz-calcite doublet l_2, (ii) refracted and dispersed by the prism p, (iii) reflected almost normally from the mirror surface M, and then (iv) retraverse the prism p and objective l_2. The prism p is placed in the position for minimum deviation with respect to the ray Hg 3660, and the ultra-violet mercury spectrum which comes

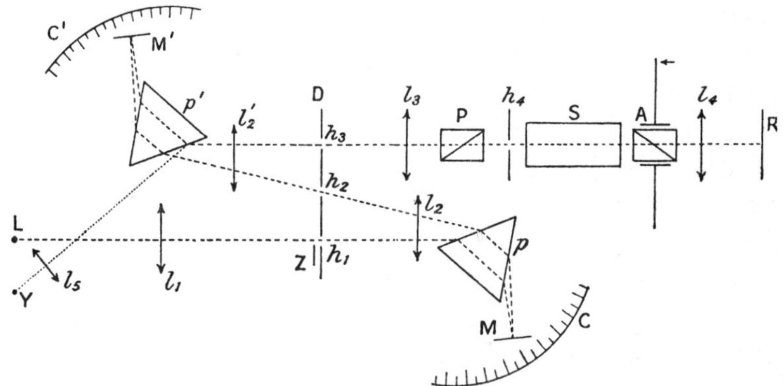

FIG. 99.—BRUHAT AND PAUTHENIER'S AUTOMATICALLY RECORDING ULTRA-VIOLET POLARIMETER.

to a focus near h_2 is adjusted, by rotating the mirror M, until the image of this ray falls on the aperture h_2, as indicated by a fluorescent screen of uranium glass. Both prism and mirror are then firmly clamped to the horizontal circular graduated mounting C. The almost completely homogeneous ultra-violet radiation which passes through the aperture h_2 is further purified by traversing the second identical dispersive system l_2' p' M′ C′. It is finally brought to a focus on the aperture h_3 and rendered plane parallel by the lens l_3 before traversing the polariser P, the optically-active medium S and the analyser A. The width of the beam is limited to 3 mm. by the stop h_4 attached to the front end of the polarimeter tube. An image of the aperture h_3 is formed by the lens l_4 on the photographic plate R; the focal length of the lens l_3 is double that of the lens l_4 so that the circular image formed on the sensitised plate is 0·5 mm. in diameter.

Any other ray from the lamp may be made to traverse the instrument by rotating the circular mountings C and C′ of the monochromators until the requisite images fall on the apertures h_2 and h_3 as shown with the help of the fluorescent uranium glass screen.

The approximate extinction positions of the analyser, with and without the optically-active medium, are found by replacing the photographic plate by the fluorescent screen and rotating the analyser until the image on the screen is as faint as possible.

By means of a carbon arc Y and a lens l_5, a beam of parallel white light, reflected from the face of the prism p', can be made to traverse the polarimeter in order to test its alignment.

The movements of the analyser and of the photographic plate between the exposures, as well as the making of the exposure itself, are effected automatically by an ingenious electrical and mechanical device. The timing mechanism consists of a chronometer, the tip of whose needle makes electrical contact every minute by passing through a drop of mercury. This operates, through an intermediate relay, a distributor to which are connected an electrically worked shutter Z in front of the aperture h_1, for intercepting the beam of light, and a mechanism for rotating the analyser and moving the photographic plate. The distributor is so constructed that, as soon as the shutter closes, the analyser and photographic plate move to the next position. At the end of two minutes the shutter automatically opens and begins an exposure. The number of minutes before the shutter closes again may be fixed at 1, 3, 5 or 11 by adjusting the connections of the contacts in the distributor. The angle through which the analyser turns between two exposures is exactly 9·5 minutes of arc, while the photographic plate moves in its own plane through 0·5 cm.

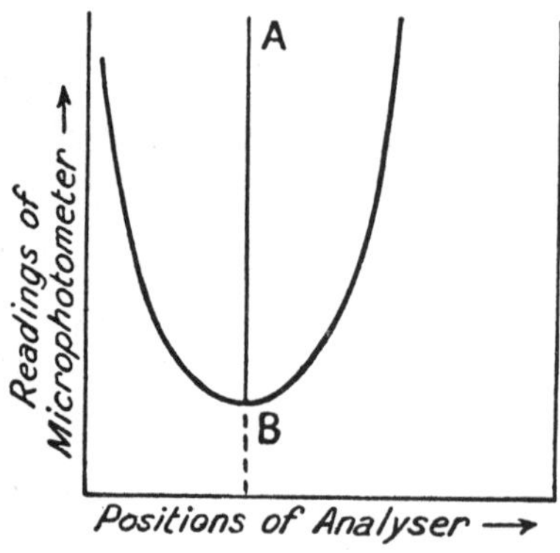

Fig. 100.—Variations of Intensity of Images in Bruhat and Pauthenier's Ultra-violet Polarimeter.

A series of exposures is taken on either side of the approximate zero position of the analyser in the absence of the optically-active medium, but with the empty polarimeter tube in position. A second series is then taken on either side of the approximate extinction position with the substance in the tube. Each series is examined by means of a Fabry and Buisson[1] microphotometer. The readings of the latter for each image are plotted as ordinates against the

[1] Fabry and Buisson, *Rev. d'Opt.*, 1922, **1**, 1. For the use of a microphotometer with a double-field polarimeter, see Lowry and Vernon, *P.R.S.*, 1928, A. **119**, 706–709.

corresponding positions of the analyser, as shown in Fig. 100. The rectilinear diameter AB of this curve indicates the exact extinction position of the analyser.

Bruhat and Pauthenier estimate that, by this method, rotations can be measured to within 0·5′ (<0·01°). This degree of accuracy, however, requires great care in selecting and using the balanced quartz end plates of the polarimeter tube. It is suggested that the tube should be filled and emptied by syphoning, so as not to disturb its position after reading the zero of the instrument through the empty tube.

Apart from the time and labour saved by the automatic working, the great advantage of this method is that only pure monochromatic light traverses the polarimeter, whereas in most other methods radiations of every wave-length pass through, and this, according to Brace, diminishes the accuracy of the observations. Bruhat and Pauthenier prevented any stray light, produced by reflection from the surfaces of the lenses of the monochromators, from entering the apertures in the screen D, by very slightly tilting the lenses about their horizontal axes.

Bruhat and Chatelain.—A photoelectric polarimeter for measuring rotatory dispersion in the ultra-violet was described in 1933 by Bruhat and Chatelain,[1] and has subsequently been improved by these authors. Their instrument has many advantages over the photoelectric polarimeters previously used by Halban and Siedentopf, Mayrhofer, Todesco, Ebert and Kortüm.[2] The general principle of these earlier forms of photoelectric polarimeter consists in measuring, by means of a photoelectric cell, the intensity of the light transmitted by a polariser and analyser whose principal sections are inclined at 45° to one another. The alteration in the intensity of the transmitted light on introducing an optically-active medium is shown by a variation in the photoelectric current. Either the polariser or the analyser is then rotated until the intensity again has its former value. This method has the drawback that the intensity of the light traversing the analyser and falling on the photoelectric cell depends not only on the rotatory power of the medium under investigation, but also on its absorptive power, and it is necessary to isolate the effects of the two phenomena.

Bruhat and Chatelain avoid this difficulty by employing a method of comparison. The general principle of their method consists in using a double-image analysing prism and setting the latter, with and without the optically-active medium in position, so that the intensity of the two beams of light transmitted by it are exactly equal. Earlier workers had found it difficult to obtain two identical photoelectric cells for making comparative measurements with a light source which was not absolutely steady. Bruhat and Chatelain,

[1] Bruhat and Chatelain, *C.R.*, 1932, **195**, 462; *Journ. de Phys.*, 1932, [vii], **3**, 501; *Rev. d'Opt.*, 1933, **12**, 1.

[2] Ebert and Kortüm, *Z. ph. C.*, 1931, B. **13**, 105.

therefore, preferred to use a single photoelectric cell, on which, as indicated by Tardy,[1] either ray from the analyser could be made to impinge by means of a quartz biprism (Fig. 101 *a*).

The polariser P is a single Glazebrook prism of calcite separated by a thin film of glycerine. The analyser A is a Wollaston prism (refracting angles 30°) giving two deviated rays, plane polarised

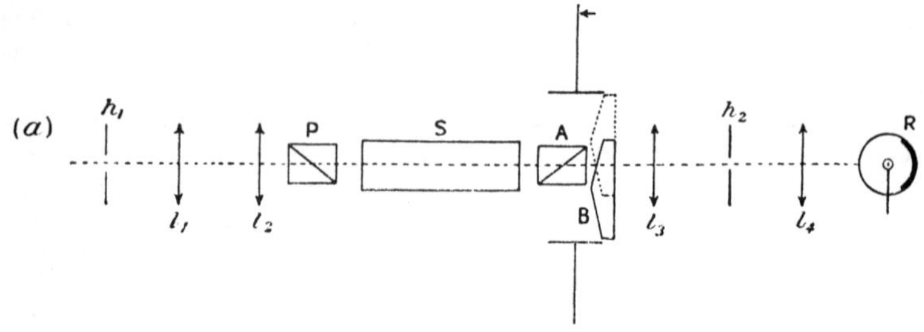

(*a*) Plan of the apparatus.

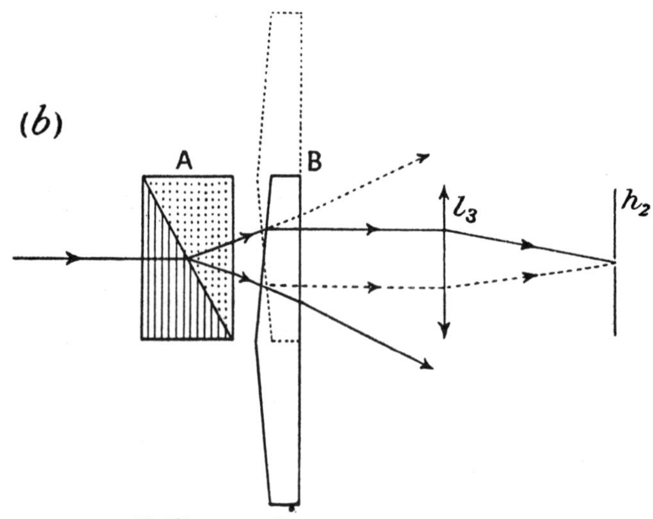

(*b*) Illustrates the action of the biprism.

FIG. 101.—BRUHAT AND CHATELAIN'S PHOTOELECTRIC POLARIMETER FOR USE IN THE ULTRA-VIOLET.

in directions mutually perpendicular. These rays will have equal intensity when the optic axis in either half of the prism is inclined at 45° to the principal section of the polariser. The arrangement is therefore equivalent to an instrument with a shadow angle of 90°. The biprism B has an angle of 35′, and its optic axis is perpendicular to the base.

[1] TARDY, *Rev. d'Opt.*, 1928, **7**, 189.

Fig. 101 (*b*) shows how one or other of the rays from the analyser can be rendered parallel to its original direction by sliding the biprism along its own length from one position to the other. As indicated by the dotted lines, either ray may then be focussed by the lens l_3 on to the aperture in the screen h_2.

Monochromatic ultra-violet radiation from a mercury vapour lamp, after traversing a monochromator of a type similar to that employed by Bruhat and Pauthenier, is focussed on the diaphragm h_1; it is rendered parallel by the lens system l_1 and l_2 before entering the polarimeter. All the lenses are achromatic quartz-fluorspar doublets. An image of the aperture h_1 is formed on the photoelectric cell R by the lenses l_3 and l_4.

The photoelectric current is amplified approximately 10^6 times by double-grid and power valves and passes through the primary of a transformer of which the secondary is connected to a ballistic galvanometer. By this means any slow variations in the photoelectric current do not make themselves apparent. It is not necessary to find directly the exact position of the analyser such that there is no " kick " in the galvanometer when the biprism is moved from one position to the other. It is sufficient to set the analyser slightly on one side of the required position and note the " kick " on moving the biprism sharply backwards and forwards; this is repeated with the analyser on the other side of the required point. The desired setting is then deduced by interpolation. This procedure is carried out with and without the optically-active medium, and with monochromatic ultra-violet radiations of different wave-length as produced by the monochromator.

By this method Bruhat and Chatelain have been able to measure rotations of the order of 45° to within 1 part in 400, and Bruhat and Guinier, by increasing the amplification, have obtained results correct to 1 part in 1000 down to 2950 A.U. and correct to 1 in 500 from 2950 to 2480 A.U.

Bruhat and Guinier.—The above method has recently been modified and improved by Bruhat and Guinier.[1] The lateral dis-

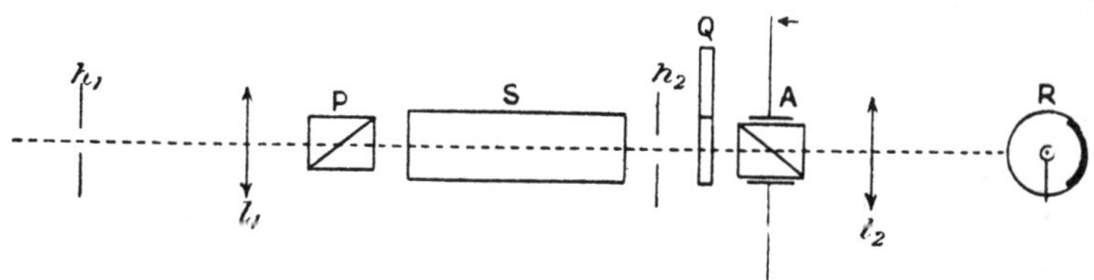

FIG. 102.—BRUHAT AND GUINIER'S PHOTOELECTRIC ULTRA-VIOLET POLARIMETER.

placement of the rays by the double-image analyser and biprism is eliminated by using, instead, a single field Glazebrook analyser A

[1] BRUHAT and GUINIER, *C.R.*, 1933, **196**, 762; *Rev. d'Opt.*, 1933, **12**, 396.

and a biquartz plate Q (Fig. 102). The latter, which is placed immediately in front of the analyser, consists of plates of dextro- and lævo-quartz of exactly the same thickness, cut perpendicular to the optic axis and united edge to edge.

The biquartz plate can be moved along its own length so that the rays traverse either the dextro- or the lævo-rotatory half. The two halves of the plate, being of equal thickness, produce equal but opposite rotations in the ray from the polariser. If, therefore, the analyser is set *exactly* to extinction with respect to the polariser the ray incident on the photoelectric cell will have the same intensity whichever half of the biquartz plate it has traversed.

It is not necessary to find this exact setting directly. The approximate extinction position is found with the biquartz plate removed altogether. The analyser is then set first on one side of the extinction position and then on the other, and the galvanometer " kick " is recorded as the biquartz plate is moved sharply backwards and forwards from one position to the other. The exact extinction position is thus found by interpolation. This is repeated with the optically-active medium in position.

CHAPTER XVIII.

MEASUREMENT OF ROTATORY DISPERSION.

(*c*) In the Infra-red.

Biot and Melloni,[1] with the help of the thermopile invented by Melloni in 1830, first showed that plane polarised heat rays have their planes of polarisation rotated in traversing a column of quartz along the optic axis. Subsequently Wartmann[2] produced qualitative evidence of the occurrence of the Faraday effect of magnetic rotation for heat rays. Heat rays, partially polarised by traversing a pile of mica plates, were made to pass through a cylinder of rock-salt between the pole-pieces of an electromagnet. The transmitted rays were analysed by a second pile of mica plates, set to extinction with respect to the polarising pile, and were finally incident on a thermopile connected to a galvanometer. A deflection of the latter, caused by switching on and off, or by reversing, the current in the electromagnet, indicated that the plane of polarisation of the heat rays was rotated in traversing the rock-salt under the influence of the magnetic field.

Desains.—The first quantitative measurements of magnetic rotatory power for heat rays were made by de la Provostaye and Desains.[3] Sunlight reflected from a heliostat was polarised by means of a rhomb of Iceland spar. The deviated extraordinary ray was cut off and the ordinary ray was allowed to traverse a piece of flint glass in a magnetic field. The analyser consisted of a second calcite rhomb set with its principal section at 45° to that of the polariser. This position gives maximum variation of the intensity of the rays transmitted by the analyser when the rotations produced by the substance are relatively small. Only the ordinary ray was allowed to fall on the thermopile.

Before the magnetic field is excited, the intensity of the rays transmitted by the analyser, as measured by the galvanometer deflection is, by the Law of Malus, $d = a \cos^2 45° = a/2$ where a is the intensity of the rays incident on the analyser. If the rotation produced by the medium on exciting the magnetic field is θ, the new galvanometer deflection, $D = a \cos^2 (45° - \theta)$, measures the

[1] Biot and Melloni, *C.R.*, 1836, **2**, 194.

[2] Wartmann, *C.R.*, 1846, **21**, 556 and 745.

[3] de la Provostaye and Desains, *A.C.P.*, 1849, [iii], **27**, 232 ; 1850, [iii], **30**, 267.

intensity of the rays now transmitted by the analyser. The rotation θ was deduced from the deflections by means of the expression

$$\frac{D - d}{d} = \frac{a \cos^2 (45 - \theta) - a \cos^2 45}{a \cos^2 45} = \sin 2\theta.$$

Later, Desains [1] modified this method in order to measure the natural rotatory power of quartz for heat rays. The inclination of the direction of vibration of the ray incident on the analyser was found by measuring the galvanometer deflection when the analyser was in one position and again after rotating it through 90°. Let OP (Fig. 103) be the direction of the vibration of intensity a incident on the analyser. Let OA and OA′ be the directions of the vibrations transmitted by the analyser when in two positions mutually at right angles, and δ be the angle OP makes with one of these directions. Then the galvanometer deflections x and y for each position of the analyser OA and OA′ are proportional to $a \cos^2 \delta$ and $a \sin^2 \delta$ respectively. Thus, since

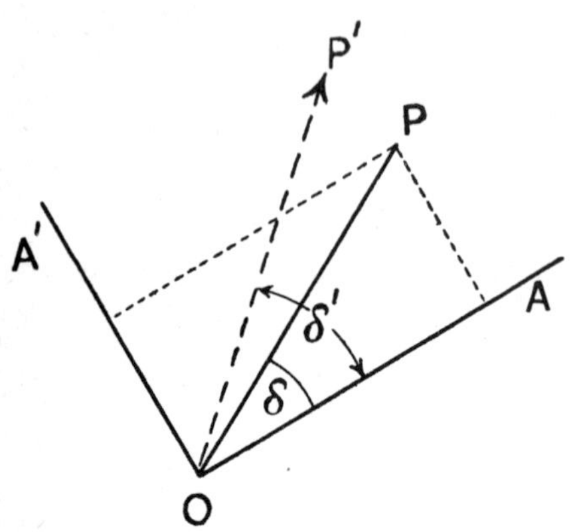

FIG. 103.—DESAINS' SECOND METHOD OF MEASURING ROTATORY DISPERSION IN THE INFRA-RED.

$$\cos^2 \delta = \frac{x}{x + y}$$

the angle δ can be found from the galvanometer deflections. The orientation δ' of the incident vibration OP′, in the absence of the quartz, was similarly found and thus the actual rotation $\theta = \delta' - \delta$.

By this means, using a calcite prism and a slit to disperse the solar radiations into a spectrum, Desains measured the rotatory power of quartz for the heat rays accompanying the yellow, green and blue components of solar light. The rotations observed were all less than the rotations for the corresponding visible light, showing that Desains was actually dealing with infra-red radiations of longer but unknown wave-length.

Carvallo.—The first observations on quartz with infra-red radiations of known wave-length were made by Carvallo.[2]

Radiations from the source L (Fig. 104) were condensed by the lens l_1 and polarised by the rhomb of Iceland spar P mounted in a

[1] DESAINS, *C.R.*, 1866, **62**, 1277; 1877, **84**, 1056.
[2] CARVALLO, *A.C.P.*, 1892, [vi], **26**, 113.

divided circle. The polarised rays traversed the block of quartz Q, and were refracted by the infra-red spectroscope S on to a thermopile T connected to a galvanometer G. The calcite prism A, with its optic axis parallel to the refracting edge, served both as a dispersive and analysing system. It was previously calibrated for certain infra-red rays so that any one of them could be made to impinge on the thermopile by rotating the prism on its graduated

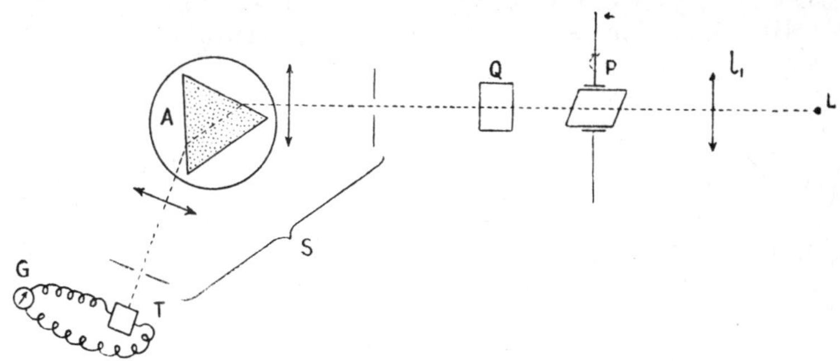

FIG. 104.—CARVALLO'S INFRA-RED POLARIMETER.

circle. By rotating the polariser the "dark" bands of Fizeau and Foucault (p. 199) were made to pass across the thermopile, and the rotatory power of quartz was thus measured for 5 wave-lengths between 10,800 and 21,400 A.U. The *ordinary* spectrum from the analysing prism gave the best results owing to its larger dispersion.

Hussel.—An ingenious method of measuring the rotatory power of quartz in the infra-red was used by Hussel.[1] A pair of Senarmont

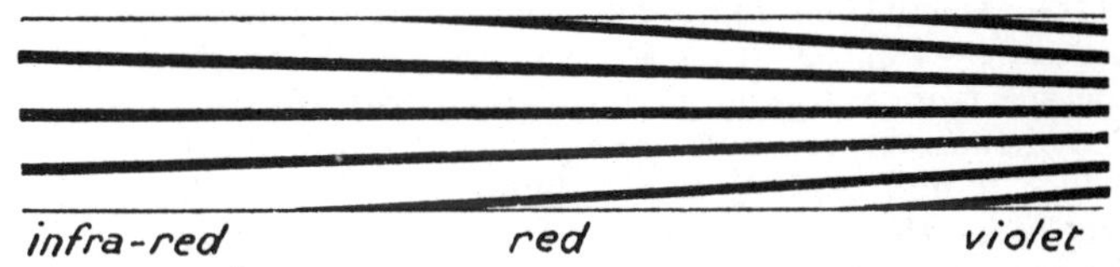

FIG. 105.—ROTATORY DISPERSION OF QUARTZ IN THE INFRA-RED (Hussel's method).

wedges (p. 184) was placed between two crossed polarising prisms, but with its refracting edges horizontal. A spectroscope placed beyond the analyser then showed a spectrum crossed by longitudinal extinction bands, which were close together in the blue and wider apart in the red as shown in Fig. 105 (compare Fig. 93, p. 216).

The distance between consecutive bands is inversely proportional to the rotation produced by quartz for the wave-length in question. The bands were rendered visible by making use of the property of

[1] HUSSEL, *Wied. Ann.*, 1891, **43**, 498.

infra-red radiations of extinguishing the phosphorescence of a screen. Hussel was thus able to measure the distance between the extinction bands and to deduce the rotatory power of quartz for various wave-lengths in the infra-red.

Moreau.—By a method analogous to those of Desains and Carvallo, but using Foucault prisms and solar radiations, Moreau [1] measured the rotatory power of quartz for 9 wave-lengths from 8420 to 17,150 A.U. with an accuracy estimated at 0·03°.

Dongier.—Further measurements with quartz were made by Dongier.[2] A double-image calcite prism was used for wave-lengths up to 21,000 A.U. At longer wave-lengths than this the two rays from calcite are unequally absorbed; the calcite prism was therefore replaced by two Wollaston prisms used in conjunction for greater separation of the ordinary and extraordinary rays.

The two spectra from the double image analyser were focussed on a differential thermopile or differential bolometer. The directions of vibration in the two rays are at right angles so that the extinction bands produced by the column of quartz were displaced in the two

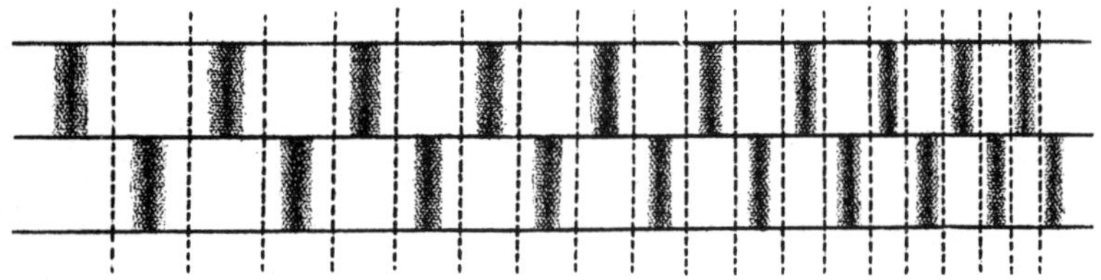

FIG. 106.—ROTATORY DISPERSION OF QUARTZ IN THE INFRA-RED (Dongier's method).

spectra as shown in Fig. 106. The rotation corresponding to each band differs by 180° from those of the bands on either side in the same spectrum. At the points marked with dotted lines the intensities of the infra-red rays are equal in the two spectra and therefore produce a null reading in the differential thermopile or bolometer. These were the points used by Dongier in measuring the rotatory power of quartz for 11 wave-lengths from 7670 to 24,000 A.U.

Meyer.—In order to extend the observations of rotatory dispersion beyond about 25,000 A.U., it is necessary to dispense with the use of calcite and glass. The lenses are therefore made of rock-salt or sylvine or are replaced by mirrors; the dispersive prism is also made of rock-salt or sylvine, while the radiations are polarised and analysed by reflection.

In Meyer's apparatus [3] (Fig. 107) infra-red radiations from a Nernst lamp L are plane polarised by reflection from the surface of a polariser

[1] MOREAU, *A.C.P.*, 1893, [vi], **30**, 433.

[2] DONGIER, *A.C.P.*, 1898, [vii], **14**, 331; *Journ. de Phys.*, 1898, [iv], **7**, 637; *Bull. Soc. fr. Phys.*, 1898, **105**, 1.

[3] U. MEYER, *Ann. der Physik*, 1909, **30**, 607.

P, and are then reflected by a concave mirror M_1 through the optically-active medium S. After traversing the rock-salt lens l_1, they are analysed by reflection from the surface of the analyser A and brought again to a focus on the slit h_1 of the infra-red monochromator. The constant deviation dispersive system D of the latter, consists of a rock-salt prism p and a plane mirror m. The concave mirror M_2 renders the radiations parallel before they enter the dispersive system and M_3 focusses them again on the exit slit h_2 of the monochromator. The radiations are finally focussed by the rock-salt lens l_2 on the sensitive bolometer B.

The reflecting surfaces of the polariser P and analyser A are of amorphous selenium obtained by melting selenium on a plane-glass surface and then removing the glass. Such a surface has a very high reflecting power and does not absorb infra-red radiations, so that it is an ideal material for producing completely plane polarised rays by reflection.

The polarising surface P and the Nernst lamp L are mounted

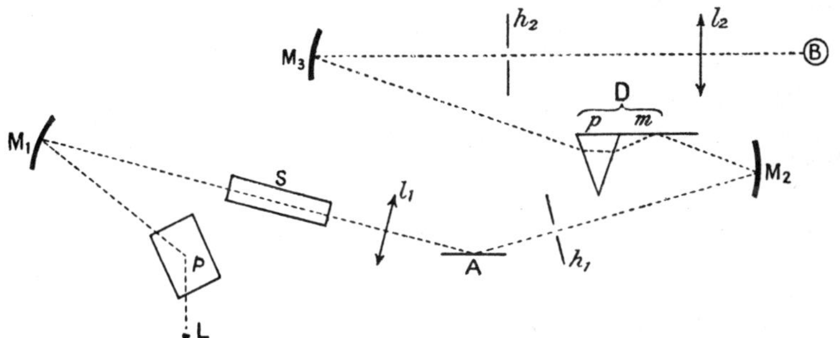

FIG. 107.—MEYER'S INFRA-RED POLARIMETER.

together so that they rotate about the reflected ray from P as axis; thus they can be set to extinction with respect to the analyser A.

By means of this apparatus Meyer was able to measure the magnetic rotatory dispersion of substances for 18 infra-red wave-lengths up to 88,500 A.U., a greater range than had previously or has since been covered. The method is applicable equally well to the measurement of natural optical activity in this region.

Ingersoll.—In 1906 Ingersoll[1] redetermined the infra-red rotatory dispersion of quartz over a larger range of wave-lengths than had been used with this medium hitherto, and in 1917[2] he modified the method and applied it to the natural and magnetic rotatory dispersion of organic substances. The arrangement of the optical parts in Ingersoll's modified apparatus (Fig. 108) bears some resemblance to that of Meyer, from which it differs, however, in the following particulars. The source L of infra-red radiations is a flat

[1] INGERSOLL, *Phil. Mag.*, 1906, **11**, 41; *Phys. Rev.*, 1906, **23**, 489.
[2] *Ibid.*, 1917, **9**, 257.

strip tungsten lamp. The polariser P and the analyser A are not reflecting surfaces but polarising prisms. The analyser is a double-image prism, and both rays which it produces are made to traverse the constant deviation infra-red spectroscope, and are finally focussed

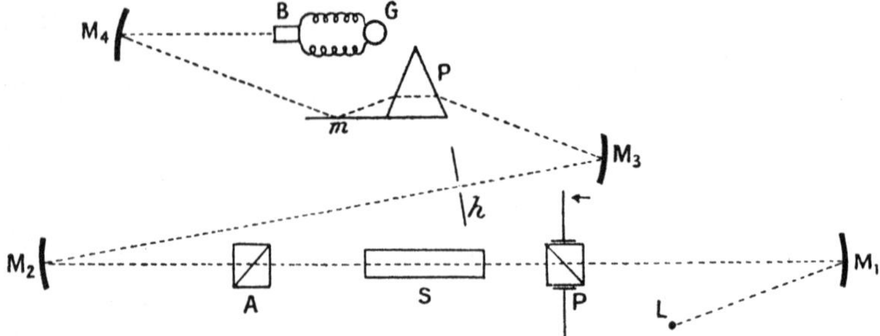

FIG. 108.—INGERSOLL'S INFRA-RED POLARIMETER.

one on each of two strips of a sensitive differential bolometer connected to a sensitive Thomson galvanometer.

In the absence of an optically-active medium the polariser is set so that the direction of the vibration which it emits is inclined at 45° to the directions of vibration of the two rays transmitted by

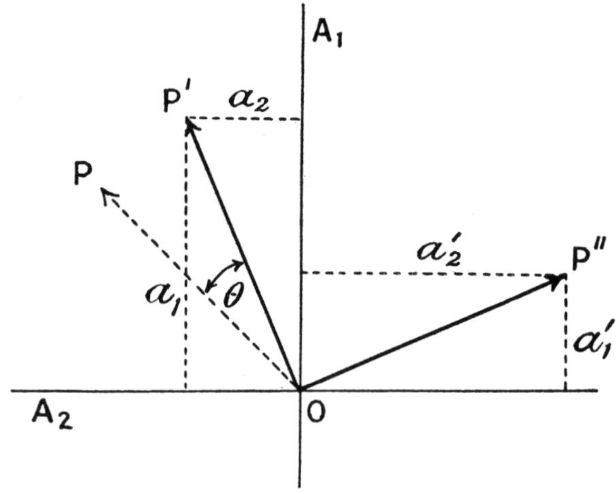

FIG. 109.—INGERSOLL'S METHOD OF MEASURING INFRA-RED ROTATORY DISPERSIONS.

the analyser. The latter are plane polarised in directions at right angles so that there is a null reading in the differential bolometer. In Fig. 109, OP and OA_1, OA_2 represent the directions of vibration of the rays transmitted by the polariser and analyser respectively. On introducing the optically-active medium the direction OP is rotated through an angle θ to OP′ and the galvanometer shows a deflection proportional to the difference between the squares of the components a_1 and a_2. The deflection from the null position is not, however, dependent only on the rotation θ because some radiation is lost by absorption and reflection when the optically-active substance is introduced. To avoid errors due to this loss the polariser is then rotated through 90° so that OP′ now moves to OP″. The intensities of the two rays from the analyser are thus reversed, and the difference in the deflection between the two positions of the polariser is the one recorded. Half this deflection is now proportional to the rotation θ if the latter is relatively small, the effect of all losses by reflection and absorption being eliminated since the substance remains in position the whole time. The ratio between deflection and rotation is obtained by noting the deflection when the polariser is rotated through a known small angle, say 3°.

Ingersoll, by this method, measured the rotatory dispersion of quartz and of five organic substances up to a wave-length of 21,400 A.U.

Lowry and Coode-Adams.—The observations of Ingersoll on quartz were repeated and extended to a wave-length of 25,170 A.U. by Lowry and Coode-Adams [1] in 1927, but by a completely different method. Radiations from a shielded Nernst lamp are made to traverse a polarimeter consisting of two Glazebrook prisms of large aperture. The emergent radiations are dispersed and examined by a constant deviation infra-red spectrometer consisting of a 60° glass or quartz prism with plane steel mirror, concave steel focussing mirrors instead of lenses and a sensitive thermopile mounted at the focus of the eyepiece of the telescope. The thermopile is connected to a sensitive Paschen galvanometer suitably shielded. The graduated drum of the spectrometer was previously calibrated with the help of the interference fringes produced by means of an etalon consisting of two lightly silvered quartz plates separated by a thin air film. The apparatus is furnished with a shutter for intercepting the radiations.

The method of working is to open the shutter and note the " kick " of the galvanometer for successive positions of the spectrometer drum. These " kicks " pass through maxima and minima on moving along the spectrum. Each maximum or minimum corresponds to a rotation, by the quartz column, of 180° more or less than that of the adjacent maximum or minimum. The positions of the maxima and minima are assigned to particular wave-lengths from the calibration curve of the spectrometer drum. By extrapolation from the known rotations in the visible spectrum the correct multiple of 180° can be assigned to each maximum and minimum.

To eliminate the effect of variations in the loss of intensity by reflection and absorption the wave-lengths of the maxima and minima were taken as the mean of the two series of values obtained

[1] LOWRY and COODE-ADAMS, *Phil. Trans.*, 1927, A. **226**, 391.

with the analyser in two positions 90° apart, when the maxima of one series correspond to the minima in the other.

Lowry and Snow.—Using an infra-red grating spectrometer described by Snow and Taylor,[1] Lowry and Snow [2] extended the measurements of rotatory dispersion of quartz up to a wave-length of 32,100 A.U. The method differed from that described above in that the analyser was rotated until the galvanometer deflection was a maximum with and without the quartz column in position. The rotation was measured for 100 wave-lengths at intervals of about 100 A.U. between 18,000 and 27,900 A.U. At this point quartz shows an intense infra-red absorption band. Four more observations were, however, obtained in a narrow region of transparency between 27,900 and 32,100 A.U., beyond which quartz again begins to absorb strongly.

Very few measurements of rotatory dispersion have been made in the infra-red, since this phenomenon shows no points of outstanding interest, the rotatory power decreasing steadily with increasing wave-length, even when passing through an infra-red absorption band.

[1] Snow and Taylor, *P.R.S.*, 1929, A. **124**, 442.
[2] Lowry and Snow, *P.R.S.*, 1930, A. **127**, 271.

CHAPTER XIX.

MEASUREMENT OF CIRCULAR DICHROISM.

Production of Circularly and Elliptically Polarised Light.—The most convenient methods of producing circularly and elliptically polarised light for measurements of circular dichroism are by means of a Fresnel rhomb or a quarter-wave plate.

(*a*) *Fresnel's Rhomb.*—Fresnel considered that, when a plane polarised ray of light is totally reflected at the surface of a less dense medium, in such a way that its plane of polarisation is inclined at an angle of 45° to the plane of incidence, then the ray is resolved into two components of equal amplitude, polarised in, and at right angles to, the plane of incidence, respectively. Moreover, in the process of reflection, these two components suffer a relative phase retardation. The magnitude of the relative phase difference Δ depends on the angle of incidence ϕ, and on the refractive index n of the less dense medium, as given by the equation

$$\tan \tfrac{1}{2}\Delta = \frac{\cos \phi \sqrt{\sin^2 \phi - n^2}}{\sin^2 \phi}.$$

When the relative phase retardation is $\pi/2$ (i.e. when $\Delta = 90°$) the reflected ray is circularly polarised.

The above equation shows that the relative difference of phase Δ is a maximum for that angle of incidence ϕ', which is given by

$$\sin^2 \phi' = \frac{2n^2}{1 + n^2}$$

and the maximum value Δ' of the difference in phase is then given by

$$\tan \tfrac{1}{2}\Delta' - \frac{1 - n^2}{2n}.$$

This equation shows that, for the relative phase retardation to be 90°, the refractive index of the denser medium must be greater than 2·4 approximately. Thus, in order to produce circularly polarised light by a single total internal reflection, one must employ a medium whose refractive index is of the order of that of the diamond.

Fresnel surmounted this difficulty by using two successive total internal reflections, so that the relative phase retardation at each reflection is only $\pi/4$ (i.e. $\Delta = 45°$). The minimum value for the refractive index of the denser medium is now only 1·5 approximately. Fresnel therefore constructed a rhomb of St. Gobain glass of refractive index 1·51. The maximum value of the relative phase retardation obtainable with this material is 45° 36′ when the angle of incidence is 51° 20′. For the phase retardation to be exactly 45° at each reflection the angle of incidence must be 48° 37½′ or 54° 37⅓′. Fresnel actually constructed a rhomb, as shown in Fig. 110, whose acute angle was 54° 37′, and thereby obtained circularly polarised light.

If three or four total internal reflections are used instead of two, the corresponding angles of incidence are respectively 69° 12′ and 74° 42′, when the refractive index of the medium is 1·51. By such multiple reflections circularly polarised light can be produced when the refractive index of the medium is considerably less than 1·5 (compare p. 253).

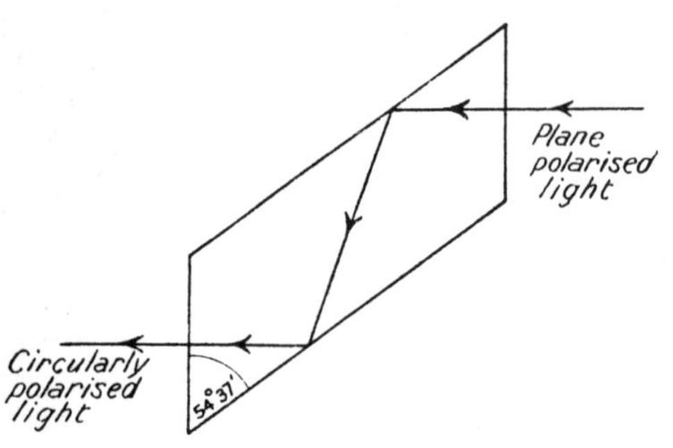

FIG. 110.—FRESNEL RHOMB.

Since the refractive index of a medium is different for different wave-lengths, a Fresnel rhomb of given angle can only give pure circularly polarised light for one particular wave-length. Thus if the rhomb is made to give circularly polarised light for the sodium *D* line, then for red light the relative phase retardation at each reflection will be less than $\pi/4$, and for violet light more than $\pi/4$. In these regions of the spectrum the emergent vibration will not be quite circular, but elliptical, the major axis of the ellipses being parallel to the direction of vibration of the incident plane polarised beam. The refractive index of glass, however, does not vary very much in the visible spectrum. A Fresnel rhomb which is constructed for sodium light will, therefore, give very nearly circularly polarised vibrations throughout the range of wave-lengths from 7000 A.U. to 4000 A.U. As an example, the following table shows the error in the relative phase retardation at the extreme ends of the visible spectrum, for a Fresnel rhomb of glass which gives pure circularly polarised sodium light.

Refractive index $\mu_D = 1·51714$. Angle of rhomb = 55° 9′.

λ.	Relative Phase Retardation. Degrees.	Error. Degrees.
7683	89·6	− 0·4
5893	90·0	± 0·0
4340	91·1	+ 1·1

If the plane of polarisation of the incident ray is rotated through 90° from its former position, the emergent ray is again circularly polarised but in the opposite direction. When the planes of polarisation and incidence are parallel or mutually at right angles the emergent ray is plane polarised. For orientations of the plane of polarisation of the incident ray other than these, the emergent vibration is elliptical. Thus all types of polarised vibration can be produced as shown in Fig. 111, where the angles refer to the mutual inclination of the planes of polarisation and incidence.

Two Fresnel rhombs of glass were used by Cotton (pp. 154 and 240) in order to measure the circular dichroism of the coloured tartrates of copper and chromium.

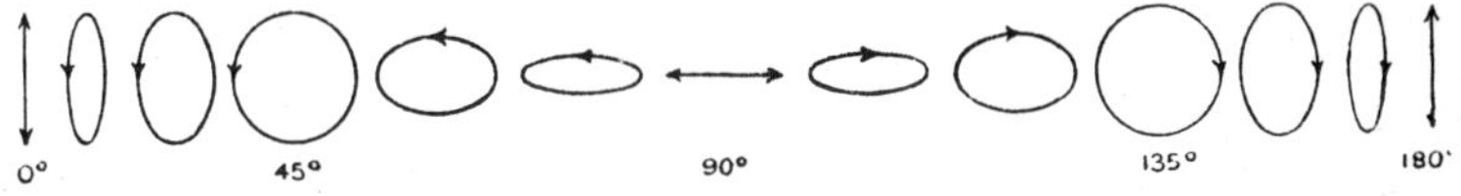

FIG. 111.—PRODUCTION OF CIRCULARLY AND ELLIPTICALLY POLARISED LIGHT BY MEANS OF A FRESNEL RHOMB.

(*b*) *Quarter-wave Plate.*—A quarter-wave plate consists of a thin lamina of a uni- or bi-axial crystal cut parallel to the optic axis. It is usually made from quartz or mica, and is of such a thickness that the ordinary and extraordinary rays suffer a relative phase difference of $\lambda/4$ in passing perpendicularly through it. For the production of circularly polarised light, a plane polarised ray is incident normally on the quarter-wave plate, its plane of polarisation being inclined at an angle of 45° to the optic axis of the plate. The ray is then resolved into two components of equal amplitude, plane polarised in directions parallel and perpendicular to the optic axis. These traverse the plate, as the ordinary and extraordinary rays, with different velocities and thus suffer a relative phase retardation. The thickness d of such a quarter-wave plate is given by

$$d = n\frac{\lambda}{4}(\mu_o \sim \mu_e),$$

where n is an integer and μ_o and μ_e are the refractive indices for the ordinary and the extraordinary ray respectively. The difference between the two refractive indices varies appreciably throughout the visible spectrum so that it is customary to use at least four plates of different thicknesses for the red, yellow, green and blue parts of the spectrum. As an example, the error in the relative phase retardation at the extreme ends of the visible spectrum is given below for a quarter-wave plate of quartz which is correct for sodium light.

λ.	Relative Phase Retardation. Degrees.	Error. Degrees.
7683	92·0	+ 2·0
5893	90·0	± 0·0
4340	87·1	− 2·9

The errors in the relative phase-retardation are thus considerably larger than those for a Fresnel rhomb.

Circularly polarised light of opposite sign, and left and right elliptical vibrations, can be produced as in the case of the Fresnel rhomb, by rotating either the plane of polarisation of the incident beam or the quarter-wave plate itself.

Cotton.—(*a*) The most direct method of measuring circular dichroism is by determining separately the absorptive power of the medium for left circularly polarised light and for right circularly polarised light. This was one of the methods used by Cotton[1] in his original researches on the phenomenon of circular dichroism (p. 149). The arrangement of the optical parts of Cotton's apparatus is shown in Fig. 112.

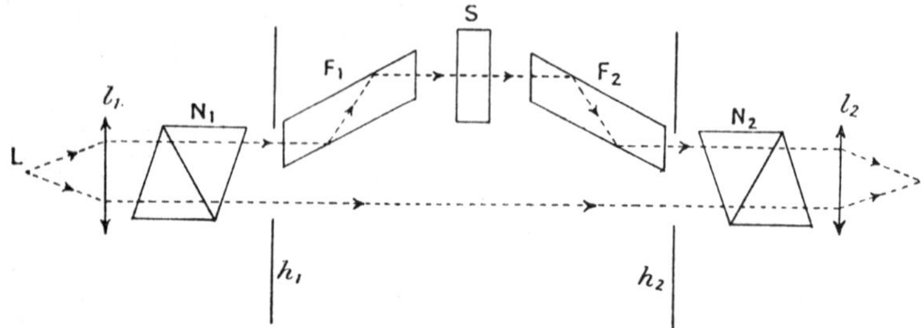

FIG. 112.—COTTON'S APPARATUS FOR MEASURING CIRCULAR DICHROISM.

Light from the source L is rendered parallel by the lens l_1 and plane polarised by the Nicol N_1. Part of this ray is incident on the Fresnel rhomb F_1, placed so that the plane of polarisation and the plane of incidence at total reflection are inclined at 45° to one another. The circularly polarised ray emerging from F_1 traverses the solution S and is rendered plane polarised again by the second rhomb F_2 and emerges with its plane of polarisation at right angles to that of the original ray. The other part of the original ray remains unaltered. Thus there enter the Nicol N_2, two rays plane polarised in directions mutually at right angles. One of these rays has, however, been reduced in intensity by being partially absorbed, as a circularly

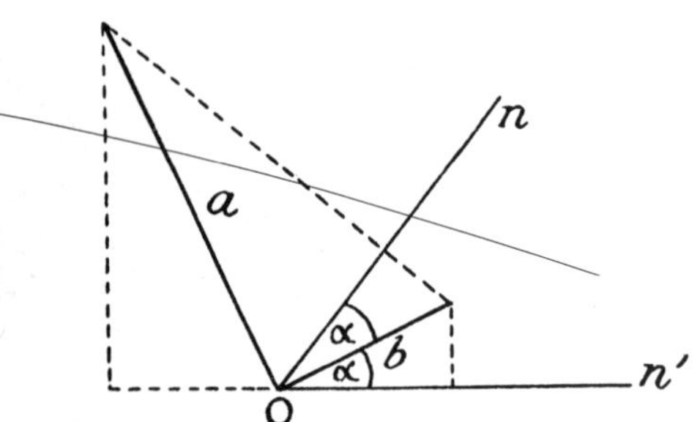

FIG. 113.—ILLUSTRATES COTTON'S METHOD OF MEASURING CIRCULAR DICHROISM.

[1] COTTON, *A.C.P.*, 1896, [vii], 8, 347.

polarised ray, by the solution S. The width of the beam traversing the apparatus is limited by apertures in the screens h_1 and h_2.

Let the amplitudes of the rays which have and have not traversed the rhombs and solution be b and a respectively (Fig. 113). There are two positions of the Nicol N_2, corresponding to On and On' in Fig. 113, which will make the intensities of the two rays traversing it equal, and thus make the two fields of vision equally bright. The tangent of half the angle 2α between these two positions gives the ratio of the amplitudes of the two rays incident on N_2, or

$$\tan \alpha = \frac{b}{a}.$$

The reduction in intensity of the ray passing through the rhombs and solution is not, however, due entirely to the absorption of the solution S; much light is also lost by reflections at the various surfaces. If these reflections were eliminated the amplitude of the ray would not be b but Kb where K is a constant larger than unity. The reduction in intensity by absorption only is given by the exponential law

$$Kb = ae^{-\delta z}$$

where δ is the absorption coefficient for, say, right circularly polarised light and z is the thickness of solution traversed, or

$$K \tan \alpha = e^{-\delta z}.$$

The observations are now repeated with a solution vessel S of a different length z'. Let α' be half the new angle between the two positions of the Nicol N_2, then

$$K \tan \alpha' = e^{-\delta z'}$$

$$\therefore \frac{\tan \alpha}{\tan \alpha'} = \frac{e^{-\delta z}}{e^{-\delta z'}} \qquad \therefore e^{\delta(z'-z)} = \frac{\tan \alpha}{\tan \alpha'}.$$

From the values of z and z', α and α', the absorption coefficient δ for right circularly polarised light may be deduced.

By rotating the Nicol N_1 through 90° left circularly polarised light is made to traverse the solution, and its absorption coefficient is determined in precisely the same way.

(*b*) Cotton used an alternative method [1] for measuring circular dichroism, in order to determine if the simultaneous passage of left and right circularly polarised rays through the same medium had any effect on the difference in the absorption. The apparatus employed by Cotton in the alternative method is shown in Fig. 114.

Light from a source L traverses a Nicol prism P and a double circular polariser C (double quarter-wave plate) orientated relative to one another so that the two halves of C produce circularly polarised rays of opposite sign as already described on page 150. The dividing line of the plate C is set horizontally. The two rays

[1] COTTON, *A.C.P.*, 1896, [vii], **8**, 347.

then traverse a cell F of fixed length and a cell V of variable length, both of which are filled with the solution to be investigated. In the upper half of the cell F is placed a thin plate of glass G whose lower edge is horizontal, and which exactly covers the upper field of the divided plate C. The two halves of the plate C are viewed through a small circular aperture in the screen h which serves to keep the eye on the axis of the apparatus.

The length of the variable vessel V is adjusted until the two halves of the plate C appear equally illuminated. Let the length of the vessel V be z when the two fields match, and let the lengths of the upper and lower parts of the vessel F be x and y respectively, i.e. the thickness of the glass plate G is $y - x$. The oppositely polarised circular vibrations produced by the upper and lower halves of the plate C therefore traverse lengths $x + z$ and $y + z$ of the solution

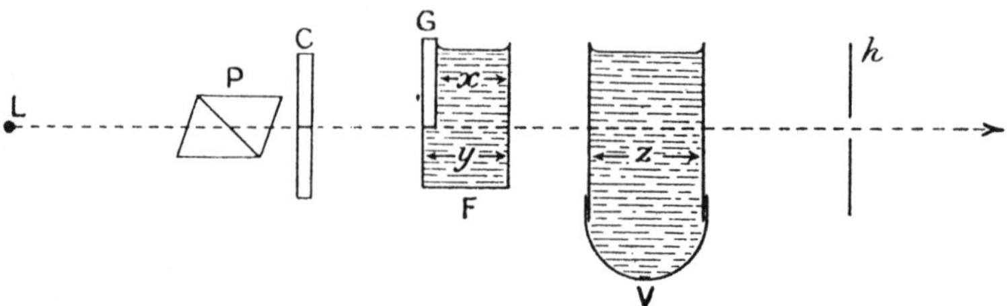

FIG. 114.—COTTON'S ALTERNATIVE METHOD OF MEASURING CIRCULAR DICHROISM.

respectively. If the absorption coefficients for the two circularly polarised rays are δ and γ, then

$$e^{-\delta(x+z)} = e^{-\gamma(y+z)}$$

or

$$\frac{\delta}{\gamma} = \frac{y + z}{x + z}.$$

If the circularly polarised ray traversing the lower half of the cell F is more strongly absorbed than the other, it will clearly be impossible to match the two halves of the plate C by altering the length of the vessel V. The Nicol polariser P must then be turned through 90° in order to reverse the sign of the circular vibrations produced by the two halves of the plate C.

If the cell F is now filled, not with the medium under investigation, but with an optically-inactive absorbing medium of absorption coefficient k, i.e. one which absorbs the two circular rays to the same extent, a different length z' of the vessel V will be required to produce equal illumination of the two fields. Then, as given above,

$$e^{-(kx+\delta z')} = e^{-(ky+\gamma z')}.$$

Therefore

$$\delta - \gamma = k\frac{y - x}{z'}.$$

If the absorption coefficient k of the inactive liquid is known, the value of $(\delta - \gamma)$ may be found and hence the absolute value of δ and γ.

(c) The method described on page 152, by which Cotton[1] first showed that a plane polarised ray is converted into an elliptically polarised ray in traversing a medium exhibiting circular dichroism, was also used to measure quantitatively the ellipticity produced. This method is of interest, because the rotatory power of the medium can also be measured without altering the apparatus. The arrangement of the latter is shown in elevation in Fig. 115.

Sunlight, reflected from a heliostat, is plane polarised by the Nicol prism P and traverses the Fresnel rhomb F, which is placed so that the plane of incidence at total reflection in the rhomb is parallel to the principal section of the polariser. In this position the ray traverses the rhomb unchanged and emerges still plane polarised. It then passes through the Bravais double plate B (p. 152), the dividing line of which is horizontal, through the analysing Nicol A set to extinction

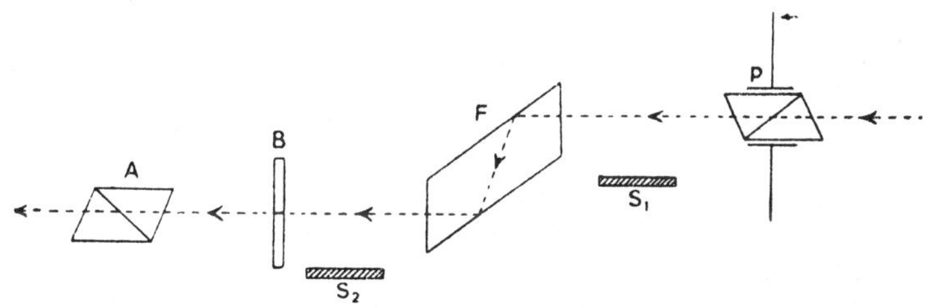

FIG. 115.—COTTON'S APPARATUS FOR MEASURING CIRCULAR DICHROISM IN TERMS OF ELLIPTICITIES.

with respect to the polariser, and finally enters the slit of a spectroscope. In the absence of a medium showing circular dichroism the dark bands are coincident in the two halves of the spectrum (p. 153).

(i) *Measurement of the Ellipticity.*—The cell containing the solution to be investigated is placed on the support S_2 between the rhomb and the Bravais double plate. The ellipticity ϕ imposed by the medium on the plane polarised ray from the rhomb causes the dark bands in the two halves of the spectrum to separate. By rotating the polariser P the vibration from the rhomb becomes elliptical, the sign of the ellipse depending on the direction of rotation of the polariser. It is therefore possible to rotate the polariser until the ellipticity produced by the rhomb is exactly equal to, but opposite in sign from, that produced by the medium. The exact counterbalance of the ellipticity is shown when the two halves of a dark band are brought back into alignment. The angle through which the polariser has been rotated is then equal to the ellipticity produced by the medium for the particular wave-length corresponding to the

[1] COTTON, *A.C.P.*, 1896, [vii], 8, 347.

centre of the dark band in question. By repeating this procedure for the other dark bands the ellipticity for various spectral regions is determined. By rotating the analyser through 90° new dark bands appear in intermediate regions of the spectrum for which the ellipticity may then be measured in like manner.

If the rotatory power of the medium is large it may be necessary to rotate the analyser slightly to make the bands as dark as possible.

A quarter-wave plate may be employed in place of the Fresnel rhomb in order to avoid the lateral displacement of the ray, but it has the disadvantage that it can only be used for a limited range of wave-lengths, as discussed on page 239.

Cotton has estimated that the experimental error involved in measuring ellipticities by this method is of the order of 0·2°.

(ii) *Measurement of the Rotation.*—The polariser is restored to its original position and the cell of solution is placed between the polariser and the rhomb on the support S_1. The elliptical vibration emerging from the solution has its major axis rotated through an angle α due to the rotatory power of the solution, so that the ray from the rhomb is also elliptical and the dark bands in the two spectra are again displaced. By rotating the polariser through an angle $-\alpha$ the major axis of the elliptical vibration incident on the rhomb is made parallel to the plane of incidence at total reflection. The rhomb converts this vibration into a plane polarised ray so that the two halves of a dark band become colinear once more. The rotation of the polariser needed to produce this effect is equal in magnitude but opposite in sign to the rotation produced by the medium. In this position the plane of polarisation of the ray emerging from the rhomb makes an angle ϕ equal to the ellipticity, with the major axis of the incident ellipse, but since the ray is plane polarised this effect does not separate the two halves of the dark bands.

Bruhat.—A number of alternative methods have been devised for measuring the ellipticity of the vibration produced when a plane polarised ray traverses an optically-active absorbing medium. Many of these methods [1] are, however, cumbersome, and employ relatively complex and delicate optical systems so that they are little used in practice. Some of these methods involve no less than eight different readings with the various optical parts of the apparatus, in order to obtain a single value for the ellipticity for a particular wave-length.

These methods will not be described here as they are similar, in general principle, to a modified and simpler method described and used by Bruhat [2] in 1915. The apparatus, which is shown in Fig. 116, consists of a polariser P, a double-field analyser A of the Lippich type, and two quarter-wave plates Q and Q′ which are accurate for the particular part of the visible spectrum in which it is wished to measure the ellipticity. In the following descriptions

[1] E.g. CHAUMONT, *Ann. de Phys.*, 1915, [ix], **4**, 61–206.

[2] BRUHAT, *ibid.*, **3**, 232.

the particular direction in a quarter-wave plate which is parallel to the vibrations which suffer retardation will be referred to as the axis of the plate. The auxiliary quarter-wave plate Q′ is pivoted at O on the mounting of the analyser, so that it rotates with the analyser but can be swung out of the path of the rays and back again to its exact former position. It is orientated so that its axis is parallel to the direction of the vibration furnished by the polariser when the analyser is set exactly to extinction.

In the absence of the optically-active medium S and the plate Q, the analyser is set to extinction, so that the two halves of its field are equally illuminated. The plate Q is then inserted in front of the analyser with its axis approximately parallel to that of the plate Q′. Q is then rotated to restore equality of the two fields. In this "zero position" the vibration from the polariser traverses both quarter-wave plates unchanged as in Fig. 117 (*a*) where the arrows p, s, q and q' represent the directions of the vibrations transmitted

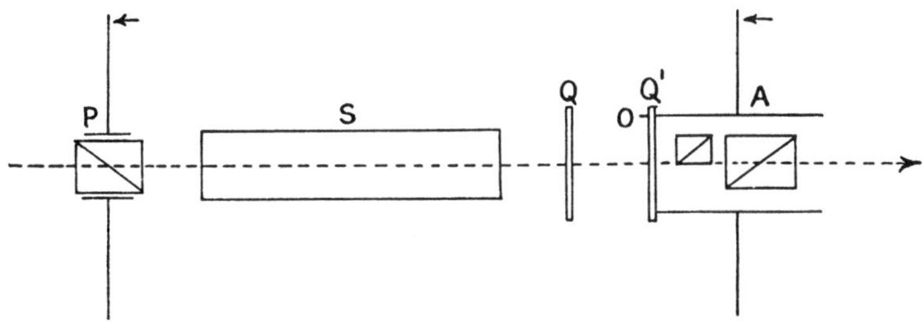

FIG. 116.—BRUHAT'S APPARATUS FOR MEASURING CIRCULAR DICHROISM IN TERMS OF ELLIPTICITIES.

by the polariser, solution, plates Q and Q′ respectively and a is the direction of the vibration *extinguished* by the analyser.

On introducing the medium S between the polariser P and the plate Q the equality of the two fields is destroyed owing to the rotation α and the ellipticity ϕ produced by the solution. The polariser P is rotated until the equality is restored, the rotation necessary being equal to $-\alpha$. This rotation makes the major axis of the ellipse produced by the solution parallel to the axes of the plates Q and Q′. The former converts the vibration into a rectilinear one whose direction makes an angle $-\phi$ with the axis of the plate. The auxiliary plate Q′ converts this back into an elliptical vibration with the direction of the major axis unchanged, i.e. parallel to the vibration extinguished by the analyser whose two fields are again equally illuminated. This condition is illustrated in Fig. 117 (*b*).

The auxiliary plate Q′ is now swung out of the path of the rays, and the analyser is rotated through an angle $-\phi$ in order to restore the equality of the two fields by extinguishing the rectilinear vibration transmitted by the plate Q. This position is shown in Fig. 117 (*c*).

If the elliptical vibration produced by the solution is not converted into an exactly linear vibration by the plate Q, i.e. if the setting of the polariser is faulty, the two fields will appear slightly unequal when the auxiliary plate Q′ is swung back into position. If this is the case, polariser and analyser are rotated slightly in succession until

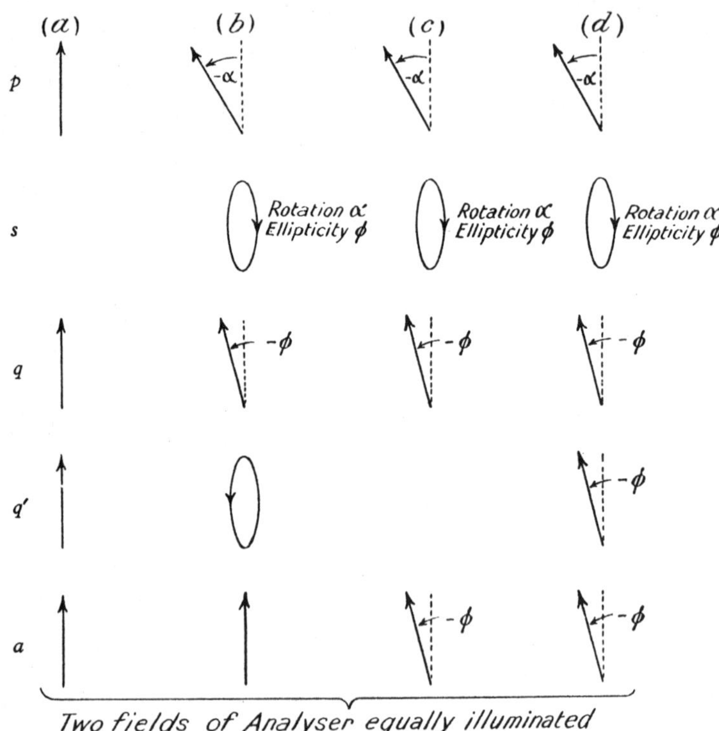

FIG. 117—BRUHAT'S METHOD OF MEASURING CIRCULAR DICHROISM IN TERMS OF ELLIPTICITIES.

the equality persists, whether the auxiliary plate Q′ is in position or not. The condition with Q′ *in* position is shown in Fig. 117 (*d*).

The rotations $-\alpha$ and $-\phi$ of the polariser and analyser respectively thus give directly the rotation and ellipticity produced by the solution, but with the signs reversed.

Bruhat [1] has subsequently described a simplified modification of this method which is applicable when the ellipticity is small. The apparatus which is shown in Fig. 118 is similar to that already

[1] BRUHAT, *Bull. Soc. Chim.*, 1930, [iv], **47**, 251; *Rev. d'Opt.*, 1929, **8**, 413.

described, except that the auxiliary quarter-wave plate Q′ is omitted and the plate Q is mounted so that it can be withdrawn and reinserted exactly in its former position.

As before, the polariser P and analyser A are set to extinction; the quarter-wave plate Q is inserted and rotated to restore equal illumination of the two fields (Fig. 119 *a*). The substance S is placed in position, the plate Q removed, and the polariser is rotated through an angle $-\alpha$ to restore the equality of illumination (Fig. 119 *b*); α is then the rotation produced by the substance. The quarter-wave plate Q is reinserted in exactly the same position as before, thereby converting the elliptical vibration (ellipticity ϕ), produced by the solution, into a linear vibration inclined at an angle $-\phi$ to the axis of the plate. Finally, the analyser is rotated through $-\phi$ to restore equal illumination.

The rotations of the polariser and analyser again give directly the rotation and ellipticity produced by the solution, but with the same reversal of signs.

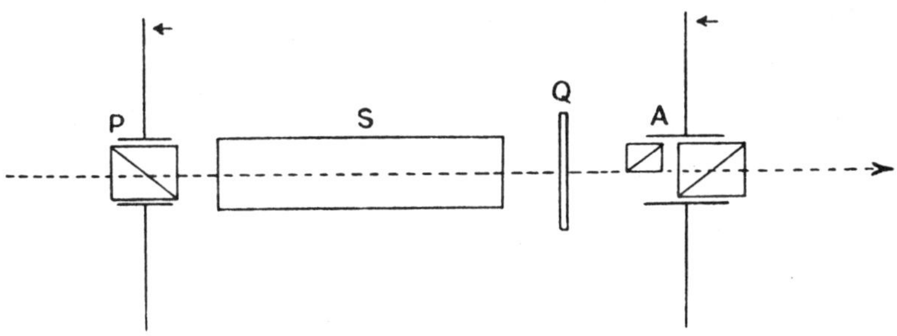

FIG. 118.—BRUHAT'S SIMPLIFIED APPARATUS FOR MEASURING CIRCULAR DICHROISM IN TERMS OF ELLIPTICITIES.

This method is not applicable when the ellipticity is very large because, with the solution in position, the setting of the polariser is made in the absence of the quarter-wave plate, so that an elliptical vibration instead of a linear one is incident on the analyser. This reduces the accuracy with which the two fields can be matched. The error involved is, however, negligible, provided that the ellipticity to be measured does not attain a value comparable with the half-shadow angle of the analyser.

By a slight modification this method of measuring ellipticity can be used with an ordinary polarimeter. The quarter-wave plate must then be mounted in a graduated circle. The double-field polariser remains stationary throughout, while the matching of the two fields is made by rotating the analyser and the quarter-wave plate. In this case the first rotation of the analyser is equal in *sign* and magnitude to the rotation produced by the substance. If the axis of the quarter-wave plate is set at right angles (instead of parallel as described above) to the vibration extinguished by the analyser, the second rotation of the latter is equal in *sign* and magnitude to the ellipticity. This setting of the quarter-wave plate should be checked with the help of a solution which produces an elliptical vibration of known sign.

For other spectral regions it may be necessary to use other quarter-wave plates correct for those regions. A slight error in phase retardation may, however, be allowed for by writing $\phi = \phi' \cos \epsilon$,

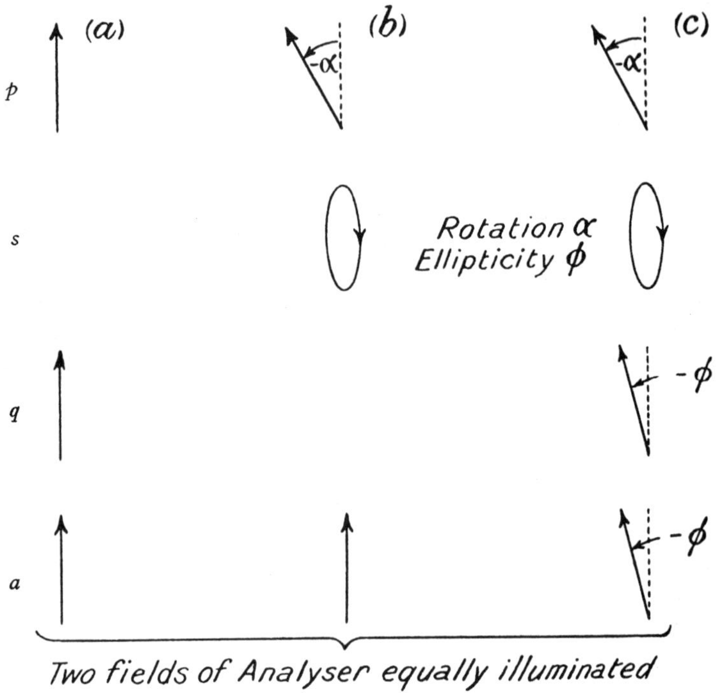

FIG. 119.—BRUHAT'S SIMPLIFIED METHOD OF MEASURING CIRCULAR DICHROISM IN TERMS OF ELLIPTICITIES.

where ϕ is the true ellipticity, ϕ' the apparent ellipticity, as measured by the rotation of the analyser, and ϵ is the error in phase retardation.

Kuhn and Braun.—A simple method of extending measurements of circular dichroism into the ultra-violet was devised in 1930 by Kuhn and Braun.[1] By employing only quartz, fluorspar and water in the optical parts of their apparatus, the method was made available for use over the whole of the visible and ultra-violet spectrum down to a wave-length of 1850 A.U. No precautions are necessary with regard to the rotatory power of the medium under investigation

[1] KUHN and BRAUN, *Z. ph. C.*, 1930, B. **8**, 445.

provided that it is not large, and that the error in phase retardation produced by the Fresnel rhomb is small.

The general principle of the method is as follows. By means of a split Rochon or Senarmont polariser and a Fresnel rhomb, two elliptically polarised rays of opposite sign and equal ellipticity are produced. The left elliptical ray may be regarded as consisting of two circularly polarised rays of opposite sign, the left component having the larger amplitude. The right elliptical ray may similarly

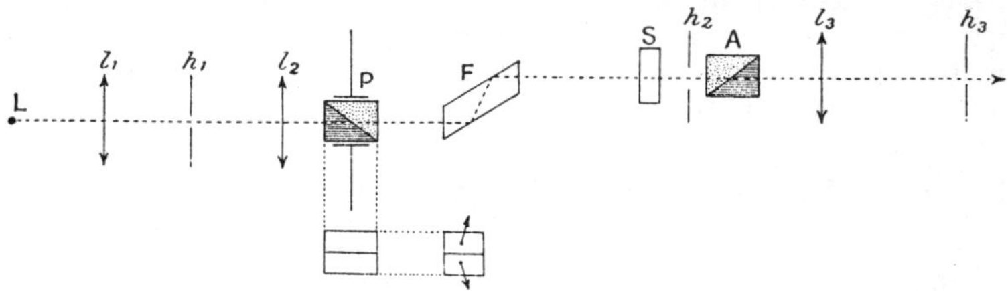

FIG. 120.—KUHN AND BRAUN'S APPARATUS FOR MEASURING CIRCULAR DICHROISM.

be regarded as consisting of opposite circular components of which the right has the larger amplitude. The two rays, on traversing a medium which absorbs left and right circular vibrations to different extents, will have their ellipticities changed, that of the one being increased and that of the other diminished. By rotating the polariser the ellipticities of the rays incident on the solution can be made unequal until those of the emergent rays are again equal. This is tested by an analyser placed so as to transmit only those vibrations corresponding to the minor axes of the two ellipses.

Except for the introduction of the Fresnel rhomb the apparatus (Fig. 120) is identical with that devised by Kuhn[1] for the measurement of rotatory dispersion (p. 219). The special feature of the apparatus is the Fresnel rhomb itself. The ideal material from which to construct the rhomb, from the point of view of transparency, would be fused silica. Until quite recently, however, it was impossible to obtain large pieces of this material which did not exhibit, at least in patches, the phenomenon of double refraction. Kuhn and Braun therefore employed a water rhomb (Fig. 121) consisting of a brass case, with four apertures closed with thin plates of fused silica, and filled with water. The acute angle of the rhomb was 62°, so that the angle of incidence at total reflection at the silica-air interface is 54°.

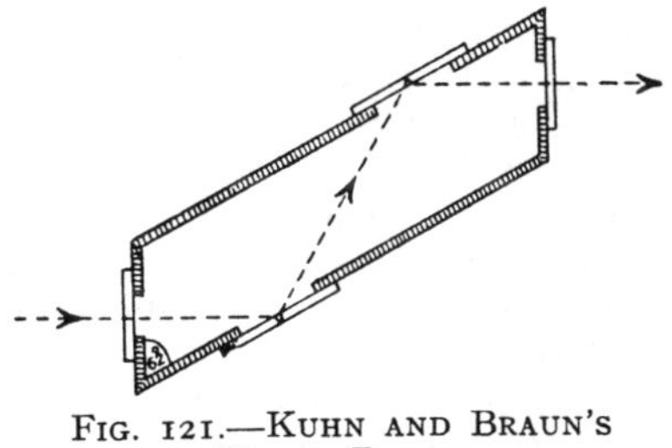

FIG. 121.—KUHN AND BRAUN'S WATER RHOMB.

[1] KUHN, *Ber.*, 1929, **62**, 1727.

Light from an iron arc L (Fig. 120) is focussed by the lens l_1 on a small circular aperture in the screen h_1, which thus serves as a steady source of illumination. An image of this aperture is formed by the achromatic quartz-fluorspar doublet l_2 on the screen h_2 which intercepts the deviated extraordinary ray from the polariser, so that only the ordinary ray enters the analyser. The rhomb F is placed so that its silica windows are vertical, and the end windows exactly perpendicular to the direction of the incident ray. The polariser P, which is mounted in a graduated circle, consists of a split Rochon prism with the dividing line approximately horizontal. The achromatic doublet l_3 focusses an image of the polariser on the slit h_3 of an ultra-violet spectrograph. The analyser A, which consists of a single Rochon prism, also produces two rays, of which only the undeviated ordinary one is allowed to enter the slit of the spectrograph. The mountings carrying the solution tube S, the screen h_2, the analyser A, and the lens l_3 are furnished with lateral slides so that they can be moved on to the optical axis of the apparatus in the absence of the rhomb F. This converts the apparatus into the ultra-violet polarimeter described on page 219.

If the directions of the plane polarised vibrations produced by the two halves of the split prism make small angles one on either side of the plane of incidence at total reflection in the rhomb, there will emerge from the latter two elliptical vibrations of opposite sign corresponding to the two halves of the polariser. If the latter is adjusted so that the plane of incidence in the rhomb bisects the shadow angle of the polariser the ellipticities of the two rays from the rhomb become equal but still opposite in sign. The major axes of both ellipses lie in the plane of incidence at total reflection. Any rotation of the polariser from this position will increase the ellipticity of one of the rays and decrease the ellipticity of the other. If the analyser is set to extinction with respect to the polariser it transmits the vibration components corresponding to the minor axes of the two ellipses. Thus any rotation of the analyser merely darkens or brightens both halves of the spectrum equally, without altering their relative intensity, while any rotation of the polariser alters this relative intensity.

The zero position of the polariser is found by rotating it slightly until the two halves of the spectrum are equally bright. This may be observed visually or photographically; the latter method is the more laborious but is the more accurate. The analyser is now rotated until the two matched fields are as faint as possible. This condition is shown in Fig. 122 (*a*), where *p*, *f*, *s* and *a* represent the vibrations transmitted by the polariser, rhomb, solution and analyser respectively. The two vibrations in each case correspond to the two halves of the field.

On introducing the optically-active absorbing medium between the rhomb and the analyser two changes in the vibrations from the rhomb occur. The major axes of the elliptical vibrations are rotated through an angle equal to the rotatory power of the medium, and the ellipticities are changed. This condition is illustrated in Fig. 122 (*b*) for the particular case of a medium which is dextrorotatory and which absorbs left circular light more than right, i.e. itself produces a right elliptical vibration. The minor axes of the two ellipses incident on the analyser are now unequal, and the two fields do not match in the spectral region where circular dichroism exists.

The polariser is now rotated in such a direction as to reduce the ellipticity of that ray which has had its ellipticity increased in

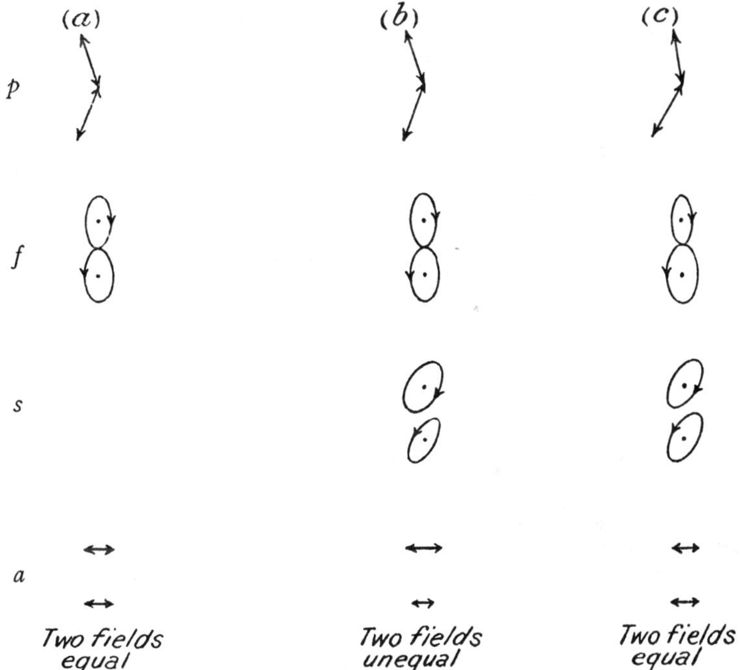

FIG. 122.—KUHN AND BRAUN'S METHOD OF MEASURING CIRCULAR DICHROISM.

traversing the solution. A series of photographs is taken with successive settings of the polariser, and the wave-lengths are read off at which the two halves of the spectrum are equal in intensity, as in measuring rotatory power. This condition is represented in Fig. 122 (*c*).

The actual angle ϕ through which the polariser is rotated is equal in magnitude to the ellipticity produced by the medium for the wave-lengths corresponding to the points of equality in the two halves of the spectrum. The difference between the molecular extinction coefficients for left and right circularly polarised light is then given by

$$(\epsilon_l \sim \epsilon_r) = \frac{4\phi}{cl} \log_{10} e \quad (\phi \text{ in radians}),$$

where c is the molar concentration and l is the length of column of the solution under investigation.

If the Fresnel rhomb produces a phase retardation between the vibrations parallel and perpendicular to the plane of incidence at total reflection, which is not exactly $\pi/2$ but $\pi/2 \pm \rho$, where ρ is small, then the major axes of the elliptical vibrations which it produces are slightly inclined. The matching of the two halves of the spectrum will then depend, not only on the position of the polariser but also on the position of the analyser and on the rotation α produced by the substance. Kuhn and Braun have shown that the ellipticity produced by the medium is then not equal to ϕ, the rotation of the polariser, but is given to a first approximation by

$$\text{ellipticity} = \phi + \alpha\rho.$$

If ρ, the error in phase retardation, is known and α, the rotation produced by the medium, is determined, this correction, which is usually very small, can be applied if necessary.

Fresnel Rhombs of Uviol Glass and of Fused Silica.—In place of the water rhomb described by Kuhn and Braun, rhombs of uviol glass and of fused silica have been used by Lowry and Hudson [1] and by Lowry and French.[2] The uviol rhomb is transparent down to a wave-length of 3100 A.U. and is therefore useful for measuring circular dichroism in the near ultra-violet. It is convenient to cut the rhomb so that it produces a phase retardation of exactly $\pi/2$ for a wave-length in the neighbourhood of 4500 A.U. in the middle of the range 7000 A.U. to 3100 A.U. for which it is to be used. The refractive index of uviol glass does not vary very much over this range; the actual error in phase retardation for a series of wave-lengths is given below:

λ.	Error in Phase Retardation. Degrees.
6563	+ 0·92
5893	+ 0·90
4861	+ 0·29
4341	− 0·16

The experimental error involved in measuring ellipticity by Kuhn and Braun's method using a uviol glass rhomb is of the order of 0·02°.

The construction of a Fresnel rhomb of fused silica presents a different problem. It has already been shown that a phase retardation of exactly $\pi/2$ can only be produced by two total internal

[1] LOWRY and HUDSON, *Phil. Trans.*, 1933, A. **232,** 117; HUDSON, WOLFROM and LOWRY, *J.*, 1933, 1179.

[2] LOWRY and FRENCH, *J.*, 1932, 2654.

reflections if the refractive index of the rhomb is greater than 1·5. Gifford and Shenstone [1] have determined the refractive index of a specimen of fused silica in the visible and ultra-violet, and their results show that the refractive index only exceeds 1·5 for wave-lengths shorter than 2650 A.U. approximately. The rhomb referred to above was cut from a specimen of fused silica, free from double refraction, whose refractive indices were slightly higher than those given by Gifford and Shenstone, so that the phase retardation was exactly $\pi/2$ for a wave-length of 2740 A.U. The error in phase retardation for a series of wave-lengths on either side of this value is given below:

λ.	Error in Phase Retardation. Degrees.
3404	+ 2·63
3034	+ 1·55
2749	+ 0·04
2573	− 1·01
2312	− 3·08

A fused silica rhomb is thus much less accurate in this part of the spectrum than is a uviol glass one in the visible and very near ultra-violet, owing to the increased refractive dispersion.

The experimental error involved in measuring ellipticity with the help of this particular fused silica rhomb was of the order of 0·05°.

Mathieu.—To avoid the necessity of using a series of quarter-wave plates (correct respectively for the red, yellow, green and blue regions of the visible spectrum), Mathieu [2] has described a method of measuring ellipticity whereby a single doubly refracting plate can be used throughout the visible spectrum. The plate, which acts as a compensator, is constructed to give a phase retardation of 30° approximately for the mercury green line (5461 A.U.), and the actual phase retardation E, which it produces for other monochromatic radiations, is first determined.

The compensator plate, placed in front of a half-shadow analyser set to extinction, is rotated until the two fields are equal, i.e. until a principal section of the compensator is parallel to that of the polariser. The optically-active absorbing medium, which produces an ellipticity ϕ to be measured, is introduced between the polariser and the compensator which is then removed. The polariser is rotated to restore equality as in Bruhat's method, through an angle $-\alpha$, α being the rotation produced by the medium.

The compensator is replaced in its former position so that the major axis of the elliptical vibration from the medium is now parallel to the principal section of the plate; but the latter does not produce a phase retardation of $\pi/2$, so that it does not convert this elliptical vibration into a linear one but into another elliptical vibration of different ellipticity, whose major axis is inclined to that of the vibration furnished by the medium. The equality of the two fields which

[1] Gifford and Shenstone, *P.R.S.*, 1904, **73**, 201.
[2] Mathieu, *C.R.*, 1931, **192**, 156; *Journ. de Phys.*, 1931, [vii], **2**, 189.

is thus destroyed is restored by rotating the compensator through an angle θ.

The required ellipticity ϕ produced by the medium is then given by

$$\tan 2\phi = \sin 2\theta \tan E/2.$$

It can be shown that the effect of the compensator, after it has been rotated, is to re-establish a vibration identical in ellipticity and orientation but opposite in sign to that produced by the solution after the polariser has been rotated.

Bruhat and Thouvenin.—An alternative method to that of Mathieu, whereby the necessity of using a series of quarter-wave plates is avoided, has been devised by Bruhat and Thouvenin.[1] These authors have described a quarter-wave plate, applicable both to the visible and ultra-violet, which can be adjusted to give a phase retardation of $\pi/2$ for any desired wave-length. It consists of two thin plates of quartz, one dextro and the other lævo, cut perpendicular to the optic axis. When placed together they produce a phase retardation of a magnitude depending on the angle at which light is incident on the plate. With plates of a suitable thickness (0·107 mm.) a phase retardation of exactly $\pi/2$ can be produced for any wave-length between 6000 A.U. and 2500 A.U. by rotating the plate so that the angle of incidence varies between 24° and 15°. Great care is required in mounting, adjusting and calibrating the plate, but once this has been done it can be readily moved from one setting to another for the different monochromatic radiations to be used.

Bruhat and Thouvenin have estimated that, with the aid of this plate, ellipticities can be measured in the visible and ultra-violet with an experimental accuracy of 0·02°.

[1] BRUHAT and THOUVENIN, *C.R.*, 1931, **193**, 727; *Rev. d'Opt.*, 1932, **11**, 49.

PART III.—*SPECIAL CASES.*

CHAPTER XX.

QUARTZ.[1]

Historical.—The optical rotatory power of quartz was discovered in 1812 by Biot[2] (p. 9) who also recorded the increase in the magnitude of the rotation on passing from the red to the violet part of the visible spectrum (p. 10), and the existence of dextrorotatory and lævorotatory forms of the mineral (p. 10). The correlation of the sign of the optical rotatory power of quartz plates with the hemihedrism of the crystals was established by Herschel[3] in 1821. Herschel, however, adopted a different convention from Biot's, since he described as *dextro* a rotation which was clockwise as viewed from the light-source, instead of from the analyser and telescope of the polarimeter. For nearly a century therefore "dextro"-quartz produced a rotation of opposite sign to *d*-camphor; but this contradiction has now been avoided by reverting to Biot's convention for quartz as well as for optically-active liquids.[4]

The rotatory dispersion of quartz was shown in 1817[5] to conform to the Law of Inverse Squares (p. 11). The use of a quartz plate of opposite sign to act as a compensator to balance the optical rotatory power of a sugar solution was described in the same memoir,[6] and the conclusion was drawn that the rotatory dispersion of the solution also obeyed the Law of Inverse Squares. Subsequent measurements of the rotatory dispersion of quartz in the visible spectrum were made by Broch in 1846,[7] by Arndtsen in 1858,[8] and by Stefan in 1864.[9] Measurements in the visible and ultra-violet were made by Soret and Sarasin in 1882,[10] by Joubin in 1889[11] and by Gumlich in 1898.[12]

[1] See also Chapter XXXIII of the A.C.S. Monograph, *The Properties of Silica*, by R. B. Sosman.
[2] Biot, *Mém. Inst.*, 1812, **1**, 1–372.
[3] Herschel, *Trans. Camb. Phil. Soc.*, 1821, **1**, 43.
[4] Cheshire, *Nature*, 1922, **110**, 807; Tutton, *ibid.*, 809.
[5] Biot, *Mém. Acad. Sci.*, 1817, **2**, 41–136. [6] *Ibid.*, 103–114.
[7] Broch, *Repert. Physik*, 1846, **7**, 113; *A.C.P.*, 1852, [iii], **34**, 119–121 (compare p. 113).
[8] Arndtsen, *A.C.P.*, 1858, [iii], **54**, 409 (p. 114).
[9] Stefan, *Sitz. Akad. Wiss., Wien*, 1864, **60**, 88–124 (p. 116).
[10] Soret and Sarasin, *Geneva Archives*, 1882, **8**, 5–59, 97–132, 201–229.
[11] Joubin, *A.C.P.*, 1889, [vi], **16**, 78.
[12] Gumlich, *Wied. Ann.*, 1898, **64**, 333–359.

Measurements in the infra-red were made by Hussel[1] in 1891, by Carvallo[2] in 1892, by Moreau[3] in 1893, by Hupe[4] in 1894, by Dongier in 1897[5] and by Ingersoll in 1917.[6] Measurements covering the range from 25,170 A.U. to 2263 A.U. were given by Lowry and Coode-Adams in 1927,[7] following an earlier series of measurements in the visible spectrum by Lowry in 1912.[8] This range has been extended in the ultra-violet to 1854 A.U. by Duclaux and Jeantet in 1926[9] and in the infra-red to 32,100 A.U. by Lowry and Snow in 1930.[10]

Rotatory Power for Standard Wave-lengths.—The rotatory power of quartz is remarkably constant when the crystals are free from twinning, etc., and have been cut accurately perpendicular to the optic axis. Thus the following values of the rotations for the yellow sodium doublet (5895·932 and 5889·965 A.U.) and for the green mercury line (5460·741 A.U.) have been tabulated by Sosman (*loc. cit.*):

TABLE 15.—OPTICAL ROTATORY POWER OF QUARTZ.

Sodium Light.		Degrees.	Mercury Green.		Degrees.
Lang	(1876)	21·724	Gumlich	(1898)	25·5317
Soret and Sarasin	(1882)	21·708	Macé de Lépinay[11]	(1900)	25·5365
Soret and Guye	(1893)	21·723	Lowry	(1912)	+ 25·5361*
Gumlich	(1896)	21·724	,,	,,	− 25·5371
Schönrock	(1898)	21·722			
Lowry	(1912)	21·728			

* Not included in Sosman's table.

All these values are probably too low, since 10 rods of quartz, cut from an exceptionally perfect crystal,[7] gave values for the green mercury line ranging from 25·5370° to 25·5399° per mm. at 20°; and the whole column, having a total length of 496·474 mm., gave in two series of readings 25·5380 and 25·5382, Mean **25·538$_1$**° per mm.

The corresponding rotations for the two sodium lines were

	Na 5895·932	= 21·7010° per mm.
	Na 5889·965	= 21·7492° ,, ,,
Optical mass-centre	5892·617	**21·729**° per mm.

Rotatory Dispersion.—The rotatory power of quartz for a series of wave-lengths in the infra-red, visible and ultra-violet regions is set out in Table 16.

[1] HUSSEL, *Ann. Phys. Chem.*, 1891, **43**, 498–508.
[2] CARVALLO, *A.C.P.*, 1892, [vi], **26**, 113–144.
[3] MOREAU, *A.C.P.*, 1893, [vi], **30**, 433–512.
[4] HUPE, *Wiss. Beilage, Charlottenburg*, 1894, pp. 1–48.
[5] DONGIER, *A.C.P.*, 1898, [vii], **14**, 331–391.
[6] INGERSOLL, *Phys. Rev.*, 1917, [ii], **9**, 257–268.
[7] LOWRY and COODE-ADAMS, *Phil. Trans.*, 1927, A. **226**, 391–466.
[8] LOWRY, *ibid.*, 1912, A. **212**, 261–297.
[9] DUCLAUX and JEANTET, *J. de Physique*, 1926, [vi], **7**, 200.
[10] LOWRY and SNOW, *P.R.S.*, 1930, A. **127**, 271–278.
[11] DE LÉPINAY, *J. de Physique*, 1900, [iii], **9**, 644–652.

TABLE 16.—OPTICAL ROTATORY POWER OF QUARTZ.

Infra-red.

(Lowry and Snow.)

A.U.	Degrees per mm. Obs.	(O–C) × 10³.
32100	0·520	+ 10
32000	0·525	+ 9
31900	0·529	+ 7
31800	0·537	+ 9
(Absorption band here.)		
27900	0·746	+ 6
27000	0·806	+ 6
26000	0·888	+ 5
25000	0·977	+ 3
24000	1·074	+ 2
23000	1·188	+ 3
22000	1·312	+ 2
21000	1·461	+ 1
20000	1·633	+ 2
19000	1·827	+ 1
18000	2·061	+ 2
(Lowry and Coode-Adams.)		
17250	2·261	+ 1
16070	2·624	−11
14800	3·168	+21
14030	3·530	+ 4
13420	3·894	+19

Visible Spectrum.

(Lowry and Coode-Adams.)

	A.U.	Degrees per mm. Obs.	(O – C) × 10⁴.
Li	6707·846	16·5359	− 8
Cd	6438·4696	18·0243	+ 4
Zn	6362·345	18·4797	−11
Na	5895·932	21·7010	− 9
Na	5889·965	21·7492	+ 4
Hg	5790·659	22·5465	+ 8
Cu	5782·159	22·6170	+10
Hg	5769·598	22·7211	+ 5
Cu	5700·248	23·3115	+ 7
Ag	5471·551	25·4328	+21
Ag	5465·489	25·4921	+13
Hg	5460·742	25·5384	+ 3
Tl	5350·65	26·6725	+ 9
Ag	5209·081	28·2451	+ 5
Cu	5153·251	28·9049	+ 7
Cu	5105·543	29·4861	− 3
Cd	5085·822	29·7323	+ 2
Zn	4810·535	33·5168	− 7
Cd	4799·909	33·6769	−11
Zn	4722·164	34·8885	−10
Zn	4680·138	35·5721	− 1
Cd	4678·163	35·6057	+ 8
Hg	4358·343	41·5506	+ 2

Ultra-violet.

(Lowry and Coode-Adams.)

	A.U.		Obs.	O–C
	4352·741	I	41·664	−4
	4315·089	I	42·469	−1
	4282·408	I	43·185	−1
	4233·615	I	44·288	−1
	4191·443	I	45·275	−3
	4147·676	I	46·340	−1
	4134·685	I	46·661	−3
	4118·552	I	47·068	−1
	4076·642	I	48·144	−5
	4021·872	I	49·616	−2
	3977·746	I	50·850	−3
	3935·818	I	52·068	−2
	3906·482	I	52·946	−3
	3865·527	I	54·212	−3
	3727·622	B	58·838	−1
	3693·999	B	60·058	−2
	3490·577	B	68·360	+5
Cd	3403·6529	MB	72·455	+2
	3383·985	B	73·434	±
	3271·003	F	79·485	−3
	3225·790	F	82·129	−2
	3075·725	F	91·968	−1
	2912·157	F	104·967	±
	2813·290	F	114·286	+3
	2739·550	F	122·123	+6
	2628·296	F	135·661	−4
	2474·818	B	158·660	−4
	2413·310	F	169·678	−6
	2359·23	E	180·426	+1
	2327·49	E	187·247	±

Ultra-violet

(Duclaux and Jeantet.)

A.U.	Obs.
2269·09	200·90
2263·34	202·27
2210·03	216·50
2174·02	226·91
1989·79	295·65
1935·18	322·76
1930·30	325·31
1862·09	365·6
1857·35	368·6
1853·98	370·9

I = International Standard.
F = Fabry and Buisson (1908).
B = Burns (1914).
MB = Meggers and Burns (1922).
E = Exner and Haschek (1897).

The preceding data from 32,000 to 2327 A.U. can be expressed very accurately by means of the equation

$$\alpha = \frac{9{\cdot}5639}{\lambda^2 - 0{\cdot}0127493} - \frac{2{\cdot}3113}{\lambda^2 - 0{\cdot}000974} - 0{\cdot}1905. \quad . \quad \text{(i)}$$

Thus for 18 wave-lengths in the visual range the average error is less than $\pm 0{\cdot}001°$ per mm. and the average deviation is less than $0{\cdot}0001°$ per mm. In the photographic range in the visible and ultra-violet spectrum the average error in 105 readings from 5371·495 to 2327·49 A.U. is $\pm 0{\cdot}002_3$ and the average deviation is only $-0{\cdot}001_6°$ per mm. In the infra-red range from 18,000 to 28,000 A.U. the average deviation is about $+0{\cdot}003°$ per mm. and even in the narrow "window," from 31,800 to 32,100 A.U. (where the light is again transmitted on the further side of a narrow absorption band), the average deviation is still only $+0{\cdot}009°$ per mm.

Since the curve of rotatory dispersion shows no deviation on either side of the infra-red band at 32,000 A.U., we may accept the conclusion which Drude reached in 1900 [1] that "the kinds of ions whose natural periods lie in the infra-red are inactive," or in more modern terms, that the infra-red absorption bands are not optically active, and in particular that the "dissymmetry factor" (p. 394) of the band at 32,000 A.U. is zero. On the other hand, the close concordance between the observed and calculated rotations indicates that the rotatory power of quartz is dominated throughout the spectrum by two characteristic frequencies in the Schumann region. These frequencies have been located provisionally at 1130 and 310 A.U. respectively; but these numbers may require some modification when measurements have been made to confirm the values given by Duclaux and Jeantet in the difficult region between 2250 and 1850 A.U. Since the deviations from his readings in this region range from 0 up to $-2°$ per mm., it is clear that the formula cited above is not valid for extrapolation to wave-lengths less than about 2250 A.U.; but it is unlikely that the characteristic frequencies will be modified at all drastically when the constants of the equation have been modified in order to cover the increased range to 1850 A.U. Beyond this range further measurements are not possible, on account of the opacity of the crystals to light of shorter wave-lengths.

Influence of Temperature.—It has been known since 1846 that the rotatory power of quartz increases with temperature,[2] and this is true even when a correction has been made for the coefficient of expansion along the optic axis, which accounts for only about 1/13 of the increment.[3] According to von Lang,[4] Sohncke [5] and

[1] DRUDE, *Physical Optics*, 1907, translation, p. 413.
[2] SOLEIL, *C.R.*, 1845, **20**, 435; DUBRUNFAUT, *C.R.*, 1846, **23**, 44.
[3] FIZEAU, *A.C.P.*, 1864, [iv], **2**, 176.
[4] VON LANG, *Sitz. Akad. Wiss., Wien*, 1875, **71**, 707.
[5] SOHNCKE, *Wied. Ann.*, 1878, **3**, 516-531.

Joubert,[1] the increment does not vary substantially with the wavelength in the visible spectrum, where the coefficient is about 0·000145, but Soret and Sarasin [2] recorded an increase to 0·000175 in the far ultra-violet at Cd 2265 A.U.

Le Chatelier [3] discovered in 1889 that there is an abrupt increment of rotatory power on passing from α or "low quartz" to β or "high quartz" at 570°, and that above this temperature the increment of rotatory power in "high quartz" is very much smaller

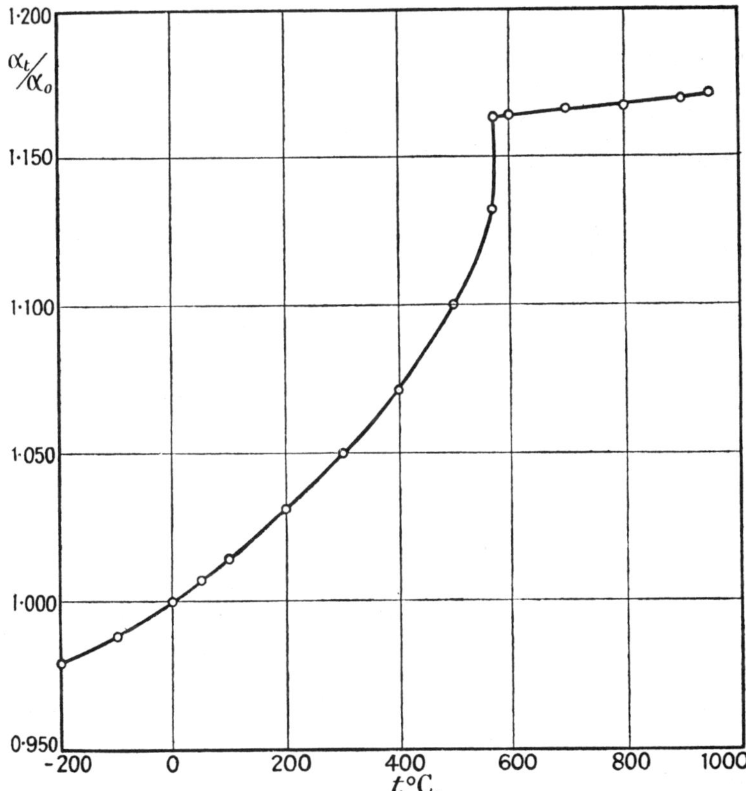

FIG. 123.—ROTATORY POWER OF α- AND β-QUARTZ.

than in "low quartz."* The data on this subject have been summarised by Sosman [4] in Table 17, and are plotted in Fig. 123.

[1] JOUBERT, *J. de Physique*, 1879, [i], **8**, 1; *C.R.*, 1878, **87**, 497.
[2] SORET and SARASIN, *Geneva Archives*, 1882, **8**, 5, 97.
[3] LE CHATELIER, *C.R.*, 1889, **109**, 264.
[4] SOSMAN, *International Critical Tables*, VI, 343.
* The crystal structure of α- and β-quartz is illustrated in Figs. 144 and 145 (pp. 344–345); the calculation of the optical rotatory power of β-quartz is referred to on p. 346.

TABLE 17.—CHANGE OF ROTATORY POWER OF QUARTZ WITH TEMPERATURE.*

Ratio of the measured rotation at temperature t to the measured rotation of the same plate at 0° C.

t.	α_t/α_0.	t.	α_t/α_0.	t.	α_t/α_0.
° C.		° C.		° C.	
−200	0·979	200	1·031	600	1·164
−100	0·988	300	1·050	700	1·166
0	1·000	400	1·071	800	1·167
+ 50	1·007	500	1·100	900	1·169
100	1·014	573	{ 1·132 1·163	1000	1·171
—	—			—	—

* Data by von Lang (1875), Joubert (1879), le Chatelier (1889), Gumlich (1895), Bates (1906), and Molby (1909).

CHAPTER XXI.

ALCOHOLS, ACIDS AND ESTERS.

Historical.—(*a*) The optical activity of *amyl alcohol*,

$$\begin{array}{l} CH_3 . CH_2 \\ \qquad\qquad \rangle \mathbf{C}H . CH_2 . OH, \\ CH_3 \end{array}$$

was discovered by Biot, who reported it to Pasteur in 1849. After purification by fractional crystallisation of *barium amyl sulphate*, $Ba(O . SO_2 . O . C_5H_{11})_2$, Pasteur [1] obtained a rotation of $- 4°$ per decimetre, whence $[\alpha]_D = - 4{\cdot}9°$. Le Bel [2] in 1874 purified it by converting the inactive amyl alcohol into chloride and then fractionating, and observed rotations of $- 4{\cdot}53°$ to $- 4{\cdot}63°$ per decimetre, whence $[\alpha]_D = - 5{\cdot}6°$ to $- 5{\cdot}7°$. Its activity is destroyed by distillation with sodium hydroxide or metal,[3] and by conversion into amylene,[3] but it gives an active chloride,[4] bromide,[4] $[\alpha]_D + 3{\cdot}68°$,[5] iodide,[4] $[\alpha]_D + 5{\cdot}64°$,[5] cyanide,[6] carboxylic acid [6] (capronic), and diamyl [6] (by electrolysis of the acid)

$$C_5H_{11} . OH \rightarrow C_5H_{11} . I \rightarrow C_5H_{11} . CN \rightarrow C_5H_{11} . COOH \rightarrow C_5H_{11} . C_5H_{11}.$$

These derivatives are all of small rotatory power but are uniformly *dextrorotatory*,[7] although their configuration is assumed to be the same as that of the *lævorotatory* alcohol.

(*b*) *Isovaleric acid*, $\begin{array}{l} CH_3 . CH_2 \\ \qquad\qquad \rangle \mathbf{C}H . CO . OH, \\ CH_3 \end{array}$ the carboxylic acid corresponding with amyl alcohol, occurs in a *dextrorotatory* form,[8] which can also be prepared by oxidation of *lævorotatory* amyl alcohol.[9] The specific rotation of the acid is higher than that of the alcohol $[\alpha]_D = 18{\cdot}2°$.[10] The rotatory dispersion of the acid in the visual region is apparently simple, since it can be expressed by the

[1] PASTEUR, *C.R.*, 1855, **41**, 296.
[2] LE BEL, *Bull. Soc. Chim.*, 1874, [ii], **21**, 542
[3] *Ibid.*, 1876, [ii], **25**, 545.
[4] *Ibid.*, 1874, [ii], **21**, 542.
[5] MARCKWALD, *Ber.*, 1904, **37**, 1038.
[6] WURTZ, *Ann.*, 1858, **105**, 295.
[7] PIERRE and PUCHOT, *C.R.*, 1873, **76**, 1332.
[8] FRANKLAND and DUPPA, *J.*, 1867, **5**, 116; *Ann.*, 1868, **145**, 92; ERLENMEYER and HELL, *Ann.*, 1871, **160**, 257–303.
[9] PEDLER, *J.*, 1868, **6**, 74; *Ann.*, 1868, **147**, 243.
[10] POWER and ROGERSON, *J.*, 1912, **101**, 406.

equation $\alpha/\alpha_{5461} = 0{\cdot}2607/(\lambda^2 - 0{\cdot}0375)$, where the dispersion constant corresponds with a characteristic frequency at 1940 A.U.[1]

These two compounds figured largely in the discussions by le Bel and van't Hoff, as to the origin of optical rotatory power, which led to the enunciation of the theory of the asymmetric carbon atom as a principal cause of optical activity. Closely related to them is *amyl valerate*,

$$\begin{matrix} CH_3 . CH_2 \\ CH_3 \end{matrix}\!\!>\!\mathbf{C}H . CO . O . CH_2 . \mathbf{C}H\!<\!\!\begin{matrix} CH_2 . CH_3 \\ CH_3 \end{matrix} \qquad [\alpha]_D = 8{\cdot}7^\circ,$$

prepared by Ley [2] and by Pierre and Puchot [3] in 1873, and shown in each case to be dextrorotatory.

Rotatory Powers of Homologous Series of Alcohols, Esters and Ethers.—(*a*) A large number of secondary alcohols were resolved, from 1911 onwards, by Pickard, Kenyon and their colleagues, by coupling the alcohol to an alkaloid by means of a dibasic acid (e.g. succinic or phthalic acid), one hydrogen of which was esterified with the alcohol whilst the other was displaced by the kation of the alkaloid. Thus *methylethylcarbinol*, $CH_3 . \mathbf{C}HOH . C_2H_5$, was resolved by crystallising the *brucine* and *strychnine* salts of the hydrogen phthalic ester, $C_4H_9 . O . CO . C_6H_4 . CO . OH$.

Many attempts were made to prepare optically-active tertiary alcohols by the same method, but the resolutions failed just as consistently as they had succeeded with the secondary alcohols. This failure suggests that the *hydrogen* of the $>\mathbf{C}HOH$ radical may play an essential part in the resolution, perhaps by checking "free rotation" in the radical

$$\begin{matrix} R_1 \\ R_2 \end{matrix}\!\!>\!\mathbf{C}\!<\!\!\begin{matrix} H \cdots O \\ \quad\;\; \| \\ O—C—R_3 \end{matrix}$$

(*b*) In the course of these investigations the following series of alcohols were investigated:

Methyl carbinols [4]	$CH_3 . \mathbf{C}HOH . R$
Ethyl carbinols [5]	$C_2H_5 . \mathbf{C}HOH . R$
iso-Propyl carbinols [6]	$CHMe_2 . \mathbf{C}HOH . R$

Table 18 (p. 265) shows that in the homologous series of *methyl carbinols* the optical rotations decrease steadily from ethyl to decyl without reaching a limiting value, but that the rotation of the *propyl* compound, $CH_3 . \mathbf{C}HOH . C_3H_7$, is abnormally high compared with the rest of the series. On the other hand, the *ethyl carbinols* show a maximum at the *amyl* compound, $C_2H_5 . \mathbf{C}HOH . C_5H_{11}$, and iso-

[1] Lowry and Dickson, *J.*, 1913, **103**, 1074. [2] Ley, *Ber.*, 1873, **6**, 1369.
[3] Pierre and Puchot, *C.R.*, 1873, **76**, 1332.
[4] Pickard and Kenyon, *J.*, 1911, **99**, 45.
[5] *Ibid.*, 1913, **103**, 1923. [6] *Ibid.*, 1912, **101**, 620 *et seq.*

propylcarbinols show a maximum at the *butyl* compound, C_3H_7 . **C**HOH . C_4H_9.

(*c*) The homologous *esters* of these optically-active aliphatic secondary alcohols were also examined,[1] as well as those of secondary alcohols containing aromatic radicals, namely

Methyl benzylcarbinol [2] CH_3 . **C**HOH . CH_2 . C_6H_5 and
l-Naphthyl-*n*-hexylcarbinol [3] C_6H_{13} . **C**HOH . $C_{10}H_7$

and of esters formed from aliphatic alcohols and aromatic acids, e.g. *benzoic* and 1- and 2-*naphthoic acids*.[4]

These esters, as well as the alcohols described above, provided remarkable illustrations of the suggestion of Frankland,[5] that "a chain of carbon atoms might be expected to form a spiral which would complete one turn at about the fifth member, and that accordingly anomalies in physical properties might be looked for at that point." [6] Thus, when the rotations of the normal esters of *γ-nonanol*,

$$\begin{matrix} C_2H_5 \\ C_6H_{13} \end{matrix} \Big\rangle \mathbf{C}H \, . \, O \, . \, CO \, . \, C_nH_{2n+1}$$

were plotted against the number of carbon atoms ($n + 1$) in the acyl group, the curve for solutions in carbon disulphide showed obvious upward bulges when $n + 1 = 5$ and 10 or 11. These bulges were repeated as slight depressions in the curve for the homogeneous esters, but could not be seen in the curve for solutions in ethyl alcohol (Fig. 124).[7] Similar irregularities were observed in the acetates of the ethyl carbinols of the series C_2H_5 . **C**HOH . R, and were again very marked (at $R = C_6H_{13}$ and $C_{11}H_{23}$) in solutions in carbon disulphide.

Further investigations (including the rotations of the formic esters [8]) confirmed these abnormalities, since exaltations may often be observed either when the growing chain R_1 or the complete chain of carbon atoms comprising the whole molecule contains 5 or 10 carbon atoms.[9]

(*d*) The rotations of the normal aliphatic *ethers* of *d-β-octanol* and of *d-benzylmethylcarbinol*

$$\begin{matrix} CH_3 \\ C_6H_{13} \end{matrix} \Big\rangle \mathbf{C}H \, . \, O \, . \, C_nH_{2n+1} \quad \text{and} \quad \begin{matrix} CH_3 \\ C_6H_5 \, . \, CH_2 \end{matrix} \Big\rangle \mathbf{C}H \, . \, O \, . \, C_nH_{2n+1}$$

were examined by Kenyon and McNicol [10] and by Phillips.[11] The rotations of the members of these homologous series showed abnormalities

[1] PICKARD and KENYON, *J.*, 1914, **105**, 830–898; KENYON, *ibid.*, 2226–2261.
[2] KENYON and PICKARD, *ibid.*, 2262–2280.
[3] *Ibid.*, 2644–2665.
[4] *Ibid.*, 1915, **107**, 115–132.
[5] FRANKLAND, *J.*, 1899, **75**, 368.
[6] KENYON, *T.F.S.*, 1930, **26**, 439.
[7] KENYON, *J.*, 1914, **105**, 2226.
[8] PICKARD, KENYON and HUNTER, *J.*, 1923, **123**, 1–14.
[9] KENYON, *T.F.S.*, 1930, **26**, 440.
[10] KENYON and MCNICOL, *J.*, 1923, **123**, 14–22.
[11] PHILLIPS, *J.*, 1923, **123**, 22–31.

which were again greatest in carbon disulphide, smaller in the homogeneous state at 120° and at 20°, and least in solutions in ethyl alcohol.

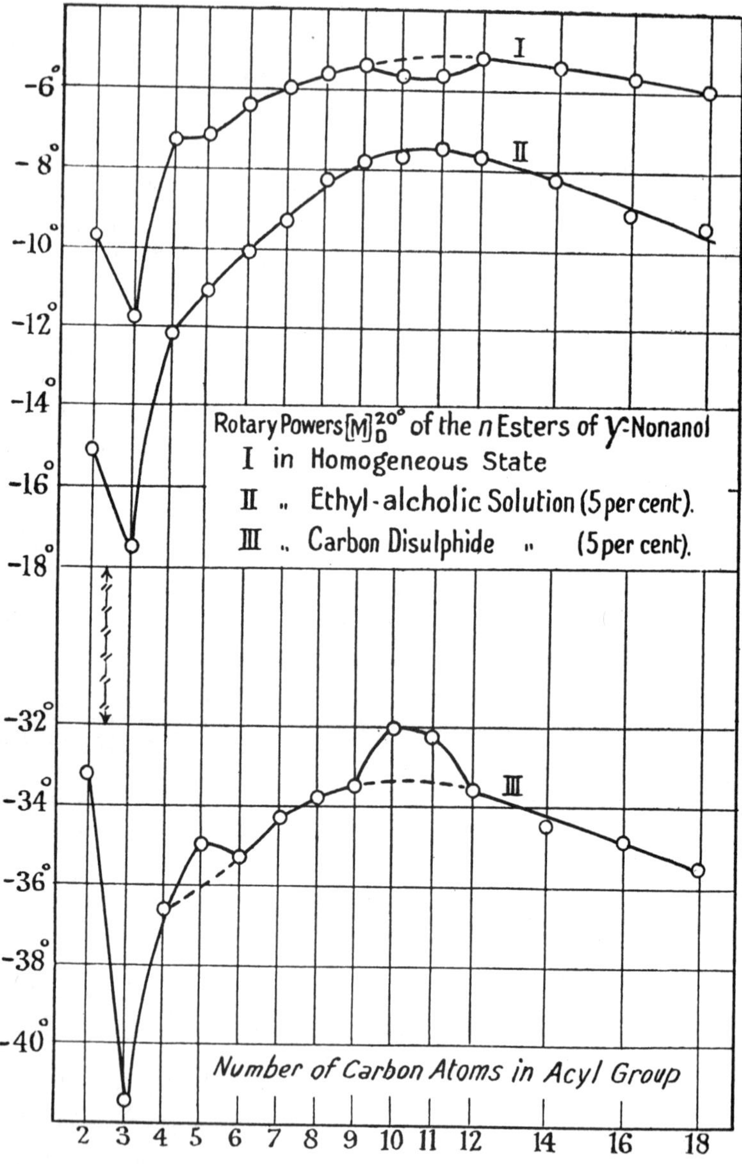

FIG. 124.—MOLECULAR ROTATIONS OF HOMOLOGOUS ESTERS (Kenyon).

TABLE 18.—ROTATIONS AND ROTATORY DISPERSIONS OF SECONDARY ALCOHOLS.

	$\frac{\alpha_{4358}}{\alpha_{5461}}$.	λ_0^2.	λ_0.	α_{5461} (100 mm.).
Methyl Carbinols.				Degrees.
$CH_3 . CHOH . C_2H_5$	1·661	0·0262	1620	12·57
$CH_3 . CHOH . C_3H_7$	1·652	0·0240	1550	13·25 (max.)
$CH_3 . CHOH . C_4H_9$	1·653	0·0242	1555	10·84
$CH_3 . CHOH . C_5H_{11}$	1·651	0·0237	1540	9·88
$CH_3 . CHOH . C_6H_{13}$	1·653	0·0242	1555	9·54
$CH_3 . CHOH . C_7H_{15}$	1·651	0·0237	1540	8·68
$CH_3 . CHOH . C_8H_{17}$	1·649	0·0232	1520	8·44
$CH_3 . CHOH . C_9H_{19}$	1·651	0·0237	1540	7·87
$CH_3 . CHOH . C_{10}H_{21}$	1·653	0·0242	1555	7·52
Ethyl carbinols.				
$C_2H_5 . CHOH . C_3H_7$	1·615 *	0·0139 *	—	1·60
$C_2H_5 . CHOH . C_4H_9$	1·650	0·0234	1530	7·94
$C_2H_5 . CHOH . C_5H_{11}$	1·639	0·0205	1430	8·07 (max.)
$C_2H_5 . CHOH . C_6H_{13}$	1·639	0·0205	1430	7·80
$C_2H_5 . CHOH . C_8H_{17}$	1·634	0·0192	1385	6·14
iso-*Propyl carbinols.*				
$C_3H_7 . CHOH . CH_3$	1·697	0·0346	1860	4·74
$C_3H_7 . CHOH . C_2H_5$	1·661	0·0262	1620	14·71
$C_3H_7 . CHOH . C_3H_7$	1·665	0·0274	1655	20·62
$C_3H_7 . CHOH . C_4H_9$	1·665	0·0274	1655	24·97 (max.)
$C_3H_7 . CHOH . C_5H_{11}$	1·663	0·0267	1630	22·46
$C_3H_7 . CHOH . C_6H_{13}$	1·661	0·0262	1620	21·16
$C_3H_7 . CHOH . C_8H_{17}$	1·661	0·0262	1620	18·38
$C_3H_7 . CHOH . C_{10}H_{21}$	1·669	0·0286	1690	15·59
Butyl carbinols.				
$Me_2CH . CH_2 . CHOH . CH_3$	1·631	0·0192	1385	19·44
†$Me_2CH . CH_2 . CHOH . C_2H_5$	1·633	0·0190	1380	9·39
†$Me_2CH . CH_2 . CHOH . C_3H_7$	1·651	0·0237	1540	4·99
$Me_3C . CHOH . CH_3$	1·707	0·0368	1920	7·87
Aromatic secondary alcohols.				
$C_6H_5 . CHOH . CH_3$	1·736	0·0429	2070	52·49
$C_6H_5 . CHOH . CH_2 . CH_3$	1·674	0·0293	1710	32·37
$C_6H_5 . CH_2 . CHOH . CH_3$	1·833	0·0613	2475	32·47
$C_6H_5 . CH_2 . CH_2 . CHOH . CH_3$	1·679	0·0306	1750	16·55

* The rotatory power was too small to give trustworthy figures for the dispersive power of this compound.

† These samples were of doubtful purity.

Rotatory Dispersion in the Region of Transparency.—(*a*) The rotatory dispersions of *amyl alcohol* and of iso-*valeric acid* in the visual region from 6708 to 4358 A.U. were examined by

Lowry and others in 1913–1914.[1] Over this range the dispersions are apparently simple and almost identical, since they can be expressed by the following equations :

$$\text{Amyl alcohol} \quad \alpha/\alpha_{5461} = 0{\cdot}2628/(\lambda^2 - 0{\cdot}0354)$$
$$\alpha_{4358}/\alpha_{5461} = 1{\cdot}700. \quad \lambda_0 = 1880 \text{ A.U.}$$
$$\textit{iso}\text{-Valeric acid} \quad \alpha/\alpha_{5461} = 0{\cdot}2607/(\lambda^2 - 0{\cdot}0375)$$
$$\alpha_{4358}/\alpha_{5461} = 1{\cdot}710. \quad \lambda_0 = 1940 \text{ A.U.}$$

The dispersion ratio $\alpha_{4358}/\alpha_{5461} = 1{\cdot}700$ of the primary alcohol is exceptionally high, as compared with the aliphatic secondary alcohols, for which $\alpha_{4358}/\alpha_{5461} = 1{\cdot}65$ (approx.), and as compared with its own magnetic rotatory dispersion for which $\alpha_{4358}/\alpha_{5461} = 1{\cdot}634$. That of the acid is in harmony with the existence of strong absorption bands on the edge of the Schumann region.[2]

(*b*) Simple dispersions in the visual range were also recorded by Lowry, Pickard and Kenyon [3] for a large number of *secondary alcohols*. After the first member the dispersion-ratios are fairly uniform at 1·65 for the *methyl* carbinols, 1·64 for the *ethyl* carbinols, and 1·66 for the iso-*propyl* carbinols, with a general average of about 1·65. Large increases of rotatory dispersion are, however, produced by the accumulation of methyl groups in the series

	$\alpha_{4358}/\alpha_{5461}$.	λ_0.
CH_3 . **C**HOH . CH_2Me	1·661	1620 A.U.
CH_3 . **C**HOH . $CHMe_2$	1·697	1860 ,,
CH_3 . **C**HOH . CMe_3	1·707	1920 ,,

Even more remarkable fluctuations were recorded in the dispersions of the aromatic secondary alcohols as shown in Table 18. These fluctuations suggest that the dispersion might prove to be complex if examined over the wider range of wave-lengths which has now become available for measurements of this kind.

(*c*) The optically-active *esters* prepared by Pickard and Kenyon, including those of the simplest possible composition, differed from the alcohols in showing anomalous rotatory dispersion, e.g. when subjected to high temperatures or when dissolved in certain solvents. This characteristic of the esters finds a ready explanation in the "induced dissymmetry" (p. 147) of the —C(=O)O— radical, since this radical has an absorption band at about 2100 A.U., which may become optically-active when coupled to a dissymmetric alkyl radical. Thus it was suggested by Lowry and Cutter in 1925 [4] "the complex dispersion, which so often appears on passing from an optically-active alcohol to its esters, may be due to the development of a partial

[1] Lowry and Dickson, *J.*, 1913, **103**, 1067–1075 (*iso*-valeric acid) ; Lowry, Pickard and Kenyon, *J.*, 1914, **105**, 94–102 (amyl alcohol).

[2] Scheibe, Povenz and Linström, *Z. ph. C.*, 1933, B. **20**, 283.

[3] Lowry, Pickard and Kenyon, *J.*, 1914, **105**, 94–102.

[4] Lowry and Cutter, *J.*, 1925, **127**, 609.

rotation of opposite sign in the carbonyl radical of the unsymmetrical molecule, just as in the case of camphor." The same phenomenon is described below in connection with the experiments of Levene, whose measurements, covering a wider range of wave-lengths, disclosed a complete series of anomalies in esters prepared from optically-active esters and inactive alcohols.

(*d*) The *alkyl ethers*, like the aliphatic alcohols, usually show simple rotatory dispersion, and their rotations are, in general, only slightly susceptible to changes of temperature and the action of solvents. On the other hand, alcohols which contain aromatic nuclei, and esters derived from both classes of alcohols, show complex rotatory dispersion, and their rotatory powers are in general very susceptible to changes in temperature and the action of solvents.[1] A striking exception is found, however, in a series of *n*-alkyl ethers of *d*-γ-nonanol,[2]

$$\begin{matrix} C_2H_5 \\ C_6H_{13} \end{matrix}\!\!\!> \mathbf{C}H \,.\, O \,.\, C_nH_{2n+1}$$

These give dispersion ratios, $\alpha_{4358}/\alpha_{5461}$, between 1·4 and 1·5, i.e. below the minimum value 1·57 for $\lambda_0 = 0$ in a one-term Drude equation. Their rotatory dispersions are therefore obviously complex, perhaps because the electrons associated with the oxygen atom of the ether have an optical activity of opposite sign to those associated with the asymmetric hydrocarbon radical; but the stereochemical factors which result in a quasi-anomalous rotatory dispersion in this case, and not in the ethers derived from methyl hexyl carbinol or methyl phenyl carbinol are still unknown. An analogous change resulting from the introduction of an additional methylene group in an optically-active acid is, however, described under (*e*) below.

(*e*) The rotatory dispersion of a large range of *fatty acids* and *esters* in the visible and ultra-violet, has been examined by Levene, Rothen and Marker.[3] In many cases the molecular rotations could be expressed by a single term of Drude's equation. Thus the dispersions of *iso*-valeric acid and its ester were apparently simple up to the limits of complete transparency at about 2500 A.U.,

l-α-*Methylbutyric Acid* (*iso*Valeric acid.) $\begin{matrix} CH_3 \\ C_2H_5 \end{matrix}\!\!\!> \mathbf{C}H \,.\, COOH$

Acid (in heptane) $[M]^{25} = -5{\cdot}8110/(\lambda^2 - 0{\cdot}03526)$ $\lambda_0 = 2360$ A.U.
Ethyl ester (no solvent) $[M]^{25} = -7{\cdot}2843/(\lambda^2 - 0{\cdot}02923)$ $\lambda_0 = 1710$ A.U.

In other cases, however, two terms of Drude's equation were required to express the rotatory dispersion. These two terms were always of

[1] KENYON, *T.F.S.*, 1930, **26**, 440.
[2] KENYON and BARNES, *J.*, 1924, **125**, 1395.
[3] LEVENE, ROTHEN and MARKER, *J. Chem. Phys.*, 1933, **1**, 662–676.

opposite signs, and a complete range of anomalies was recorded in β-methyl valeric acid and its ethyl ester (Fig. 125).[1]

l-β-Methylvaleric acid

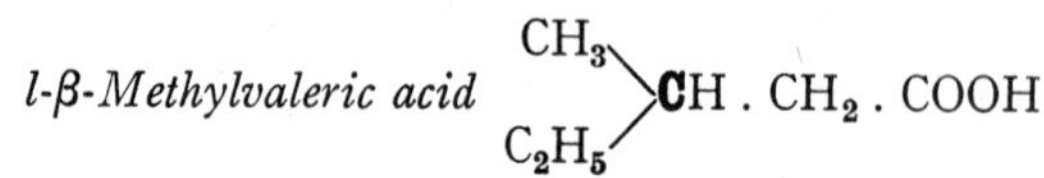

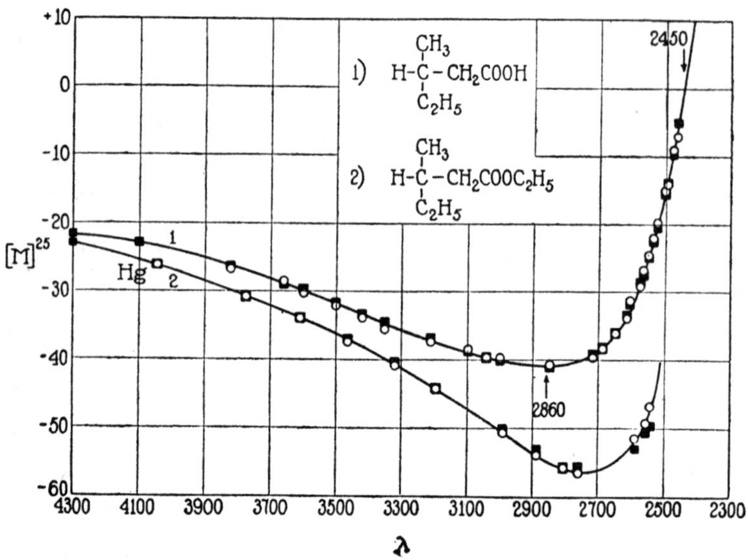

FIG. 125.—ROTATORY DISPERSION OF β-METHYLVALERIC ACID AND ITS ETHYL ESTER (Levene, Rothen and Marker).

Acid (in heptane)

$$[M]^{25} = \frac{8{\cdot}088}{\lambda^2 - 0{\cdot}042} - \frac{11{\cdot}68}{\lambda^2 - 0{\cdot}034}.$$

Inflection at 3212 A.U.
Maximum at 2860 A.U.
Reversal at 2450 A.U.

Ethyl ester (no solvent)

$$[M]^{25} = \frac{11{\cdot}614}{\lambda^2 - 0{\cdot}0420} - \frac{15{\cdot}364}{\lambda^2 - 0{\cdot}0373}.$$

Inflection at 3000 A.U.
Maximum at 2704 A.U.
Reversal at 2378 A.U. (calc.).

It is remarkable that the change from the simple rotatory dispersions of *iso*-valeric acid and its ester to the anomalous dispersion

[1] LEVENE, ROTHEN and MARKER, *J. Chem. Phys.*, 1933, **1**, 667.

recorded above should have been brought about merely by interposing a single methylene radical between the asymmetric carbon atom and the carboxyl-group.

In two further papers Levene and Rothen[1] describe the rotatory dispersion of (i) *a*-substituted fatty acids containing the sulphide and sulphonic groups

$$R.\mathbf{C}H(SH).CO.OH \quad \text{and} \quad R.\mathbf{C}H(SO_3H).CO.OH$$

and (ii) *a*-derivatives of propionic and caproic acids and their salts

$$R.\mathbf{C}HX.CO.OH \text{ where } R = CH_3 \text{ or } C_4H_9$$
$$X = Cl, Br, I, OH, \text{ or } OCH_3$$

They conclude that "in members of a homologous series, the total rotation of consecutive members may differ in sign, but the signs of the partial rotations remain constant." In the series of halogen compounds, the well-marked absorption band of iodopropionic acid at 2840 A.U. and of its sodium salt at 2660 A.U. are attributed to the iodine atom, since iodobutane gives a band at 2525 A.U.; but, unlike Kuhn and his colleagues (see below), they attribute the low-frequency partial rotation of the chloro- and bromo-derivatives to the carboxyl-group, and not to the halogen.

Rotatory Dispersion in the Region of Absorption.—W. Kuhn and his colleagues[2] have plotted and analysed the optical rotations in the region of absorption of a number of derivatives of propionic acid

$$CH_3.\mathbf{C}HX.CO.Y \text{ where } X = Cl, Br, N_3$$
$$Y = OCH_3, OC_2H_5 \text{ or } N(CH_3)_2$$

An absorption band at 2830 A.U. in $N_3.\mathbf{C}H(CH_3).COOMe$ and at 2900 A.U. in $N_3.\mathbf{C}H(CH_3).CO.NMe_2$, and partial rotations passing through a zero value at these wave-lengths, were attributed to the azido-group. Similarly an absorption band at 2370 A.U. in $Br.\mathbf{C}H(CH_3).CO.OEt$, although reduced to a mere step-out, gave rise to a partial rotation, passing through a zero value at this wave-length, which was attributed to the bromide radical. In the dimethyl amide, however, the absorption was so great that neither the maximum nor the reversal of sign could be recorded.

[1] LEVENE and ROTHEN, *J. Chem. Phys.*, 1934, **2**, 681–688; *J. Biol. Chem.*, 1934, **107**, 533-553.

[2] KUHN, *T.F.S.*, 1930, **26**, 293; KUHN and BRAUN, *Z. ph. C.*, 1930, B. **8**, 281–313; KUHN, FREUDENBERG and WOLF, *Ber.*, 1930, **63**, 2367; see also pp. 433–440 and 439–440.

CHAPTER XXII.

SUGARS.

Historical.—The optical rotatory power of *cane sugar* and of *beet sugar* in aqueous solutions was discovered by Biot in 1818.[1] He showed[2] that the dextrorotation of a sugar solution could be compensated throughout the spectrum (i) by plates of quartz of opposite sign or (ii) by tubes of lævorotatory turpentine. The conclusion was drawn that the rotatory dispersion of the sugar solution conformed to the Law of Inverse Squares, which had already been established (in the same Memoir) for plates of quartz. The existence of optical rotatory power in crystals of sugar cut perpendicular to the optic axis was also recorded.[3] In 1832 he showed that the specific rotatory power of cane sugar was approximately constant at different concentrations[4] and persisted without substantial change in a block of amorphous sugar, 37·3 × 24·6 mm., which give, for the light transmitted through a red (copper) glass the observed rotations + 24·23° and + 15·54°, as compared with the calculated rotations + 24·97° and + 16·47°.[5] He also recorded the dextrorotation of *milk-sugar*, *grape-sugar* and *honey-sugar;* but the syrupy residues from grape-sugar and honey-sugar and the sugars extracted from apples, currants, and various types of grapes and wine, were found to be lævorotatory.[6]

In the same Memoir, the name DEXTRIN was applied to the syrup prepared from starch, on account of its large dextrorotation, which was second only to the rotatory power of quartz;[7] but *mannitol* and *citric acid* were reported to be inactive.[8] Finally, in an Appendix[9] to the Memoir he records the important discovery of the INVERSION of cane sugar by acids (compare p. 20). The term "inverted," to describe the product of the action of acids on cane sugar, was introduced by Biot in a footnote to a later paper[10] in which he describes the complete compensation of dextrorotatory cane sugar by the lævorotatory product of inversion. For this purpose, solutions of these two sugars were examined in tubes about 147 mm. in length and then mixed in the proportions required to produce equal and opposite rotations. The mixture was then examined with white

1 BIOT, *Mem. Acad. Sci.*, 1817, **2**, 47.
2 *Ibid.*, pp. 103–114.
3 *Ibid.*, p. 113.
4 *Ibid.*, 1835, **13**, 113–126.
5 *Ibid.*, pp. 126–132.
6 *Ibid.*, pp. 160–169.
7 *Ibid.*, p. 171.
8 *Ibid.*, Tables facing pp. 167 and 169.
9 *Ibid.*, pp. 174–175.
10 *A.C.P.*, 1844, [iii], **10**, 35.

light in a tube 501 mm. in length, but showed no signs of optical rotatory power.

Specific Rotation of Cane Sugar.—A normal sucrose solution containing 26 grams of sugar (weighed in air of density 0·0012 against brass weights) in 100 ml. at 20° C., gives a rotation of 34·617° at 5892·5 A.U., and 40·763° at 5461 A.U. The ratio of these two rotations is 0·84922 as compared with 0·85085 for quartz. The thickness of a quartz plate which will produce an equal rotation is 1·5934 mm. The corresponding specific rotations, based upon weights *in vacuo*, are $[\alpha]_D^{20} = 66{\cdot}53°$ and $[\alpha]_{5461}^{20} = 78{\cdot}34$.[1] The variations of specific rotation with concentration are given by the equations [2]

$$[\alpha]_D^{20} = 66{\cdot}435 + 0{\cdot}00870c - 0{\cdot}000235c^2 \quad (c = 0\text{–}65 \text{ per cent.})$$
$$= 66{\cdot}412 + 0{\cdot}01267_3p - 0{\cdot}000376_5p^2 \quad (p = 0\text{–}50 \text{ per cent.}),$$

where c = grams of sucrose in 100 ml. of solution,
p = weight per cent. of sucrose.

Other data used in saccharimetry, and similar data for other sugars, may be found in the article "Saccharimetry" in the *International Critical Tables*, vol. II, pp. 334–355.

Mutarotation.*—(i) Dubrunfaut in 1846[4] discovered that freshly-prepared solutions of *glucose* in water had a rotation which was greater than that of older solutions, in the ratio of 2 : 1 approximately.[5] He therefore proposed to describe the two states of the sugar as MONOROTATORY and BIROTATORY.[5] A similar phenomenon was observed in *lactose* by Erdmann[6] (1855), in *galactose* by Pasteur[7] (1855), and in *maltose* by Soxhlet[8] (1880). In none of these sugars, however, was the ratio of the initial and final rotations equal to 2. Schmoeger[9] in 1880 therefore introduced the term "half-rotation" to describe a form of lactose of which the initial rotation was only half as great as the final rotation; and Wheeler and Tollens in 1889[10] introduced the terms MULTIROTATION and PAUCIROTATION to cover downward and upward changes of rotatory power. Finally, the term MUTAROTATION was introduced by the author in 1899[11] to describe the general phenomenon of change of rotatory power with time, which in the case of *nitrocamphor* was often accompanied by a change in the *sign* as well as in the *magnitude* of the rotation.

* A fuller account of this phenomenon is given in a report on Mutarotation presented to the Liège meeting of the *Union International de Chimie* in 1931.

[1] BATES and JACKSON, *Bureau of Standards, Scientific Papers*, 1916, 268; *Bulletin*, 1916, **13**, 67; see also *International Critical Tables*, II, 336.
[2] See LANDOLT, *Das Optische Drehungsvermögen*, 1898, p. 420.
[3] LOWRY and SMITH, *Rapport sur les Hydrates de Carbone*, pp. 79–121.
[4] DUBRUNFAUT, *C.R.*, 1846, **23**, 38. [5] *Ibid.*, 1856, **42**, 228.
[6] ERDMANN, *Dissertatio de sacharo lactico et amylaceo*: cited in *Jahresbericht* für 1855, p. 671.
[7] PASTEUR, *C.R.*, 1856, **42**, 347.
[8] SOXHLET, *J. pr. Chem.*, 1880, **21**, 283.
[9] SCHMOEGER, *Ber.*, 1880, **13**, 2130.
[10] WHEELER and TOLLENS, *Ann.*, 1889, **254**, 304.
[11] LOWRY, *J.*, 1899, **75**, 211.

The changes of rotatory power observed in some of the principal sugars are shown in Table 19.

TABLE 19.—MUTAROTATORY SUGARS.

		Degrees.	Degrees.	Degrees.
Glucose	$[\alpha]_D^{20}$	$+ 110 \cdot 12 \rightarrow$	$+ 52 \cdot 1 \leftarrow$	$+ 19 \cdot 26$ [1]
	$[\alpha]_{5461}^{25}$	$+ 129 \cdot 0 \rightarrow$	$+ 61 \cdot 5 \leftarrow$	$+ 23 \cdot 8$ [2]
Lactose	$[\alpha]_D$	$+ 90 \cdot 0 \rightarrow$	$+ 55 \cdot 3 \leftarrow$	$+ 35$ [3]
Galactose	$[\alpha]_D^{20}$	$+ 145 \rightarrow$	$+ 81 \leftarrow$	$+ 54$ [4]
	$[\alpha]_{5461}^{20}$	$+ 173 \rightarrow$	$+ 94 \leftarrow$	$+ 63 \cdot 5$ [5]
Maltose	$[\alpha]_D^{20}$		$+ 136 \leftarrow$	$+ 118$ [6]
Fructose	$[\alpha]_D^{20}$	$- 130 \cdot 8 \rightarrow$	$- 91 \cdot 0$ [7]	

(ii) The slow mutarotation of *gluconic acid* and of its *lactone* was attributed by Emil Fischer in 1890 [8] to reversible hydrolysis,

$$C_6H_{10}O_7 + H_2O \rightleftharpoons C_6H_{12}O_8$$

Gluconic lactone. Gluconic acid.

$[\alpha]_D + 55 \cdot 9^\circ \rightarrow + 18 \cdot 8^\circ$ $\quad + 12^\circ \leftarrow - 1 \cdot 7^\circ$.

He postulated a similar reversible hydration as a cause of the mutarotation of glucose; but this explanation was rendered improbable by the observations of Erdmann [9] and of Schmoeger [10] that lactose exists in three *anhydrous* forms of *high*, *medium* and *low* rotation; and it was rendered even less likely when Tanret in 1895 [11] confirmed these observations on lactose and extended them to glucose and galactose. Finally, since the mutarotation of nitrocamphor in many anhydrous solvents could only be attributed to REVERSIBLE ISOMERIC CHANGE, the view was advanced that the mutarotation of glucose was due to an interconversion of the isomeric α and β sugars. The stable product was therefore not a third isomer, as had formerly been supposed, but an equilibrium-mixture of α and β-glucose.[12]

(iii) These two sugars are the parents of the α- and β-glucosides.[13] They are now formulated as 6-ring (pyranose) oxides,[14] but the presence of an open-chain aldehyde or aldehydrol in the equilibrium mixture

[1] RIIBER, *Ber.*, 1923, **56**, 2185.
[2] ANDREWS and WORLEY, *J.P.C.*, 1927, **31**, 1880.
[3] HUDSON and JANOSKY, *J.A.C.S.*, 1917, **39**, 1032.
[4] RIIBER, MINSAAS and LYCHE, *J.*, 1929, 2173.
[5] SMITH and LOWRY, *J.*, 1928, 666.
[6] HUDSON and JANOSKY, *J.A.C.S.*, 1917, **39**, 1032.
[7] NELSON and BEEGLE, *J.A.C.S.*, 1919, **41**, 559.
[8] FISCHER, *Ber.*, 1890, **23**, 2625.
[9] ERDMANN, *Fortschr. der Physik*, 1855, 13; *Ber.*, 1880, **13**, 2180.
[10] SCHMOEGER, *Ber.*, 1880, **13**, 1915, 1922, 2130.
[11] TANRET, *C.R.*, 1895, **120**, 1060; *Bull. Soc. Chim.*, 1896, [iii], **15**, 195, 349.
[12] LOWRY, *J.*, 1899, **75**, 211.
[13] SIMON, *C.R.*, 1901, **132**, 487; E. F. ARMSTRONG, *J.*, 1903, **83**, 1305.
[14] GOODYEAR and HAWORTH, *J.*, 1927, 3136.

must also be postulated in order to provide a mechanism for the inversion of the terminal >**C**HOH radical.[1] The presence of 5-ring (furanose) sugars in the mixture must also be admitted, since the difference of stability of the 5- and 6-atom rings is not sufficient to exclude them altogether from the equilibrium. The final equilibrium must therefore include, in addition to the open-chain aldehydic sugar, which has been isolated in the form of a penta-acetate by Wolfrom,[2] *four* glucosidic sugars (α- and β- pyranose, α- and β- furanose)[3] corresponding with the four ethyl glucosides which have now been isolated:[4]

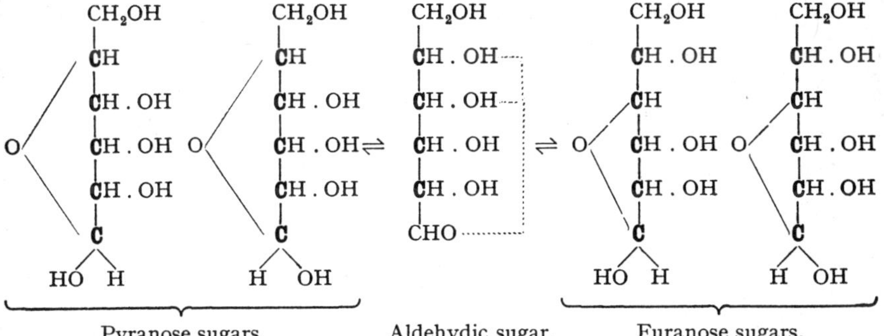

The dotted lines have been added to suggest that, even in the open-chain compound, the carbonyl group is probably co-ordinated with a hydroxyl group in the γ or δ position.

(iv) Mutarotation is characteristic of all the natural reducing sugars in freshly-prepared solutions in water. It persists in the tetramethyl and tetra-acetyl derivatives of the hexoses, in which ring formation is still possible, but is inhibited when the alcoholic hydrogen of the terminal >**C**HOH group is replaced by methyl in the α- and β- methylglucosides, >**C**H . OMe, and also when the sugar is fixed in the open-chain form by methylating, acetylating or benzoylating the *five* >**C**HOH groups of the chain. It can, however, be arrested in *non-aqueous solutions* of methylated or acetylated sugars, e.g. 2 : 3-*dimethylglucose* in *acetone*,[5] although mutarotation proceeds rapidly on adding a catalyst. The reversible isomeric changes which give rise to mutarotation are therefore not spontaneous, but proceed readily only in the presence of an amphoteric solvent. Thus the mutarotation of *tetramethylglucose* can be arrested both in *pyridine* and in *cresol;* but in a mixture of 1 part of pyridine with 2 parts of

[1] Lowry, *J.*, 1903, **83**, 1314; 1904, **85**, 1551.
[2] Wolfrom, *J.A.C.S.*, 1929, **51**, 2188 *et seq.*
[3] Lowry and Smith, Report on *Mutarotation*, Liège, 1931, p. 92.
[4] Haworth and Porter, *J.*, 1929, 2796.
[5] Irvine and Scott, *J.*, 1913, **103**, 575.

cresol it proceeds 20 times faster than in pure water,[1] the efficiency of which is also obviously due to its amphoteric character. The action of acid and basic catalysts in mutarotation, and the mechanism of prototropic change in compounds which exhibit mutarotation, are discussed more fully in a paper on *An Electrolytic Theory of Catalysis by Acids and Bases.*[2]

Optical Superposition.—Van't Hoff's PRINCIPLE OF OPTICAL SUPERPOSITION [3] suggests that the partial rotation contributed by a given asymmetric carbon atom, to the total rotation of a molecule containing several such atoms, is independent of the configuration of these other atoms Thus, the configurations of the four pentose sugars in an open-chain form may be shown as follows :

CH_2OH	CH_2OH	CH_2OH	CH_2OH
HO . **C** . H	HO . **C** . H	HO . **C** . H	H . **C** . OH
HO . **C** . H	HO . **C** . H	H . **C** . OH	HO . **C** . H
H . **C** . OH	HO . **C** . H	H . **C** . OH	H . **C** . OH
CHO	CHO	CHO	CHO
Arabinose.	Ribose.	Lyxose.	Xylose.
$+A$	$+A$	$+A$	$-A$
$+B$	$+B$	$-B$	$+B$
$+C$	$-C$	$+C$	$+C$

We assign to the three asymmetric carbon atoms of arabinose the partial rotations $+A+B+C$. The sign of C, B, and A is then reversed in the three sugars which follow. Since

$$(A+B-C)+(A-B+C)+(-A+B+C)=(A+B+C)$$

van't Hoff predicted that "the rotation of arabinose (probably the highest) should be equal to the rotations of xylose, ribose and the expected fourth type [lyxose] taken together." [3] Guye and Gautier [4] expressed these rules as follows :

FIRST PRINCIPLE.—*In a molecule containing several asymmetric carbon atoms, each of them behaves as if the rest of the molecule were inactive.*

SECOND PRINCIPLE.—*The optical effects of different asymmetric carbon atoms in the same molecule can be added algebraically.*

Consequently : *The rotatory power of a substance with several asymmetric carbon atoms may be deduced by evaluating the optical effect of each asymmetric carbon atom as if the rest of the molecule were inactive, and forming the algebraic sum of these different effects.*

[1] LOWRY and FAULKNER, *J.*, 1925, **127**, 2883.
[2] LOWRY, *Reunion Int. de Chim. Physique*, 1928, pp. 219–232.
[3] VAN'T HOFF, *The Arrangement of Atoms in Space*, London, 1898, p. 160. The name was introduced by GUYE and GAUTIER, *C.R.*, 1894, **119**, 740, 953.
[4] GUYE and GAUTIER, *C.R.*, 1894, **119**, 741.

This method was used to predict the specific rotations of amyl ether and of amyl valerate as follows:

$\alpha_D{}^*(l = 0{\cdot}5 \text{ dm.})$.

Amyl ether:

$$\begin{matrix} C_2H_5 \\ CH_3 \end{matrix}\!\!>\mathbf{C}H\,.\,CH_2\,.\,O\,.\,CH_2\,.\,\mathbf{C}H\!<\!\!\begin{matrix} C_2H_5 \\ CH_3 \end{matrix} \quad +0{\cdot}50° \text{ (obs.)} +0{\cdot}49° \text{ (calc.)}.$$

Amyl valerate:

$$\begin{matrix} C_2H_5 \\ CH_3 \end{matrix}\!\!>\mathbf{C}H\,.\,CH_2\,.\,O\,.\,CO\,.\,\mathbf{C}H\!<\!\!\begin{matrix} C_2H_5 \\ CH_3 \end{matrix} \quad +5{\cdot}32° \text{ (obs.)} +5{\cdot}34° \text{ (calc.)}$$

$$+5{\cdot}62° \text{ (calc.)}.$$

C. S. Hudson's Isorotation Rules.—Van't Hoff's "Principle of Optical Superposition" applies strictly only to stereoisomers which differ only in the *configurations* of a series of asymmetric carbon atoms. It has been extended to a wider range of compounds by C. S. Hudson,[1] who has put forward two ISOROTATION RULES as set out below:

Rule 1.—"*The rotation of Carbon 1 in the case of many substances of the sugar group is affected in only a minor degree by changes in the structure of the remainder of the molecule.*"

Thus, the effect of reversing the configuration of the terminal $>\mathbf{C}\!<\!\!\begin{matrix} H \\ OMe \end{matrix}$ group in a series of methyl glycosides is seen to be almost exactly the same for (i) a pentose, (ii) a hexose, (iii) a biose.

(i) α-Methyl *d*-xyloside	$[M]_D =$	252°	Difference 359°
β- ,, ,,	$=$	−107°	
(ii) α-Methyl *d*-glucoside	$[M]_D =$	308·3°	Difference 374·6°
β- ,, ,,	$=$	−66·3°	
(iii) α-Methyl *d*-gentibioside	$[M]_D =$	233·2°	Difference 361·2°
β- ,, ,,	$=$	−128·0°	

Rule 2.—"*Changes in the structure of Carbon 1 in the case of many substances of the sugar group affect in only a minor degree the rotation of the remainder of the molecule.*"

This rule has no direct relation to van't Hoff's principle, since we have no right to expect that the partial rotation of an asymmetric carbon atom will be unaffected by an alteration of *structure* (and not merely of *configuration*) in one of the four different radicals to which it owes its asymmetry. Nevertheless Simon in 1901 [2] was able to show that the average specific rotation of α and β glucose is equal to that of the α and β methyl glucosides, thus:

[1] C. S. HUDSON, *J.A.C.S.*, 1909, **39**, 66 *et seq.*; Union International de Chimie, *sur Rapport les Hydrates de Carbone*, 1931, pp. 59–78.
[2] SIMON, *C.R.*, 1901, **132**, 487.

	$[\alpha]_D$.		$[\alpha]_D$.
α-Methylglucoside	+ 157°	α-Glucose	+ 105°
β-Methylglucoside	− 32°	β-Glucose	+ 22°
Mean	+ 62·5°	Mean	+ 63·5°

In accordance with Hudson's second rule of isorotation, these observations show that, when the partial rotation of the terminal carbon atom is eliminated by averaging the rotations of the α and β forms, the *specific* rotation of the remainder of the molecule is nearly the same for glucose itself and for the methyl glucosides derived from it. Hudson applied his rule to *molecular* rotations, and cited the following data as illustrations :

(i) α-Glucose	$[M]_D$ =	203·0°	Sum = 237·2°
β- ,,	=	34·2°	
(ii) α-Methylglucoside	$[M]_D$ =	308·3°	Sum = 242·0°
β- ,, ,,	=	−66·3°	
(iii) α-Glycol glucoside	$[M]_D$ =	303·5°	Sum = 235·1°
β- ,, ,,	=	−68·4°	

Other examples of the same rule have been given by Maltby,[1] who attributes to the terminal carbon atom of a series of glucosides the following molecular partial rotations :

Hydrogen.	*Methyl.*	*Ethyl.*	*Propyl.*	*Butyl.*	*Amyl.*	*Allyl.*	*Benzyl.*	*Phenyl.*
−84°	−184°	−194°	−200°	−210°	−210°	−210°	−270°	−324°

These values are obviously one half of the differences tabulated by C. S. Hudson.

By making use of these rules C. S. Hudson has been able to predict the molecular rotations of a large number of unknown sugars, and in fourteen cases the values thus predicted have been verified within a few degrees.[2] Conversely, in order to account for the observed rotations of certain sugars, he has suggested new structures, which are sometimes in conflict with the chemical evidence. Thus, when classified according to their response to the rules of isorotation, the six common aldoses fall into two categories as follows : [3]

Glucose.	Gulose.	Xylose.	Gluco-heptose.	Mannose.	Galactose.
CHO	CHO	CHO	CHO	CHO	CHO
-|-OH	-|-OH	-|-OH	-|-OH	HO-|-	-|-OH
HO-|-	-|-OH	HO-|-	-|-OH	HO-|-	HO-|-
-|-OH	HO-|-	-|-OH	HO-|-	-|-OH	HO-|-
-|-OH	-|-OH	CH_2 . OH	-|-OH	-|-OH	-|-OH
CH_2 . OH	CH_2 . OH		-|-OH	CH_2 . OH	CH_2 . OH
			CH_2 . OH		
Category B.				Category A.	

[1] MALTBY, *J.*, 1922, **121**, 2608 ; 1923, **123**, 1404.
[2] C. S. HUDSON, *Rapport sur les Hydrates de Carbone*, Liege, 1931, p. 76.
[3] HAWORTH and HIRST, *J.*, 1930, p. 2616.

The α and β sugars in category B show a constant difference which amounts in the case of α- and β-glucose to 94°, in the specific rotations of the two forms, whereas the specific rotations of α- and β-mannose differ only by 47°:

α-Glucose	+ 113°	α-Mannose	+ 30°
β-Glucose	+ 19°	β-Mannose	− 17°
	+ 94°		+ 47°

In order to explain this contrast, Hudson assumed that the ring-structures are different. In particular he assigned to α-mannose a 5-atom (furanose) ring, whilst admitting the presence of a 6-atom (pyranose) ring in all the sugars of the category B, as well as in β-mannose. He therefore regarded α-mannose and α-methylmannoside as having a different structure from β-mannose and β-methylmannoside as well as from the α and β glucosides.

This assignment led Hudson to postulate a change of ring-structure during methylation, since α- and β-methylmannosides yield crystalline tetramethyl derivatives, which on hydrolysis give the same tetramethyl mannose, and the same δ-lactone when this sugar is oxidised. Since this hypothetical ring-change is not supported by any independent chemical evidence, Haworth and Hirst attribute the irregularities of rotatory power to the influence of dissimilarities of configuration, in accordance with the view of Rosanoff [1] who holds that "the optical rotatory power of an asymmetric carbon atom depends upon the composition, constitution and *configuration* of each of its four groups." In support of this conclusion they point out that all the sugars of category B exhibit a similarity of configuration of three asymmetric carbon atoms (indicated by a bracket) which does not occur in the sugars of category A, and suggest that abnormalities may perhaps be found "where aggregates of *cis*-hydroxyl groups occur in spacial proximity."

The view that the rotatory powers of the mannose series are less regular than those of the glucose series is supported by observations on the completely methylated lactones, as set out in Table 20.[2]

In the glucose series the specific rotations of a given lactone cover a range of only about 25°; but in the galactose series the range is increased to a maximum of 60°, and in the mannose series to over 150°, accompanied by a reversal of sign in four lactones out of six.

In general it may be concluded that each asymmetric centre in a molecule gives rise to a partial rotation with an appropriate characteristic frequency. Van't Hoff and C. S. Hudson were probably right in supposing that this partial rotation need not be influenced very greatly by the configuration, or even by the structure, of radicals which are not related very closely to the asymmetric centre. Rules such as those cited above can, however, only be valid when the

[1] Rosanoff, *J.A.C.S.*, 1906, **28**, 525; 1907, **29**, 536.
[2] Haworth and Hirst, *J.*, 1930, p. 2619.

TABLE 20.—SPECIFIC ROTATIONS $[\alpha]_D$ OF COMPLETELY METHYLATED LACTONES IN DIFFERENT SOLUTIONS.

		Solvent.			
	Lactones.	Water.	Chloroform.	Ether.	Benzene.
		Degrees.	Degrees.	Degrees.	Degrees.
Mannose series.	δ-Mannono (*d*-)	+150	+ 59·5	+ 35	+ 20
	γ-Mannono (*d*-)	+ 65	− 10	− 36	− 50
	δ-Lyxono (*d*-)	+ 35·5	− 60	− 87	−102
	γ-Lyxono (*d*-)	+ 82·5	− 28	− 70	− 70
	δ-Rhamnono (*l*-)	−130	− 68	− 39	− 15
	γ-Rhamnono (*l*-)	− 56·5	+ 13	+ 65	+ 87
Glucose series.	δ-Glucono (*d*-)	+ 98	+103	+123	+121
	γ-Glucono (*d*-)	+ 62	+ 42	+ 67	+ 68
	δ-Xylono (*d*-)	0	+ 9	+ 12	+ 17
	γ-Xylono (*d*-)	+ 88	+ 81	+ 84	+106
Galactose series.	δ-Galactono (*d*-)	+153	+101	+ 96	+128
	γ-Galactono (*d*-)	− 34	− 13	− 11	− 11
	δ-Arabono (*l*-)	+181	+125	+105	+166
	γ-Arabono (*l*-)	− 44	− 9	− 3	+ 16

mutual influence of the radicals (which has been invoked in interpreting the phenomenon of induced dissymmetry, pp. 147 and 408) can be ignored. If, however, a strong coupling exists between adjacent asymmetric carbon atoms, deviations may result, which will make the rules invalid. The magnitude of these deviations cannot yet be foretold, but it is probable that they are at a maximum when the radicals possess a low-frequency period, as in unsaturated and chromophoric groups, and at a minimum when the characteristic frequencies are all in the Schumann region. Regularities are therefore most likely to occur in series of compounds showing simple rotatory dispersion, but are likely to be much rarer and more fortuitous in compounds which exhibit complex or anomalous rotatory dispersion.

Rotatory Dispersion of Cyclic Sugars.—The cyclic sugars, and the polysaccharides generally, are characterised by the presence of the group $>$CH . OH, which may be repeated many times over in the molecule.

β-Glucose.

β-Fructose.

Thus β-glucose, the sugar from which cellulose is built up, contains this asymmetric secondary alcohol group repeated *four* times, but once with oxygen instead of carbon as a neighbour. The molecule also contains an inactive H>CHOH group in the side chain and an active ether-group >**C**H . O . C< in the ring. Similarly β-fructose contains *three* asymmetric >**C**HOH groups in the ring, an inactive H>CHOH group in the side chain, an inactive $>CH_2$ group in the ring and an active ether-group >**C**(OH) . O . C< also in the ring. It may therefore be expected that the rotatory dispersion of the sugars would be similar to that of the optically-active secondary alcohols, prepared in such large numbers by Pickard and Kenyon [1] (p. 262) which contain the same radical linked to two alkyl radicals R_1 and R_2, thus

$$\begin{matrix} R_1 \\ R_2 \end{matrix} > \mathbf{C}\text{H . OH}$$

These secondary alcohols have a simple or pseudo-simple rotatory dispersion [2] with dispersion constants ranging from 0·0190 to 0·0286, corresponding with wave-lengths from 1044 to 1690 A.U. Sucrose contains the group >CHOH repeated eight times in the molecule, three times as H(C)>CHOH (inactive) and five times as C(C)>**C**HOH (active), together with two active >**C**H . OC< groups and two active C(C)>**C**<(O)(O) groups. These variants in the type of the radicals from which the molecule is built up may be expected to give rise to unequal frequencies, but all lying within a relatively narrow range in the Schumann region, since no absorption band has been observed in the more readily accessible part of the ultra-violet spectrum. It is, therefore, not surprising to find that the rotatory dispersion of sucrose is apparently simple, not only in the visible and early ultra-violet spectrum [3] (see p. 131), but also over a more extended range in the ultra-violet, where the simple dispersion law is obeyed as far as 2356 A.U.[4]

[1] PICKARD and KENYON, *J.*, 1911, **99**, 45; 1912, **101**, 620; 1913, **103**, 1923 *et seq.*

[2] LOWRY, PICKARD and KENYON, *J.*, 1914, **105**, 94–102.

[3] LOWRY and RICHARDS, *J.*, 1924, **125**, 2523.

[4] HARRIS, HIRST and WOOD, *J.*, 1932, 2115.

The dispersion-constant of the equation

$$[\alpha] = 21{\cdot}648/(\lambda^2 - 0{\cdot}0213)$$

for cane sugar corresponds with a characteristic wave-length, $\lambda_0 = 1460$ A.U., lying within the normal range of the secondary alcohols. The range of frequencies of the constituent radicals of the sugars is, however, sufficiently wide to give rise in other cases to marked deviations from the law of simple rotatory dispersion; thus, of twelve sugar-derivatives examined by Harris, Hirst and Wood, only three conformed closely to this law, whilst the other nine showed deviations ranging up to 13 per cent.

Mutarotation and Rotatory Dispersion of Acids and Lactones.—Emil Fischer discovered in 1889 that the conversion of *mannonic acid* into its lactone could be followed by observing the changing rotatory power of the solutions. In 1890 [1] he showed that changes of opposite sign occur when *gluconic acid* and its lactone are converted reversibly into one another in aqueous solutions. Similar changes were observed by Allen and Tollens [2] in 1890 in *arabonic* and *xylonic acids* and by Schnelle and Tollens [3] in 1892 in *rhamnonic acid* and its lactone as well as in *gluconic* and *galactonic acids* and their lactones. Their figures for gluconic acid and its lactone have already been cited, but complete equilibrium (as evidenced by identity of rotatory power of solutions prepared from the acid and lactone) had not been reached at the end of their observations. More recently the mutarotation curves of a series of methylated sugar acids and their lactones have been studied by Haworth and his colleagues [4] who have shown that the γ-lactones undergo hydrolysis much more slowly than the isomeric δ-lactones, although the final equilibrium of the fully methylated lactones with

TABLE 21.—EQUILIBRIUM BETWEEN METHYLATED SUGAR ACIDS AND LACTONES.

	γ-Lactones.			δ-Lactones.		
	Acid.	Lactone.	Time, Hours.	Acid.	Lactone.	Time, Hours.
Tetramethyl gluconolactone	87	13	520	95	5	5
Tetramethyl galactonolactone	—	—	—	98·5	1·5	18
Tetramethyl mannonolactone	11	89	>900	64	36	140
Trimethyl xylonolactone	31	69	>500	65	35	70
Trimethyl arabonolactone	45	55	>500	99·5	0·5	4

[1] FISCHER, *Ber.*, 1890, **23**, 2625.
[2] ALLEN and TOLLENS, *Ann.*, 1890, **260**, 289, 396.
[3] SCHNELLE and TOLLENS, *Ann.*, 1892, **271**, 61.
[4] DREW, GOODYEAR AND HAWORTH, *J.*, 1927, 1237; HAWORTH and PORTER, *J.*, 1928, 611.

the corresponding methylated acids (which must contain a free hydroxyl radical in the appropriate γ or δ position) is not always very different.

As an illustration, Fig. 126[1] shows in the case of the isomeric trimethylxylonic acids :

I. The slow hydrolysis of the γ-lactone.
II. The slow lactonisation of the " γ "-acid.
III. The rapid hydrolysis of the δ-lactone.
IV. The rapid lactonisation of the " δ "-acid.

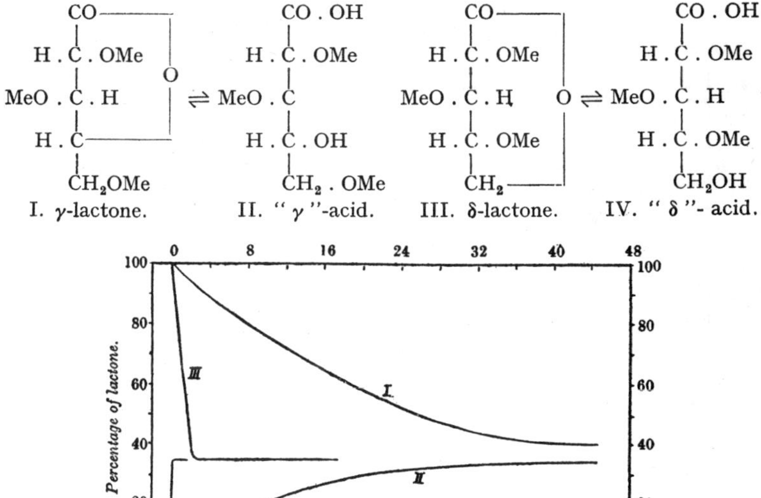

FIG. 126.—MUTAROTATION OF TRIMETHYLXYLONIC ACIDS AND LACTONES (Haworth and Porter).

A very full study of the mutarotation and rotatory dispersion of galactonic acid and its lactone was made by Lowry and Krieble in 1931.[2] The lævorotatory *lactone* showed in the visible spectrum a *simple rotatory dispersion*, as expressed by the equation

$$\alpha/\alpha_{5461} = 0{\cdot}2389/(\lambda^2 - 0{\cdot}0586).$$

The dispersion constant of this equation corresponds with a characteristic frequency at 2420 A.U. as compared with 1460 A.U. for cane sugar (p. 131). On the other hand, the dextrorotatory *sodium salt* provided the first example of *anomalous rotatory dispersion* in the sugar series, since the dextrorotation reached a maximum in the

[1] HAWORTH and PORTER, *J.*, 1928, p. 613.
[2] LOWRY and KRIEBLE, *Z. ph. C.*, Bodenstein Festband, 1931, 881–889.

blue region at about 4700 A.U. The rotatory power of the acid was intermediate between those of the lactone and of the sodium salt, since it was feebly lævorotatory throughout the visible spectrum, the dispersion being *complex but normal.* The mutarotation curves for the conversion of the acid into the lactone were recorded for eleven wave-lengths and shown to be unimolecular, with a velocity-constant ranging only from 0·00086 to 0·00089 in presence of 0·109 N HCl and 0·0653 N NaCl. Ten corresponding values were obtained for the equilibrium-proportions, the average being 25·2 per cent. acid, 74·8 per cent. lactone, on the assumption that the equilibrium included only the acid and the γ-lactone and was not complicated

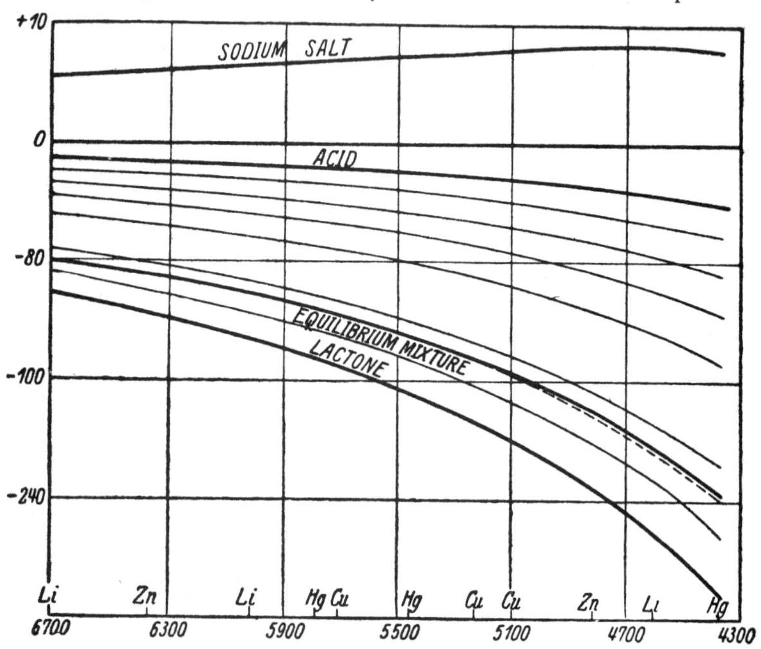

FIG. 127.—ROTATORY DISPERSION AND MUTAROTATION OF GALACTONIC ACID AND LACTONE (Lowry and Krieble).

by the formation of smaller quantities of the δ-lactone. Curves showing the rotatory dispersion of the lactone, acid and sodium salt are reproduced in Fig. 127,[1] together with intermediate curves showing the progressive change in dispersion as the acid and lactone were gradually converted into the equilibrium-mixture. Fig. 127 may be compared with the diagram of Austin and Park,[2] which shows the progressive conversion by hydrolysis of the *simple, positive* rotatory dispersion of *diacetyltartaric anhydride,* through

[1] LOWRY and KRIEBLE, *loc. cit.*, p. 888.
[2] AUSTIN and PARK, *J.*, 1925, **127**, Fig. 1, p. 1929.

a series of *anomalous* intermediate curves, into the apparently * *simple negative* dispersion of *diacetyltartaric acid* (p. 290).

Rotatory Dispersion of Aldehydic and Ketonic Sugars.—Special interest attaches to the optical properties of open-chain sugars, in which the formation of a ring has been prevented by eliminating all the free hydroxyl groups by methylation,[2] acetylation [3] or benzoylation.[4] These sugars may be expected to show the normal properties of aldehydes and ketones, and this prediction has been verified in the five following compounds [5]:—

CHO	CHO		CH_2 . OAc	CHO
H . **C** . OAc	H . **C** . OAc	CHO	CO	AcO . **C** . H
AcO . **C** . H	AcO . **C** . H	H . **C** . OAc	AcO . **C** . H	H . **C** . OAc
H . **C** . OAc	AcO . **C** . H	AcO . **C** . H	H . **C** . OAc	H . **C** . OAc
H . **C** . OAc	H . **C** . OAc	AcO . **C** . H	H . **C** . OAc	AcO . **C** . H
CH_2 . OAc	CH_2 . OAc	CH_2 . OAc	CH_2 . OAc	CH_3
I. Penta-acetyl μ-*d*-glucose.	II. Penta-acetyl μ-*d*-galactose.	III. Tetra-acetyl μ-*l*-arabinose.	IV. Penta-acetyl μ-*d*-fructose.	V. Tetra-acetyl μ-*d*-fucose.

In particular, maxima of absorption and of circular dichroism have been recorded as follows :—

log ε (max.).	Half width.	$(\varepsilon_r - \varepsilon_l)$ (max.).	Half width.
I. 1·505 at 2920 A.U.	423 A.U.	1·00 at 2920 A.U.	420 A.U.
II. 1·460 at 2900 A.U.	500 A.U.	2·25 at 2910 A.U.	400 A.U.
III. 1·580 at 2900 A.U.	470 A.U.	1·80 at 2910 A.U.	400 A.U.
IV. 1·591 at 2830 A.U.	472 A.U.	1·00 at 2806 A.U.	392 A.U.
V. 1·380 at 2935 A.U.	520 A.U.	2·15 at 2924 A.U.	446 A.U.

The rotatory dispersions of the first two compounds in the region of transparency were obviously complex, but those of III and IV were at least approximately simple. Thus the rotatory dispersion of the arabinose acetate III could be expressed by the equation

$$[\alpha] = -\ 16{\cdot}1571/(\lambda^2 - 0{\cdot}08575),$$

where $\lambda_0 = 2928$ A.U. as compared with 2910 for the maximum of circular dichroism and 2900 for the maximum of absorption.

* By using ether instead of acetone as a solvent, the rotatory dispersion of the acid was shown to be obviously complex in the ultra-violet region of the spectrum.[1]

[1] AUSTIN, *J.*, 1928, p. 1828.
[2] LEVENE and MEYER, *J. Biol. Chem.*, 1926, **69**, 175 ; 1927, **74**, 695.
[3] WOLFROM, *J.A.C.S.*, 1929, **51**, 2188 *et seq.*
[4] BRIGL and MÜHLSCHLEGEL, *Ber.*, 1930, **63**, 1551.
[5] HUDSON, WOLFROM and LOWRY, *J.*, 1933, 1179–1192 *et seq.*

In the region of absorption, negative and positive maxima of specific rotation, on either side of a reversal of sign, were observed as follows :

	First maximum.	Reversal.	Second maximum.
I.	$-464°$ at 3113 A.U.	2969 A.U.	$+755°$ at 2600 A.U.
II.	$-1090°$ at 3113 A.U.	2925 A.U.	$+1420°$ at 2650 A.U.
III.	$-1145°$ at 3122 A.U.	2909 A.U.	$+1145°$ at 2678 A.U.
IV.	$500°$ at 3020 A.U.	2830 A.U.	$-500°$ at 2580 A.U.
V.	$1130°$ at 3150 A.U.	2980 A.U.	$-1755°$ at 2702 A.U.

In compounds I, II and V the second maximum is seen to be much larger than the first maximum, showing that the (anomalous) partial rotation due to the aldehydic group is superposed on a partial rotation of opposite sign and of high frequency, due to the more saturated parts of the molecule. In the third and fourth sugars, however, the negative and positive maxima are seen to be exactly equal, although their signs are interchanged in III and IV.

This remarkable result implies that the partial rotations due to the three asymmetric carbon atoms of arabinose or fructose have cancelled out, so that they make no appreciable contribution to the rotatory power of the molecule. Within the limits of the errors of observation and of method, therefore, the whole of the rotatory power must be attributed to the induced dissymmetry of the carbonyl group. It is of interest to notice that the two compounds in which this phenomenon occurs contain the enantiomorphous radicals

```
        X                      X
        |                      |
        CO                     CO
        |                      |
  H . C . OAc          AcO . C . H
        |                      |
AcO . C . H              H . C . OAc
        |                      |
AcO . C . H              H . C . OAc
        |                      |
      CH2OAc                 CH2 . OAc
```

where in III, $X = H$, and in IV $X = .\,CH_2\,.\,O\,.\,CO\,.\,CH_3$. The radical $H[CHOAc]_4$ in these two compounds therefore appears to be optically inactive by internal compensation of the partial rotations of the three asymmetric carbon atoms, at least when it is linked to a $>$CO group ; but, since the $>$CO group is mainly under the influence of the contiguous asymmetric carbon atom, it exhibits an induced dissymmetry, which is not destroyed, or even weakened, by the neutralisation of the partial rotation of this contiguous asymmetric atom by those of the two more distant asymmetric atoms. The partial rotations for the *D*-line are distributed as follows in these sugars :

	I.	II.	III.	IV.	V.
Partial rotation of >CO group	−109°	−238°	−188°	+119°	+239·0°
Partial rotation of other groups . . .	+ 89°	+145°	[− 9°]	[+ 15°]	−105·5°
Total molecular rotation .	− 20°	− 93°	−197°	+ 134°	+133·5°

The residual rotations — 9° and + 15° may be attributed to the failure of the theoretical formula to give *equal* maxima of opposite sign, and should not be regarded as real.

CHAPTER XXIII.

HYDROXY-ACIDS.

A. Tartaric Acid.

HO . CO . **C**H(OH) . **C**H(OH) . CO . OH.

Historical.—The optical rotatory power of tartaric acid was discovered in 1832 by Biot,[1] who devoted one of his longest memoirs [2] to a detailed account of its properties when mixed with water, with alcohol, and with wood spirit. In this memoir it was shown that the rotatory power of the acid, instead of increasing progressively from red to blue, in approximate agreement with the Law of Inverse Squares, reached a maximum in the green.[3] This anomaly persisted in alcoholic solutions, but disappeared in the alkali salts, which showed a normal type of rotatory dispersion throughout the visible spectrum. In a sealed communication to the Academy on August 25, 1832, it was shown that the addition of boric acid to aqueous solutions of tartaric acid also gave rise to strongly dextrorotatory solutions with a normal type of dispersion.[4] The properties of the amorphous acid, both alone [5] and mixed with tartaric acid, were also studied by Biot, who recorded the occurrence of a *negative* rotation in pure amorphous *dextro*-tartaric acid, in agreement with an earlier prediction [6] that the rotatory power for red light would become negative below 23° C.

The changes in the properties of tartaric acid resulting from the action of solvents, of alkalis and of boric acid were all regarded by Biot as chemical rather than physical in character; but this point of view was stated in a more definite form by Arndtsen, who suggested in 1858 [7] that tartaric acid could be regarded as *a mixture of two substances*, which differed only in their optical properties, including the sign of their rotatory powers and the magnitude of their rotatory dispersions.

Rotatory Dispersion of Tartaric Acid and its Derivatives.—(*a*) The anomalous rotatory dispersion of aqueous solutions of *tartaric*

[1] Biot, *Mém. Acad. Sci.*, 1835, **13**, Table G, p. 168; paper read November 5, 1832.
[2] *Ibid.*, 1838, **15**, 93–279; paper read January 11, 1836.
[3] *Ibid.*, p. 236.
[4] Biot, *C.R.*, 1835, **1**, 457–460; *Mém. Acad. Sci.*, 1838, **16**, 271.
[5] Biot, *A.C.P.*, 1850, [iii], **28**, 366.
[6] Biot, *Mém. Acad. Sci.*, 1838, **16**, 269.
[7] Arndtsen, *A.C.P.*, 1858, [iii], **54**, 421.

acid is shown in Fig. 35 (p. 108).[1] Up to the limits of solubility of the acid in water, the *inflection* and the *maximum* on each dispersion curve lie within the visual range between Li 6708 and Hg 4358, whilst the *reversal of sign* falls within the violet or ultra-violet range beyond Hg 4358; but in the amorphous acid the reversal is in the green and the maximum and inflection have disappeared into the infra-red.

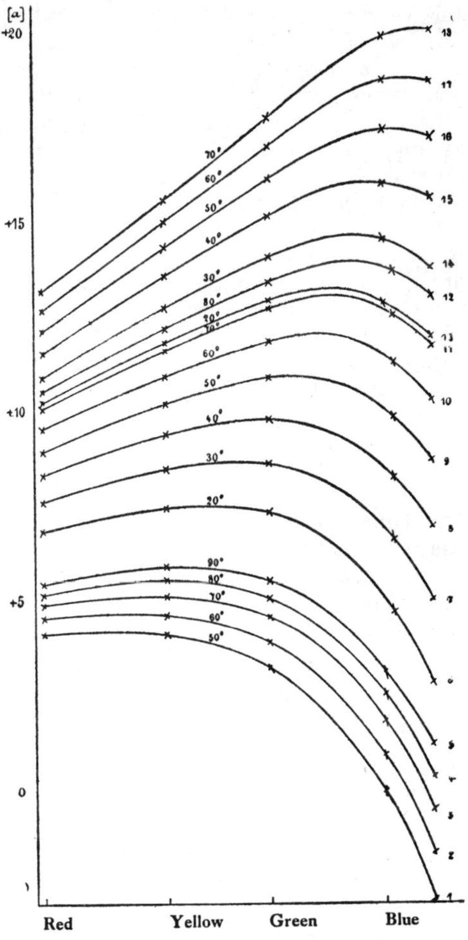

FIG. 128.—ROTATORY DISPERSION OF TARTARIC ESTERS (Winther, 1902).

1–5 Methyl tartrate.
6–12 Ethyl tartrate.
13–18 Propyl tartrate.

(*b*) The rotatory dispersion of the *tartrates* and *borotartrates* is *normal* in the visible spectrum,[2] but the observations of Descamps[3] and of Lowry and Vernon[4] have shown that dilute solutions of sodium tartrate exhibit anomalous rotatory dispersion in the ultra-violet, e.g. an aqueous solution of sodium tartrate containing 0·963 g. per 100 grams of solution shows a *maximum* at 3580 A.U. and a *reversal of sign* at 3275 A.U.[4] The borotartrates, however, show no anomalies in the ultra-violet when examined in presence of a sufficient excess of boric acid to prevent the dissociation of the complex acid;[3] and when this complex acid is stabilised by converting it into potassium or

[1] LOWRY and AUSTIN, *Bakerian Lecture, Phil. Trans.*, 1922, A. **222,** Fig. 4, p. 275.
[2] *Ibid.*, Fig. 5, p. 277.
[3] DESCAMPS, *Thesis*, Brussels, 1928.
[4] LOWRY and VERNON, *P.R.S.*, 1928, A. **119,** 706–709.

ammonium borotartrate, the dispersion is not only *normal* but *simple*.[1] Tartar emetic also appears to give a simple rotatory dispersion both in aqueous solutions, where it is strongly dextrorotatory, and in presence of an excess of alkali, when it is almost equally strongly lævorotatory.[2]

(*c*) The *alkyl tartrates* show similar anomalies to the acids. The anomalous rotatory dispersion of *methyl*, *ethyl* and *propyl* tartrates was discovered in 1902 by Winther, whose diagram [3] is reproduced as Fig. 128. The data for the rotatory dispersion of pure *ethyl tartrate* [4] have already been given in Table 11 (p. 138) and Fig. 43 (p. 137), but the variations of rotatory dispersion with changes of solvent and temperature are illustrated in Fig. 129.[5] In the upper range of rotatory powers (e.g. in formamide) the anomalies, which are quite obvious in the pure ester and in its solutions in *acetone*, have been displaced into the ultra-violet, exactly as in the case of sodium tartrate (p. 287). Conversely, in the halogenated hydrocarbons the anomalies are displaced into the infra-red, but do not necessarily disappear. Thus calculation shows that the rotatory dispersion of a solution of ethyl tartrate in *carbon tetrachloride* can be expressed by the equation [6]

$$[\alpha] = 21{\cdot}65/(\lambda^2 - 0{\cdot}03) - 20{\cdot}13/(\lambda^2 - 0{\cdot}058).$$

This equation, in which the two numerators are nearly equal, gives rise to relatively small rotations in the visible spectrum, but enables us to calculate the wave-lengths of the three anomalies, of which only the reversal can be observed experimentally, since the maximum and inflection have now been displaced into the infra-red

Reversal of sign . . .	6550 A.U.
Maximum	9020 A.U.
Inflection	10940 A.U.

On the other hand, the rotations of a solution of ethyl tartrate in *ethylene dibromide* can be expressed by the equation

$$[\alpha] = 18{\cdot}08/(\lambda^2 - 0{\cdot}03) - 19{\cdot}335/(\lambda^2 - 0{\cdot}061).$$

The numerator of the negative term is now greater than that of the positive term, and since this term also has the greater dispersion, the rotations are negative at all wave-lengths and the dispersion is normal, but complex.

(*d*) In marked contrast to the alkyl tartrates many of the *cyclic derivatives* of tartaric acid exhibit rotatory dispersions which are not only normal but are apparently *simple*, since they can be expressed

[1] Lowry, *J.*, 1929, 2853.
[2] Lowry and Austin, *Phil. Trans.*, 1922, A. **222**, 304–305.
[3] Winther, *Z. ph. C.*, 1902, **41**, 161.
[4] Lowry and Cutter, *J.*, 1922, **121**, 532–544.
[5] Lowry and Dickson, *J.*, 1915, **107**, 1183.
[6] Lowry, *J.*, 1929, 2858.

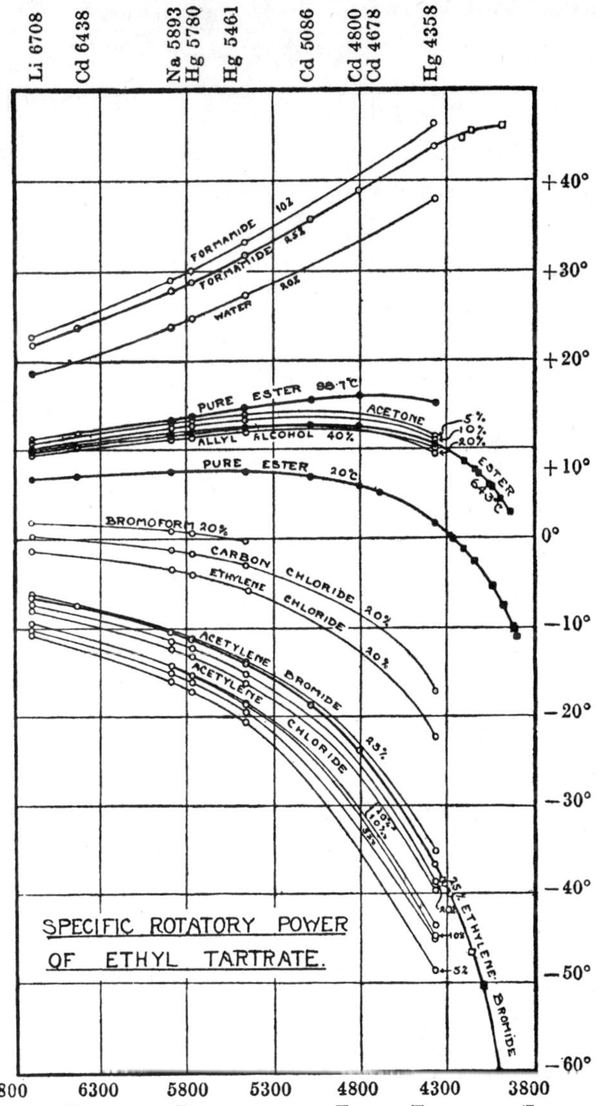

FIG. 129.—ROTATORY DISPERSION OF ETHYL TARTRATE (Lowry and Austin, 1922).

by a single term of Drude's equation. Thus the rotatory dispersion of *methylene tartrate* (I) can be represented by the equation [1]

$$[\alpha] = -24{\cdot}637/(\lambda^2 - 0{\cdot}0446)$$

[1] AUSTIN and CARPENTER, *J.*, 1924, **125**, 1939–1946.

and that of *dimethylene tartrate* (II) by the equation [1]

$$[\alpha] = + 34{\cdot}168/(\lambda^2 - 0{\cdot}03132).$$

$$CH_2\begin{cases} O\,.\,\mathbf{C}H\,.\,COOH \\ \qquad | \\ O\,.\,\mathbf{C}H\,.\,COOH \end{cases}$$

I. Methylene tartrate.

$$\begin{array}{l} CH_2\,.\,O\,.\,\mathbf{C}H\,.\,CO\,.\,O \\ |\qquad\qquad |\qquad\qquad | \\ O\text{---}CO\,.\,\mathbf{C}H\,.\,O\text{---}CH_2 \end{array}$$

II. Dimethylene tartrate.

In the same way, Austin and Park [2] have shown that *diacetyltartaric anhydride* (III) exhibits simple rotatory dispersion, whilst that of *diacetyltartaric acid* (IV) is only pseudo-simple (p. 283, footnote).

$$\begin{array}{l} CH_3\,.\,CO\,.\,O\,.\,\mathbf{C}H\,.\,CO \\ \qquad\qquad\qquad | \qquad\qquad \rangle O \\ CH_3\,.\,CO\,.\,O\,.\,\mathbf{C}H\,.\,CO \end{array} \quad \underset{}{\overset{\pm H_2O}{\rightleftharpoons}} \quad \begin{array}{l} CH_3\,.\,CO\,.\,O\,.\,\mathbf{C}H\,.\,CO\,.\,OH \\ \qquad\qquad\qquad | \\ CH_3\,.\,CO\,.\,O\,.\,\mathbf{C}H\,.\,CO\,.\,OH \end{array}$$

III. Diacetyltartaric anhydride (dextrorotatory).

IV. Diacetyltartaric acid (lævorotatory).

The reversal of sign may be attributed to an abrupt change in the shape of the molecule when the ring is broken, and the atoms are thus enabled to take up new positions in space. Austin and Park also showed that the hydrolysis of the anhydride is accompanied by a mutarotation from *d* to *l*, in the course of which the mixture of anhydride and acid exhibits anomalous rotatory dispersion (p. 283).

The dispersion ratios, and dispersion constants of eleven cyclic compounds which exhibit simple or pseudo-simple rotatory dispersion are set out in Table 22. These all contain a ring of 5 or 6 atoms and in most of them free rotation about the central bond is impeded by ring-formation. The four types of rotatory dispersion which stand out most clearly in the Table are distinguished as + A, — B, + C, — D. The optical properties of the borotartrates belong to types + A and — D, but their structures are too dissimilar from those of the alkyl, acyl and anhydro derivatives to justify any close correlation with them. On the other hand, tartar emetic and its bismuth analogue in neutral and alkaline solutions appear to establish two new optical types, which are distinguished in the Table as + E and — F.

The relation of these different optical types to the absorption spectra of the compounds is still obscure, since no maximum of selective absorption has yet been observed experimentally, although Bruhat [3] has extrapolated the absorption curve of aqueous tartaric acid to a hypothetical maximum at 2330 A.U. This wave-length can be inserted in the negative term of his equation for the rotatory dispersion of tartaric acid and is substantially identical with the

[1] Austin and Carpenter, *J.*, 1924, **125**, 1939–1946.
[2] Austin and Park, *J.*, 1925, **127**, 1926–1934.
[3] Bruhat, *T.F.S.*, 1930, **26**, 400–411.

TABLE 22.—ROTATORY DISPERSION OF CYCLIC DERIVATIVES OF TARTARIC ACID.

Substance.	Solvent.	Formula.	$[\alpha]_{5461}$.	$[M]_{5461}$.	$\frac{\alpha_{4358}}{\alpha_{5461}}$.	λ_0.	Type.
			Degrees.	Degrees.			
I. Methylene tartaric acid[2] . . .	Water (31·919/100 c.c.)	$H_2C<$ O . CH . COOH O . CH . COOH	−97·1	−157	1·745	2112	−B
	Water (6·299/100 c.c.)		−96·84	−157	1·739	2087	−B
II. Methyl benzylidene tartrate[6] . .	Ethyl acetate	C_6H_5 . HC< O . CH . $COOCH_3$ O . CH . $COOCH_3$	−44·92	−120	1·739	2087	−B
	Benzene		−40·46	−108	1·702	1883	
III. Ethyl benzylidene tartrate[6] . . .	Methyl alcohol	C_6H_5 . HC< O . CH . $COOC_2H_5$ O . CH . $COOC_2H_5$	−53·97	−159	1·739	2087	−B
	Ethyl acetate		−46·20	−136	1·727	2022	−B
IV. Diacetyl tartaric anhydride[3] . .	Acetone	CH_3CO . O . CH . CO CH_3CO . O . CH . CO >O	+74·15	+160	1·777	2251	+C
	Ethyl acetate		+50·48	+109	1·814	2387	+C
V. Dimethylene tartrate[2]	Ethyl acetate	H_2C —O— CH —CO— O O —CO— CH —O— CH_2	+128·0	+223	1·682	1770	+A
VI. Dibenzylidene tartrate[4] . . .	Acetone	C_6H_5 . HC —O— CH —CO— O O —CO— CH —O— CH . C_6H_5	+142	+463	1·696	1860	+A
VII. Potassium borotartrate[5] . . .	Water	CHOH . COOH [CO . O >B< O . CH CH . O / \ O . CO] K, H_2O CHOH . COOH	+59·7	+206	1·683	1778	+A

VIII. Ammonium borotartrate[5] . . .	Water	CHOH . COOH [CO . O, CH . O]>B<[O . CH, O . CO] NH4, H2O CHOH . COOH	+61·37	+211	1·683	1778	+A
IX. Potassium ethyl borotartrate[7] . .	Methyl alcohol	[C_2H_5COO . HC . O, C_2H_5COO . HC . O]>B<[O . CH . $COOC_2H_5$, O . CH . $COOC_2H_5$] K, H_2O	−24·63	−117	1·970	2790	−D
X. Acetylmalic anhydride[6]	Acetone	CH_3 . CO . O . CH . CO>O CH_2 . CO	−31·02	− 49	1·963	2786	−D
	Acetic anhydride		−34·63	− 55	1·960	2778	−D
XI. Ethyl β′β′β′-trichloro-α′-hydroxyethyl α-anhydrotartrate[6] .	Ethyl alcohol	CH(OH) . $COOC_2H_5$ CH . O, CO . O>CH . CCl_3 (probably)	+44·68	+137	1·615	—	—
XII. Tartar emetic[1] .	Water	KSbO($C_4H_4O_6$), ½H_2O	—	+573·7	1·771	2184	+E
	KOH aq		—	−380·4	1·846	2505	−F
XIII. Potassium bismuthyl tartrate[1] . .	KOH aq	KBiO($C_4H_4O_6$) (?)	—	−159·9	1·869	2563	−F

Note 1.—The formulæ V and VI are the most probable, but 5-atom rings are not excluded. Conversely, a 6-atom ring is a possible alternative to the 5-atom ring in formula XI.

Note 2.—The dispersion ratio 1·615 for XI is suspiciously low: since the acetyl derivative gives a ratio, $\alpha_{4358}/\alpha_{5461}$ = 1·531, below the theoretical minimum, 1·57, it is probable that *neither* dispersion is simple.

[1] Lowry and Austin, Bakerian Lecture, *Phil. Trans.*, 1922, A. **222,** 249-308.
[2] Austin and Carpenter, *J.*, 1924, **125,** 1939-1946.
[3] Austin and Park, *J.*, 1925, **127,** 1926–1934.
[4] Austin (private communication).
[5] Lowry, *J.*, 1929, 2853–2858.
[5] Brynmor Jones, *J.*, 1933, 788–796.
[7] *Ibid.*, 951–955.

characteristic wave-length, 2370 A.U., of the negative term of the equation which represents with remarkable accuracy the rotatory dispersion of ethyl tartrate; but these rotations are of opposite sign to that of the compound (diacetyltartaric anhydride) which exhibits most nearly the same dispersion-ratio, and cannot therefore be used as evidence of identity of structure.

Origin of Anomalous Rotatory Dispersion of Tartaric Acid. —The anomalous rotatory dispersion first observed by Biot in aqueous solutions of tartaric acid cannot be due to (i) hydration or (ii) ionisation, since it persists both in the glassy amorphous acid and in its esters. Thus Bruhat's curves for amorphous tartaric acid (Fig. 34, p. 103) correspond fairly closely with those of aqueous solutions of different concentrations; and his curve for the specific rotatory power of glassy tartaric acid at 44·6° can actually be superposed upon Winther's curve for methyl tartrate at 80° (curve 4, Fig. 128, p. 287).

Empirically the anomalous rotatory dispersion of aqueous solutions of tartaric acid is due to the existence of *two partial rotations of opposite sign and unequal dispersion.* The two characteristic frequencies (which Bruhat has located at 2330 and 1750 A.U.) may be identified with those of two optically-active absorption bands. Similarly the anomalous rotatory dispersion of ethyl tartrate may be attributed to the existence of two optically-active absorption bands, which have been located by Lowry and Cutter at 2370 and 1730 A.U. The outstanding problem is therefore to discover why the acid and the ethyl ester have *two* optically-active absorption bands at about 2350 and 1740 A.U., whilst, for instance, the rotatory dispersions of the methylene and dimethylene derivatives are controlled by a single absorption band at 2100 and 1770 A.U. respectively.

In this connection great importance attaches to the observations of Longchambon [1] on the rotatory dispersion of crystalline tartaric acid in plates cut perpendicular to an optic axis. These observations confirm the statement of Dufet [2] (who discovered the rotatory power of the crystalline acid in 1904), that the dispersion is *normal*, the rotations recorded by Longchambon being as follows:

$\lambda =$ 650	600	578	550	546	500	480	436 $\mu\mu$.
$\alpha =$ 9·1	10·3	11·2	12·6	13·0	16·9	19·4	24°/mm.

The rotations are negative and the dispersion is apparently simple. The dispersion-ratio 1·85 is comparable with that of tartar emetic in alkaline solutions. It corresponds with an absorption band at 2500 A.U., not far from the wave-length, 2370 A.U., attributed by Bruhat to the low-frequency absorption band of tartaric acid in aqueous solutions. There can therefore be little doubt that the lævorotatory "α-acid" (p. 111) when dissolved in water, contributes

[1] LONGCHAMBON, *C.R.*, 1924, **178**, 951.
[2] DUFET, *Bull. Soc. fr. Min.*, 1904, **27**, 156.

a negative partial rotation to the aqueous solutions, since it is unlikely that this acid disappears completely on dissolution. The hypothetical β-acid, to which Longchambon attributes the positive rotations of these solutions, must then be formed from the α-acid by a process of reversible isomeric change, since hydration and ionisation have already been excluded as fundamental factors in the development of anomalous rotatory dispersion in this series of compounds, and the hypothesis of a reversible polymeric change, suggested by Astbury in 1923,[1] is equally untenable. Since the β-acid has not been isolated, and is still a mere hypothesis, we may assume provisionally that it has a simple rotatory dispersion, like dimethylene tartrate, since the sign of the rotation and the magnitude of the dispersion constant are almost identical in the two cases ($\lambda_0 = 1750$ for tartaric acid and 1770 for dimethylene tartrate).

Lucas [2] has suggested that tartaric acid can exist in three configurations, with different optical properties, as a result of rotation about the central bond. This form of Arndtsen's hypothesis as to the origin of the different partial rotations of tartaric acid receives support from the empirical observations that simple rotatory dispersion is associated (i) with ring-formation,[3] and (ii) with the impedance of free rotation about the central bond of the molecule, which may result from ring-formation between the two halves of the molecule [4] or perhaps from the introduction of bulky substituents, as in diacetyltartaric acid.[5] Since, however, the characteristic frequencies are quite sharply defined, both in the complex dispersions of the free acid and its esters, and in the simple dispersions of their cyclic derivatives, it appears likely that, as Wolf has suggested,[6] the " free rotation " of the acid and its alkyl esters may be impeded by the presence of minima and maxima of energy, so that certain configurations of maximum stability are favoured at the expense of others. This would give rise to equilibrium-mixtures of acids of definite configuration, as postulated by Lucas, and two of these might perhaps be identified with Longchambon's α- and β-tartaric acid. On the other hand, internal co-ordination cannot be excluded in compounds which are so richly endowed with hydroxyl and carboxyl groups. It is, therefore, *a priori* almost certain that the acid and its alkyl esters must exist in co-ordinated forms [7] such as

[1] ASTBURY, *P.R.S.*, 1923, **102**, 506; compare LANDOLT, *Handbook of the Polariscope* (1882), p. 62; PRIBRAM, *Monatsh.*, 1888, **9**, 401; HAMMERSCHMIDT, *Z. Rübenzuckerind.*, 1890, **40**, 939; WYROUBOFF, *C.R.*, 1892, **115**, 832, for a similar, but equally incorrect, explanation of the mutarotation of the sugars.

[2] LUCAS, *Thesis*, Paris, 1927, p. 46; *Ann. de Physique*, 1928, [x], **9**, 425.

[3] AUSTIN, *J.*, 1924, **125**, 1941; 1925, **127**, 1926; *T.F.S.*, 1930, **26**, 415.

[4] AUSTIN, *loc. cit.*; BRYNMOR JONES, *J.*, 1933, 793.

[5] See footnote *, p. 283.

[6] WOLF, *T.F.S.*, 1930, **26**, 315.

[7] LOWRY and BURGESS, *J.*, 1923, **123**, 2118; LOWRY and CUTTER, *J.*, 1924, **125**, 1468.

```
        OH
        |                                   OH
        C     O                              |
     //   \  / \                       O=C          O—H
    O      CH    H                       :  \      /    :
    :      |     :                        :  CH—CH      :
    H      CH    O                       H—O      \    :
     \    /  \  //                                 C=O
       O       C                                   |
               |                                   OH
               OH
```

These definite chemical compounds may therefore replace, in certain instances, the less precise structures postulated by Lucas and by Wolf, although fixity of structure, which appears to be essential for the production of *simple* rotatory dispersion, is not essential to any theory of the production of *complex* rotatory dispersion by superposition of different optical types.

In conclusion, it may be pointed out that the optical activity associated with an absorption band is probably much more sensitive to changes of configuration than is the absorption band itself. In other words the "dissymmetry factor" $(\epsilon_l \sim \epsilon_r)/\epsilon$ (p. 394) is likely to vary much more than the coefficient ϵ. Thus if the same group occurs twice in the same molecule, as it obviously does in tartaric acid and the tartrates, it may be assumed as a first approximation that the absorptive powers of the two groups will be additive; but their contributions to the dissymmetry of the molecule may be directly opposed to one another, as they certainly are in meso-tartaric acid. It is therefore possible that even the "fixed" derivatives of tartaric acid, which exhibit simple rotatory dispersion, may possess the two absorption bands which give rise to the opposite partial rotations of tartaric acid and the alkyl tartrates, but that the optical activity of one or other of these bands disappears when the rotatory dispersion becomes simple. This phenomenon has already been observed in nicotine, where the principal absorption band of the base becomes inactive in the salts, and it is unfortunate that the shortness of the wave-lengths in question makes a similar complete diagnosis in the case of tartaric acid so much more difficult.

B. Malic Acid.

HO . CO . **C**H(OH) . CH_2 . CO . OH.

Malic Acid and the Malates.—The rotatory power of *malic* acid I, and of its derivatives II and III, was first described by Pasteur[1] in 1851 in a Memoir, *Sur les relations qui peuvent exister entre la forme cristalline, la composition chimique et le phénomène de la polarisation rotatoire.*

[1] Pasteur, *A.C.P.*, 1851, [iii], **31**, 81.

$HO.CH.CO.OH$	$NH_2.CH.CO.OH$	$NH_2.CH.CO.NH_2$		OH
$\vert$	$\vert$	$\vert$	or	
$CH_2CO.OH$	$CH_2.CO.OH$	$CH_2.CO.OH$		NH_2
I. Malic acid.	II. Aspartic acid.	III. Asparagin.		

In this memoir Pasteur recorded the lævorotation for the "neutral tint," $[\alpha]_j$, of *asparagin*, III, the amide of dextrorotatory *aspartic acid*, II, when dissolved in water or in ammonia, and its dextrorotation when dissolved in nitric or hydrochloric acid.

Asparagin in soda	$[\alpha]_j = -\ 7{\cdot}3°$;	in nitric acid	$[\alpha]_j = +35{\cdot}09°$
,, in ammonia	$[\alpha]_j = -11{\cdot}18°$;	in hydrochloric acid	$[\alpha]_j = +34{\cdot}4°$
Aspartic acid in soda	$[\alpha]_j = -\ 2{\cdot}22°$;	in nitric acid	$[\alpha]_j = +38{\cdot}86°$
,, ,, in ammonia	$[\alpha]_j = -11{\cdot}67°$;	in hydrochloric acid	$[\alpha]_j = +27{\cdot}68°$

More recent determinations have given the following values:

	In Water.	In HCl.
l-Asparagine [1] . . .	$[\alpha]_D^{20} = -4{\cdot}9°$	$+28{\cdot}5°$
l-Aspartic acid [2] . . .	$[\alpha]_D^{20} = +4{\cdot}3°$	$+25{\cdot}7°$

Malic acid and *ammonium bimalate* were lævorotatory; but a dextrorotatory solution was obtained by dissolving calcium malate in strong hydrochloric acid, and this remained dextrorotatory on adding an excess of ammonia.

Malic acid (33 per cent.) in water $[\alpha]_j^{10}$	$= -5{\cdot}00°$
Ammonium bimalate (23 per cent.) in water $[\alpha]_j^{20}$	$= -7{\cdot}22°$
,, ,, (27 per cent.) in nitric acid of density 1·2	$= +5{\cdot}60°$

A very high dextrorotation was also recorded in *ammonium antimonyl malate*, which gave $[\alpha]_j^{17} = +115{\cdot}47°$ as compared with $[\alpha]_r = +119{\cdot}75°$ for tartar emetic.

The rotatory power of *l*-malic acid and its salts for sodium light was studied in 1880 by H. Schneider,[3] who found that the *l*-acid became inactive at 35 per cent. concentration and dextrorotatory in very strong aqueous solutions. The solutions were therefore lævorotatory only at concentrations below 35 per cent., the influence of concentration being expressed by the equation

$$[\alpha]_D = 5{\cdot}891 - 0{\cdot}0895929\,(100 - p).$$

Esters of *l*-Malic Acid.—A large number of derivatives of malic acid were included in an investigation by Walden (p. 357) of the *optically-active derivatives of succinic acid*,[4] in which he sought to discover how the sign of the rotation was affected by changes of structure. These observations were, however, limited to the yellow

[1] E. Fischer and Koenigs, *Ber.*, 1904, **37**, 4585.
[2] Fischer and Fiedler, *Ann.*, 1910, **375**, 181.
[3] Schneider, *Ann.*, 1881, **207**, 257.
[4] Walden, *Z. ph. C.*, 1895, **17**, 245–266.

sodium line, and Purdie and Williamson [1] have shown that some of the substances were partially racemised. The specific rotations recorded by the latter were as follows :

Methyl malate $[\alpha]_D$ − 7·34°	Propyl malate $[\alpha]_D$ − 13·7°
Ethyl malate $[\alpha]_D$ − 12·4°	*n*-Butyl malate $[\alpha]_D$ − 12·2°

Rotatory Dispersion.—The rotatory dispersion of a concentrated (dextrorotatory) solution of *l-malic acid* was recorded by Arndtsen [2] and found to be normal ; but a reversal of sign, from negative in the red to positive in the violet, was discovered with the help of light filters by Nasini and Gennari [3] in 1896. This anomaly disappeared into the dark red at concentrations greater than 41 per cent. by volume or 49 per cent. by weight, and into the dark blue at concentrations less than 17 per cent. by volume or 18 per cent. by weight. Higher concentrations of the *sodium salt* were needed, however, in order to give solutions which were dextrorotatory, even at the blue end of the spectrum, where the reversal of sign first appears.

The rotatory dispersion of the *l*-acid for a series of colours was also determined by Walden [4] for solutions in several solvents and mixtures of solvents. In some of these (benzyl alcohol alone and mixed with benzene or carbon disulphide) it was dextrorotatory, whilst in others (e.g. aldehyde, acetone, *iso*butyl alcohol and pyridine) it was lævorotatory throughout the visible spectrum ; in formic acid dextrorotations appeared in the blue region when the concentration was increased to $c = 38{\cdot}3$ per cent., and this anomaly moved into the yellow on cooling to 0° C.

The rotatory dispersion of *methyl malate* was studied for the first time in 1910 by Grossmann and Landau,[5] who plotted a series of curves showing the rotations of the homogeneous ester and of solutions in twelve different solvents, for six colour-filters (Fig. 130).[6] Data thus obtained have no exact value, but the curves are reproduced for comparison with the dispersion curves for ethyl tartrate (Figs. 43 and 129, pp. 137 and 289). Solutions in $AsCl_3$, $C_2H_2Br_4$ and $C_6H_4Cl_2$ were dextrorotatory, and those in formamide, quinoline, acetaldehyde, formic acid and pyridine were lævorotatory, throughout the visible spectrum from red to violet, but a reversal of sign was recorded for a solution in glycerine, which was lævorotatory in the red, yellow and green, inactive in the light blue, and dextrorotatory in the dark blue and violet.

Solutions in chloral, bromal and phosphorus trichloride exhibited mutarotation and gave strongly lævorotatory products. Two condensation products formed in this way were isolated by Patterson

[1] Purdie and Williamson, *J.*, 1896, **69**, 818–839.
[2] Arndtsen, *A.C.P.*, 1858, [iii], **54**, 417.
[3] Nasini and Gennari, *Z. ph. C.*, 1896, **19**, 111–129.
[4] Walden, *Ber.*, 1899, **32**, 2849–2862.
[5] Grossmann and Landau, *Z. ph. C.*, 1910, **75**, 129–218.
[6] *Ibid.*, p. 214.

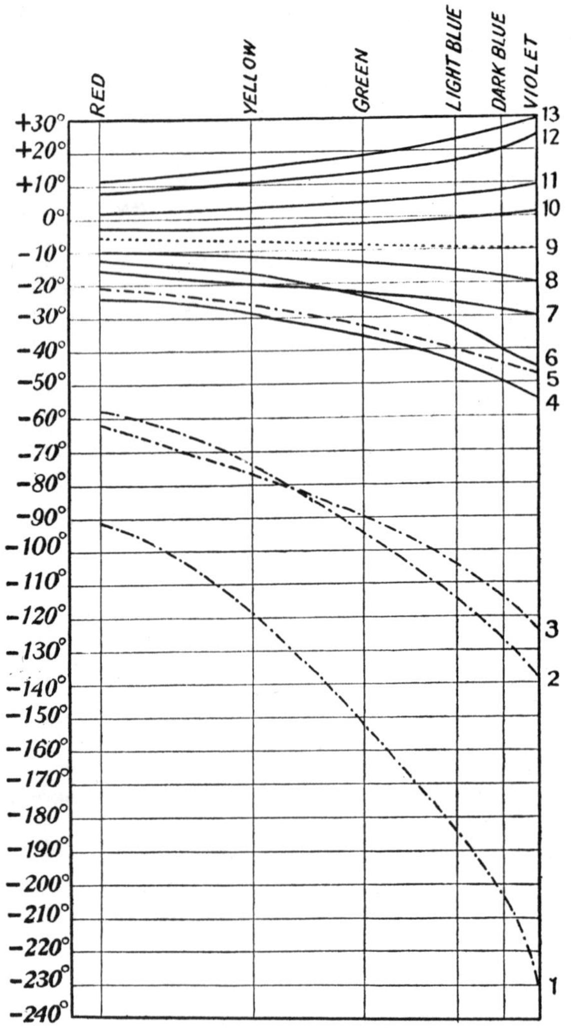

FIG. 130.—ROTATORY DISPERSION OF METHYL MALATE 209 (Grossmann and Landau, 1910).

1. Phosphorus trichloride.
2. Chloral.
3. Bromal.
4. Pyridine.
5. Formic acid.
6. Acetaldehyde.
7. Quinoline.
8. Formamide.
9. Pure ester.
10. Glycerine.
11. Dichlorobenzene.
12. Tetrabromoethane.
13. Arsenic trichloride.

and McMillan [1] from mixtures of chloral with ethyl tartrate and ethyl malate and formulated as IV and V. Their rotatory dispersion was examined by Brynmor Jones,[2] together with that of the acetyl-derivative VI of the condensation product of ethyl tartrate with chloral (Table 22, p. 290).

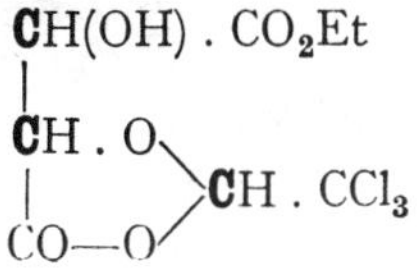

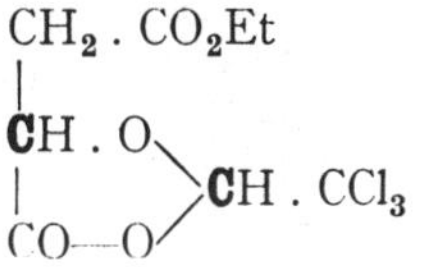

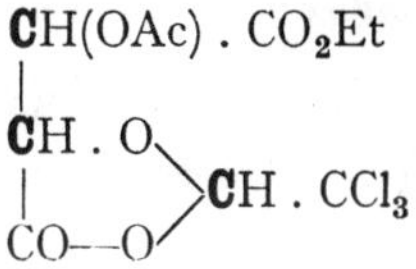

```
CH(OH) . CO2Et        CH2 . CO2Et          CH(OAc) . CO2Et
|                     |                    |
CH . O                CH . O               CH . O
|      \              |      \             |      \
|       >CH . CCl3    |       >CH . CCl3   |       >CH . CCl3
|      /              |      /             |      /
CO—O                  CO—O                 CO—O
    (IV)                  (V)                  (VI)
```

The rotatory dispersions of the two condensation products IV and V are apparently simple in the visible spectrum, since they can be expressed by the following equations :

$$\text{IV.}\quad \alpha/\alpha_{5461} = 0{\cdot}2860/(\lambda^2 - 0{\cdot}01230)\,;\quad \lambda_0 = 1100 \text{ A.U.},$$
$$\text{V.}\quad \alpha/\alpha_{5461} = 0{\cdot}2845/(\lambda^2 - 0{\cdot}01366)\,;\quad \lambda_0 = 1170 \text{ A.U.},$$

but the dispersion ratios, $\alpha_{4358}/\alpha_{5461} = 1{\cdot}610$ and $1{\cdot}613$ (corresponding with absorption bands at 1100 and 1170 A.U.), are suspiciously low for a simple dispersion, just as in the case of octyl oxalate (p. 133), and this suspicion was confirmed by the fact that the dispersion ratio of the acetyl derivative VI, $\alpha_{4358}/\alpha_{5461} = 1{\cdot}531$, is below the minimum value, $\alpha_{4358}/\alpha_{5461} = 1{\cdot}57$, for a simple dispersion. It is therefore probable that the dispersion of IV and V is also complex, although the deviations from the simple formula are too small to be detected in the visual range from 6708 A.U. to 4358 A.U. In the same way, the rotatory dispersion of *potassium boromalate*, which can be expressed, over the range from 6708 A:U. to 4358 A.U. by the equation

$$\alpha/\alpha_{5461} = 0{\cdot}2897/(\lambda^2 - 0{\cdot}008542),\quad \lambda_0 = 924 \text{ A.U.}$$

is probably complex, since the dispersion ratio, $\alpha_{4358}/\alpha_{5461} = 1{\cdot}597$, and the characteristic wave-length $\lambda_0 = 924$ A.U. are too small for a genuine simple dispersion in a compound of this structure.

***d*-Malic Acid.**—A real *d*-malic acid, enantiomorphous with the *l*-acid, was prepared in 1875 by Bremer [3] by partial reduction of *d*-tartaric acid with hydrogen iodide. The same acid was also separated by a process of resolution from the inactive (racemic) form of the acid. Piutti [4] in 1886 prepared the *d*-acid by the action of nitrous acid on *d*-asparagin, whilst Walden in 1899 [5] obtained it by the action of potassium hydroxide on lævorotatory *d*-chloro-succinic acid or of silver oxide on dextrorotatory *l*-chlorosuccinic

[1] Patterson and McMillan, *J.*, 1912, **101**, 788–803.
[2] Brynmor Jones, *J.*, 1933, p. 792.
[3] Bremer, *Ber.*, 1875, **8**, 861, 1594 ; 1880, **13**, 351.
[4] Piutti, *Ber.*, 1886, **19**, 1691.
[5] Walden, *Ber.*, 1899, **32**, 1855.

acid. Since the same *d*-malic acid was obtained from the *d*- and *l*-chloro-acids, one of these substitutions must obviously be accompanied by a WALDEN INVERSION. By taking advantage of this inversion, the *d*-acid can therefore be prepared from the ordinary *l*-acid, or conversely, by the successive action of phosphorus pentachloride and of silver oxide; but independent evidence is required to show in which of these processes inversion takes place, since the configuration of the compounds of this series cannot be deduced merely from the sign of their rotatory powers.

C. LACTIC ACID.

$$CH_3 \,.\, \mathbf{C}H(OH) \,.\, CO \,.\, OH.$$

Sarcolactic Acid.—The acid of sour milk was amongst the organic acids discovered by Scheele, who described it in 1780 as LACTIC ACID. This acid is optically inactive, as is also the acid prepared by the LACTIC FERMENTATION of glucose and fructose

$$C_6H_{12}O_6 = 2C_3H_6O_3.$$

On the other hand, SARCOLACTIC ACID or PARALACTIC ACID, which Berzelius obtained from flesh, and which Liebig[1] in 1847 prepared from meat extract, is dextrorotatory and is therefore now distinguished as *d*-lactic acid. Its optical activity was discovered in 1869 by Wislicenus,[2] who showed that the *dextrorotatory acid*, $[\alpha]_D = +3{\cdot}5°$, yields a *lævorotatory zinc salt*, $Zn(C_3H_5O_3)_2, 2H_2O$, for which $[\alpha]_D = -8{\cdot}37°$, and also a strongly *lævorotatory anhydride* or LACTIDE,[3]

$$CH_3 \,.\, \mathbf{C}H\left\langle\begin{array}{c} O \,.\, CO \\ CO \,.\, O \end{array}\right\rangle \mathbf{C}H \,.\, CH_3,$$

for which $[\alpha]_D = -85{\cdot}9°$ in alcohol, or $-298°$ in benzene.[4] At the close of this investigation Wislicenus (preceding le Bel and van't Hoff) made the important suggestion that paralactic acid and fermentation lactic acid, which are *structurally identical*, might differ in the *arrangement of the atoms in space.*[5]

Separation of *d*- and *l*-Lactic Acid.—The resolution of *lactic acid* into the *d* and *l* forms was effected in 1892 by Purdie and Walker.[6] Lewkowitch in 1883[7] had obtained a dextrorotatory solution by the selective fermentation of ammonium lactate; but eight years later Linossier[8] obtained a lævorotatory solution, which gave a lævorotatory *zinc salt* and a dextrorotatory acid. On the

[1] LIEBIG, *Ann.*, 1847, **62**, 326–331.
[2] WISLICENUS, *Ber.*, 1869 **2**, 550 and 620.
[3] WISLICENUS, *Ann.*, 1873, **167**, 321.
[4] JUNGFLEISCH and GODCHOT, *C.R.*, 1905, **141**, 111.
[5] WISLICENUS, *Ann.*, 1873, **167**, 343.
[6] PURDIE and WALKER, *J.*, 1892, **61**, 754–765; PURDIE, *J.*, 1893, **63**, 1143–1157.
[7] LEWKOWITCH, *Ber.*, 1883, **16**, 2720.
[8] LINOSSIER, *Bull. Soc. Chim.*, 1891, [iii], **6**, 10–12.

other hand, Schardinger,[1] by the bacterial decomposition of cane sugar, obtained a lævorotatory acid, $[\alpha]_D - 4{\cdot}3°$, which gave a zinc salt, crystallising with $2H_2O$ like the sarcolactate, but combining with it to give the ordinary inactive zinc salt with $3H_2O$.

Purdie's resolution was effected by crystallising the *strychnine salt*. It gave a dextrorotatory zinc salt $[\alpha]_D + 5{\cdot}63°$ (as compared with $- 6{\cdot}36°$ for the sarcolactate) and a lævorotatory *l*-lactic acid, isomeric with the dextrorotatory sarcolactic acid. In the following year the resolution was repeated by crystallisation of super-saturated solutions of *zinc ammonium lactate*, Zn $(C_3H_5O_3)_2$, Am $(C_3H_5O_3)$, $2H_2O$, on the addition of a nucleus of active salt. Two crops of active zinc ammonium lactate prepared in this way gave $[\alpha]_D = + 6{\cdot}06°$ and $- 6{\cdot}06°$, as compared with a maximum value $[\alpha]_D + 6{\cdot}49°$ for resolution with the help of strychnine, and $- 6{\cdot}44°$ for the lævorotatory fraction after conversion into the zinc salt. The zinc lactates prepared from these two fractions gave $[\alpha]_D + 6{\cdot}81°$ and $- 6{\cdot}83°$, as compared with Wislicenus' value $[\alpha]_D - 6{\cdot}83°$ for (hydrated) zinc sarcolactate.

The rotatory powers of the salts, which are opposite in sign to those of the acids, were described in a subsequent paper.[2]

Lactic Esters.—Klimenko in 1880 [3] prepared lævorotatory *ethyl* d-*lactate* from silver sarcolactate and ethyl iodide, and obtained $[\alpha]_D - 14{\cdot}19°$. Le Bel [4] in 1893 prepared the methyl ester, and other derivatives were prepared by Frankland and Henderson [5] and by Walden [6] in 1895. From the material which had been resolved by Purdie and Walker, J. W. Walker in 1895 [7] prepared

Methyl *d*-lactate $[\alpha]_D - 11{\cdot}1°$
Ethyl *l*-lactate $[\alpha]_D + 14{\cdot}52°$
Propyl *d*-lactate $[\alpha]_D - 17{\cdot}06°$.

For the *acetyl derivative* of ethyl *d*-lactate,

$$CH_3 . CO . O . \mathbf{C}H(CH_3) . CO . O . C_2H_5,$$

Purdie and Williamson [8] found $[\alpha]_D - 49{\cdot}9°$.

Esters of lactic acid were also prepared by Patterson and Forsyth,[9] who separated the dextrorotatory acid with the help of morphine [10] and by crystallisation of the zinc ammonium salt. They studied the rotatory power of the methyl ester and of its methoxy and acetoxy derivatives, both in the homogeneous state and in solution

[1] Schardinger, *Monatschefte*, 1891, **11**, 545–559.
[2] Purdie and Walker, *J.*, 1895, **67**, 616–640.
[3] Klimenko, *J. Russ. Chem. Soc.*, 1880, **12**, 25–28; *Bull. Soc. Chim.*, 1880, [ii], **34**, 321.
[4] Le Bel, *Bull. Soc. Chim.*, 1893, [3], **9**, 674.
[5] Frankland and Henderson, *Proc. Chem. Soc.*, 1895, **11**, 54.
[6] Walden, *Ber.*, 1895, **28**, 1287–1297.
[7] J. W. Walker, *J.*, 1895, **67**, 914–918.
[8] Purdie and Williamson, *J.*, 1896, **69**, 828.
[9] Patterson and Forsyth, *J.*, 1913, **103**, 2263-2271.
[10] Irvine, *J.*, 1906, **89**, 935.

in nitrobenzene and in acetylene tetrabromide, at temperatures ranging from — 75° to 141°. The rotations tend to converge as the temperature rises. Thus methyl *l*-lactate became increasingly dextrorotatory whilst a solution in acetylene tetrabromide, which was lævorotatory at lower temperatures, became dextrorotatory above 70°; conversely, the large dextrorotations of the methoxy and acetoxy propionates diminished as the temperature was raised.

A more extensive series of observations of acyl derivatives of methyl lactate, at temperatures from 0° to 140° and for six different wave-lengths from 6716 A.U. to 4358 A.U., has also been recorded by Patterson and Lawson.[1] The rotations show the same general tendency to converge as the temperature rises, but many of the curves intersect one another. An interesting feature of these observations is a *reversal of sign* in solutions of methyl *d*-lactate * in ethylene dibromide. This showed dextrorotations from the violet almost into the red at 21°; but the range of these dextrorotations decreased with rising temperature, and above 65° the solution was lævorotatory from the red to the violet at 4358 A.U. or beyond.

Two other researches on the lactic esters have been described by C. E. Wood, Such and Scarf.[2] In the first of these it is shown that the molecular rotations of the homologous series of (dextrorotatory) normal esters, $CH_3 . CHOH . CO . OC_nH_{2n+1}$, of *l*-lactic acid, both at 20° and at 110°, increase to a value which is almost constant from amyl to nonyl for four wave-lengths from 6563 A.U.

TABLE 23.—MOLECULAR ROTATIONS OF LACTIC ESTERS.

Alkyl *l*-lactate.	$[M]^{20}_{5893}$.	$\alpha_{4358}/\alpha_{5461}$.	$[M]^{110}_{5893}$.	$\alpha_{4358}/\alpha_{5461}$.
	Degrees.		Degrees.	
Methyl	+8·59	1·280	+11·26	1·375
Ethyl	13·37	1·302	16·47	1·413
n-Propyl	17·44	1·461	19·56	1·491
n-Butyl	19·58	1·445	22·05	1·501
n-Amyl	20·16	1·461	22·51	1·481
n-Hexyl	21·25	1·472	25·40	1·490
n-Heptyl	21·33	1·472	23·54	1·495
n-Octyl	18·91	1·496	20·83	1·570
n-Nonyl	19·85	1·486	23·27	1·513

* The acid used in this research was dextrorotatory like sarcolactic acid, but the authors described it as *l*-lactic acid, following WOHL and FREUDENBERG (*Ber.*, 1923, **56**, 309), who showed that it is configuratively related to *l*-tartaric acid (p. 301) and therefore proposed to describe it as *l*(+) lactic acid. In the opinion of the writer, the configurative relationship of lactic acid to tartaric acid is not of sufficient importance to justify the confusion which must inevitably result when the meaning attached to a symbol is reversed at an arbitrary date. In the present text the traditional nomenclature is maintained, and sarcolactic acid is described throughout as *d*-lactic acid.

[1] PATTERSON and LAWSON, *J.*, 1929, 2042–2051.

[2] WOOD, SUCH and SCARF, *J.*, 1923, **123**, 600–616; 1926, 1928–1938.

to 4455 A.U., although the highest values are given by the *n*-hexyl or *n*-heptyl ester. The rotatory dispersions of the esters are *normal* between these limits of wave-length, but are obviously *complex* since the dispersion ratios are below the minimum value for a *simple* rotatory dispersion.

The second paper describes the isomeric butyl lactates. The *specific* rotations and dispersions were as follows :—

TABLE 24.—SPECIFIC ROTATIONS OF BUTYL LACTATES.

Temperature 20°.	$[\alpha]_{6708}$.	$[\alpha]_{5893}$.	$[\alpha]_{5461}$.	$[\alpha]_{5086}$.	$[\alpha]_{4358}$.
	Degrees.	Degrees.	Degrees.	Degrees.	Degrees.
n-Butyl *l*-lactate .	+ 10·41	+ 13·45	+ 15·50	+ 17·62	+ 22·40
iso-Butyl *l*-lactate .	+ 11·72	+ 15·18	+ 17·55	+ 20·08	+ 25·97
tert.-Butyl *l*-lactate .	+ 7·76	+ 9·45	+ 10·93	+ 12·44	+ 15·90
dl-sec.-Butyl *l*-lactate	+ 7·50	+ 9·44	+ 10·74	+ 12·03	+ 14·80
d-sec.-Butyl *l*-lactate	+ 16·02	+ 20·67	+ 23·91	+ 27·42	+ 35·37
d-sec.-Butyl *d*-lactate	− 1·53	− 1·83	− 1·83	− 1·54	+ 0·75

The dispersions are all *complex but normal*, except in the case of *d-sec*.-butyl *d*-lactate, where a reversal of sign was produced by preparing the ester from *d*-lactic acid. In this case the lævorotatory *d*-lactic ester radical was almost completely neutralised by the dextrorotatory *d*-butyl radical. The small residual rotations were therefore negative, except in the violet at low temperatures, where a positive rotation was recorded. The dispersion curve for this ester at 20° therefore includes a *negative maximum* in the *orange* and a *reversal of sign* in the *violet*.

D. CONFIGURATION OF HYDROXY ACIDS.

Chemical Methods.—The relative configurations of a related series of optically-active compounds can be deduced with almost absolute certainty by means of transformations which do not involve those radicals which are attached directly to the asymmetric atom. In this way Freudenberg[1] has shown that the relative configuration is the same in *l*-lactic acid, *l*-glyceric acid, *d*-malic acid, *d*-tartaric acid.

COOH		COOH		COOH		COOH	COOH		COOH
H**C**OH		H**C**OH		H**C**OH		H**C**OH	H**C**OH		H**C**OH
HOCH	→	CH_2	→	CH_2NH_2	→	CH_2OH	CH_2Br	→	CH_3
COOH		COOH							
d-Tartaric acid.		*d*-Malic acid.		*d*-Iso serin.		*l*-Glyceric acid.	*l*-Bromo lactic acid.		*l*-Lactic acid.

(A curved arrow leads from *d*-Iso serin to *l*-Bromo lactic acid.)

[1] FREUDENBERG, *Ber.*, 1914, **47**, 2027–2037.

Physical Methods.—The changes of sign which so often accompany changes of temperature, solvent and concentration, especially amongst the hydroxy acids and their derivatives, make it quite impossible to deduce the relative configurations from the sign of their optical rotations, even if there were a sound theoretical basis for doing so in the absence of these irregularities. Nevertheless many attempts have been made to deduce these configurations from the *direction* in which the rotatory power varies with a given alteration in the physical and chemical conditions. Thus, in conformity with Freudenberg's conclusions, Clough[1] has shown that the changes of rotatory power produced by the addition of *inorganic salts* to aqueous solutions of the preceding acids, and of their esters, are similar in sign. He has also pointed out that changes of *temperature* produce similar changes in the rotatory powers of these related acids, since the dextrorotations are increased and the lævorotations are diminished when the temperature is raised. Moreover, the rotations of the acids are all increased by *dilution*, and the *molecular rotations of the sodium and potassium salts* are more positive than those of the free acids in aqueous solutions.

Similarly, the *temperature-rotation curves* for the esters[2] of these four acids are similar in character, their *molecular rotations* are higher than those of the acids and increase with increase of molecular weight, and the *influence of solvents* is generally similar. Thus the rotatory power of ethyl *d*-tartrate is raised by the use of formamide as a solvent and depressed by the use of tetrabromoethane (Fig. 129, p. 289), whilst that of methyl *l*-malate (which is of opposite configuration) is depressed by formamide and raised by tetrabromoethane (Fig. 130, p. 296).

C. E. Wood[3] considers that "anomalous rotatory dispersion is a trustworthy criterion of relative configuration." He therefore assigns similar configurations to all those substances in which a reversal of sign takes place in the same direction, e.g. from left to right as the wave-length is decreased. This test is obviously limited to those substances in which a high-frequency partial rotation is accompanied by a low-frequency partial rotation of opposite sign. The test requires that *both* terms should be of similar character before similar configurations are assigned.

With the same purpose in view, Freudenberg[4] has directed attention to the changes of rotation which result from substitution in radicals not attached directly to an asymmetric carbon atom, so that Walden inversions are unlikely to occur, e.g.

[1] Clough, *J.*, 1914, **105**, 49; 1915, **107**, 96, 1509; 1918, **113**, 526.
[2] Patterson and Forsyth, *J.*, 1913, **103**, 2263; Frankland and McGregor, *J.*, 1894, **65**, 760–771; Clough, *J.*, 1915, **107**, 96; Pictet, *Jahresbericht*, 1882, 855.
[3] Wood and Nicholas, *J.*, 1928, 1712–1727.
[4] Kuhn, Freudenberg and Wolf, *Ber.*, 1930, **63**, 2367; Freudenberg, Kuhn and Bumann, *ibid.*, 2380; Freudenberg, *Ber.*, 1933, **66**, 177-194.

$CO.OH$	$CO.NH_2$	$CO.OH$	$CO.NH_2$
$H.C.OH$	$H.C.OH$	$H.C.OH$	$H.C.OH$
C_6H_5	C_6H_5	C_6H_{11}	C_6H_{11}
$-240°$	$-137°$	$-40°$	$+76°$

The differences are

$$NH_2 - OH = +103° \text{ and } +116°$$
$$C_6H_{11} - C_6H_5 = +200° \text{ and } +213°$$

Untrustworthiness of Polarimetric Methods.—In spite of the many regularities recorded above, it is very doubtful whether any method based upon the study of optical rotations can give decisive evidence in reference to molecular configuration, when direct chemical evidence is not available. The argument which leads to this conclusion may be set out as follows:

Let us suppose that $\genfrac{}{}{0pt}{}{CH_3}{C_2H_5}\!>\!C\!<\!\genfrac{}{}{0pt}{}{X}{Y}$ is an optically-active compound containing a single asymmetric carbon atom, linked to methyl and ethyl and to two other radicals X and Y. It is then certain that, however complex the rotatory dispersion of the molecule may be, or however complicated the influence of temperature or solvent, the rotatory power (under any conditions whatever which are not themselves dissymmetric) (i) will be reversed in sign but unaltered in magnitude if the methyl and ethyl radicals are interchanged, and (ii) will disappear completely if the methyl is replaced by a second ethyl radical, or conversely. If (iii) this process is carried further, and methyl is replaced, not by ethyl but by propyl, it is generally admitted that the sign of the rotation will be reversed, i.e. that

$$\genfrac{}{}{0pt}{}{CH_3}{C_2H_5}\!>\!CXY \quad \text{and} \quad \genfrac{}{}{0pt}{}{C_3H_7}{C_2H_5}\!>\!CXY$$

will have opposite rotations, although there has been *no change of configuration* in the univalent radical $\genfrac{}{}{0pt}{}{}{C_2H_5}\!>\!C\!<\!\genfrac{}{}{0pt}{}{X}{Y}$. This reversal of sign applies in the first instance to the partial rotations associated with the hydrocarbon radical $\genfrac{}{}{0pt}{}{CH_3}{C_2H_5}\!>\!C\!<$ or $\genfrac{}{}{0pt}{}{C_3H_7}{C_2H_5}\!>\!C\!<$. It is unlikely that the partial rotations associated with X and Y (which may be unsaturated or chromophoric radicals) will behave in identically the same way when the sign of this radical is reversed in passing from $\genfrac{}{}{0pt}{}{CH_3}{C_2H_5}\!>\!C\!<$ to $\genfrac{}{}{0pt}{}{C_2H_5}{CH_3}\!>\!C\!<$ on the one hand and to $\genfrac{}{}{0pt}{}{C_3H_7}{C_2H_5}\!>\!C\!<$ on the other ; but it is more likely that the signs of these subsidiary rotations will be reversed in both cases than that they will remain unchanged in one of them. If, however, this reversal is admitted even as a possibility, it follows that the whole complex system of partial rotations in a molecule like Levene's α-methylvaleric acid or ester, $C_2H_5 . \mathbf{C}H(CH_3) . CH_2 . COOH$ or $C_2H_5 . \mathbf{C}H(CH_3) . CH_2 . CO . OC_2H_5$, is liable to be inverted *without any change of configuration* merely by replacing CH_3 by C_3H_7. *A fortiori* it follows that the replacement of —H by —COOH on passing from lactic to malic acid (which is much more drastic that the replacement of CH_3 by C_3H_7), might produce a complete reversal of the complex system of partial rotations in the molecule $X . CH_2 . \mathbf{C}HOH . COOH$.

The conclusion appears to be inevitable that the various lines of evidence cited above are of value mainly in eliminating the complications resulting from the complex or anomalous dispersion of molecules which contain unsaturated or chromophoric groups. Thus it is easy to show, by methods such as these, that the partial rotations of natural dextrorotatory tartaric acid and of dextrorotatory malic acid (the enantiomorph of the natural acid) are fundamentally similar, and in particular that both the low-frequency and the high-frequency terms are of similar sign in these two acids. This coincidence, however, carries us no further than in the case of two substances with simple rotatory dispersion which both happen to be dextrorotatory, like *d*-amyl alcohol and the chloride of the *l*-alcohol, and does not therefore throw any light on the relative configuration of the two acids.

Deductions from optical rotatory power are therefore only of value in exceptional cases, such as (i) the homologous series of secondary alcohols prepared by Pickard and Kenyon, where the relations between chemical constitution and rotatory power are much simpler than in Levene's fatty acids, (ii) compounds like the sugars, where the configuration of one asymmetric carbon atom may be used as a check upon the configuration of another. In general, however, the configuration of a hydroxy acid, like the ring structure of a sugar, must be deduced by chemical methods, and cannot be based securely upon speculations in reference to its optical rotatory power.

CHAPTER XXIV.

CAMPHOR AND CAMPHORQUINONE.

Historical.—The optical rotatory power of *d*-camphor was discovered by Biot in 1815.[1] As early as 1818 [2] he used it to compensate the rotation of *l*-turpentine, and concluded that both media obeyed the Law of Inverse Squares; but in 1852 [3] he made use of the unequal dispersions of these two materials in order to prepare mixtures (comparable with an achromatic doublet) which had an almost constant rotatory power over a range of wave-lengths (p. 109). The variability of specific rotation, which accompanies its high dispersion, was also studied by Biot, who expressed, by means of linear formulæ,[4] the decrease of specific rotation on progressive dilution with alcohol or acetic acid (p. 93). The decrease in the specific rotation of camphor on dilution with seven different solvents was also described and plotted (Fig. 29, p. 94) by Landolt,[5] who extrapolated his curves to 0 and 100 per cent. concentration and recorded the differences between these limiting values (p. 95). Winther in 1906 [6] sought to correlate Landolt's data with the changes of density of the solutions, and made some additional measurements of the influence of concentration on the specific rotations of solutions of camphor in phosphorus trichloride, carbon disulphide, ethylene bromide, benzene, chloroform, and acetic acid.[7] He also made some rough determinations (with the help of Landolt's filters) of the dispersion, for five bands of visible colour, of solutions of camphor in benzene, ethylene bromide, chloroform and acetic acid,[8] and concluded that the rotatory dispersion is independent of the concentration and of the changes of internal pressure produced by changes of solvent and concentration. *l*-Camphor was discovered by Chautard in 1853.[9]

[1] Biot, *Bull. Soc. Philomatique,* 1815, 190–192.
[2] Biot, *Mém. Acad. Sci.*, 1817, **2,** 117.
[3] Biot, *A.C.P.*, 1852, [iii], **36,** 405–489.
[4] *Ibid.*, pp. 306–309.
[5] Landolt, *Ann.*, 1877, **189,** 241–337.
[6] Winther, *Z. ph. C.*, 1906, **56,** 703–718.
[7] *Ibid.*, 1907, **60,** 613–615.
[8] *Ibid.*, pp. 685–705.
[9] Chautard, *C.R.*, 1853, **37,** 166.

Rotatory Dispersion in the Region of Transparency.—

```
              H                                   H
              |                                   |
              C                                   C
           /  |  \                             /  |  \
     H2C  π   |   π  CH2αα'              H2C      |      C=O
      |   Me.C.Me    |                    |   Me.C.Me    |
(?)βH2C       |      C=O                 H2C      |      C=O
           \  |  /                             \  |  /
              C                                   C
              |                                   |
            CH3β(?)                              CH3
```

I. Camphor. II. Camphorquinone.

(*a*) In the region of transparency, the rotatory dispersion of *camphor*, I, can be expressed by the equation

$$[\alpha] = \frac{29 \cdot 384}{\lambda^2 - 0 \cdot 08720} - \frac{20 \cdot 138}{\lambda^2 - 0 \cdot 05428}$$

for a concentrated solution in benzene (33·827 g. in 100 grams of solution), over the range from 6708 A.U. to 3600 A.U.[1] Although the two partial rotations are of opposite sign, the rotatory dispersion is *normal*, since the numerator of the low-frequency term is already larger than that of the high-frequency term. The wave-length of the low-frequency term for which $\lambda_n = 2950$ A.U. corresponds with that of the ketonic absorption band at 2880 A.U., and this term is therefore attributed to the 'induced dissymmetry' (p. 147) of the carbonyl group.

(*b*) The rotatory dispersion of *camphorquinone*, II, in the narrow range of transparency at the red end of the visible spectrum, can be expressed by the equation [2]

$$[\alpha] = -13 \cdot 170/(\lambda^2 - 0 \cdot 22352).$$

The frequency of this term $\lambda_0 = 4730$ A.U., corresponds with that of the quinonoid absorption band at 4650 A.U. (Fig. 169, p. 406), and is attributed to the induced dissymmetry of the two closely-coupled carbonyl groups. The molecule of camphorquinone is, however, nearly symmetrical, like that of camphoric anhydride, $C_8H_{14}\langle{}^{CO}_{CO}\rangle O$, for which $[\alpha]_D = -1°$ only [3] in chloroform, or of camphorimide, $C_8H_{14}\langle{}^{CO}_{CO}\rangle NH$, for which $[\alpha]_D = 1 \cdot 55°$ only [4] in chloroform (5 per cent.). The partial rotations of the two asymmetric carbon atoms

[1] LOWRY and CUTTER, *J.*, 1925, **127**, 612. [2] *Ibid.*, p. 614.
[3] LOWRY, *J.*, 1903, **83**, 968. [4] TAFEL and ECKSTEIN, *Ber.*, 1901, **34**, 3274.

are therefore too small to call for an additional term in the dispersion equation for wave-lengths longer than that of the absorption band.

(*c*) The rotatory dispersion of several of the *π-sulphonic derivatives* of camphor in the region of complete transparency can be expressed by a single term of the Drude equation,[1] e.g.

Ammonium α-bromocamphor π-sulphonate $[\alpha] = \frac{22{\cdot}015}{\lambda^2 - 0{\cdot}09668}$ $\lambda_0 = 3110$ A.U.

Ammonium α-chlorocamphor π-sulphonate $[\alpha] = \frac{17{\cdot}677}{\lambda^2 - 0{\cdot}09946}$ $\lambda_0 = 3150$ A.U.

As in the case of camphorquinone, the low-frequency term again dominates the dispersion curve ; but the dispersion remains apparently simple over a much wider range of wave-lengths. This effect can be interpreted as a result of the increased " longitudinal " or " lateral " dissymmetry of the molecule, III, enhancing the induced dissymmetry of the carbonyl group. On the other hand, in *camphor β-sulphonic acid*, IV, only the " latitudinal " dissymmetry is increased ; the low-frequency partial rotation is therefore (as in camphor itself) not much larger than the high-frequency partial rotation and gives rise to small complex (but not anomalous) positive total rotations :

H
|
C
H_2C CHX >CMe.CH_2.SO_2.OH
H_2C CO
C
|
CH_3

III. π-Sulphonic acids (X = Cl or Br).

H
|
C
H_2C CH_2 >CMe_2
H_2C CO
C
|
CH_2
|
SO_2.OH

IV. β-Sulphonic acid.

Rotatory Dispersion in the Region of Absorption.—(*a*) In the region of absorption the rotatory dispersion of *camphor* exhibits a reversal of sign and very large positive and negative maxima [2] (Figs. 131 and 132).[3]

	Positive Maximum.	Reversal.	Negative Maximum.
Camphor vapour			
$[\alpha]^{180°}$	+ 2000° at 3200 A.U.	± 0° at 3020 A.U.	− 1860° at 2800 A.U.
$[M]^{180°}$	+ 3040° " 3200 "	± 0° " 3020 "	− 2830° " 2800 "
Solution in *cyclo*hexane			
$[\alpha]^{20°}$	+ 2600° at 3200 A.U.	± 0° at 3020 A.U.	− 2200° at 2720 A.U.
$[M]^{20°}$	+ 3950° " 3200 "	± 0° " 3020 "	− 3340° " 2720 "

1 Richards and Lowry, *J.*, 1925, **127**, 1503–1514.
2 Lowry and Gore, *P.R.S.*, 1932, A. **135**, 13–22.
3 *Ibid.*, p. 15.

These loops cannot be represented by the equations which have been found effective in other cases (p. 441); but, since the positive maximum of lower frequency is larger than the negative maximum of higher frequency, it appears likely that the residual rotation, due to active absorption bands of shorter wave-length, is *positive* in sign, whereas the equations which represent the rotatory dispersion in the visual and early ultra-violet regions down to 3600 A.U. include a *negative* high-frequency term. The unusual complexity of the dispersion suggests that not less than *three* dissimilar partial rotations must be included in the total rotations, although two terms of Drude's equation suffice to express the data in the region of complete transparency up to the limits of accuracy which can now be attained by careful working.

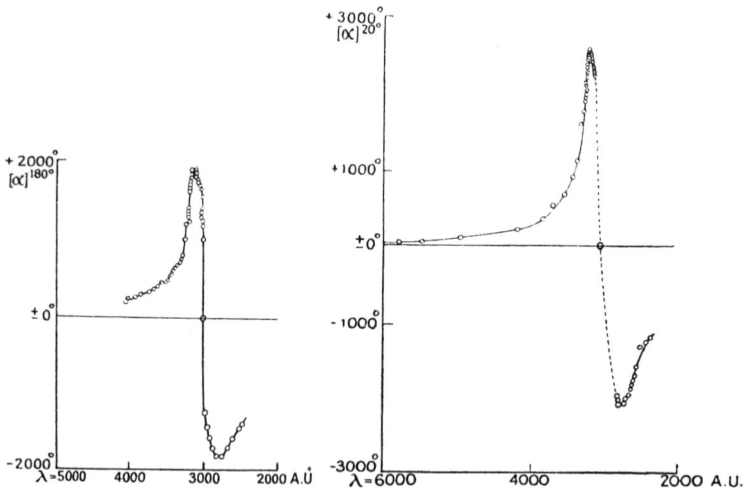

FIG. 131.—ROTATORY DISPERSION OF CAMPHOR VAPOUR AT 180° C. (Lowry and Gore).

FIG. 132.—ROTATORY DISPERSION OF CAMPHOR IN CYCLOHEXANE AT 20° (Lowry and Gore).

A further indication of the complexity of the rotatory dispersion is provided by the fact that the circular dichroism of the ketonic absorption band, instead of being distributed uniformly throughout the band, with an approximately constant dissymmetry factor, is concentrated on the side of longer wave-lengths (Fig. 169, p. 406).[1] The absorption band is therefore composite in character, including (i) a strongly-dextrorotatory component of longer wave-length, and (ii) a component of shorter wave-length, the optical activity of which is too small to produce a measurable circular dichroism in a region of

[1] KUHN and GORE, *Z. ph. C.*, 1931, B. **12**, 389–397; LOWRY and GORE, *P.R.S.*, 1932, A. **135**, 22.

such strong absorption. If, however, it were feebly dextrorotatory, this second component could be invoked to account for the positive residual rotation, which has been cited above.

(*b*) Reychler's *camphor β-sulphonic acid*, IV, $C_{10}H_{15}O \,.\, SO_2 \,.\, OH$, resembles camphor itself in many of its optical properties, although it differs from it in being readily soluble in water. Its rotatory dispersion in the region of transparency down to 3969 A.U. can be expressed by the equation [1]

$$[\alpha] = \frac{22{\cdot}17}{\lambda^2 - 0{\cdot}08515} - \frac{18{\cdot}32}{\lambda^2 - 0{\cdot}04945}.$$

The rotatory dispersion in the region of absorption has the following characteristics: [2]

Positive Maximum.	Reversal.	Negative Maximum.
$[\alpha] = 2000°$ at 3090 A.U.	$\pm 0°$ at 2950 A.U.	$-2450°$ at 2690 A.U.

Since the negative maximum is larger than the positive maximum the residual rotation is therefore still negative, like the high fre-

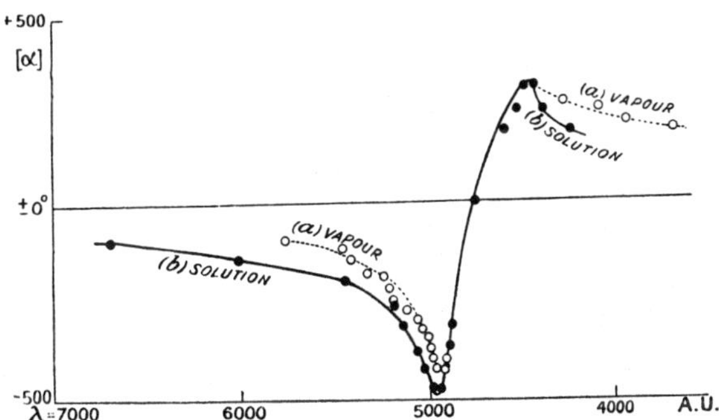

FIG. 133.—ROTATORY DISPERSION OF CAMPHORQUINONE (Lowry and Gore).

quency term in the preceding equation; but the dispersion in the region of absorption is too complex to be expressed by any of the usual equations.

The composite character of the absorption band is shown by the fact that the circular dichroism reaches a maximum at a longer wave-length than the absorption for unpolarised light, exactly as in the case of camphor (Fig. 169 p. 406).[3]

(*c*) The rotatory dispersion of *camphorquinone* in the region of absorption was traced by Lifschitz [4] as far as the first (negative)

[1] RICHARDS and LOWRY, *J.*, 1925, **127**, 1511.
[2] LOWRY and FRENCH, *J.*, 1932, 2657.
[3] *Ibid.*, p. 2655.
[4] LIFSCHITZ, *Z. ph. C.*, 1923, **105**, 27-54.

maximum. Dispersion curves for the vapour and for a solution in *cyclo*-hexane are reproduced in Fig. 133.[1] They show two maxima and a reversal of sign, but these are all in the visible region of the spectrum. The characteristic features of these curves are set out below :

	Negative Maximum.	Reversal.	Positive Maximum.
Vapour $[\alpha]^{200°}$ =	$-$ 500° at 4950 A.U.	± 0° at 4740 A.U.	+ 300° at 4440 A.U.
Solution $[\alpha]^{20°}$ =	$-$ 450° ,, 4940 A.U.	± 0° ,, 4740 A.U.	+ 300° ,, 4440 A.U.

The negative maximum is substantially larger than the positive maximum of higher frequency. The residual rotation (due to active absorption bands of higher frequency) is therefore negative ; but the curves are again too complex to be expressed by the usual equations.

Optical Superposition in the Camphor Series.—When a hydrogen atom in the α or α′ position is replaced by another radical, the camphor-molecule acquires an additional asymmetric carbon atom, which may be either *d* or *l* ; but this disappears when two identical substituents occupy the α and α′ positions, e.g.

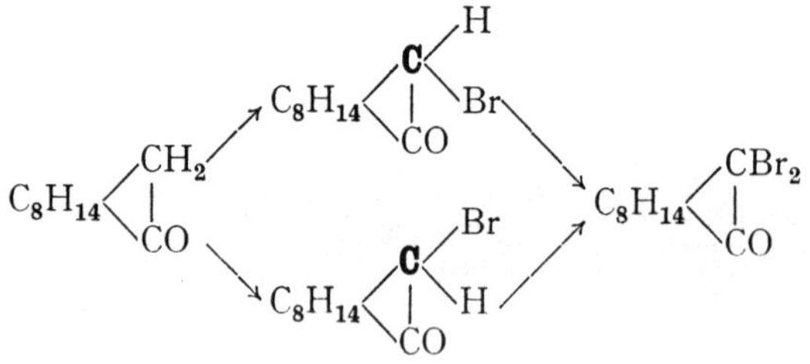

$$C_8H_{14}\left\langle\begin{matrix}CH_2\\ |\\ CO\end{matrix}\right. \nearrow C_8H_{14}\left\langle\begin{matrix}\mathbf{C}\begin{matrix}H\\ Br\end{matrix}\\ |\\ CO\end{matrix}\right. \searrow C_8H_{14}\left\langle\begin{matrix}CBr_2\\ |\\ CO\end{matrix}\right.$$

$$C_8H_{14}\left\langle\begin{matrix}CH_2\\ |\\ CO\end{matrix}\right. \searrow C_8H_{14}\left\langle\begin{matrix}\mathbf{C}\begin{matrix}Br\\ H\end{matrix}\\ |\\ CO\end{matrix}\right. \nearrow C_8H_{14}\left\langle\begin{matrix}CBr_2\\ |\\ CO\end{matrix}\right.$$

The principle of optical superposition (p. 274) suggests that the new asymmetric carbon atom in the α and α′-derivatives should produce equal increments and decrements of rotatory power. This simple relation is unlikely, however, to be fulfilled, since the introduction of a halogen on the α-carbon atom cannot be without influence on the partial rotation of the contiguous asymmetric carbon atom, or on the "induced dissymmetry" of the contiguous carbonyl group on the other side. It is, however, not unreasonable to suppose that the secondary changes thus produced by the two successive substitutions might be similar in magnitude, so that the average molecular rotation of the two monobromo-derivatives would be equal to the average molecular rotation of camphor and of the di-derivative.

For the purpose of checking this relation the following data are available : [2]

[1] LOWRY and GORE, *P.R.S.*, 1932, A, **135**, 15. The zero line, which was drawn incorrectly in the original publication, is here inserted correctly.

[2] CUTTER, BURGESS and LOWRY, *J.*, 1925, **127**, 1260–1274.

TABLE 25.—ROTATORY DISPERSION OF CAMPHOR AND ITS BROMO-DERIVATIVES.

Camphor in benzene at 20°

$$[M] = \frac{44{\cdot}702}{\lambda^2 - 0{\cdot}08720} - \frac{30{\cdot}636}{\lambda^2 - 0{\cdot}05428}. \quad . \quad . \quad . \quad (1)$$

α-Bromocamphor in benzene at 20°

$$[M] = \frac{30{\cdot}681}{\lambda^2 - 0{\cdot}10855} - \frac{45{\cdot}134}{\lambda^2 - 0{\cdot}0562}. \quad . \quad . \quad . \quad (2)$$

α′-Bromocamphor in benzene at 20°

$$[M] = \frac{23{\cdot}646}{\lambda^2 - 0{\cdot}10630} - \frac{60{\cdot}994}{\lambda^2 - 0{\cdot}03525}. \quad . \quad . \quad . \quad (3)$$

αα′-Dibromocamphor in benzene at 20°

$$[M] = \frac{45{\cdot}465}{\lambda^2 - 0{\cdot}11296} - \frac{25{\cdot}331}{\lambda^2}. \quad . \quad . \quad . \quad . \quad . \quad (4)$$

The dispersion curves corresponding with these equations are plotted in Fig. 134. The two averages agree quite closely in the

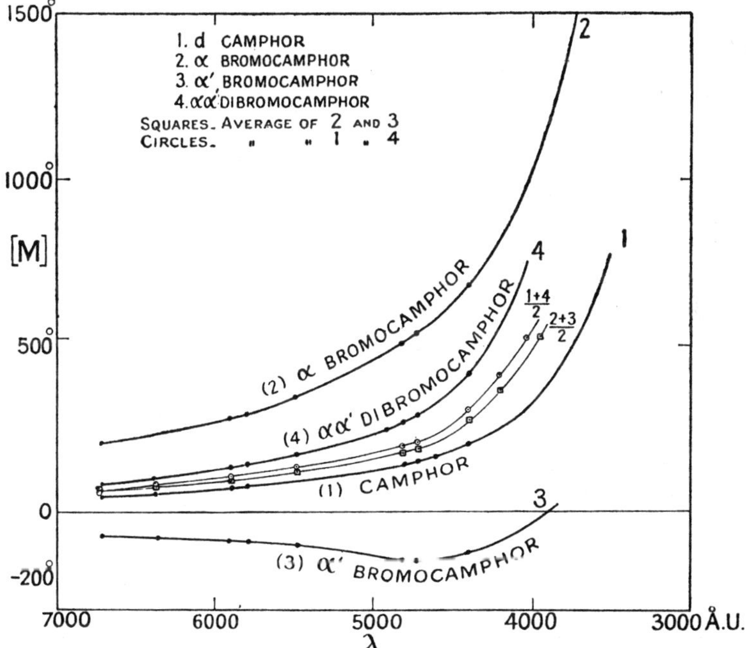

FIG. 134.—OPTICAL SUPERPOSITION IN THE CAMPHOR SERIES.

visible spectrum, both in this case and in that of the corresponding chloro-compounds.

	$[M]_{5461}$.		$[M]_{5461}$.	
Camphor .	+ 79·7°	α-Chlorocamphor .	+153·7°	Diff.
αα′-Dichlorocamphor	+147·3°	α′-Chlorocamphor .	+ 69·9°	83·8°
Mean .	+113·5°	Mean .	+111·8°	
Camphor .	+ 79·7°	α-Bromocamphor .	+324·2°	Diff.
αα′-Dibromocamphor	+147·6°	α′-Bromocamphor .	−101·1°	425·3°
Mean .	+113·6°	Mean .	+111·5°	

This concordance is remarkable, in that the *secondary* effects of the two halogens, as indicated by the mean values, are seen to be identical for chlorine and bromine. On the other hand, the *primary* effects of the additional asymmetric carbon atom, as shown by the differences in the last column, amount to ± 42° for >**C**HCl and to ± 213° for >**C**HBr. At shorter wave-lengths, however, the two averages diverge progressively, until they differ by about 60° at 4000 A.U. This divergence is obviously due to the increased dispersion of the positive partial rotation of the halogen derivatives, as shown in the following table:

	λ_1.	λ_0.		λ_1.	λ_0.
Camphor . .	2950	2880	Camphor . .	2950	2880
α-Chlorocamphor .	*3140**	3050	α-Bromocamphor .	3290	3120
α′-Chlorocamphor .	*3300**	3100	α′Bromocamphor .	3260	3140
			αα′Dibromocamphor	*3770**	3230

This causes the dispersion curve for αα′-dibromocamphor to turn up very steeply, and produces a corresponding elevation of the average curve $\frac{1+4}{2}$ to which it contributes a half-share.

* Italicised wave-lengths under λ_1 are derived from 3-constant equations; values shown under λ_0 are the wave-lengths of maximum absorption of solutions in *cyclo*-hexane (Lowry and Owen, *J.*, 1926, 606–622).

CHAPTER XXV.

DERIVATIVES OF MENTHOL AND BORNEOL.

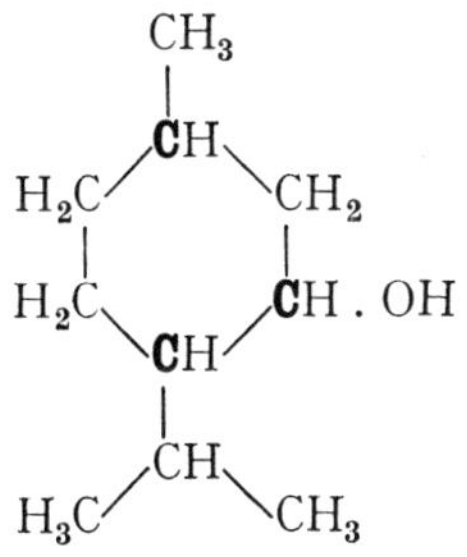

```
            CH3                                  CH3
            |                                    |
            CH                                   C
     H2C/      \CH2                          /   |   \
      |          |                    H2C/       |    \CHOH
     H2C\      /CH . OH                |      MeCMe      |
            CH                        H2C\       |      /CH2
            |                                \   |   /
            CH                                   C
     H3C/      \CH3                              |
                                                 H
```

I. Menthol. 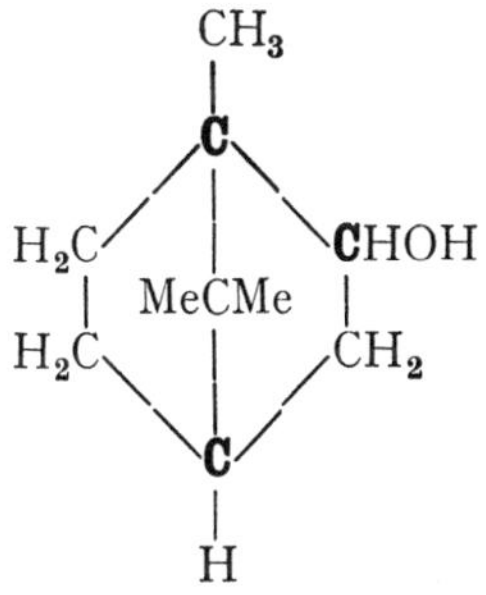 II. Borneol.

Menthol and its Esters.—A comprehensive study of the optical rotations of *l*-menthol and its esters for four wave-lengths in the visible spectrum and at temperatures ranging from 20° to 180° was made by Kenyon and Pickard in 1915 ; [1] and Hall [2] (from the same laboratory) has given data for the dimenthyl esters of normal dicarboxylic acids from $CH_2(CO_2H)_2$ to $C_{10}H_{20}(CO_2H)_2$.

Menthol itself gives a dispersion ratio $\alpha_{4358}/\alpha_{5461} = 1{\cdot}664$, which is practically constant from 20° to 180°, and corresponds with the simple rotatory dispersion of a secondary alcohol. This ratio is preserved in the *acetate* (1·661) and in the *oxalate* (1·654) ; but in the *phenylacetate* it is raised to 1·693 at 20°, falling to 1·669 between 100° and 180°, and in the o-*nitrobenzoate* to 2·367 at 20°, falling to 2·316 at 100°. The carboxyl radical of the esters therefore enhances the rotation without altering the rotatory dispersion ; but the aromatic nuclei enhance both the rotatory power and the rotatory dispersion in a way that points clearly to the presence of a large partial rotation of low frequency. In this connection it is of interest to notice that

$$l\text{-Menthylamine} \quad [M]^{20}_{5461} = -72{\cdot}91^\circ \quad \alpha_{4358}/\alpha_{5461} = 1{\cdot}624$$

resembles menthol itself, whilst in

$$l\text{-Menthone} \quad [M]^{20}_{5461} = -49{\cdot}39^\circ \quad \alpha_{4358}/\alpha_{5461} = 1{\cdot}781$$

[1] KENYON and PICKARD, *J.*, 1915, **107**, 35–62.
[2] HALL, *J.*, 1923, **123**, 105–113.

TABLE 26.—ROTATORY DISPERSION OF MENTHYL ESTERS.

	$t°$ C.	$[M]_{5461}$.	$\alpha_{4358}/\alpha_{5461}$.
		Degrees.	
l-Menthol	20	− 90·33	1·668
,, ...	180	− 86·72	1·659
l-Menthyl acetate . .	20	−186·4	1·659
,, ,, . .	180	−181·6	1·662
l-Menthyl oxalate . .	20	−404·0	1·649
,, ,, . .	180	−390·1	1·661
l-Menthyl phenylacetate .	20	−222·3	1·693
,, ,, .	180	−200·0	1·669
l-Menthyl *o*-nitrobenzoate .	20	−492·0	2·367
,, ,, .	100	−450·2	2·316

an increased rotatory dispersion is accompanied by a decreased lævorotation, which is converted into a dextrorotation in *iso*-menthone, when the asymmetric carbon atom contiguous to the ketonic group is inverted by means of an alkali. A fuller investigation of these substances, covering the region of absorption, as well as the region of transparency, is obviously called for, in order that the component partial rotations of the saturated and unsaturated radicals may be effectively separated.

Stereoisomeric Menthylamines.—Read[1] has prepared and examined the series of stereoisomeric menthylamines, of which four pairs may be derived from the two pairs of menthones

```
   H      CH3              H      CH3
     \   /                   \   /
      C                        C
     /   \                   /   \
 H2C       CH2           H2C       CH2
  |         |             |         |
 H2C       C=O           H2C       CH . NH2
     \   /                   \   /
      C                        C
     /   \                   /   \
   H      CH(CH3)2         H      CH(CH3)2
```

III. Menthone. IV. Menthylamine.

These resemble the cyclic sugars in containing a series of asymmetric carbon atoms >CHX in a closed ring. They can be distinguished by showing the relative positions of H and X on either side of the ring. Thus the four bases prepared by Read are represented as follows :—

```
 Me|H           Me|H            Me|H           Me|H
  H|NH2        H2N|H           H2N|H             H|NH2
  H|Prβ          H|Prβ         Prβ|H           Prβ|H
```

(i) *l*-Base. (ii) *d*-*neo*-Base. (iii) *d*-*iso*-Base. (iv) *d*-*neoiso*-Base.

The bases (i) and (ii) are prepared from menthone, in which the methyl and *iso*-propyl groups are on opposite sides of the ring. The bases (iii) and (iv) are prepared from *iso*-menthone, in which the

[1] READ, *T.F.S.*, 1930, **26**, 441.

$>\mathbf{C}<^{H}_{Pr}$ group has been inverted, by taking advantage of the enolisation of the contiguous $>CO$ group in presence of alkali

$$\begin{matrix} -C{=}O \\ | \\ -\mathbf{C}HPr \end{matrix} \rightleftharpoons \begin{matrix} -COH \\ \| \\ -CPr \end{matrix}$$

On the other hand, since there is no mechanism available for inverting the $>\mathbf{C}<^{H}_{CH_3}$ group, none of the mirror-image series is known. The conditions of isomeric change are indeed very similar to those which prevail in camphoric acid,

$$\begin{matrix} & \mathbf{C}H\,.\,CO\,.\,OH \\ CH_2 \diagup & | \\ | & CMe_2 \\ CH_2 \diagdown & | \\ & \mathbf{C}Me\,.\,CO\,.\,OH \end{matrix}$$

where the carbon atom which carries a hydrogen atom can be inverted by enolisation of the carboxyl group

$$>\mathbf{C}H\,.\,CO\,.\,OH \rightleftharpoons\ >C{=}C(OH)_2,$$

but not the carbon atom in which the corresponding atom of hydrogen has been replaced by methyl.

The molecular rotations of the four bases and of some of their derivatives are set out in Table 27 below, whilst the specific rotations (usually in chloroform) are shown in Fig. 135, with the signs of (ii) and (iii) reversed, so that the configuration of the $>\mathbf{C}<^{H}_{NH_2}$ group is the same throughout.

TABLE 27.—OPTICAL SUPERPOSITION IN THE MENTHYLAMINES.

	(i) *l.*	(ii) *d-neo.*	(iii) *d-iso.*	(iv) *d-neoiso.*	Total.
	Degrees.	Degrees.	Degrees.	Degress.	Degrees.
1. Base (no solvent) .	$[M]_D$ − 67	+ 23	+ 45	+ 0·4	=+ 1
2. Base (chloroform) .	− 59	+ 14	+ 44	+14	=+13
3. Hydrochloride (in water) . .	− 70	+ 41	+ 45	+40	=+56
4. Formyl der. .	− 153	+ 99	+ 57	− 7	=− 4
5. Acetyl der. . .	− 161	+104	+ 61	− 5	=− 1
12. Chloroacetyl der. .	− 166	+117	+ 69	−23	=− 3
14. Phenacetyl der. .	− 165	+ 94	+ 91	− 9	=+11
15. Benzoyl der. .	− 163	+ 59	+ 47	−27	=−84
17. Benzylidene der. .	− 322	+150	+220	−83	=−35

Since *l*-menthylamine has the largest (negative) rotatory power, the four stereoisomers would be represented in van't Hoff's nomenclature (p. 274) as follows:

(i) − A − B − C. (ii) − A + B − C.
(iii) − A + B + C. (iv) − A − B + C.

Read, however, assigns a *negative* value to the rotation A (but not to the configuration of the carbon atom) when B and C are of similar sign (*cis* configuration) and a *positive* value to A when B and C are

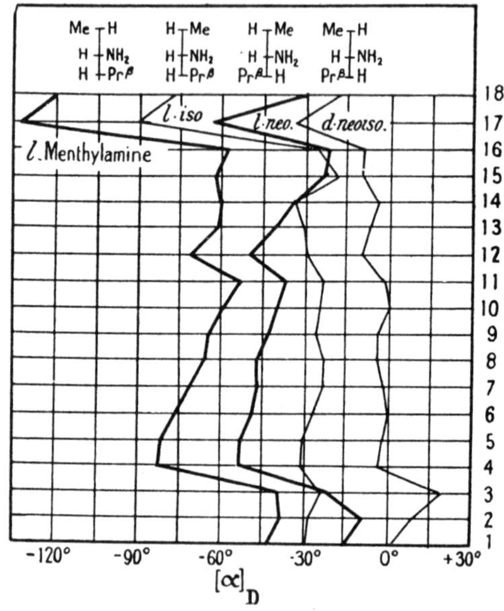

FIG. 135.—SPECIFIC ROTATIONS OF STEREOISOMERIC MENTHYLAMINES AND DERIVATIVES (Read).

1. Base (no solvent). 2. Base (chloroform). 3. Hydrochloride aq.
4. Formyl der. 5. Acetyl der. 6. Propionyl der.
7. Butyryl der. 8. *iso*-Butyryl der. 9. *iso*-Valeryl der.
10. Caproyl der. 11. Capryl der. 12. Chloroacetyl der.
13. Bromoacetyl der. 14. Phenacetyl der. 15. Benzoyl der.
16. Anisoyl der. 17. Benzylidene der. 18. Salicylidene der.

of opposite signs (*trans* configuration). For the purpose of calculating rotations, therefore, his assignment is as follows:

(i) − A − B − C. (ii) + A + B − C.
(iii) − A + B + C. (iv) + A − B + C.

The sum of these four rotations should therefore be zero. This is approximately true for the base and for its acyl derivatives, but large deviations occur in the hydrochloride, where the whole molecule is perhaps disturbed by the positive charge on the $—\overset{+}{N}H_3$ radical, and in the benzoyl and benzylidene derivatives, where the induced dissymmetry of the substituent would be at a maximum.

When the four rotations add up to zero, the usual algebraic relations necessarily follow, i.e. numerical values can be assigned to A, B and C, and the rotation of any one stereoisomer can be predicted from those of the other three, in accordance with van't Hoff's principle. The fundamental assumption of a reversal of sign of the partial rotation of an asymmetric carbon atom, resulting from a change of configuration of other asymmetric carbon atoms, without any change in the configuration of the atom itself, is, however, very difficult to justify in terms of current conceptions, when applied to the fixed dissymmetry of a saturated radical, although it is perfectly natural when applied to the induced dissymmetry of an unsaturated radical, which has no inherent rotatory power of its own when forming part of an otherwise symmetrical molecule.

Optically-Active Xanthates.—In the course of his studies of terpenes and camphor-derivatives, Tschugaeff prepared a number of optically-active xanthates and related compounds, containing the group >C=S, which were coloured and gave rise to anomalous rotatory dispersion in the visible spectrum. The optically-active radicals used in these experiments included the following:

```
        CH3                      CH3                      CH3
         |                        |                        |
        CH                        C                        C
  H2C /    \ CH2          H2C /   |   \ CH—        H2C /   |   \ CH—
   |         |             |   MeCMe   |            |     CH2  |  / CH3
  H2C \    / CH—          H2C \   |   / CH2        H2C \   |   C
        CH                        C                        C      \ CH3
         |                        |                        |
        CH                        H                        H
  CH3 /    \ CH3
```

Menthyl radical. Bornyl radical. Fenchyl radical.

These were linked to radicals containing the group >CS in compounds such as the following:

(i) Alkyl xanthates, etc.	R . O . CS . S . CH_3	RO . CS . S . CH_2 . C_6H_5
	R . O . CS . S . C_2H_5	RO . CS . S . $CH(C_6H_5)_2$
		RO . CS . S . $C(C_6H_5)_3$
(ii) Thioanhydrides	RO . CS . S . CS . OR	
(iii) Dixanthides	RO . CS . S . S . CS . OR	
(iv) Methylene esters, etc.	RO . CS . S . $[CH_2]_n$. S . CS . OR (n = 1, 2, 3)	
(v) Monothiourethanes	RO . CO . $N(C_6H_5)$. CS . C_6H_5	
(vi) Dithiourethanes	RO . CS . $N(C_6H_5)$. CS . C_6H_5	

Although many of these compounds are coloured, they do not as a rule exhibit an absorption band in the visible spectrum. The red dithiourethanes, however, show an absorption band in the middle of the visible spectrum at about 5200 A.U., i.e. at a wave-length a little less than that of the green mercury line Hg 5461. The other compounds show a strong absorption band, with a maximum, $\log \epsilon = 4$, in the ultra-violet at about 2800 A.U., and a weaker band of longer wave-length, with a maximum not recorded by Tschugaeff, $\log \epsilon < 2$ at about 3600 A.U. The extension of the foot of this band into the visible spectrum was noted by Tschugaeff and is responsible for the yellow tint of many of the compounds.

Anomalous Rotatory Dispersion of Xanthates.—(*a*) In his first paper[1] Tschugaeff announced the discovery of anomalous rotatory dispersion in solutions in toluene of the following three compounds:

I. *l*-Menthyl dixanthide
$(C_{10}H_{19} . O . CS)_2S_2$ (straw yellow).

II. *l*-Bornyl dixanthide
$(C_{10}H_{17} . O . CS)_2S_2$ (bright yellow).

III. Thioanhydride of *l*-menthyl xanthic acid.
$(C_{10}H_{19} . O . CS)_2S$ (greenish-yellow).

The rotations of these three compounds showed *negative* maxima at 4810, 5480 and 6400 A.U., corresponding with the increasing depth of the colour. In the two darker compounds, the dispersion curves were followed beyond the maxima until zero values were recorded at 4520 and 5020 A.U. respectively, followed by *positive* rotations at shorter wave-lengths.* These anomalies were observed in the region of transparency, since the maxima of absorption are in the ultra-violet, and only the foot of the band extends into the visible region. They must therefore be ascribed, like those of the tartaric esters, to the influence of optically-active absorption bands of opposite sign, lying beyond the range of observation.

(*b*) Tschugaeff and Ogorodnikoff in 1911[2] showed that negative maxima occurred in *superfused melts* (as well as in solutions in toluene) of

IV. Methyl *l*-menthylxanthate $C_{10}H_{19} . O . CS . S . CH_3$
V. Ethyl *l*-menthylxanthate $C_{10}H_{19} . O . CS . S . C_2H_5$
II. *l*-Bornyl dixanthide $[C_{10}H_{17} . O . CS . S .]_2$

The dispersion curves for the melts lie very close together, but dissolution in toluene *raises* the rotatory power of IV and V, and *lowers* the rotatory power of VI, without producing any important displacement of the maxima, which range from 5100 to 5500 A.U. only. These experiments, like those of Bruhat on glassy tartaric acid (p. 103), proved that anomalous rotatory dispersion is not dependent on the presence of a solvent.

[1] TSCHUGAEFF, *Ber.*, 1909, **42**, 2244.
[2] TSCHUGAEFF and OGORODNIKOFF, *A.C.P.*, 1911, [viii], **22**, 137–144.
* Compare Fig. 136, where the dispersion curves are of the same type but of opposite sign.

Anomalous Rotatory Dispersion in Transparent and in Absorbing Media.—(*a*) A more extensive investigation by the same authors [1] included the absorption and rotatory dispersion of the following colourless or yellow compounds:

VI. Methyl *d*-bornylxanthate $C_{10}H_{17} . O . CS . S . CH_3$
VII. Ethyl *d*-bornylxanthate $C_{10}H_{17} . O . CS . S . C_2H_5$
II. *l*-Bornyl dixanthide $[C_{10}H_{17} . O . CS . S .]_2$

These three compounds showed positive maxima of rotation at 5300 to 5500 A.U., whilst the absorption became appreciable only beyond about 4400 A.U. (Fig. 136).

(*b*) Three coloured dithiourethanes were also examined, namely:

VIII. Diphenyl *l*-fenchyl dithiourethane
$C_{10}H_{17} . O . CS . NPh . CS . C_6H_5$
IX. Diphenyl *d*-bornyl dithiourethane
$C_{10}H_{17} . O . CS . NPh . CS . C_6H_5$
X. Diphenyl *l*-menthyl dithiourethane
$C_{10}H_{19} . O . CS . NPh . CS . C_6H_5$

The first two compounds showed maxima of absorption at 5300 A.U., and the third at 5200 A.U. (Fig. 137). Since the maximum of absorption occurs near the brightest part of the visible spectrum, it was possible, by making use of dilute solutions (about 0·14 per cent.), to follow the rotatory power of these three compounds into the region of absorption, and thus to make a visual study of the Cotton phenomena. Thus, the first two compounds gave *positive* maxima and the third compound a *negative* maximum, at 5800, 5700 and 5600 A.U. approximately. These maxima were followed by reversals of sign at 5500, 5400 and 5300 A.U. approximately. At still shorter wavelengths, large rotations of opposite sign were observed, but without reaching a maximum of opposite sign.

(*c*) These two groups of three compounds provide an interesting contrast, since in the first group of colourless or pale yellow xanthates the anomalies are of the type discovered by Biot in tartaric acid and by Winther in the colourless tartaric esters (pp. 21 and 287), whilst in the second group they are of the type discovered by Cotton in the coloured tartrates of chromium and copper (p. 154). Thus, in the two colourless xanthates VI and VII a *positive* rotation of higher frequency is overpowered by a *negative* rotation of lower frequency as the region of absorption in the near ultra-violet (in which this negative rotation has its origin) is approached. On the other hand, the *positive* rotations of the two coloured urethanes VIII and IX are due primarily to a low-frequency.term, corresponding with an optically-active absorption band in the green. This absorption band is so close that the low-frequency partial rotation entirely masks any high-frequency partial rotation that may be present, exactly as

[1] Tschaugaeff and Ogorodnikoff, *Z. ph. C.*, 1910, **74**, 503-512.

in the case of camphorquinone (p. 134). Since, however, the absorption is not confined to a single wave-length, but is spread out

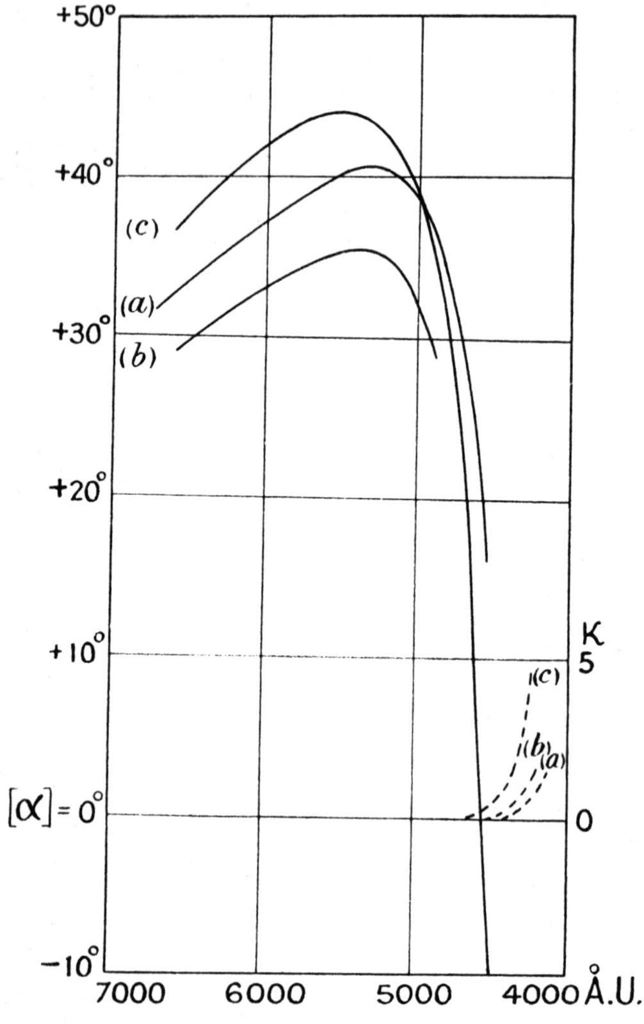

FIG. 136.—ABSORPTION AND ROTATORY DISPERSION OF COLOURLESS XANTHATES.

(*a*) VI. Methyl *d*-bornylxanthate.
(*b*) VII. Ethyl *d*-bornylxanthate.
(*c*) II. *l*-Bornyldixanthide (sign of rotation reversed).
Broken curves represent absorption.

into an absorption band of finite width, the positive rotation does not increase indefinitely, but reaches a maximum soon after entering

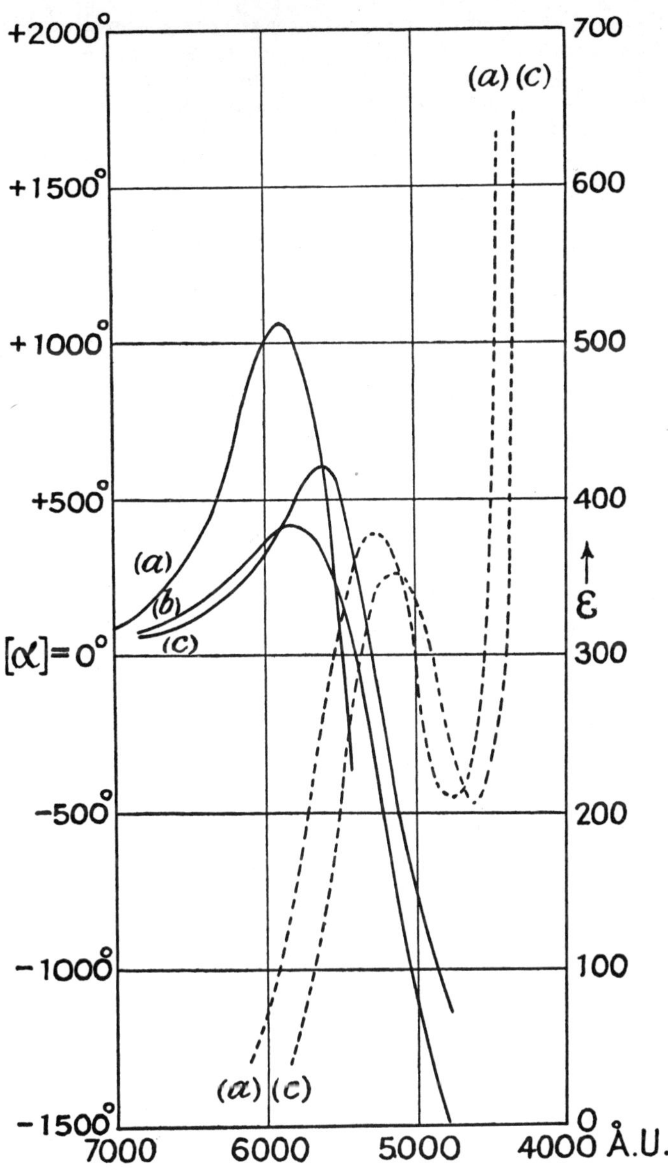

FIG. 137.—ABSORPTION AND ROTATORY DISPERSION OF COLOURED XANTHATES.

(a) VIII. Phenyl *l*-fenchyl dithiourethane.
(b) IX. Phenyl *d*-bornyl dithiourethane.
(c) X. Phenyl *l*-menthyl dithiourethane (sign of rotation reversed).
Broken curves represent absorption.

the region of absorption. This *positive maximum* at 5800 to 5600 A.U. is followed by a *reversal of sign* at 5500 to 5300 A.U., i.e. at a wavelength not far removed from the maximum of absorption at 5400 to 5200 A.U.; and a *negative* maximum might then have been recorded if the decreasing visual intensity of the light and the increasing absorption due to the second absorption band had not made the observations too difficult.

The two types of behaviour described above were repeated in the compounds II and X, but with rotations of opposite sign, as is indicated in Figs. 136 and 137.

(*d*) In two subsequent papers,[1] the absorption and rotatory dispersion of twenty-six optically-active derivatives of this type were recorded, but without disclosing any new type of dispersion.

Circular Dichroism and Rotatory Dispersion in the Region of Absorption.—A comprehensive series of measurements of absorption, circular dichroism, and rotatory dispersion in the regions of transparency and of absorption in nine of Tschugaeff's xanthates, was described by Lowry and Hudson in 1933.[2] The nine compounds examined had the following formulæ:—

Substance.	Formula.
A. *Bornyl xanthates.*	
I. Methyl *d*-bornyl xanthate	$C_{10}H_{17}O . CS . S . CH_3$
II. Ethyl *d*-bornyl xanthate	$C_{10}H_{17}O . CS . S . C_2H_5$
III. *d*-Bornyl dixanthide	$C_{10}H_{17}O . CS . S . S . CS . O . C_{10}H_{17}$
IV. Methylene *d*-bornyl xanthate	$C_{10}H_{17}O . CS . S . CH_2 . S . CS . O . C_{10}H_{17}$
B. *Menthyl xanthates.*	
V. Methyl *l*-menthyl xanthate	$C_{10}H_{19}O . CS . S . CH_3$
VI. *l*-Menthyl dixanthide	$C_{10}H_{19}O . CS . S . S . CS . O . C_{10}H_{19}$
VII. Methylene *l*-menthyl xanthate	$C_{10}H_{19}O . CS . S . CH_2 . S . CS . O . C_{10}H_{19}$
C. *Dithiourethanes.*	
VIII. Diphenyl *d*-bornyl dithiourethane	$C_{10}H_{17}O . CS . N(C_6H_5) . CS . C_6H_5$
IX. Diphenyl *l*-menthyl dithiourethane	$C_{10}H_{19}O . CS . N(C_6H_5) . CS . C_6H_5$

(*a*) *Absorption.*—Each compound was shown to have two absorption bands, in addition to a general absorption. The parameters of the bands are set out below.

TABLE 28.—MAXIMA OF ABSORPTION.

		First Maximum. log ε.	λ_1	Second Maximum. log ε.	λ_2.
A.	I.	1·565	at 3540 A.U.	4·12	at 2760 A.U.
	II.	1·570	at 3580 A.U.	3·99	at 2790 A.U.
	III.	1·72	at 3660 A.U.	3·90	at 2880 A.U.
	IV.	1·92	at 3550 A.U.	4·44	at 2820 A.U.
B.	V.	1·590	at 3530 A.U.	4·24	at 2760 A.U.
	VI.	1·82	at 3630 A.U.	4·20	at 2850 A.U.
	VII.	2·030	at 3550 A.U.	4·48	at 2810 A.U.
C.	VIII.	2·270	at 5200 A.U.	4·10	at 3300 A.U.
	IX.	2·250	at 5150 A.U.	4·05	at 3300 A.U.

[1] TSCHUGAEFF and OGORODNIKOFF, *Z. ph. C.*, 1912, **79**, 471–480; 1913, **85**, 481–511.

[2] LOWRY and HUDSON, *Phil. Trans.*, 1933, A. **232**, 117–154.

The band at 2800 A.U. may be attributed to the >C=S group, since Purvis, Jones and Tasker[1] and Hantzsch and Scharf[2] found a strong band between 2750 and 3050 A.U. in a number of thion-, thionthio- and trithio-carbonates, all of which contain the >C=S group. The much weaker band at 3600 A.U. (approx.) may be attributed to the group —S . CS—, since Hantzsch and Scharf found a weak band of similar wave-length (3509 A.U.) in diethyl thion-thio-carbonate, C_2H_5 . O . CS . S . C_2H_5, but not in diethyl thiocarbonate C_2H_5 . O . CS . O . C_2H_5.

In some of the compounds the two bands were separated by well-defined minima as in the curve for methylene *l*-menthyl xanthate (Fig. 162, p. 401); but in others, e.g. the dixanthides, these minima disappeared, and the maxima were sometimes reduced to a "step-out." The bands with well-defined maxima were found to be *symmetrical on a scale of wave-lengths*, instead of on a scale of frequencies as postulated by Kuhn and Braun (p. 397). A new equation was, therefore, devised to express this important relation (p. 399), and was made the basis of new equations for circular dichroism and rotatory dispersion.

(*b*) *Circular Dichroism.*—Measurements of circular dichroism were made in the region covered by the first weak band, but could not be extended into the second band, where the absorption soon became too strong. In general the "dissymmetry factor" $(\epsilon_l \sim \epsilon_r)/\epsilon$ increased slightly with increasing frequency, in accordance with theoretical indications (p. 402). The circular dichroism, however, always fell off more rapidly than the calculated values on the side of shorter wave-lengths, and in three cases crossed over to the negative side. It follows that the second absorption band is also optically-active, but of opposite sign to the first and weaker absorption band. The parameters of the curves of circular dichroism are set out below.

TABLE 29.—MAXIMA OF CIRCULAR DICHROISM.

		$(\epsilon_l \sim \epsilon_r)$ λ_0.
A.	I. $C_{10}H_{17}O$. CS . S . Me	0·90 at 3550
	II. $C_{10}H_{17}O$. CS . S . Et	0·80 at 3580
	III. $[C_{10}H_{17}O . CS . S]_2$	1·80 at 3660
	IV. $[C_{10}H_{17}O . CS . S]_2CH_2$	1·30 at 3560
B.	V. $C_{10}H_{19}O$. CS . S . Me	1·72 at 3540
	VI. $[C_{10}H_{19}O . CS . S]_2$	4·80 at 3630
	VII. $[C_{10}H_{19}O . CS . S]_2CH_2$	2·70 at 3560
C.	IX. $C_{10}H_{19}O$. CS . NPh . CS . Ph	1·75 at 5150

(*c*) *Rotatory Dispersion in the Region of Transparency.*—With the exception of the methylene esters, where the principal anomalies (including the reversal of sign) had disappeared in one case into the red, and in the other into the infra-red, the rotatory dispersion of all these compounds in the region of transparency was obviously

[1] PURVIS, JONES and TASKER, *J.*, 1910, **97**, 2287.
[2] HANTZSCH and SCHARF, *Ber.*, 1913, **46**, 3570.

anomalous, but could be expressed by two terms of Drude's equation with opposite signs. The positions of the anomalies as deduced from these equations are set out below.

TABLE 30.—OBSERVED AND CALCULATED POSITION OF ANOMALIES FOR BORNYL AND MENTHYL XANTHATES.

	Inflexion.		Maximum.		Reversal.
	[α].	A.U.	[α].	A.U.	A.U.
	Degrees.		Degrees.		
I. (calc.) . .	31·13	6386	36·39	5374	4416
(obs.) . .	—	—	36·35	5370	4440
II. (calc.) . .	27·91	6342	31·45	5471	4451
(obs.) . .	—	—	31·42	5460	4460
III. (calc.) . .	56·31	5925	63·05	5198	4400
(obs.) . .	—	—	62·91	5219	4398
IV. (calc.) . .	1·60	11,430	1·79	9454	6992
(obs.) . .	—	—	—	—	7000
V. (calc.) . .	− 78·91	5919	−88·36	5146	4288
(obs.) . .	—	—	−88·38	5153	4285
VI. (calc.) . .	−277·31	5363	−308·66	4818	4279
(obs.) . .	—	—	−308·71	4800	4170

In VII, VIII and IX the corresponding anomalies were not observed, as they had disappeared into the infra-red region of the spectrum.

(*d*) *Rotatory Dispersion in the Region of Absorption.*—The first maximum in the region of absorption occurs on the near side of the first absorption band, and was easily observed, since $\epsilon = 20$ to 40 only. The reversal of sign near the head of the band, where $\epsilon = 40$ to 80, was also observed without difficulty, since the rotations

TABLE 31.—ANOMALIES IN THE REGION OF ABSORPTION.

		First Maximum.	Reversal of Sign.	Second Maximum.
		[α]. Degrees. A.U.	A.U.	[α]. Degrees. A.U.
A.	I.	− 490 at 3765	3600	not reached
	II.	− 445 at 3815	3637	,, ,,
	III.	− 486 at 3900	3728	,, ,,
	IV.	− 490 at 3826	3586	,, ,,
B.	V.	+ 659 at 3760	3648	−2450 at 3330*
	VI.	+ 407 at 3923	3800	not reached
	VII.	+ 904 at 3795	3593	−1450 at 3350*
C.	VIII.	+ 410 at 5780	5395	not reached
	IX.	− 590 at 5700	5285	,, ,,

* Deduced from observations on both sides of the maximum.

change extremely rapidly with changes of wave-length. The second maximum, however, was much more difficult to record, on account of overlapping by the very strong absorption of the second band, which may increase to $\epsilon = 10,000$ at the maximum. This maximum of rotation was therefore only reached in two menthyl compounds, where the rotatory power was 2 to 4 times greater than in the corresponding bornyl compounds. The positions of the anomalies in the region of absorption are set out on the opposite page.

The partial rotations due to the first absorption band were calculated by means of the equation of Lowry and Hudson (p. 441) and were subtracted from the observed total rotations. The residual rotations, due to bands of shorter wave-length, were then expressed by quasi-hyperbolic curves, disturbed by ripples due to the imperfect elimination of the anomalies produced by the absorption band. This elimination was particularly good in the menthyl compounds.

CHAPTER XXVI.

NITROGENOUS BASES.

General.—The optical activity of a number of alkaloids was investigated by Bouchardat [1] and others from 1843 onwards for the dark red light used by Biot. A much more detailed study was made about 1875 by Hesse,[2] who measured the specific rotations, for the *D* line, of anhydrous quinine in several solvents, as well as of the hydrate, hydrochloride, sulphate, oxalate, etc. These were all found to be lævorotatory, as also were morphine and its salts, whilst cinchonine and its salts were dextrorotatory. Lævorotatory narcotine, however, gave a dextrorotatory hydrochloride.

A similar reversal of sign occurs in the salts of nicotine (see below, p. 329), coniine and *d*-amylamine, as well as in the sodium salts of lactic acid (see above, p. 298), glyceric acid and methoxysuccinic acid. These are cited by Rule [3] as extreme examples of the influence of polar substituents on the optical rotatory power of organic compounds.

Most of the observations on the rotatory power of nitrogenous bases have been limited to the influence of solvents and concentration on rotations for a single wave-length, and have had but little theoretical interest. Read's studies of the stereoisomeric menthylamines have, however, already been described (pp. 314–317), and in the following pages summaries are given of Betti's investigations of the derivatives of the optically-active base *β-naphthyl-phenyl-aminomethane*,

$$\begin{array}{ccc} HO\,.\,C_{10}H_6 & & H \\ & \mathbf{C} & \\ C_6H_5 & & NH_2 \end{array}$$

and of more detailed work on the rotations and rotatory dispersion of nicotine and its derivatives.

Condensation Products of Naphthyl-phenyl-aminomethane.—The base, prepared and resolved by Betti,[4] contains only one asymmetric

[1] BOUCHARDAT, *A.C.P.*, 1843, [iii], **9**, 213; BOUCHARDAT and BOUDET, *Journ. Pharm. Chim.*, 1853, [iii], **23**, 288; BUIGNET, *ibid.*, 1861, [iii], **40**, 252–276; DE VRIJ and ALLUARD, *C.R.*, 1864, **59**, 201.

[2] HESSE, *Ann.*, 1873, **166**, 217–278; 1875, **176**, 89–128, 189–239; **178**, 241–266. See also OUDEMANS, *Ann.*, 1873, **166**, 65–77; 1876, **182**, 33–69.

[3] RULE, *J.*, 1927, 54–59; *T.F.S.*, 1930, **26**, 321–336.

[4] BETTI, *Gazzetta*, 1901, **31**, I, 384; 1906, **36**, II, 392; 1907, **37**, I, 62, II, 5; 1916, **46**, I, 200 and 220; 1920, **50**, II, 276; *T.F.S.*, 1930, **26**, 337–347.

carbon atom; it has a very high rotatory power, $[\alpha]_D + 58{\cdot}9°$, $[M]_D + 146{\cdot}6°$, and can be substituted in close proximity to the asymmetric carbon atom. On condensing the base with a series of aldehydes, to produce compounds of the type

$$HO . C_{10}H_6 . \mathbf{C}HPh . N{=}CHR,$$

a remarkable parallelism was found between the molecular rotations of the condensation products of the aldehydes and the dissociation constants of the corresponding acids, R . CO . OH.

TABLE 32.—ALDEHYDIC DERIVATIVES OF NAPHTHYL-PHENYL-AMINOMETHANE (Betti).

$[M]_D$ of the Aldehydic Compounds.	Aldehydes.	$K \times 10^5$ at 25° of the corresponding Acids.
+ 2676·0	*p*-Dimethylaminobenzoic	0·94
+ 1049·5	*p*-Oxybenzoic	2·9
+ 648·0	3-Bromo-*p*-oxybenzoic	—
+ 588·8	Protocatechuic	3·3
+ 559·6	3-Nitroanisic	—
+ 504·5	*m*-Toluic	5·6
+ 373·1	**Benzoic**	**6·6**
+ 362·6	*m*-Oxybenzoic	8·33
+ 311·8	*p*-Chlorobenzoic	9·3
+ 280·9	*m*-Bromobenzoic	13·7
+ 255·9	*m*-Chlorobenzoic	15·5
+ 167·6	*m*-Nitrobenzoic	34·8
− 85·7	Salicylic	106
− 128·4	*o*-Chlorobenzoic	132
− 308·2	*o*-Bromobenzoic	145
− 990·7	*o*-Nitrobenzoic	657

Further researches by Betti and Bonino, in part unpublished, showed that a quantitative relationship existed between the molecular rotatory powers $[M]_D$ of the aldehydic condensation products and the activity constant of the acid, plotted as $p_K = - \log K$. Two slightly different curves were obtained for the *m*- and *p*-substituted aldehydes, but a close agreement was found between eleven values of p_K calculated from the molecular rotatory powers of the aldehydes, and the values determined by direct observation of the acid. When, however, the molecular rotations for the *D*-line were replaced by the ABSOLUTE MOLECULAR ROTATIONS, extrapolated to a wave-length at which $\lambda^2 - \lambda_0^2 = 1$,[1] the data fell on a single curve, which could be expressed by the equation

$$[M]_{abs.} = 83{\cdot}32x^2 + 17{\cdot}38x + 34{\cdot}63,$$

where $x = p_K - 3{\cdot}4$. The observed and calculated values are shown on page 328.

[1] LOWRY and DICKSON, *J.*, 1913, **103**, 1068.

Acid.	p_K Calculated.	p_K Determined directly.
Protocatechuic	4·469	4·481
m-Toluic	4·329	4·252
Benzoic	4·116	4·180
p-Chlorobenzoic	4·002	4·031
m-Bromobenzoic	3·883	3·863
m-Nitrobenzoic	3·465	3·458

These observations showed clearly that it is the *polarity* of the substituents, rather than their *masses*, which determines the magnitude of the molecular dissymmetry. The absolute molecular rotations of six of the condensation products were therefore compared with their own dipole moments, as given by Debye's equation,

$$\frac{\epsilon - 1}{\epsilon + 2}\frac{1}{\rho} = \frac{4\pi}{3}\frac{N}{M}\left(\gamma + \frac{\mu^2}{3kT}\right)$$

and with the " mean " moment

$$\bar{\mu} = \frac{\mu^2}{3kT}.$$

The results are shown in Table 33, and are plotted in Fig. 138.

TABLE 33.—RELATION BETWEEN MOLECULAR ROTATORY POWER AND DIPOLE MOMENTS.

Compound.	Absolute Rotatory Power. $[M]_{abs}$.	Dipole Moment.		$-\log K$.
		$\mu \cdot 10^{18}$.	$\mu^2 \cdot 10^{36}$.	
Protocatechuic	148·52	3·27	10·7	4·48
m-Toluic	122·78	2·87	8·25	4·25
Benzoic	89·76	2·31	5·34	4·18
p-Chlorobenzoic	75·32	2·18	4·75	4·03
m-Bromobenzoic	62·54	2·02	4·08	3·86
m-Nitrobenzoic	36·13	1·41	1·99	3·46

The relation between $[M]_{abs.}$ and μ^2 is almost linear. Moreover, the curve on which the six points lie appears to pass through the origin. It therefore appears that in this series of compounds a zero dipole moment would be accompanied by a zero value for the absolute rotation, even although there were still a great dissymmetry of mass.[1]

[1] Compare WALDEN's experiments (p. 357) in which equality of two masses, e.g. in $CH_3 . CO . O . CHMe . CO . O . CH_3$, did not give rise to a zero rotation.

Nicotine.—(*a*) *Rotatory Power.*—The optical properties of nicotine are of special interest because of the high lævorotatory power of the base [1] and the reversal of sign which occurs in its salts [2] and some other derivatives. Thus a six-decimetre column of the base, I, gave $\alpha_{5461} = -1237 \cdot 38°$ in the green, rising to $\alpha_{4358} = -2266 \cdot 31°$ in the violet; [3] the specific rotations in the yellow and green are $[\alpha]_D^{20} = -169 \cdot 3°$, $[\alpha]_{5461} = -204 \cdot 1°$ and the molecular rotations are $[M]_D^{20} = -274 \cdot 6°$, $[M]_{5461} = -331 \cdot 0°$.

Winther's data for the specific rotation of nicotine in a series of fourteen solvents, at concentrations from 0 to 100 per cent., are reproduced in Fig. 139.[4] They may be compared with Patterson's curves for ethyl tartrate (Fig. 32, p. 98) where the solvents follow a very similar sequence.

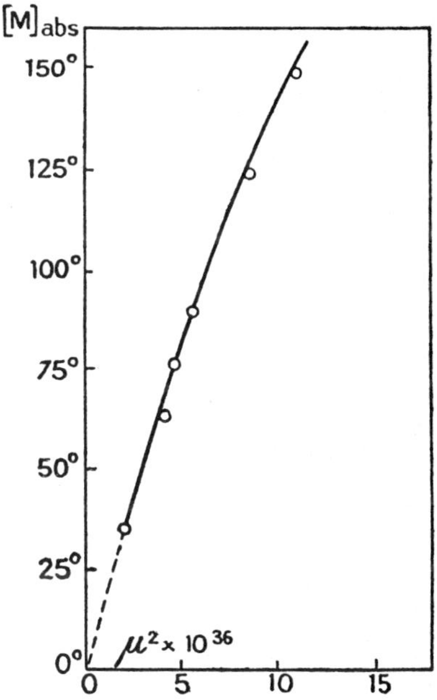

FIG. 138.—RELATION BETWEEN MOLECULAR ROTATIONS AND DIPOLE MOMENTS (Betti).

(*b*) *Rotatory Dispersion.*—The rotatory dispersion of nicotine is nearly simple,[5] since the rotations in the visual region can be expressed approximately by the formula

$$\alpha/\alpha_{5461} = 0 \cdot 2383/(\lambda^2 - 0 \cdot 060);$$

but complete concordance in the visual and photographic readings can only be obtained by introducing an additional partial rotation as follows:—

$$\frac{\alpha}{\alpha_{5461}} = \frac{0 \cdot 21415}{\lambda^2 - 0 \cdot 06376} + \frac{0 \cdot 0257}{\lambda^2}$$

This partial rotation is so small that its dispersion constant can be ignored, as in Drude's equation for quartz, but it is exceptional in being of similar sign to the low-frequency term, which has a characteristic frequency at 2525 A.U. On the other hand, the nicotinium ion, II, in the form of its zinc chloride compound is feebly

[1] LANDOLT, *Ann.*, 1878, **189**, pp. 317–328.
[2] GENNARI, *Z. ph. C.*, 1896, **19**, 130–134.
[3] LOWRY and LLOYD, *J.*, 1929, 1771.
[4] WINTHER, *Z. ph. C.*, 1907, **60**, p. 621.
[5] LOWRY and SINGH, *C.R.*, 1925, **181**, 909.

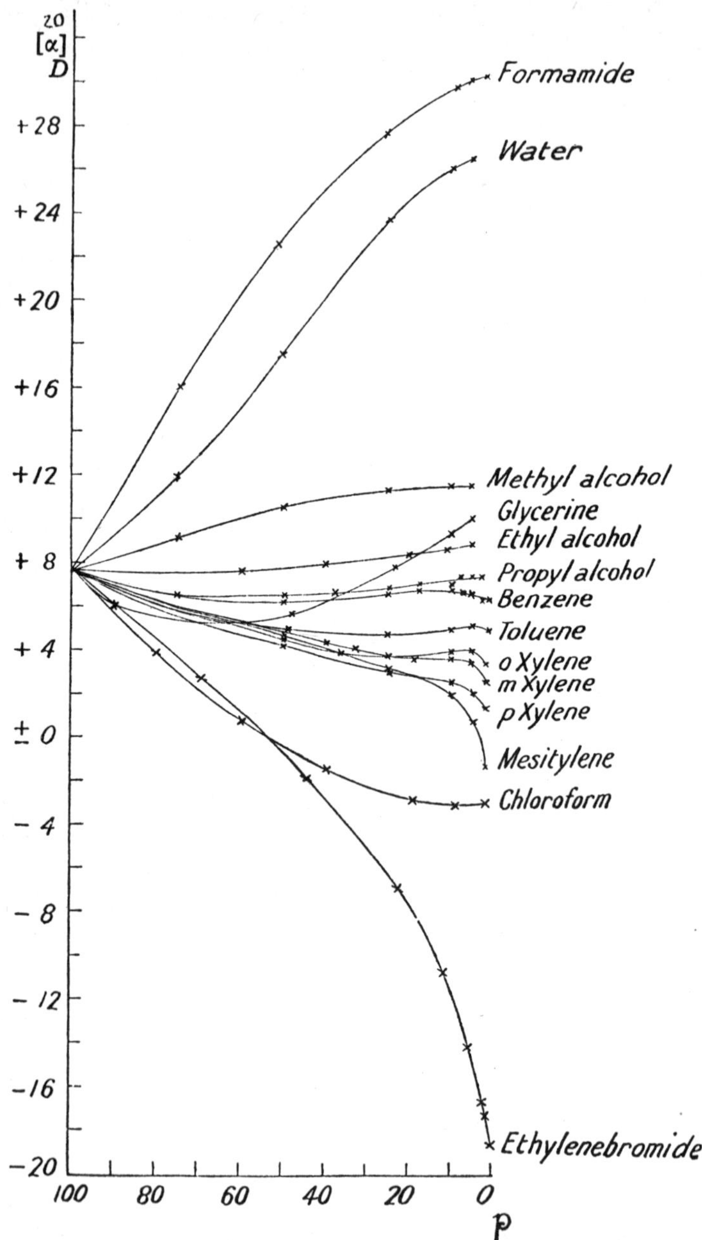

Fig. 139.—Specific Rotation of Nicotine in Different Solvents (Winther).

dextrorotatory and gives a simple rotatory dispersion with a characteristic frequency at 1480 A.U.

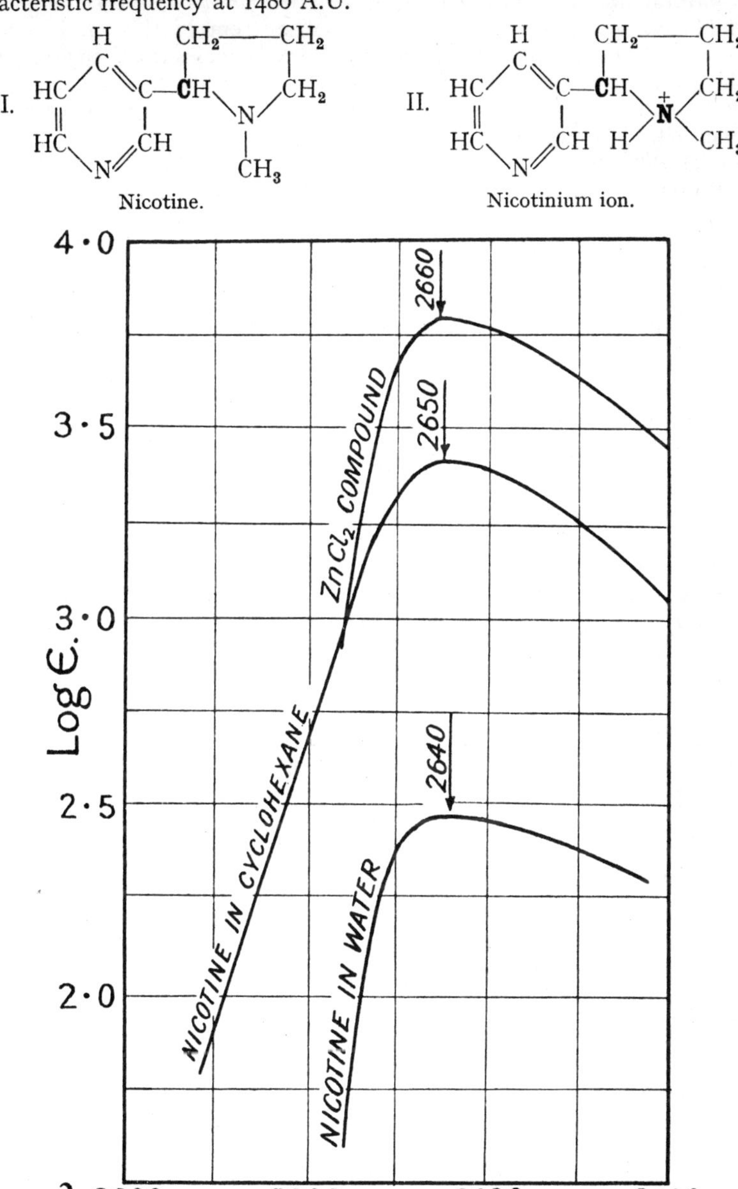

FIG. 140.—ABSORPTION SPECTRUM OF NICOTINE.

The nicotinium ion is of interest as an almost unique example of a natural product containing an *asymmetric atom of nitrogen* ;[1] but the development of a second asymmetric atom in the ion is not responsible for the reversal of sign which accompanies the change from the free base to its salts, since a similar reversal is observed in the methiodide, IV, where the positively charged nitrogen atom carries two methyl groups and is therefore not asymmetric.

(*c*) *Absorption.*—In spite of the wide contrast in their optical rotations, the base and its ion both give rise to an intense absorption band at 2650 A.U. (Fig. 140). This band is weakened nearly ten-fold when the base is dissolved in water, instead of in *cyclo*-hexane, perhaps by reason of hydration of the pyridinium nucleus,

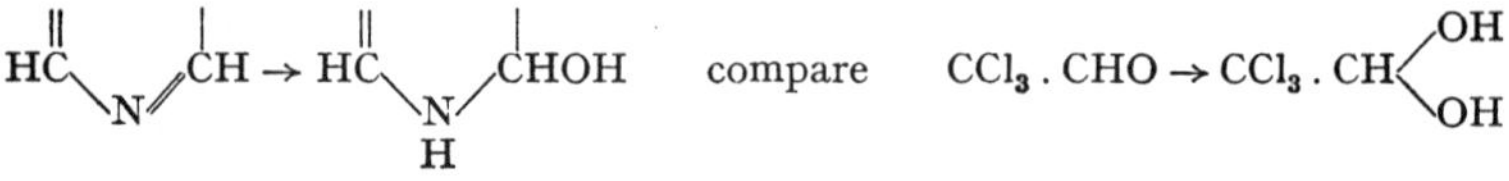

$$HC{=}N{-}CH \rightarrow HC{-}NH{-}CHOH \quad \text{compare} \quad CCl_3.CHO \rightarrow CCl_3.CH(OH)_2$$

On the other hand, it is intensified three-fold in the ion. It is therefore remarkable that, whilst the strong lævorotation of the *base* is controlled almost exclusively by a frequency in the near neighbourhood of this absorption band, the simple dextrorotation of the *ion* is controlled by a frequency in the Schumann region at 1480 A.U. This must mean that the absorption band of the unsaturated pyridine nucleus becomes inactive (or, in other words, that its "induced dissymmetry" disappears completely) when *the pyrrolidine nitrogen is ionised*, leaving the dispersion under the control of a hypothetical band in the Schumann region. This appears to be a general result,

III. 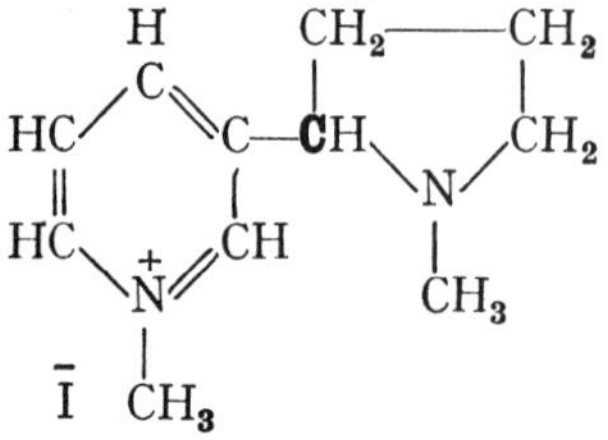 *iso*-Methiodide (lævorotatory).

IV. 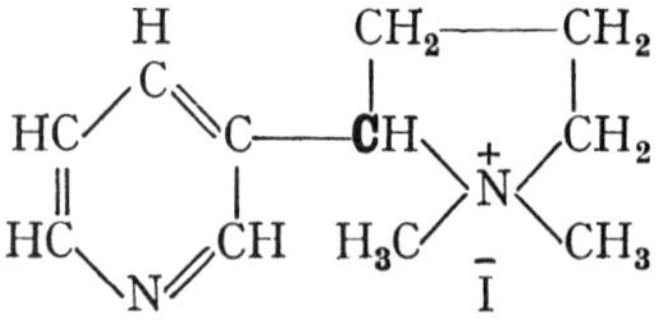 Methiodide (dextrorotatory).

V. 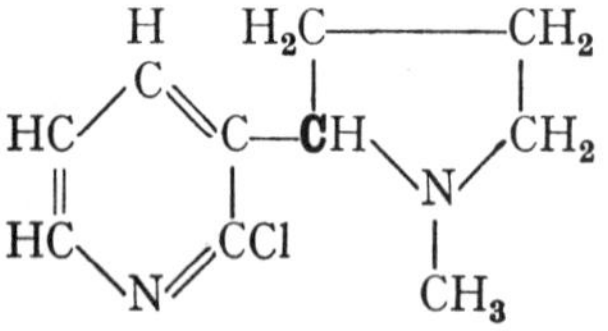 Chloronicotine.

VI. 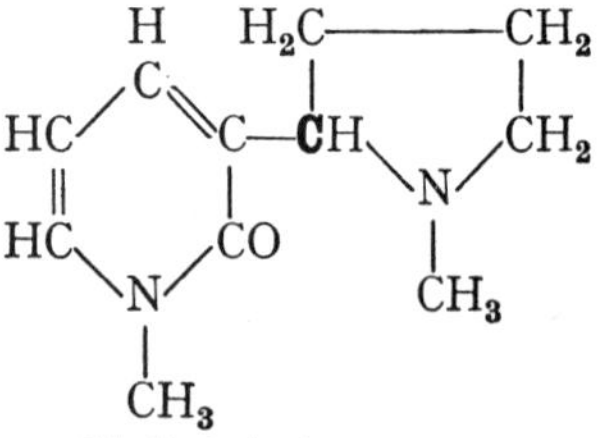 Methylnicotone.

[1] LOWRY, *Nature*, 1926, **117**, 417.

since the methiodides, which give rise to absorption bands of similar wave-length and intensity to the zinc chloride compound, are lævorotatory in the case of the *iso*methiodide, III, where *the pyrrolidine nitrogen is tervalent*, although *the pyridyl nitrogen is ionised*, but dextrorotatory in the case of the mono and dimethiodides where the pyrrolidine nitrogen is ionised as in IV.

Chloronicotine, V, in alcohol and methylnicotone,[1] VI, in *cyclo*-hexane give absorption bands at 2700 and 2950 A.U., the latter being a normal value for a ketone. The two compounds are lævorotatory, like nicotine itself,

$$[\alpha]_{5461}^{20} = -145^\circ \text{ for chloronicotine in alcohol}$$

$$\text{and } [\alpha]_{5461}^{20} = -109^\circ \text{ for methylnicotone in ethyl acetate.}$$

Their rotatory dispersions can be expressed accurately by Drude equations with a single term, with characteristic frequencies at 2550 and 2668 A.U. respectively. The latter number is interesting, since it suggests that whilst the principal *absorption* is that of a ketone, the *rotatory dispersion* is still controlled by the same unsaturated system as in nicotine. It is therefore particularly unfortunate that the circular dichroism of this compound cannot be measured, on account of the intensity of the ordinary absorption and the consequent minuteness of the dissymmetry factor $(\epsilon_l \sim \epsilon_r)/\epsilon$.

[1] Lowry and Gore, *J.*, 1931, 319.

CHAPTER XXVII.

CENTROASYMMETRIC AND AXIALLY-SYMMETRIC MOLECULES.

Methylcyclohexylideneacetic Acid.—

$$\begin{matrix} CH_3 \\ H \end{matrix} > C \begin{matrix} \diagup CH_2 . CH_2 \diagdown \\ \diagdown CH_2 . CH_2 \diagup \end{matrix} C \Longleftrightarrow C \begin{matrix} \diagup H \\ \diagdown CO . OH \end{matrix}$$

Special interest attaches to the rotatory dispersion of "centroasymmetric" compounds (p. 63) which (according to the ordinary definition) contain no asymmetric atom. The classical example of this type is 1-methyl*cyclo*hexylidene-4-acetic acid, of which optically-active *d* and *l* forms were prepared in 1909 by Pope, Perkin and Wallach.[1] This compound, when dissolved in methylal, showed a simple rotatory dispersion,[2] which could be expressed over the range from 6708 to 4072 A.U. by the equation

$$\alpha/\alpha_{5461} = 0{\cdot}2422/(\lambda^2 - 0{\cdot}056).$$

The dispersion constant corresponds with a wave-length $\lambda_0 = 2364$ A.U., which may be identified with the characteristic frequency of the conjugated system

$$>C{=}CH{-}C(OH){=}O$$

The acid has a very intense absorption in this region, but the maximum lay beyond the last observation at 2327 A.U., where, however, the intensity, log $\epsilon = 4{\cdot}15$, was so great as to suggest that the maximum had been very nearly reached.

Co-ordination Compounds.—The dissymmetric co-ordination compounds of the transition elements (Chap. VI) are of interest (i) on account of their simple structure, which in the case of the complex oxalates includes an ion which is dissymmetric but not asymmetric, since it has not less than four axes of symmetry, (ii) on account of their colour, which brings the Cotton phenomenon within the range of visual observation, e.g. in the complex chromitartrates on which he made his first observations, (iii) on account of their exceptionally large optical rotatory power, which in the case of the cobaltioxalate ion reaches a maximum value of about [M] = 10,000°.

On account of their colour, Werner did not always record the

[1] POPE, PERKIN and WALLACH, *J.*, 1909, **95**, 1789.
[2] RICHARDS and LOWRY, *J.*, 1925, **127**, 238.

rotatory power of his optically-active co-ordination compounds for sodium light. Thus, for the chromioxalate he gives for the *G*-line (presumably Fe 4307), observed with the help of a filter of copper sulphate and methyl violet, the following values:[1]

In water $p = 0{\cdot}5$ per cent., $l = 5$ cm., $t = 9°$,
$\alpha_G + 3{\cdot}25°$, $[\alpha]_G = + 1300°$ $[M]_G + 5637°$.
In acetone $p = 0{\cdot}1$ per cent., $l = 5$ cm., $t = 16°$,
$\alpha_G + 0{\cdot}68°$ $[\alpha]_G = + 1360°$ $[M]_G + 5897°$.

The rotatory dispersions of several anions and kations of this series was investigated about 1915 by F. M. Jaeger[2] throughout the visible spectrum, including in several cases the region covered by an absorption band.* Both anions and kations were examined,

e.g. $[Cr\,.\,3C_2H_4(NH_2)_2]^{+++}$ as the triiodide
and $[Cr\,.\,3(C_2O_4)]^{---}$ as the tripotassium salt,

with similar derivatives of rhodium and cobalt. The purpose of the paper was largely crystallographic; but special interest attaches to the anomalous curves for the rotatory dispersion in the region of absorption of (i) the *dark blue chromioxalate* which shows a loop on either side of a reversal of sign at 5700 A.U., (ii) the *dark green cobalti-oxalate*, which shows a loop on either side of a reversal of sign at 6250 A.U., and (iii) the *blood-red rhodioxalate*, which shows a reversal of sign at 6000 A.U., as contrasted with (iv) the *pale-yellow iridio-oxalate*, the rotatory dispersion of which increases progressively with the frequency, although apparently tending to a maximum in the blue.

Attention may also be directed to papers by Lifschitz[3] in 1923, and by Jaeger and Blumendal[4] in 1928, in which the effects produced by co-ordinating an optically-active acid or base with a metallic ion were studied.

The molecular rotations recorded by Jaeger for *potassium cobaltioxalate*, $K_3[Co\,.\,3C_2O_4]$, H_2O were as follows:

	First Maximum.	*Reversal.*	*Second Maximum.*
[M]	$\pm$ 543° at 6380 A.U.	6250 A.U.	$\mp$ 1266° at 6020 A.U.

Johnson and Mead in 1933,[5] by a much more drastic resolution, raised these figures about ten-fold. Thus their best sample of the *d*-salt gave the following molecular rotations:

	First Maximum.	*Reversal.*	*Second Maximum.*
[M]	= + 7700° at 6540 A.U.	6280 A.U.	− 10,000° at 5820 A.U.

* His molecular rotations, expressed in c.g.s. units, must be divided by 10 in order to make them comparable with those given by Werner or (later) by C. H. Johnson.

[1] WERNER, *Ber.*, 1912, **45**, 3070.
[2] F. M. JAEGER, *Proc. Akad. Wet., Amsterdam*, 1915, **17**, 1217 *et seq.*; *Rec. trav. chim.*, 1919, **38**, 171–316.
[3] LIFSCHITZ, *Rec. trav. chim.*, 1922, **41**, 427–436; *Z. ph. C.*, 1923, **105**, 27–54.
[4] JAEGER and BLUMENDAL, *Proc. Akad. Wet., Amsterdam*, 1928, **31**, 637–650; *Z. anorg. Chem.*, 1928, **175**, 161–230.
[5] JOHNSON and MEAD, *T.F.S.*, 1933, **29**, 626.

From the inequality of the two maxima they concluded that the loop of anomalous rotatory dispersion in the red absorption band is associated with a rotation of higher frequency and of opposite sign.

This conclusion was confirmed in the following year by Kuhn and Bein.[1] The salt which they prepared had only about three quarters of the rotatory power of Johnson's sample, but their measurements of absorption and rotatory dispersion were extended into the ultra-violet to about 2800 A.U., and they also recorded the circular dichroism of the red and blue absorption bands, as set out in Fig. 141.[2] Their measurements of rotatory dispersion showed that the red and

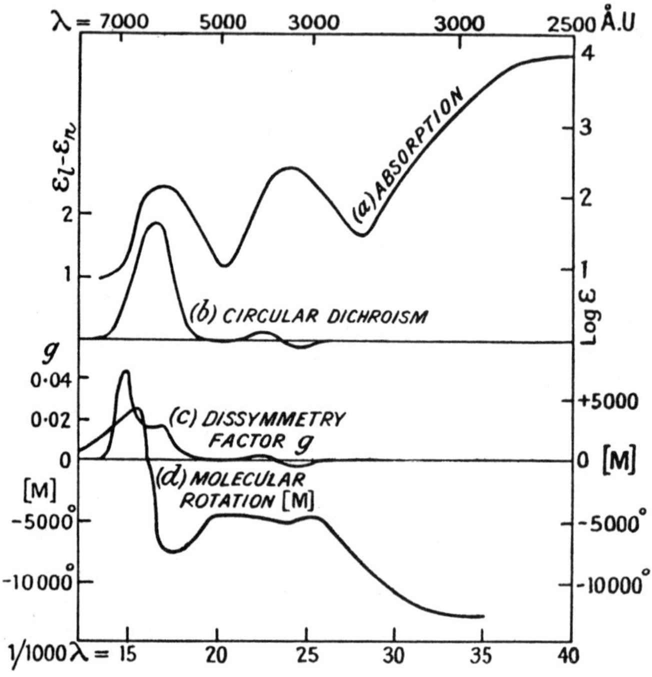

FIG. 141.—ABSORPTION AND ROTATORY DISPERSION OF POTASSIUM *d*-COBALTIOXALATE (Kuhn and Bein, 1934).

blue absorption bands are both optically-active; but both bands appear to be composite in character, like the ketonic band of camphor (p. 406), since the dissymmetry factor, $g = (\epsilon_l \sim \epsilon_r)/\epsilon$, of the first band includes two unequal positive maxima, whilst that of the second is represented with two shallow maxima of opposite sign. The same authors[3] also tabulated[4] the absorption and rotatory dispersion of sixteen complex salts of similar type, but the data are too complex to be easily analysed.

[1] KUHN and BEIN, *Z. ph. C.*, 1934, B. **24**, 335–369.
[2] *Ibid.*, p. 362.
[3] *Ibid.*, *Z. anorg. Chem.*, 1934, **216**, 321–348.
[4] *Ibid.*, pp. 329–330.

PART IV.—*THEORETICAL CONSIDERATIONS.*

CHAPTER XXVIII.

THE ORIGIN OF OPTICAL ROTATORY POWER.

Optical Rotatory Power of Crystals.—Crystals which exhibit optical activity fall naturally into two main classes (i) those which are optically active only in the crystalline state and lose their activity when fused or dissolved, and (ii) those which are optically-active in the crystalline, amorphous and gaseous states and in solution. The optical activity of the former class is due entirely to a structural dissymmetry of the crystal, whereas that of the latter class is due to molecular dissymmetry, on which a structural dissymmetry may or may not be superposed.

Optically active crystals such as quartz, sodium chlorate, sodium ammonium tartrate, etc., are often hemihedral and not superposable on their mirror images (pp. 25–31); but this relationship is not universally true, since many optically-active crystals show no trace of hemihedral faces. This is hardly surprising since the "habit" of the crystal is influenced by many secondary factors, and it may be only by a happy chance that the facets which disclose the hemihedrism of the crystal are developed in any particular instance.

Optical activity has been observed in isotropic, uniaxial and biaxial crystals. Crystals of these three types, of which the rotatory power, and in many cases the rotatory dispersion, has been measured, are set out in Table 34.

Isotropic Crystals.—Optical activity in isotropic crystals was first observed in 1854 by Marbach [1] in hemihedral crystals of sodium chlorate, which are not superposable on their mirror images. The rotatory power is identical in all directions, and is the same for lævo- and dextro-crystals. Marbach also observed optical activity in the other three substances included in Table 34 which form isotropic crystals, namely, sodium bromate, sodium sulphantimoniate and sodium uranyl acetate. Measurements of the rotatory power of crystals of sodium chlorate have also been made by Sohncke,[2]

[1] MARBACH, *Pogg. Ann.*, 1854, **91**, 482; 1855, **94**, 412; 1856, **99**, 451; *C.R.*, 1855, **40**, 793; *A.C.P.*, 1855, [iii], **43**, 252; 1855, [iii], **44**, 41.
[2] SOHNCKE, *Wied. Ann.*, 1878, **3**, 529–531.

TABLE 34.—OPTICAL ACTIVITY IN CRYSTALS.*

(*a*) *Optically-active in Crystalline State only.*

Isotropic.	Uniaxial.	Biaxial.
Sodium chlorate	Quartz	Hydrazine sulphate
Sodium bromate	Cinnabar	Strontium formate
Sodium sulphantimoniate	Sodium periodate	Barium formate
Sodium uranyl acetate	Potassium dithionate	Lead formate
	Rubidium dithionate	Iodic acid
	Cæsium dithionate	Ammonium oxalate
	Calcium dithionate	Ammonium potassium oxalate
	Strontium dithionate	Sodium arsenate
	Lead dithionate	Lithium sulphate
	Ethylenediamine sulphate	Zinc sulphate
	Benzil	Magnesium sulphate
	Potassium lithium sulphate	Nickel sulphate
	Potassium lithium sulphochromate	Magnesium chromate
	Guanidine carbonate	Sodium dihydrogen phosphate
	Diacetylphenolphthalein	
	Potassium silicomolybdate	
	Potassium silicotungstate	

(*b*) *Optically-active in Crystalline State and in Solution.*

Uniaxial.	Biaxial.
Camphor	Cane sugar
Matico-camphor	Tartaric acid
Patchouli-camphor	Sodium tartrate
Strychnine sulphate	Potassium tartrate
Rubidium tartrate	Ammonium tartrate
Cæsium tartrate	Rochelle salt
Zinc malate	Sodium ammonium tartrate
Potassium iridium trioxalate	Ammonium antimonyl tartrate
Potassium rhodium trioxalate	Acid strontium tartrate
Corydine	Tartramide
Hydrocinchonine sulphate	Asparagine
Cinchonine antimonyl tartrate	Benzilidene-camphor
Apocinchonine succinate	Anisylidene-camphor
	Camphoric acid
	Camphor oxime
	Quercitol
	Methyl α-glucoside
	Calcium bimalate
	Ammonium bimalate
	Ammonium molybdomalate
	Barium molybdomalate

* The rotatory power of many of the crystals included in Table 34 are given in *International Critical Tables*, Vol. 7, pp. 353–354.

Guye[1] and Voigt,[2] and of sodium bromate by Traube[3] and by Rose.[4]

Uniaxial Crystals.—Optical activity in doubly refracting crystals can in practice be observed and measured readily only along the direction of an optic axis, since the rotation in all other directions is largely masked by the effects of ordinary double refraction. In uniaxial crystals, such as quartz, optical activity can be observed without difficulty in plates cut perpendicularly to the principal axis of the crystal (Biot, 1812); but the phenomenon was not observed again until 1857, when Descloizeaux[5] found that a millimetre plate cut perpendicular to the optic axis of a crystal of cinnabar rotated the plane of polarisation of red light through 325°. The *rotatory dispersion* of crystals of cinnabar was studied in 1911 by Becquerel,[6] who showed that the increase of rotation on approaching the region of strong absorption in the green part of the spectrum is precisely analogous to the corresponding effect in optically-active absorbing solutions, as described in Chapter XI (p. 154). Optical activity in uniaxial crystals has also been observed and measured by Groth,[7] Pape,[8] von Lang,[9] Bodewig,[10] Hintze,[11] Bodländer,[12] Traube,[13] Wyrouboff,[14] Rose[15] and others.

Biaxial Crystals.—Optical rotatory power in biaxial crystals is much more difficult to observe, since the optic axes do not coincide with axes of crystal symmetry, and moreover vary in direction with the wave-length of the light. Thus, although Biot[16] in 1817 obtained indications that crystalline cane sugar was optically active, his observations with white light were too complex to establish the existence of optical rotatory power in the crystal. Landolt[17] therefore in 1879 concluded quite wrongly that substances such as cane sugar, tartaric acid, asparagine, camphor, etc., which are optically active when in solution, do not exhibit optical activity in the crystalline state. In 1901, however, Pocklington[18] measured the rotatory power of a crystal

[1] GUYE, *Arch. Sc. Phys. Nat., Genève*, 1889, **22**, 130.
[2] VOIGT and HONDA, *Phys. Zeit.*, 1908, **9**, 585.
[3] TRAUBE, *Jahrb. Min.*, 1894, **1**, 171.
[4] ROSE, *Jahrb. Min.*, Beilage Band, 1910, **29**, 53–105.
[5] DESCLOIZEAUX, *C.R.*, 1857, **44**, 876 and 909.
[6] J. BECQUEREL, *C.R.*, 1908, **147**, 1281.
[7] GROTH, *Pogg. Ann.*, 1869, **137**, 433.
[8] PAPE, *Pogg. Ann.*, 1870, **139**, 224.
[9] VON LANG, *Sitz. Akad. Sci., Wien*, 1872, **65**, II, 30–32.
[10] BODEWIG, *Pogg. Ann.*, 1876, **157**, 122.
[11] HINTZE, *Pogg. Ann.*, 1876, **157**, 127.
[12] BODLÄNDER, *Inaug. Diss., Breslau; Zeit. Krist.*, 1884, **9**, 309.
[13] TRAUBE, *Jahrb. Min.*, 1894, **1**, 171; *Sitz. preuss. Akad. Wiss.*, 1895, p. 195; *Jahrb. Min.*, Beilage Band, 1897, **11**, 623.
[14] WYROUBOFF, *Journ. de Phys.*, 1894, **3**, 451; *A.C.P.*, 1894, [vii], **1**, 5–90; *Bull. Soc. fr. Min.*, 1896, **19**, 219–354; 1901, **24**, 76.
[15] ROSE, *Jahrb. Min.*, Beilage Band, 1910, **29**, 53–105.
[16] BIOT, *Mem. Acad. Sci.*, 1817, **2**, 113.
[17] LANDOLT, *Handbook of the Polariscope*, English translation, 1882, p. 15.
[18] POCKLINGTON, *Phil. Mag.*, 1901, [vi], **2**, 368.

of cane sugar in the direction of both optic axes, and thus for the first time established conclusively the existence of optical activity in a biaxial crystal. All the other crystals cited by Landolt have also been shown to be optically active, and at the present time no exception is known to the rule that *all crystals which yield optically-active solutions are themselves optically active.*

The difficult problem of *measuring* the rotatory power of biaxial crystals was solved by Dufet,[1] Wallerant [2] and Longchambon [3] (p. 205) who evolved a technique by which it is possible to obtain an accuracy of 0·1°

The rotatory power of a crystal is usually expressed as the rotation produced by 1 mm. thickness; the rotatory power of a solution may be expressed in comparable units by means of the function $[\alpha]d/100$ where $[\alpha]$ is the specific rotation and d is the density of the solution; conversely the rotatory power of a crystal, $\rho°$ per mm., can be expressed like that of a solution as $[\rho] = 100\,\rho/d$. Some of Longchambon's observations are thus set out in Table 35.

Table 35.—Rotatory Power of Crystals and Solutions (Longchambon, 1923).

Compound.	Rotatory Power of Solution. $[\alpha]d/100$.	Rotatory Power of Crystal ρ.
	Degrees.	Degrees per mm.
Matico-camphor	− 0·31	− 1·92
Patchouli-camphor	− 1·26	− 1·325
Ordinary camphor	+ 0·55	+ 0·65
Cane sugar	+ 1·05	+ 5·38 1st optic axis
		− 1·61 2nd ,, ,,
Rubidium tartrate	+ 0·69	− 10·24
Ammonium tartrate	+ 0·61	− 8·9
Ammonium molybdomalate	+ 5	+ 32·3

These observations show that the rotatory power of the crystal is usually larger than that of the solution; but there is no correspondence in magnitude or sign, since the rotation in the crystal may even be of opposite sign to that given by the same substance in solution. Moreover, the rotatory power of a crystal of cane sugar along the two optic axes is different both in sign and magnitude, although magnesium sulphate has the same rotatory power along both its axes.

Longchambon has suggested three possible explanations of the divergence between the rotatory powers of crystals and of solutions:

(i) The rotatory power of the molecule is the same in all directions, but is supplemented by a structural rotatory power which is different in different directions.

[1] Dufet, *Journ. de Phys.*, 1904, **3**, 757.

[2] Wallerant, *C.R.*, 1914, **158**, 91.

[3] Longchambon, *Bull. Soc. fr. Min.*, 1922, **45**, 161–252; *Thesis*, Paris, 1923; *C.R.*, 1921, **172**, 1187; 1921, **173**, 89; 1922, **175**, 174; 1924, **178**, 951, 1828.

(ii) The rotatory power of the molecule is different in different directions and there is no structural rotatory power.

(iii) Both the structural rotatory power and the rotatory power of the molecule vary with the direction.

Explanation (ii) seems hardly tenable in view of the rotatory power of crystals which are inactive in solution.

Rotatory Dispersion of Crystals.—Although the rotatory powers of crystals and solutions differ very widely, Longchambon showed that their rotatory dispersions are almost identical (Table 36):

TABLE 36.—ROTATORY DISPERSION OF CRYSTALS AND SOLUTIONS (Longchambon, 1923).

(*a*) *Cane Sugar.*

		Rotatory Power.			Dispersion Ratio.		
	$\lambda=$	5790	5460	4360	5790	5460	4360
Crystal 1st axis	$[\rho]=$	+ 336°	+ 366°	+ 631°	1	1·15	1·88
Crystal 2nd ,,	$[\rho]=$	− 100·6°	− 111·2°	− 191°	1	1·11	1·90
Solution . .	$[\alpha]=$	+ 69°	+ 79°	+ 129°	1	1·14	1·87

(*b*) *Ammonium Molybdomalate.*

		Rotatory Power.		Dispersion Ratio.	
	$\lambda=$	5790	4360	5790	4360
Crystal . .	$[\rho]=$	+ 1410°	+ 3182°	1	2·36
Solution . .	$[\alpha]=$	+ 220°	+ 499·4°	1	2·38

(*c*) *Camphor.*

	Dispersion Ratio.				
$\lambda=$	5890	5780	5460	4920	4360
Crystal	1	1·06	1·27	1·8	2·7
Molten	1	1·05	1·28	1·8	2·8
Vapour	1	1·06	1·27	1·8	2·9
Solution in hexane . .	1	1·055	1·270	1·790	2·930
,, ,, ether . .	1	1·053	1·262	1·785	2·918

Many of these dispersions can be represented by one term of Drude's equation $[\alpha]=k/(\lambda^2-\lambda_0^2)$ (p. 120). The "rotation-constants," k, of the crystals differ from those of the solutions, but the "dispersion constants," λ_0, are approximately the same. The rotatory dispersion of cinnabar, however, can be represented only very approximately by a Drude equation on account of the presence of an intense absorption band in the visible spectrum.[1] Longchambon's values for *potassium rhodium trioxalate* in the region of an absorption band are set out in Table 37.

TABLE 37.—ROTATORY DISPERSION OF POTASSIUM RHODIUM TRIOXALATE.

$\lambda =$	7000	6800	6600	6400	6200	6000	5800	5600	5400	5200	5000	4900
Crystal $\rho =$	−83′	−88′	−91′	−89′	−84′	−74′	−64′	−48′	−26′	−1′	+28′	+51′
Solution $\alpha =$	−2′	−3′	−4′	−4′	−2′	±0′	+3′	+6′	+10′	+15′	+18′	+28′

The magnitudes of the rotations for a given wave-length differ very widely, and the reversal of sign is displaced from 6000 A.U. for the solution to 5200 A.U. for the crystal. This phenomenon may be explained, however, by differences in the "rotation constants" of an equation containing *two* partial rotations (one of the Drude type, p. 120, and one of the Natanson type, p. 425) and does not necessarily imply any alteration of the "dispersion constants" of the equation.

Origin of Optical Rotation in Crystals.—It was first suggested by Fresnel[2] that the structural dissymmetry of a uniaxial crystal such as quartz, was due to the arrangement of the molecules of silica in a left or right-handed spiral. The same idea was suggested by Pasteur (p. 27). Reusch[3] in 1869 found that optical activity could be produced artificially by superposing thin mica sheets, each of which was turned through a definite angle with respect to its neighbours. Mallard[4] therefore suggested that optically-active crystals are built up from successive layers, each of which is rotated to the right or left with respect to the layer beneath it; but analysis by means of X-rays has shown that a spiral arrangement of MOLECULAR PLATES is not necessary for the production of optical activity in crystals. Thus the optically-active crystals of *sodium chlorate* consist of an IONIC AGGREGATE of sodium and chlorate ions, arranged dissymmetrically on a cubic lattice (Fig. 142),[5] in which no molecular plates can be recognised. On the other hand, crystals of *quartz* consist of GIANT MOLECULES, in which the silica units —O—Si—O— (with bonds above and below Si) are built

[1] ROSE, *Jahrb. Min.*, Beilage Band, 1910, **29,** 53–105.
[2] FRESNEL, *A.C.P.*, 1824, [ii], **28,** 279 (quoted on p. 19).
[3] REUSCH, *Pogg. Ann.*, 1869, **138,** 628.
[4] MALLARD, *Ann. des Mines*, 1876, **10,** 175–192; 1881, **19,** 256–313; *Bull. Soc. fr. Min.*, 1884, **7,** 349–401.
[5] KOLKMEIER, BIJVOET and KARSSEN, *Proc. Akad. Wet., Amsterdam*, 1921, **23,** 644 (Fig. 1, p. 649); *Zeit. Phys*, 1923, **14,** 291.

up into a continuous three-dimensional network of indefinite extent. In this network every atom of silicon is linked to four atoms of oxygen, whilst every atom of oxygen unites two atoms of silicon to one another. The complex aggregate which is thus formed has, however, a spiral character, which can be seen clearly in Fig. 143 although this figure is misleading in that the silicon atoms are represented by large black spheres, and oxygen atoms by small white ones, whereas in reality the oxygen atoms are so large as almost to conceal the tiny silicon atoms. Moreover, a lamellar structure can also be claimed in view of the fact that the lattice repeats itself at intervals of 1·8 A.U. along the optic axis. Successive layers, however, are rotated through 120° relatively to their neighbours, so that it is only the *fourth* layer which is superposed above the *first* with an iden-

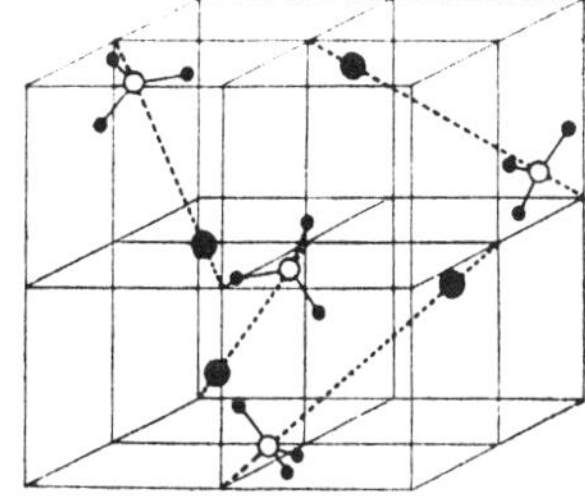

FIG. 142.—CRYSTAL STRUCTURE OF SODIUM CHLORATE (Kolkmeier, 1921).

The hollow circles represent chlorine atoms, each joined by covalent bonds to three oxygen atoms to form a chlorate ion. The sodium ions are shown as larger black circles. In reality the space is almost filled with oxygen atoms, the diameter of which is much greater than that of the sodium ions.

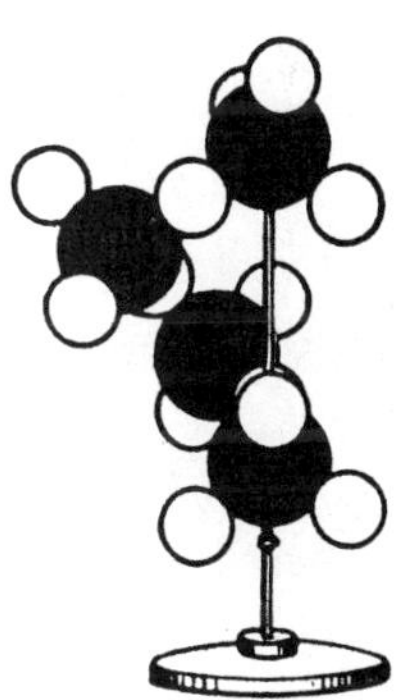

FIG. 143.—SPIRAL STRUCTURE OF QUARTZ (Bragg).

tical orientation. The major spacing along the optic axis is therefore 5·4 A.U.[1]

In this connection it may also be noted that Wyrouboff[2] has definitely established lamellar growth in crystals of ammonium lithium sulphate, where the plates make an angle of 60° with one another, whilst Burgers[3] was not able to prove the absence of lamellæ in $KLiSO_4$.

Structure of α- and β-Quartz.—The positions of the silicon atoms in α- and β-quartz are illustrated in Fig. 144 (*a*) and (*b*).[4] In both forms of quartz the atoms of silicon occupy successive layers at intervals of 1·8 A.U., which are distinguished by shading the

[1] BRAGG, *P.R.S.*, 1913, A. **89**, 575.

[2] WYROUBOFF, *Bull. Soc. fr. Min.*, 1890, **13**, 215–233.

[3] BURGERS, *P.R.S.*, 1927, A. **116**, 553–586.

[4] BRAGG and GIBBS, *P.R.S.*, 1925, A. **109**, 405–427 (Figs. 10*a* and 13, pp. 415 and 421).

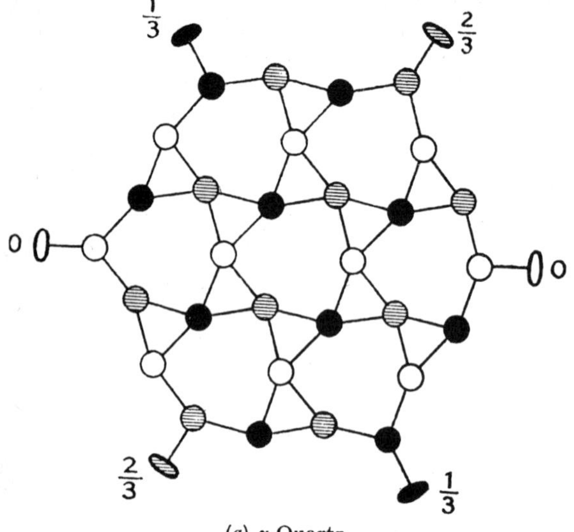

(*a*) α-Quartz.

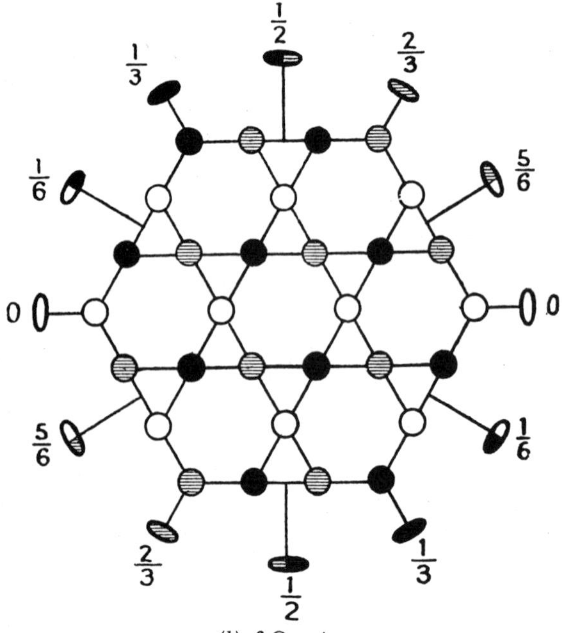

(*b*) β-Quartz.

FIG. 144.—ARRANGEMENT OF SILICON ATOMS IN QUARTZ.

circles. α-Quartz (Fig. 144 *a*) is the less symmetrical form. It has one three-fold axis (the optic axis) perpendicular to the plane of the paper, and three two-fold axes, lying respectively in the planes of the white, grey and black circles. β-Quartz (Fig. 144 *b*) is much more symmetrical, since the principal axis is now an axis of six-fold symmetry (compare Fig. 8, p. 25), and there are six axes of two-fold symmetry, lying alternately in and half-way between the planes which contain the silicon atoms.

The pattern in Fig. 144*b* consists of regular hexagons and equilateral triangles. Each equilateral triangle is the plan of a spiral of silicon atoms (compare Fig. 143) represented by the white, grey and black

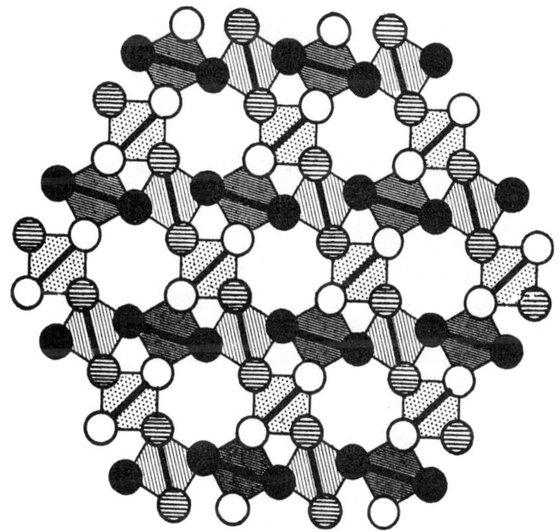

FIG. 145.—ARRANGEMENT OF OXYGEN ATOMS IN β-QUARTZ (Bragg and Gibbs).

circles. In the same way the hexagons represent *twin* spirals of white, grey and black circles, which are twisted in the opposite direction to the smaller single spirals. These spirals are all reversed on passing from *d*- to *l*-quartz. It is probable that the light is twisted in the same direction as the small spirals, but no categorical statement has been made to show in which direction the spirals turn in the two forms of quartz. The lower symmetry of α-quartz is due to a twisting of the equilateral triangles, which reduces the symmetry of the hexagons from hexagonal to trigonal. This twisting is due to the greater stability at low temperatures of the distribution of four oxygens round each silicon atom at the apices of a regular tetrahedron, since this arrangement is strained on passing over into β-quartz. The arrangement of the tetrahedra in β-quartz is shown in Fig. 145.[1]

[1] Compare BRAGG and GIBBS, *P.R.S.*, 1925, A. **109**, Fig. 12, p. 420.

In general agreement with van't Hoff's method of representation (Fig. 15, p. 43), each atom of bivalent oxygen occupies the apices of two contiguous tetrahedra, and thus forms a link between two silicon atoms. The oxygen atoms, like the silicon atoms, are arranged in layers, which are distinguished by different shading, and the regular hexagons and equilateral triangles of Fig. 144*b* can again be recognised. This symmetrical structure is twisted as before on passing to the lower symmetry of α-quartz, but this is not shown in Fig. 145.

Calculation of Rotatory Power of Crystals.—The refractive index and rotatory power of crystals of sodium chlorate and bromate for light of various wave-lengths were calculated by Hermann [1] in 1923 from the crystal structure with the help of theoretical considerations given by Born.[2] The theoretical values are of the right order of magnitude, but are not now cited, on account of a numerical error in one of the equations.[3]

Similar calculations of the double refraction and rotatory power of β-quartz were made by Hylleraas [4] in 1927. The lattice is almost certainly that of a giant molecule, held together by a network of covalent bonds; but, for the purposes of calculation, Hylleraas postulated, in accordance with a general procedure which often appears to be successful,[5] that the crystals were an aggregate of ions held together by electrostatic forces. By deducing from an observed index of refraction an arbitrary constant repressing the intensity of an inaccessible absorption band (which was thus assumed to involve *five* electrons), a close agreement was obtained between the observed and calculated values of n_e and n_o at 580° and of α at 600°; but the numerical values of these quantities were not tabulated, since they depend on the *eighth* power of the linear dimensions. The results of the calculation were therefore expressed by deducing, from the double refraction and from the rotatory power of the crystal, independent values for two of the lattice constants. These agreed closely with those given by Bragg and Gibbs,[6] although they differed substantially from the values of Wyckoff.[7]

Rotatory Power of "Liquid Crystals."—Reinitzer's [8] original researches on "crystalline liquids" included the observation of optical activity in what are now generally known as ANISOTROPIC LIQUIDS or MESOMORPHIC STATES of matter.[9] This phenomenon is always associated with the presence of dissymmetric molecules, but it can

[1] HERMANN, *Zeits. f. Physik.*, 1923, **16**, 103.

[2] BORN, *Dynamik der Krystalgitter*, Leipzig (1915); *Zeits. f. Physik.*, 1922, **8**, 390; *Atomtheorie des festen Zustandes, Encyklopödie der math. Wiss.*, 1923, **5**, 25.

[3] BORN, footnote, p. 386 below.

[4] HYLLERAAS, *Zeits. f. Physik.*, 1927, **44**, 871.

[5] BERNAL, *Chem. and Ind.*, 1934, **53**, 249.

[6] BRAGG and GIBBS, *P.R.S.*, 1925, A. **109**, 405–427.

[7] WYCKOFF, *Zeitschr. f. Krist.*, 1926, **63**, 507.

[8] REINITZER, *Monatshefte*, 1888, **9**, 421.

[9] FRIEDEL. *Ann. de Phys.*, 1922, [ix], **18**, 273–474, p. 275.

be produced by the addition, to an inactive mesomorphic medium, of dissymmetric molecules which do not normally form "liquid crystals."

Rotations of the order of 50,000° or 100,000° per millimetre may be observed in the mesomorphic state, but if the temperature is raised until the anistropic liquid becomes isotropic, the rotatory power falls at once to the almost immeasurably small rotation per millimetre which is associated with ordinary optically-active liquids and solutions. The large rotatory powers referred to above must therefore be due to a structural dissymmetry induced by and superposed upon a relatively insignificant molecular dissymmetry.

Optical activity in mesomorphic media was first observed in derivatives of cholesterol, and media which show this phenomenon are therefore termed CHOLESTERIC MEDIA, to distinguish them from the "nematic" and "smectic" types.[1] Cholesteric media can exist also in a form which does not exhibit enhanced optical activity, but the remarkable phenomena described above have been observed in mesomorphic states of the following substances.[2]

TABLE 38.—OPTICALLY-ACTIVE (CHOLESTERIC) ANISOTROPIC LIQUIDS.

Cholesteryl chloride	Cholesteryl *o*-nitrobenzoate
,, acetate	,, *m*- ,,
,, propionate	,, *p*- ,,
,, crotonate	,, phenylpropiolate
,, *n*-butyrate	,, *p*-nitrocinnamate
,, allylacetate	,, cinnamate
,, sorbate	,, hydrocinnamate
,, benzoate	,, *p*-azoxycinnamate
Amyl cinnamylideneaminocinnamate	
,, *p*-nitrobenzylideneaminocinnamate	
,, cyanobenzylideneaminocinnamate	
,, anisylideneaminocinnamate	
,, anisylideneamino-α-methylcinnamate	

The very large optical rotatory power of cholesteric media is associated with a selective transmission of circularly-polarised light of opposite signs, giving rise to ELLIPTIC POLARISATION in the emergent ray (Fig. 46, p. 152). This selective action may be of such intensity that one ray is almost totally *reflected*, whilst the other is almost totally *transmitted*, at the wave-length at which the selective action is at a maximum.[3] The effect is equivalent to a circular dichroism of almost infinite magnitude, but it is remarkable that the reflected circular component has the same sign before and after reflection, i.e. *the direction of the spin is reversed by reflection.*

The rotatory power of mesomorphic media is very sensitive to changes of temperature. Thus, variations of temperature (between the limits within which the mesomorphic state is stable) may displace

[1] FRIEDEL, *Ann. de Physik*, 1922, [ix], **18**, 276, 277, 385.
[2] VORLANDER and HUTH, *Z. ph. C.*, 1911, **75**, 641.
[3] GIESEL, *Phys. Zeit.*, 1910, **11**, 192.

the wave-length of maximum selective reflection over the whole range of the visible spectrum and cause it to disappear into the infra-red or the ultra-violet.

On either side of the wave-length of maximum selective reflection the rotatory power of the medium varies in a manner somewhat

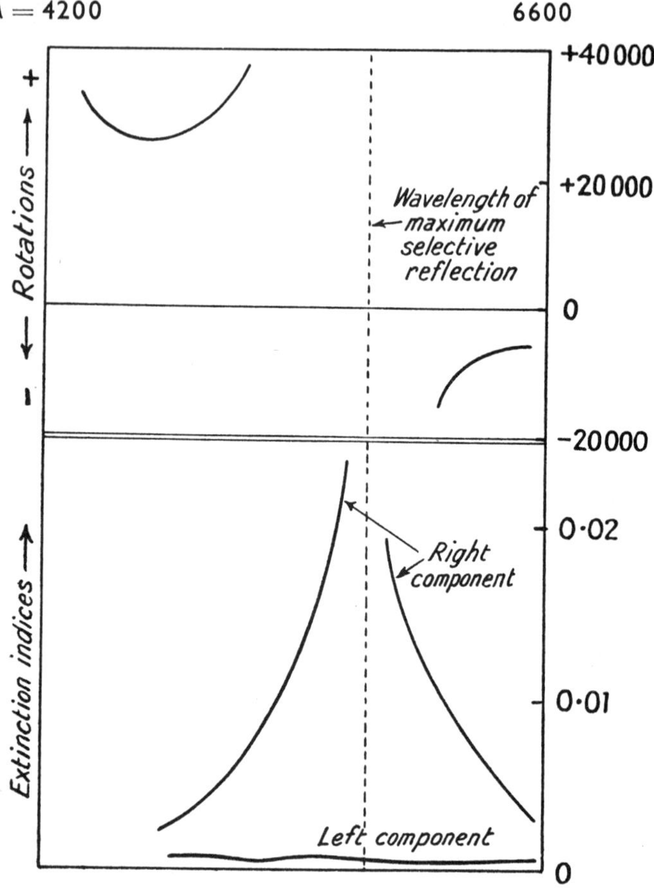

FIG. 146.—OPTICAL ACTIVITY IN "LIQUID CRYSTALS" OF AMYL CYANOBENZYLIDENEAMINOCINNAMATE (Stumpf, 1910).

The right circularly-polarised component is selectively reflected, whilst the left-component is transmitted. The wave-lengths increase from left to right.

similar to that observed by Cotton (Fig. 48, p. 154) on either side of an optically-active absorption band. Thus, the rotatory power is opposite in sign on opposite sides of the wave-length of maximum selective reflection, and increases as this wave-length is approached;

but no observations have been recorded near the centre of the region of selective reflection, where Cotton's observations suggest that a zero rotation might be expected to occur; and, as far as can be observed, the rotation of the cholesteric medium passes discontinuously from a high negative value on one side to a high positive value on the other.* These results are illustrated in Fig. 146,† for the particular case of *amyl cyanobenzylideneaminocinnamate* investigated by Stumpf.[1]

Optical Rotatory Power of Solutions.—Although the *existence* of optical rotatory power in amorphous media can always be traced back to the dissymmetry of the molecules, its *magnitude* depends very largely on the influence of solvents, of concentration and of temperature. This effect has been known, and its importance has been fully recognised, since the appearance in 1832 of Biot's memoir on Tartaric Acid (p. 90). The phenomena are necessarily complicated, since chemical as well as physical influences are at work. Thus, the reversible hydration of a sugar or of a lactone, the reversible polymerisation of a nitroso-compound, or the reversible isomeric change of nitrocamphor, will obviously affect the rotatory power of the medium in which these chemical changes take place; and the displacement of equilibrium resulting from changes of temperature or of concentration will be accompanied by corresponding changes in the specific rotation of the medium. Of a more subtle character are the distortions which may be produced in the molecule itself, when free rotation is assisted or impeded by changes of temperature,[2] or when the strength and stiffness of the bonds between the atoms are modified by an increase or decrease of thermal energy.

Superposed on the chemical factors of alterations of molecular structure and configuration are the physical effects of thermal expansion. This acts primarily as a form of dilution, which can sometimes be carried to the extreme limit of vaporisation without affecting very seriously the rotatory power of the molecule (pp. 100–104). In other cases, however, the increasing distance between the molecules may produce important changes of rotatory power by decreasing their mutual influence upon one another; and similar changes may be produced by the intervention of molecules of a solvent. Evidence is now available to show that the polarity of the molecules is a dominant factor in this interaction (compare Betti, p. 328), and this point of view is developed more fully in the following paragraphs.

Finally, since optical rotatory power depends directly on circular double refraction, and therefore less directly on circular dichroism, changes of temperature may produce important alterations of rotatory power by altering the energy levels of the molecule. Thus, the

* Compare R. W. Wood's observations on the magnetic rotatory power of sodium vapour in the neighbourhood of the D-lines (p. 461).

† The diagram is based upon curves plotted by Bruhat (*Traité de Polarimétrie*, Fig. 234, p. 351).

[1] Stumpf, *Phys. Zeit.*, 1910, **11**, 780; *Ann. der Physik*, 1912, **37**, 351; see p. 206 for methods of observation.

[2] Wolf, *T.F.S.*, 1930, **26**, 315–320.

distribution of the vibrational levels corresponding with a given electronic level may be modified both in the ground state and in the excited state, i.e. at both ends of the electronic jump which is responsible for a given partial rotation. In some cases the effects thus produced may be altered profoundly by the mutual influence of the molecules upon one another; but this will not necessarily be the case, since the changes of energy-levels do not depend on this interaction, and may therefore occur even when the molecules are almost completely isolated from one another in a dilute vapour. The primary effect of these changes is to alter the form and position of the curves of absorption and circular dichroism; but these primary effects will give rise to secondary alterations in the curve of rotatory dispersion in the region of absorption, and to corresponding alterations of rotatory power in the region of transparency.

In view of the complicated character of the changes thus produced by alterations of solvent, concentration and temperature, it is not surprising that, over a period of nearly a century, only empirical relations were forthcoming. During this long period, therefore, the data that were accumulated had but little theoretical significance, except as raw materials for future analysis. In recent years, however, definite relations have been established by H. G. Rule,[1,2] between the dipole moment of (i) a substituent and (ii) a solvent, and its influence on the rotatory power of an optically-active solute. The influence of the polarity of substituents has already been discussed in the case of *β-naphthylmethylaminomethane* (p. 328); the influence of the polarity of the solvent is discussed in the following paragraphs.*

Fatty and Aromatic Solvents.—Rule observed regular alterations in the rotatory power of an optically-active compound when dissolved in a series of closely related solvents, such as $C_6H_5 . X$ or $CH_3 . X$, in which X is a substituent group of variable polarity. In each series the influence of the solvent tends to be related closely to its dipole moment. Thus in the majority of cases an *increase* in the polarity of the solvent *decreases* the rotatory power, although for some optically-active compounds the reverse occurs. Provided that a clear agreement exists between the polarity of the solvent and the rotatory power of the solution, the rotation is found to be displaced in the same direction by highly polar solvents, whatever the type of hydrocarbon from which they are derived.

[1] H. G. Rule, "Optical Activity and the Polarity of Substituent Groups," I, *J.*, 1924, **125**, 2155; II, 1925, **127**, 2188; III, 1926, 553; IV, 1926, 2116; V, 1926, 3202; VI, 1927, 54; VII, 1928, 178; VIII, 1928, 1347; IX, 1928, 1493; X, 1929, 401; XI, 1929, 2274; XII, 1929, 2516; XIII, 1929, 2524; XIV, 1930, 1887; XV, 1930, 1894; XVI, 1930, 2319; XVII, 1931, 669; XVIII, 1931, 1478; XIX, 1931, 1482; XX, 1931, 1929.

[2] H. G. Rule, "Studies in Solvent Action," I, *J.*, 1931, 674; II, 1931, 2652; III, 1932, 1400; IV, 1932, 1409; V, 1932, 2332; VI, 1933, 376; VII, 1933, 1217; VIII, A. McLean, *J.*, 1934, 351.

* Unless otherwise indicated, references on pp. 350 to 356 refer to papers cited under ref. [2] above.

It is suggested that the influence of the solvent is mainly transmitted in the following ways. (i) According to Debye, *dipole-association* occurs between polar molecules, e.g.

$$\begin{array}{l} C_6H_5\,.\,\overset{\delta+}{C}\equiv\overset{\delta-}{N} \\ \qquad\quad \overset{\delta-}{N}\equiv\overset{\delta+}{C}\,.\,C_6H_5 \end{array}$$

When association of this kind occurs between a dipole in the optically-active solute and another in the solvent, the field of force within the active molecule will be weakened, and the contribution of the dipolar radical to the total optical activity of the molecule will be reduced. The more powerful the dipoles in the solvent, the greater will be the degree of dipole association between solute and solvent and the greater the observed change in optical rotation. With weakly polar solvents, solvation may lead only to a tendency for the solvent dipoles to become loosely oriented towards the solute molecules, giving rise to minor effects of the same kind.

(ii) The *position of an optically-active absorption band* may be displaced by the influence of the solvent. Thus Scheibe[1] has noted a progressive displacement of the $>$ CO absorption band of ketones towards the further ultra-violet region as the polarity of the solvent is increased. This displacement will obviously affect the partial rotation associated with the absorption band.

(iii) The *hydrocarbon radical* of the solvent will also exert a minor influence upon the rotation. Thus among *n*-alkyl derivatives, R . X, the frequency with which the polar group X can come into action falls off as the bulk of R increases. For this reason the extent of dipole association between solute and solvent is greater with CH_3I than with C_4H_9I, and the former behaves as the more polar solvent (I, 678, 679; III, 1406). In the same way, branched chains may screen the polar group, and minimise its influence by steric hindrance. Thus, in the series of *n*-, *sec*- and *tert*-butyl chlorides (I, 679) and alcohols (III, 1406) the *tert*-compounds tend to behave like pure hydrocarbons. Similar effects are observed on comparing an aldehyde with the corresponding methyl ketone (I, 676, 679; II, 2653, 2655).

(iv) The contrast between *aryl* and *aliphatic* derivatives (I, 676) is probably connected with the unsaturated nature of the aromatic nuclei and their greater tendency to acquire an *induced polarisation* under the influence of adjacent or attached polar groups. A minor solvent influence, which becomes the main effect in the case of non-polar optically-active solutes such as *d*-pinane, is related directly to the refractive index of the medium.[2]

The above influences were first demonstrated in solutions of *methyl* l-*menthyl naphthalate*, the molecular rotation of which in over fifty solvents varied from $[M] = -219°$ in CH_3NO_2 to $-788°$ in

[1] SCHEIBE, *Ber.*, 1926, **59**, 2617–2628.
[2] RULE and CHAMBERS, *Nature*, 1934, **133**, 910.

dekalin. They are illustrated in Figs. 147 and 148, which show a progressive increase of molecular rotation [M] with decreasing dipole moment μ in aliphatic and in aromatic solvents respectively. Similar

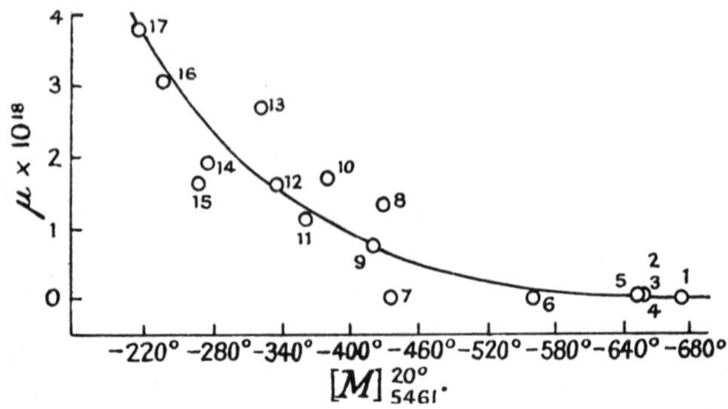

FIG. 147.—MOLECULAR ROTATION OF *l*-MENTHYL METHYL NAPHTHALATE IN ALIPHATIC SOLVENTS.

1. *Cyclo*-hexane.
2. Heptane.
3. Hexane.
4. Pentane.
5. Tetranitromethane.
6. Carbon tetrachloride.
7. Carbon disulphide.
8. Bromoform.
9. Acetic acid.
10. Methyl alcohol.
11. Chloroform.
12. Methyl iodide.
13. Acetaldehyde.
14. Methylene bromide.
15. Methylene chloride.
16. Acetonitrile.
17. Nitromethane.

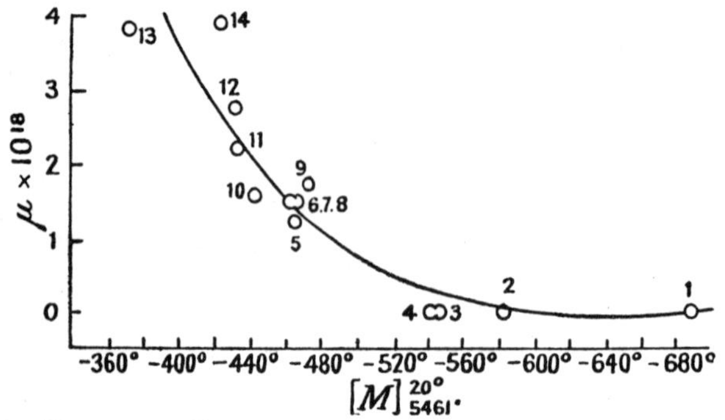

FIG. 148.—MOLECULAR ROTATION OF *l*-MENTHYL METHYL NAPHTHALATE IN AROMATIC SOLVENTS.

1. *Cyclo*-hexane.
2. Mesitylene.
3. Toluene.
4. Benzene.
5. Anisol.
6. Chlorobenzene.
7. Iodobenzene.
8. Bromobenzene.
9. *o*-Chlorotoluene.
10. Aniline.
11. *o*-Dichlorobenzene.
12. Benzaldehyde.
13. Benzonitrile.
14. Nitrobenzene.

effects, however, have since been found in more than thirty other optically-active compounds.

Solvents Containing Two or More Polar Substituents.—The optical effect of these solvents does not necessarily correspond with their dipole moments. Thus, derivatives of methane of the type

CS_2 and CCl_4 ($\mu = 0$) behave as hydrocarbons, as a result of the close contiguity of the compensating dipoles, e.g. $\overset{\delta-}{S}{=\!=}\overset{\delta+\delta+}{C}{=\!=}\overset{\delta-}{S}$. On the other hand, molecules of the type p-$NO_2 . C_6H_4 . NO_2$, which also have no dipole moment, behave as polar media, since the two CX dipoles are so far apart that their electrical influences may be exerted individually upon adjacent molecules, even though the solvents are completely non-polar when examined in a uniform electrical field. They therefore usually produce an effect which is comparable with that of the monosubstituted compound C_6H_5X (I, 680, 681; II, 2656; IV, 1414). Probably for a similar reason, solvents such as $CHCl_3$, CH_2I_2, etc., are commonly subject to minor displacements when [M] is compared with μ.

Influence of Concentration in Non-polar Media.—If the influence of the solvent is due to association between solvent and solute, it should be possible to predict the direction in which the rotatory power will move when we increase the concentration of an optically-active polar solute dissolved in a *non-polar solvent.* At high dilutions, the active solute molecules (unless they have a strong tendency towards self-association), are separated from one another by the non-polar molecules of the solvent. On increasing the concentration, the molecules of the solute are brought into closer proximity, favouring solute-solute association, or at least promoting a definite orientation of the active molecules towards one another. The result is therefore equivalent to an increase in the polarity of the solvent, and the rotation is displaced in the same direction as if the non-polar medium were replaced by a more polar one.

The accuracy of this prediction has been verified in a large number of cases, such as *d-amyl alcohol* (VI, 383), *β-octyl alcohol* (VI, 378), *octyl methyl ether* (VI, 378), *chloro-* and *bromo-octanes* (VI, 379), and *ethyl tartrate* (VII, 1219). In other cases, where changes of concentration have not been examined, a displacement in the above direction is obvious from the fact that the value of [M] for the active compound in the homogeneous state lies between the extreme values found for solutions in non-polar and highly polar solvents respectively.

Influence of Temperature.—A rise in temperature probably influences the rotatory power of an active compound mainly by increasing the rotational mobility of groups joined by single bonds and thus modifying the spatial configuration of the molecule. The rotatory power of a solution must, however, obviously be affected also by changes in the degree of association resulting from increase in molecular vibration with rise of temperature.

If we take the temperature-rotation curve for an active compound at high dilution in a non-polar solvent as a standard for comparison, the corresponding diagram for a highly polar solvent will in general be represented by a curve lying either above or below the standard. At low temperatures, however, the degree of dipole association between solvent and solute will be greater than at high temperatures:

consequently the displacement in rotatory power due to the association will also be greater at low temperatures. It thus follows that the characteristic influence of the polar solvent diminishes as the temperature rises, and in general it may be concluded that the temperature-rotation curves for highly polar and non-polar solvents respectively will converge at higher temperatures, as may be seen in Table 39.

TABLE 39.—INFLUENCE OF TEMPERATURE ON MOLECULAR ROTATION.*

Data showing the convergence of temperature-rotation curves for highly polar and non-polar solvents respectively with rise of temperature.

p = per cent. by weight. c = grams per 100 cc.

1. *Ethyl tartrate.*[1]

Highly polar solvents raise the dextro-rotation.

Temperature.	In Nitrobenzene ($p=2$).	Homogeneous Ester.	In Mesitylene ($p=10$).
Degrees.	Degrees.	Degrees.	Degrees.
20	$[M]_D = +79{\cdot}1$	$+16{\cdot}7$	$+3{\cdot}71$
100	$[M]_D = +58{\cdot}7$	$+28{\cdot}0$	$+22{\cdot}8$
Δ 80	Fall of 20·4	Rise of 11·3	Rise of 19·1

2. *Dimethyl methylenetartrate.*

Highly polar solvents raise the lævo-rotation.

Temperature.	In Nitrobenzene ($c=2$).	In Bromobenzene ($c=3$).	In Mesitylene ($c=4$).
Degrees.	Degrees.	Degrees.	Degrees.
20	$[M]_{5461} = -226$	$-214{\cdot}7$	$-209{\cdot}5$
100	-209	-204	-205
Δ 80	Fall of 17	Fall of 10·7	Fall of 4·5

3. *l-Menthyl o-nitrobenzoate.*

Highly polar solvents depress lævo-rotation.

Temperature.	In Nitrobenzene ($c=4$).	Homogeneous Ester.[2]	In Dekalin ($c=4$).
Degrees.	Degrees.	Degrees.	Degrees.
20	$[M]_{5461} = -439$	-492	-572
100	-429	-450	-517
Δ 80	Fall of 10	Fall of 42	Fall of 55

* Unless otherwise stated the molecular rotations are from RULE's measurements for Hg 5461.

[1] PATTERSON, *J.*, 1908, **93**, 1849.

[2] PICKARD and KENYON, *J.*, 1915, **107**, 48.

TABLE 39.—*Continued.*

4. *l-Menthyl hydrogen phthalate.*

Highly polar solvents depress lævo-rotation.

Temperature.	In Nitrobenzene (c=4).	In Dekalin (c=4).
Degrees.	Degrees.	Degrees.
20	$[M]_{5461}$ = − 296	− 404
100	− 290	− 390
Δ 80	Fall of 6	Fall of 14

5. *l-Menthyl* 2 : 4-*dinitrobenzoate.*

Polar solvents depress lævo-rotation (VIII, 355, 356).

Temperature.	In Nitrobenzene (c=3).	In Dekalin (c=2).
Degrees.	Degrees.	Degrees.
20	$[M]_{5461}$ = − 398	− 676
80	− 399·5	− 623
Δ 60	Rise of 1·5	Fall of 53

6. *l-Benzoin.*

Highly polar solvents depress lævo-rotation.

Temperature.	In Nitrobenzene (c=1).	In Toluene (c=1).
Degrees.	Degrees.	Degrees.
20	$[M]_{5461}$ = − 460	− 567
100	− 387	− 466
Δ 80	Fall of 73	Fall of 101

7. *l-Benzoin methyl ether.*

Highly polar solvents give lævo-rotation and non-polar solvents give dextro-rotation.

Temperature.	In Nitrobenzene (c=4).	In Bromobenzene (c=4).	In Mesitylene (c=4).
	Degrees.	Degrees.	Degrees.
20	$[M]_{5461}$ = − 117	+ 35	+ 268
100	− 40	+ 74	+ 187
Δ 80	Fall in negative rotation of 77°	Rise in positive rotation of 39°	Fall in positive rotation of 81°

(These three curves apparently tend to a common meeting-point in the neighbourhood of + 80° to + 90° at high temperatures.)

This effect of temperature was discovered in a series of compounds studied by Rule, but it is also valid in the case of ethyl tartrate (VII, 1222), for which very full data have been given by Patterson.[1] Such a relationship can, however, only be expected to hold when a definite relationship can also be traced between the values of [M] and μ.

Abnormal Effects of Solvents.—When the solvent can form a chemical compound with the optically-active solute, it is not to be expected that its influence will be in agreement with its polar properties (VI, 385–386). Examples of this kind have been found in *octyl iodoacetate* dissolved in *pyridine*,[2] and in *l-menthyl hydrogen naphthalate* dissolved in basic solvents (IV, 1412).

Apparent irregularities will also occur if the optically-active compound has a strong tendency to exist in the associated state. When a solvent of zero or low polarity fails, even at high dilutions, to disrupt the association complexes of such a solute, it is clear that the observed rotatory power is that of the more or less associated compound. The solvent under consideration will therefore appear to behave as a more highly polar medium than is actually the case. This factor has been examined in the case of *ethyl tartrate* (VII, 1221), and *β-nitrooctane* (VI, 381), and the explanation has been supported by determinations of molecular weight. It is probable that the different effects produced by various non-polar solvents, in cases where a close relationship between [M] and μ is otherwise observed, may also be largely due to this cause.

[1] Patterson, *J.*, 1904, **85**, 765 *et seq.*
[2] Rule and Mitchell, *J.*, 1926, 3202.

CHAPTER XXIX.

MOLECULAR THEORIES OF OPTICAL ROTATORY POWER.

Influence of Mass.—(*a*) *Crum Brown and Guye.* The first attempts to predict the optical rotatory power of a liquid from its molecular structure were made in 1890 by Crum Brown [1] and by Guye.[2]

Since the rotatory power of a molecule containing an asymmetric carbon atom depends on the dissimilarity of the four attached radicals, and vanishes when two of them become equal, it was evident that the observed rotation must depend on a function involving the *differences* between the radicals, and vanishing when any one of these differences becomes zero. For this purpose they proposed to use a PRODUCT OF ASYMMETRY of the form

$$(a-b)(b-c)(c-d)(a-c)(a-d)(b-d).$$

The quantities a, b, c, d, in this function were identified with the *masses* of the radicals.

(*b*) *Walden.* This identification was proved to be incorrect by Walden,[3] who showed that substantial optical rotations were given by molecules in which two of the radicals have equal *masses*, but dissimilar *structures*. The compounds which he studied were derivatives of *malic acid* of the type

$$\begin{array}{l} \mathrm{X.CO.O.\mathbf{C}H.CO.OY} \\ \qquad\qquad\quad | \\ \qquad\qquad\quad \mathrm{CH_2.CO.OZ} \end{array}$$

The masses of two of the radicals attached to the asymmetric carbon atom could be made equal (i) by making $X = Y$, (ii) by making $X = Z + CH_2$. In every case a large rotatory power was developed by making use of the isomeric pairs of radicals :

(i) $\mathrm{O.CO}$— and —$\mathrm{CO.O}$—
or (ii) —$\mathrm{O.CO.CH_2}$— and —$\mathrm{CH_2.CO.O}$—

The results are shown in Table 40.

It is also noteworthy that in this series of homologues the specific rotation is (with one exception) almost stationary between $-22°$

[1] CRUM BROWN, *Proc. Roy. Soc. Edin.*, 1890, **17**, 181.
[2] GUYE, *C.R.*, 1890, **110**, 714.
[3] WALDEN, *Z. ph. C.*, 1895, **17**, 245–246.

TABLE 40.—ROTATORY POWER OF DERIVATIVES OF MALIC ACID (Walden).

(i) *Case I.* $X = Y$.					(ii) *Case II.* $X = Z + CH_2$.				
X.	*Y.*	*Z.*	$[\alpha]_D$.	$[M]_D$.	*X.*	*Y.*	*Z.*	$[\alpha]_D$.	$[M]_D$.
			Degrees.	Degrees.				Degrees.	Degrees.
CH_3	CH_3	CH_3	− 22·92	− 46·76	C_2H_5	CH_3	CH_3	− 22·94	− 50·01
C_2H_5	C_2H_5	C_2H_5	− 22·20	− 54·62	n-C_3H_7	C_2H_5	C_2H_5	− 22·22	− 57·78
n-C_3H_7	n-C_3H_7	n-C_3H_7	− 22·40	− 64·50	i-C_3H_7	C_2H_5	C_2H_5	− 21·99	− 57·17
i-C_4H_9	i-C_4H_9	i-C_4H_9	− 19·91	− 65·70	i-C_4H_9	n-C_3H_7	n-C_3H_7	− 21·68	− 65·47

and − 23°, whilst the molecular rotation ranges from − 47° to − 66°. This result is in accordance with the experience of Pickard and others that *specific rotations* often show regularities which would be expected to occur only when *molecular rotations* were calculated.

(*c*) *Pope.* In confirmation of Walden's conclusions, the optical activity of methyl*cyclo*hexylideneacetic acid (p. 66) may also be cited, since in this compound a mere difference of *configuration* in two of the four unlike radicals attached to one of the carbon atoms suffices to produce optical activity.

Gray's Molecular Theory of Optical Rotatory Power.—An attempt to unite the chemical and physical interpretations of optical activity was made by Gray [1] in 1916.

> "The natural activity of isotropic substances has been an interesting subject both to the chemist and to the physicist. Each has attacked the problem according to his own particular method; but unfortunately, the two lines of attack have not met. On the chemical side Pasteur, le Bel and van't Hoff developed the present conception of molecular structure and proposed the theory connecting optical activity with an asymmetric structure of the molecule, a theory confirmed by the enormous amount of experimental work carried out in recent years. In the domain of physics the subject has been treated more from the optical view-point, and Drude and Voigt have shown just what form of electro-magnetic equations must hold in an active medium in order to produce rotation. However, the two theories have not been tied together; it has not been shown why the asymmetric molecule should give rise to these particular equations."

In an attempt to bring about this alliance, Gray considered the effect on a wave of plane polarised light of a hypothetical molecule consisting of a central atom surrounded by four dissimilar atoms at the apices of a tetrahedron. Each atom is supposed to be electrically polarised by the application of an external electric field. This in-

[1] GRAY, *Phys. Rev.*, 1916, **7**, 472.

duces electric moments in each of the atoms, and these moments interact with one another in a complicated way. The problem reduces to the calculation of the net polarisation in an incident harmonic plane polarised field, in which the molecules have a random orientation. The calculations were too complicated to lead to an explicit formula for rotation in terms of the geometrical relations of the molecule, but the following general results were deduced:

(i) Optical activity results if the four outer atoms are all different.

(ii) The rotation vanishes if the total number of atoms is less than four.

(iii) If two of the atoms are interchanged, giving a mirror image, the only result is a change in the *sign* of the rotation.

(iv) Any absorption band in the molecule may be optically active, and thus give rise to anomalies in the curve of rotatory dispersion, whereas on Drude's theory only electrons vibrating along helical paths can give rise to the Cotton effect.

Gray did not consider that the observations of Cotton [1] and of Tschugaeff and Ogorodnikoff,[2] on anomalous rotation in the region of an absorption band, provided sufficient proof of his theory as distinct from that of Drude, since the absorption bands in question were known to be strongly influenced by the rest of the molecule. In order to test his theory more fully, he therefore measured the rotatory dispersion of a *tartrate* of *neodymium*, since this element has a group of absorption bands between 5600 and 5900 A.U., which are known to be very little influenced by the remainder of the molecule. The curve of rotatory dispersion showed small anomalies in the region of bands at 5837, 5819 and 5800 A.U., a large anomaly near a band at 5770 A.U., and a tiny anomaly near a very weak band at 5642 A.U., whereas another weak band at 5733 A.U. produced no anomaly. These results supported his theory, but this was not developed further, since his experiments were equally in harmony with the more general theories of Stark, Born and Oseen, which were published whilst his own experimental work was still in progress.

de Mallemann's Molecular Theory.—Since the equal *masses* of two radicals do not cancel out in the product of asymmetry, some alternative basis must be found for predicting the magnitude of the molecular rotatory power in terms of the dissymmetry of the molecule. For this purpose de Mallemann [3] made use of the POLARISABILITY of the radicals as measured by their *refractivities*. Thus each of the four radicals attached to an asymmetric carbon atom could be represented by an *ellipsoid* with three PRINCIPAL AXES OF POLARISATION at right angles to one another. The electrons of these radicals were assumed to be held in their equilibrium positions by quasi-elastic forces, such as those by which the four radicals are attached to the central asymmetric atom; but no coupling forces were supposed to

[1] COTTON, *A.C.P.*, 1896, **8**, 347–433.
[2] TSCHUGAEFF and OGORODNIKOFF, *Z. ph. C.*, 1910, **74**, 503; 1913, **85**, 481.
[3] DE MALLEMANN, *Rev. Gen. Sci.*, 1927, **38**, 453.

exist between the electrons of the different radicals, which were assumed to be completely independent of one another. By taking into account the changes of phase and of amplitude of light waves passing through the molecule, it was possible to calculate the circular double refraction and hence the rotatory power of the molecule, in terms of the refractivities of the radicals and the linear dimensions of the molecule. Thus, four anisotropic radicals at the corners of an irregular tetrahedron were shown to give rise to optical rotatory power when each was coupled independently to a central carbon atom; but, since the 3×4 partial polarisabilities of the anisotropic radicals are still entirely unknown, it was not possible to calculate the rotatory power of such a molecule in terms of quantities which could be taken from existing data or easily be determined by *ad hoc* measurements.

In order to reduce his MOLECULAR THEORY [1] to a more practicable form, de Mallemann introduced two simplifying assumptions,[2] as follows:

(i) That the four dissimilar atoms which are attached to the asymmetric carbon atom in the molecule CHClBrI can be treated as *isotropic spheres*, of which the MEAN POLARISABILITY can be deduced from the known ATOMIC REFRACTIONS of the atoms.

(ii) That the three halogen atoms can be regarded as lying on the rectilinear axes of x, y and z, at distances from the origin corresponding with their known ATOMIC RADII.

He also appears at this stage to have abandoned the idea of four radicals coupled independently to a central carbon atom in favour of a direct mutual coupling of the four radicals of the same type as that postulated by Born in 1915 (p. 374), by Gray in 1916 and by Boys in 1934 (p. 361). On this basis he found that the specific rotatory power of his model was given by the formula

$$[\alpha] = \pm \frac{4\pi^3}{27\lambda^2}\frac{N}{M}(n^2+2)^2 A_1A_2A_3A_4 f(a, b, c, p, q, r), \quad \text{(i)}$$

where M is the molecular weight of the molecule,

λ is the wave-length of the light,

N is the number of molecules in one gram molecule,

n is the observed refractive index of the medium.

$A_1A_2A_3A_4$ are the *reduced refractivities* of the four atoms as deduced from the molecular refractivities of *gases*, by means of the formula $A = (n^2 - 1)\frac{M}{d}$, instead of from the formula $R = \frac{n^2-1}{n^2+2}\frac{M}{d}$ of Lorenz and Lorentz, which is now generally used for *liquids* whose refractive indices differ substantially from unity.

$f(a, b, c, p, q, r)$ is a complicated function of the linear dimensions of the molecule, of which a, b, c are co-ordinates, showing the distances

[1] DE MALLEMANN, *T.F.S.*, 1930, **26**, 281–292.
[2] *Ibid.*, *C.R.*, 1925, **181**, 298.

of the centres of the halogen atoms from the origin of the rectilinear axes, and p, q, r are the lengths of the inclined edges of the tetrahedra.

By means of this formula de Mallemann calculated the specific rotatory power of the compound

$$\mathbf{C}HClBrI\ [\alpha]_D = \pm 3{\cdot}2^\circ.$$

This compound has not yet been prepared, and indeed no compound containing only five atoms has yet been resolved into optically-active forms. The calculation cannot, therefore, be checked against any experimental data; but the value deduced is similar to that observed in the simplest asymmetric molecules, and is therefore obviously of the right order of magnitude. We may therefore conclude that Pasteur's model of an irregular tetrahedron, with four dissimilar radicals at the apices, provides a sound theoretical basis for the origin of optical rotatory power, and that by the use of this model it is possible to predict not only the existence, but also the approximate magnitude of the optical rotatory power of the simplest dissymmetric molecules, in terms of the refractivities of the radicals and the linear dimensions of the tetrahedron.

Boys' Theory of Optical Rotatory Power.—The molecular theory of optical rotatory power put forward by S. F. Boys in 1934 [1] differs from de Mallemann's original theory, and resembles his later point of view, in that the development of optical rotatory power in an asymmetric molecule of the type **C**RR′R″R‴, is attributed to a mutual interaction between the radicals RR′R″R‴, which was at first ignored by de Mallemann, whilst on the other hand the coupling of these radicals to the central atom of carbon, which de Mallemann postulated in his first theory, does not now enter into the calculations. Boys' theory is based on the initial hypothesis, already used by Gray and by Born (p. 374), that every vibrator in the molecule has a natural position of rest and a characteristic polarisability, which must be taken into account in considering its interaction with a light wave passing through the medium, but that it also responds to every other vibration in the molecule. This idea is illustrated by a simple mechanical model (Fig. 148A) which is described as follows:

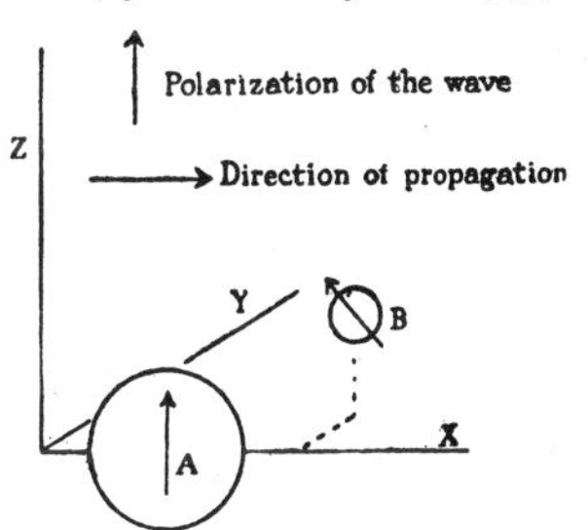

FIG. 148A.—S. F. BOYS' MOLECULAR MODEL.

> " It is possible to construct a medium which shows a rotatory power for an ordinary transverse elastic wave. This is formed when we have an elastic solid with heavy masses embedded in it; as an example we may consider a gelatine gel with ball bearings as heavy particles. These balls are arranged in sets

[1] S. F. Boys, *P.R.S.*, 1934, A. **144**, 655–691.

of four; all these sets are exactly alike, but the balls in each separate set may be different and may be at the corners of an irregular tetrahedron. These tetrahedral sets are distributed entirely at random, so that the gel as a whole is isotropic. There are two kinds of model, related to each other as an object to its mirror image; and hence a gel containing only one 'enantiomorph' corresponds to an optically active liquid. Each ball vibrates under the influence of a passing wave and has an effect on neighbouring balls, e.g. if a certain ball A vibrates up and down, then a smaller ball B a little farther along the path of the wave, a little to one side, and a little above A, is dragged to and fro in a horizontal direction by the motion of A. The secondary wavelets caused by all these vibrators have to be added to the original wave, and, if the total effect of these is not exactly in the same direction as the original wave, the plane of polarisation is altered. For a medium containing a symmetrical model the resultant of these secondary wavelets would be in the same direction as the inducing wave, but with an unsymmetrical model this condition does not hold and the medium is 'mechanically active.'

"This model bears the same relation to the natural optically-active molecule as the Sellmeier vibrators in the early theory of optical dispersion do to the modern electron structure. Replace the ball bearings by atoms containing electrons free to move under the influence of an electric field; replace the elasticity of the gel by the equations of electricity; and the model becomes an asymmetric molecule under the influence of a light wave. Under the action of the electric field of a light wave each atom becomes an oscillating electric doublet. Inside the molecule the electric field of the wave is altered by the fields of the doublets themselves, and, making allowance for this, the exact polarisations of the molecule can be calculated. When the polarisations in a medium can be expressed in terms of the electrical fields, it is possible to find the velocity of a light wave in this medium by use of Maxwell's equations. Fresnel has shown that an active medium can be regarded as a medium in which the velocities of circularly polarised waves of opposite sense are different, and the rotation can be calculated from this difference. Hence, having found the expression for the total polarisation, we can derive the formula for the rotation" (*ib.*, pp. 656–658).

According to Boys' calculations the specific rotation of a medium containing molecules composed of four isotropic spheres at the corners of an irregular tetrahedron can be expressed by means of the formula

$$[\alpha] = \frac{72900}{32\pi^2\lambda^2 MN^3}(n^2+2)(n^2+5)R_A R_B R_C R_D I \quad . \quad . \quad \text{(ii)}$$

where N is the Avogadro number,

M is the molecular weight of the molecule,

n is the observed refractive index of the medium,

$R_A R_B R_C R_D$ are the refractivities of the radicals, deduced by means of the formula of Lorenz and Lorentz,

$$R = \frac{n^2 - 1}{n^2 + 2} \frac{M}{d}.$$

I is a complicated function of the radii of the spheres.

The specific rotation $[\alpha]$ in formula (ii) above represents the sum of all the secondary effects resulting from the primary polarisation due to the incident light. These may be pictured as being relayed, for instance, from particle A to particle B, and thence to C and to D, or back to A, as represented symbolically by $[AB]$, $[ABC]$, $[ABCD]$, $[ABA]$, etc. Ultimately it is shown that these can all be expressed in terms of the binary interactions $[AB]$, $[BC]$, $[CD]$, etc., and that these in turn are dependent merely on the distances between the centres of the particles.

Application of Boys' Formula.—The formula of Boys differs from that of de Mallemann (i) in the form of the simple model on which it is based, (ii) in the use of refractivities derived from the formula of Lorenz and Lorentz, instead of the less exact formula of Newton, (iii) in that magnetic as well as electrostatic interactions are postulated, as a result of which the factor $(n^2 + 2)^2$ is replaced by $(n^2 + 2)(n^2 + 5)$, and the rotations are approximately doubled when n is nearly equal to unity.

In order to compare the rotations calculated from his formula (ii) with those observed experimentally, Boys made two additional simplifying assumptions:

(i) That *radicals* such as OH, NH_2, CH_3, C_2H_5, CH_2OH (as well as *atoms* such as the halogens) can be treated as *isotropic spheres* of known refractive index and radius, in close-packed contact with one another.

(ii) That the volume and refractivity of the central asymmetric carbon atom may be distributed equally amongst the four surrounding radicals.

The former assumption may be justified, in the case of the simpler radicals, OH, NH_2, CH_3, by Langmuir's theory of ISOSTERISM,[1] since these three radicals contain the same number of electrons and may be credited with the same electronic configuration as an atom of fluorine; but it is obviously invalid in the case of more complex radicals such as $—CH_2 . CH_3$ and $—CH_2 . OH$, which cannot be either spherical or isotropic.

The latter assumption, whereby the four pairs of shared valency electrons (which form the outer octet of the asymmetric carbon atom) are assigned to the four attached radicals, is a reversion to the scheme

[1] LANGMUIR, *J.A.C.S.*, 1919, **41**, 1543–1559.

proposed by Kossel when he formulated carbon tetrachloride as an ionic aggregate of the type

$$\begin{matrix} \overset{-}{\mathrm{Cl}} & & \overset{-}{\mathrm{Cl}} \\ & \overset{++++}{\mathrm{C}} & \\ \overset{-}{\mathrm{Cl}} & & \overset{-}{\mathrm{Cl}} \end{matrix}$$

It can be justified by the general rule [1] that the properties of covalent compounds can often be deduced from those of a corresponding ionic aggregate ; and it finds a close precedent in Hylleraas' calculation (p. 346) of the rotatory power of quartz as an aggregate of silicon and oxygen ions Si^{+4} and O^{-2}, although the lattice is certainly held together by bonds.

With the help of these simplifying assumptions Boys deduced the following values for the specific rotations of four of the simplest asymmetric molecules.

TABLE 41.—SPECIFIC ROTATIONS OF SIMPLE MOLECULES.

		$[\alpha]_D$ (calc.).	$[\alpha]_D$ (obs.).
act-Amylamine . .	$\begin{matrix} C_2H_5 & & H \\ & >C< & \\ CH_3 & & CH_2 . NH_2 \end{matrix}$	Degrees. 3·6	Degrees. 5·86
act-Amyl alcohol . .	$\begin{matrix} C_2H_5 & & H \\ & >C< & \\ CH_3 & & CH_2 . OH \end{matrix}$	4·0	5·90
sec-Butylamine . .	$\begin{matrix} C_2H_5 & & H \\ & >C< & \\ CH_3 & & NH_2 \end{matrix}$	7·4	7·44
sec-Butyl alcohol . .	$\begin{matrix} C_2H_5 & & H \\ & >C< & \\ CH_3 & & OH \end{matrix}$	9·3	13·9

In spite of the impossible character of the assumptions made in reference to the more complex radicals, the agreement is close enough to show that the model, and the method of coupling of the radicals which was postulated in deducing the formula, are sufficient to account, not merely for the existence of optical rotatory power in the system, but also for its approximate magnitude.

Boys points out that the factor I in his formula contains the term

$$(a-b)(a-c)(a-d)(b-c)(b-d)(c-d)/(a+b+c+d)^{14}$$

and is very sensitive to small variations in the radii, a, b, c, d, of the four radicals. The variations in rotatory power resulting from

[1] BERNAL, *Chem. and Ind.*, 1934, **53**, 249.

progressive substitution in a series of compounds are therefore due mainly to alterations in the *volumes* of the radicals. Thus, in the amyl series, the only radicals which have smaller volumes than $—CH_3$ are $—NH_2$ and —OH. The replacement of hydroxyl in the radical $—CH_2 . OH$ by another heavier substituent, as in *amyl chloride*, therefore makes the $—CH_2X$ group larger than the adjacent $—CH_2 . CH_3$ group instead of smaller. The rotations of all the amyl derivatives, except *amylamine*, are therefore of opposite sign to that of *amyl alcohol*, e.g. *lævorotatory* amyl alcohol gives *dextrorotatory* amyl chloride.

$$\begin{matrix} CH_3 . CH_2 \searrow \\ CH_3 \nearrow \end{matrix} \mathbf{C}H . CH_2 . OH \longrightarrow \begin{matrix} CH_3 . CH_2 \searrow \\ CH_3 \nearrow \end{matrix} \mathbf{C}H . CH_2Cl$$

Amyl alcohol (lævorotatory) — Amyl chloride (dextrorotatory)

Rotatory Dispersion of Simple Molecules.—The molecular theory of optical rotatory power, as expressed in the formulæ of de Mallemann and of Boys, can be tested in a very simple way by using these formulæ to predict the rotatory dispersions of some of the simple optically-active compounds for which data are already available.[1] This test has the advantage of eliminating all hypotheses as to the size and shape of the molecules, since these remain fixed when only the wave-length of the incident light is changed. In particular, the errors inherent in the assumption that the four radicals attached to an asymmetric carbon atom may be treated as spheres are avoided, since it is only necessary to calculate the values of the product

$$(n^2 + 2)^2 A_1 A_2 A_3 A_4 / \lambda^2 \text{ (de Mallemann)},$$
$$\text{or } (n^2 + 2)(n^2 + 5) R_A R_B R_C R_D / \lambda^2 \text{ (Boys)}$$

for a series of wave-lengths, from the refractive indices of the medium and the refractivities of the radicals, and then to compare the *ratios* of these products with the ratios of the observed rotations for the same wave-lengths.

For the purpose of this test, it is convenient to use the DISPERSION RATIOS, $\alpha_{4358}/\alpha_{5461}$, for the violet and green mercury lines and to select for comparison the data for the two simple alcohols, for which the absolute rotations were calculated by Boys, namely:

		$\dfrac{\alpha_{4358}}{\alpha_{5461}}$
sec-Butyl alcohol	$\begin{matrix} C_2H_5 \searrow \\ CH_3 \nearrow \end{matrix} \mathbf{C}H . OH$	1·661
sec-Amyl alcohol	$\begin{matrix} C_2H_5 \searrow \\ CH_3 \nearrow \end{matrix} \mathbf{C}H . CH_2 . OH$	1·700

[1] LOWRY and DICKSON, *J.*, 1913, **103**, 1067.

The dispersion ratio of *sec*-butyl alcohol differs but little from the average value for several series of secondary alcohols which were examined in 1915 by Lowry, Pickard and Kenyon,[1] e.g. 1·652 for the other methyl carbinols, 1·660 for the ethyl carbinols, and 1·663 for the *iso*-propyl carbinols, the lowest member of each homologous series being excluded. *act*-Amyl alcohol, on the other hand, gives a disperson ratio which is abnormally high, but agrees closely with those of

		$\frac{\alpha_{4358}}{\alpha_{5461}}$
Methyl *iso*-propyl carbinol	$(CH_3)_2 > CH \cdot \mathbf{C}HOH \cdot CH_3$	1·697
Methyl *ter*-butyl carbinol	$(CH_3)_3 \geq C \cdot \mathbf{C}HOH \cdot CH_3$	1·707
iso-Valeric acid	$C_2H_5, CH_3 > \mathbf{C}H \cdot CO \cdot OH$	1·710

The atomic refractivities of carbon, hydrogen and oxygen for the two mercury lines are not included in the usual tables, but the following values have been deduced by Mr. H. F. Willis from the refractivities tabulated by Eisenlohr [2] for the sodium and hydrogen lines, using in each case the formula of Lorenz and Lorentz, in which the refractivity is expressed by the function

$$R_L = \frac{n^2 - 1}{n^2 + 2} \frac{M}{d}.$$

TABLE 42.—REFRACTIVITIES OF ATOMS AND RADICALS.

	[H].	[O].	[C].	[OH].	[CH_3].	[C_2H_5].	[$CH_2 . OH$].
Hg 4358 .	1·122	1·540	2·465	2·662	5·831	10·540	7·371
Hg 5461 .	1·106	1·527	2·423	2·633	5·741	10·376	7·268
Ratio .	1·015	1·009	1·018	1·011	1·016	1·016	1·014

The principal factor in determining the dispersion ratio is the term $1/\lambda^2$, which for the two wave-lengths in question gives rise to a ratio 1·570. In the case of the carbinols, this minimum ratio, which corresponds with Biot's Law of Inverse Squares, has to be increased by 4 to 8 per cent. with the help of four refractivities, which provide multipliers ranging from 1·011 for the hydroxyl radical to 1·016 for

[1] LOWRY, PICKARD and KENYON, *J.*, 1914, **105**, 94.

[2] EISENLOHR, *Spektrochemie Organischer Verbindungen*, Stuttgart, 1912, p. 48.

the alkyl radicals. As de Mallemann has pointed out,[1] this is sufficient to bring the calculated ratio within the range of the values observed in compounds of this class. The results of this comparison for the product $R_A R_B R_C R_D/\lambda^2$ are set out below:

	Dispersion (observed).	Ratio (calculated).
sec-Butyl alcohol .	1·661	1·661 or 1·664
act-Amyl alcohol .	1·700	1·666 or 1·668

In deducing the first pair of values for the calculated ratio, the central atom of carbon was ignored; in deducing the second pair of values, the shared electrons of the carbon atom were assigned to the octets of the four surrounding radicals, after the manner of Kossel, as used by Boys.

The lower value, 1·661, for *sec*-butyl alcohol is correct to the third decimal, whilst the higher value, 1·664, is perhaps within the limits of experimental error; but the further increase to 1·677, produced by the factor $(n^2 + 2)(n^2 + 5)$, would make the product definitely incorrect. On the other hand, the dispersion ratio of *act*-amyl alcohol, 1·700, is too high to be expressed by this formula, since the maximum value given by the product $R_A R_B R_C R_D/\lambda^2$ for any saturated compound of carbon, hydrogen and oxygen is $1{\cdot}57 \times (1{\cdot}016)^4 = 1{\cdot}673$, and the factor $(n^2 + 2)(n^2 + 5)$ only increases this to 1·686. Whilst, therefore, the formula deduced by Boys can be used to account for the rotatory dispersion of the simpler secondary alcohols, the experimental values for the primary alcohols, and for the secondary alcohols with branched chains, are outside the range of the formula.

In this connection it is of interest to note that the rotatory dispersion of a simple molecule, which can be expressed by *one* term of Drude's equation, with a *single* characteristic frequency at about 1500 A.U., may be expressed by the molecular theory as a product of *four* refractivities with frequencies at about 1200 A.U. The contrast between these two wave-lengths may perhaps serve as a means of checking the relative merits of the molecular and electronic theories, when the requisite frequencies in the Schumann region become known.

Application of the Molecular Theory of Rotatory Power to Rotatory Dispersion in a Region of Absorption.—The molecular theory of optical rotatory power can be used to predict the *sign* (p. 369) and the *order of magnitude* of the specific rotation of some of the simplest dissymmetric molecules, namely, molecules containing only one asymmetric carbon atom and no unsaturated or chromophoric radical; it can also be used to account (to a first approximation) for the rotatory dispersion of those simple molecules of which the

[1] DE MALLEMANN, *T.F.S.*, 1930, **26**, 292.

characteristic frequencies are most remote from the visible spectrum. It is therefore of interest to inquire to what extent and under what conditions the molecular theory may be applied to compounds which exhibit selective absorption within the range of experimental observations.

For this purpose, only that part of the expression for specific rotation which varies with wave-length is in question, namely,

$$(n^2 + 2)^2 A_1 A_2 A_3 A_4/\lambda^2 \text{ (de Mallemann)},$$
$$\text{or } (n^2 + 2)(n^2 + 5) R_A R_B R_C R_D/\lambda^2 \text{ (Boys)}.$$

Consider a molecule containing, round a central carbon atom, three hydrocarbon radicals of refractivities R_A, R_B, and R_C, whose absorption bands lie in the Schumann or Lyman regions, and one chromophoric radical, refractivity R_D, with an absorption band in the ordinary ultra-violet at wave-length λ_0. The variations of R_A, R_B and R_C at wave-lengths greater than λ_0 will be small compared with the variation of R_D. These three refractivities may therefore be taken as constant to a first approximation, whereas the variation with wave-length of R_D in the region of transparency may be expressed by a term of the type $R_D = k\lambda^2/(\lambda^2 - \lambda_0^2)$. Since $(n^2 + 2)(n^2 + 5)$ and $R_A R_B R_C$ are approximately constant, the rotatory dispersion of the molecule will depend on R_D/λ^2 and can therefore be expressed by a term of the type $\alpha = k/(\lambda^2 - \lambda_0^2)$. This result agrees exactly with experimental observations in compounds such as camphorquinone (p. 134), where the rotatory dispersion in the region of transparency is dominated by a contiguous absorption band.

In the region of absorption, the rotatory dispersion is dominated even more completely by the optically-active absorption band. The anomaly in the curve of rotatory dispersion as the absorption band is penetrated is therefore often extremely large, and in a large number of cases the rotation passes through a zero value near the middle of the band, with maxima of opposite sign on either side (Figs. 48, p. 154; 131–133, pp. 308–309; 141, p. 336; 182–184, pp. 446–448). A zero rotation, however, can only be deduced from the preceding formulæ if one of the refractivities also passes through a zero value. This phenomenon has been realised by R. W. Wood [1] in sodium vapour, where refractive indices less than unity were recorded on the short wave-length side of an absorption line, but the absorption bands of organic compounds are much too diffuse to produce any analogous effect. Thus Allsopp [2] has shown that the ultra-violet absorption band of the >C=O radical, which gives rise to such remarkable anomalies in the rotatory dispersion of camphor, produces only a tiny ripple on the curve showing the refractive indices of *cyclo*hexanone:

[1] R. W. Wood, *Phil. Mag.*, 1904, [vi], **8**, 324.
[2] Allsopp, *P.R.S.*, 1934, A. **146**, p. 307.

$$\begin{array}{c} O \\ \| \\ H_2C\diagup C \diagdown CH_2 \\ | \qquad\quad | \\ H_2C\diagdown C \diagup CH_2 \\ H_2 \end{array}$$

Moreover, even when the refractivities of the five methylene radicals were eliminated, the refractivity of the carbonyl radical was found to vary only by about 1 per cent. ($\pm$ 0·05 on a value of + 5·5), and therefore made no approach to a zero value. [1]

The impossibility of realising a zero value for the refractivity of a chromophoric radical, such as the >C=O group, is easily understood, since *all* the outer electrons of the radical contribute to its refraction. Thus in the case of the carbonyl group, the major contribution appears to be made by electrons with natural frequencies in the Schumann region at about 1200 A.U., including probably the two lone pairs of the oxygen atom and the shared electrons of the single bonds which link the carbon atom to the two radicals with which it is associated in the carbonyl compound. The shared electrons of the double bond, which are responsible for the ketonic absorption band, therefore contribute only a part of the refractivity of the radical, and do not produce an anomaly of sufficient magnitude to carry the curve across the axis of zero refractivity. If therefore the very attractive conception of optical rotatory power as depending on a product of four refractivities is to be extended to the region of absorption, we must make use of smaller units than the complete atoms or radicals used in the formulæ of de Mallemann and of Boys, and segregate for this purpose the actual electrons which produce the optically-active absorption band. This condition obviously implies the presence of a new centre of asymmetry *within the chromophoric radical* and is therefore in accord with the view put forward by Lowry and Walker in 1924 that the carbonyl radical in a ketone may acquire an *induced dissymmetry* when coupled sufficiently closely to the asymmetric atoms (pp. 146 and 410).

The Absolute Configuration of Optically-Active Molecules.— Ever since Pasteur showed that the molecules of the two forms of an optically-active compound are related to each other in the same way as an object and its mirror image, the question has been asked "Which of the two forms gives rise to a dextro-rotation and which to a lævo-rotation?"

Boys claims that his theory gives an unambiguous answer to the question so far as the simple tetrahedral molecule is concerned. The specific rotation of such a molecule can be expressed by a function which contains the product $(a - b)(a - c)(a - d)(b - c)(b - d)(c - d)$

[1] LOWRY and ALLSOPP, *P.R.S.*, 1934, A. **146**, 317.

of the differences of the radii, a, b, c, d of the four radicals A, B, C, D, which occupy the apices of the tetrahedron, but *this result is only true for that configuration in which, when A is placed nearest the observer, the groups B, C, D appear in clockwise order* (Fig. 149). When

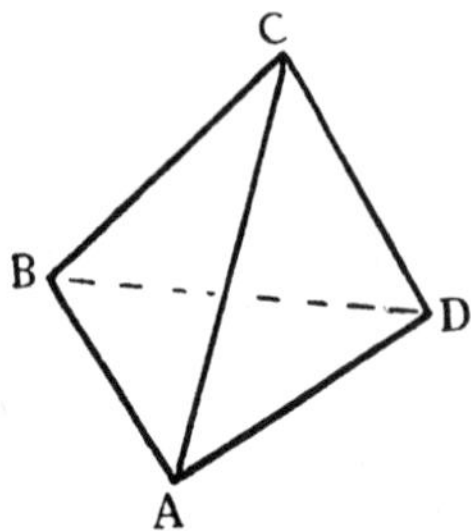

Fig. 149.—Boys' Tetrahedral Model.

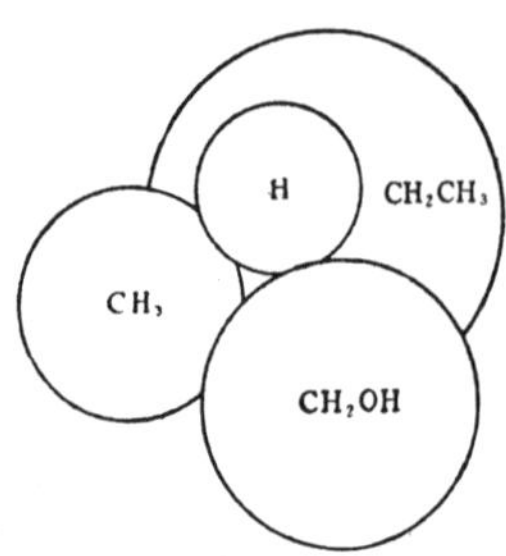

Fig. 150.—Absolute Configuration of *l*-Amyl Alcohol (Boys).

$a > b > c > d$ this product is necessarily positive, and the molecule must be dextrorotatory. Boys therefore concludes that

> " According to this analysis, we can say with absolute certainty that a dextro-compound has the configuration such that, when the largest group is nearest to the hypothetical observer, the other groups in order of diminishing size appear in a clockwise rotation " (*loc. cit.*, p. 682).

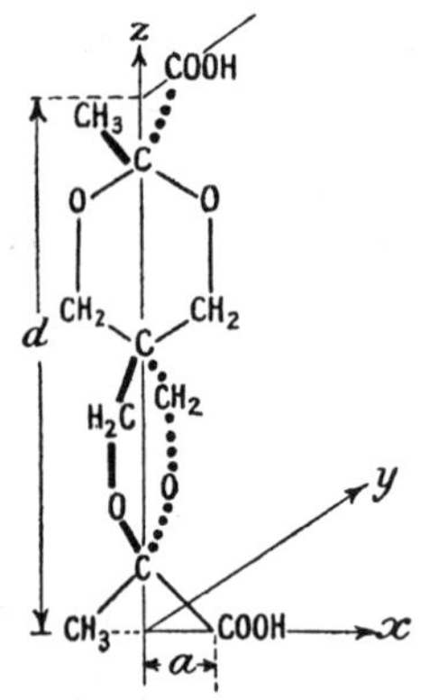

Fig. 151.—Absolute Configuration of *d*-Dipyruvic Erythritol (Kuhn and Bein).

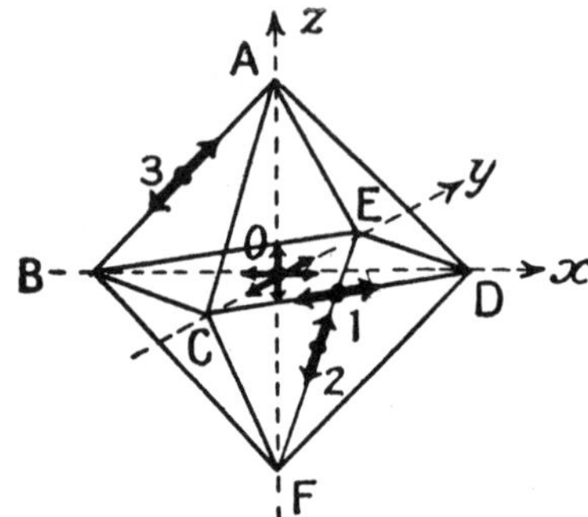

Fig. 152*a*.—Absolute Configuration of *d*-Cobaltioxalates (Kuhn and Bein, 1934).

As an illustration, the absolute configuration of lævorotatory amyl alcohol is represented by Fig. 150, in which " when the CH_2OH group is placed nearest to the observer the other groups taken in clockwise order are C_2H_5, CH_3, and H " (*ib.*, p. 686).

Almost simultaneously, a much more tentative determination of absolute configuration was made by Kuhn and Bein [1] in the case of Boeseken's *dipyruvic erythritol.*[2] Their investigations indicated that the configuration represented by Fig. 151 was *dextrorotatory*, but this deduction is directly opposite to the conclusion reached by Born (p. 391) from the study of a much more plausible model of molecules of this type. In a subsequent paper,[3] the same authors studied the rotatory dispersion of complex salts, and came to the conclusion that the form of the complex salts

$$K_3[Co . 3C_2H_4] \text{ and } [Co . 3C_2H_4(NH_2)_2]Br_2$$

which is *dextrorotatory for red light*, but *lævorotatory for the yellow sodium D-lines* has the configuration shown in Fig. 152*a*. This figure is not easy to read, but by a simple geometrical translation it can be changed into the projection shown in Fig. 152*b*, from which it appears that *the form which is dextrorotatory at long wave-lengths has the configuration of a right-handed screw.* This form has in fact the symmetry of *a three-bladed propeller which when rotated in a clockwise direction, tends to move away from the observer.* Since according to Biot's convention (p. 26) a dextrorotatory medium rotates the light in a clockwise direction *as viewed by the observer*, and in a counter-clockwise direction as viewed in the direction of propagation, Kuhn's assignment implies that *the spiral structure of the molecule is opposite in sign to the spiral described by the line showing the direction of the plane of polarisation in the medium.* As in the spiro-compounds referred to above, this deduction appears to be opposed to the general principles developed by Born, and should therefore be regarded as tentative and liable to be reversed when a more realistic model of the ion is investigated.

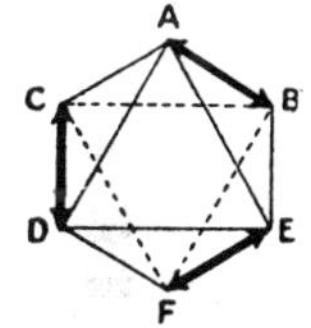

FIG. 152*b*.—ABSOLUTE CONFIGURATION OF *d*-COBALTIOXALATES.

[1] KUHN and BEIN, *Z. ph. C.*, 1934, B. **24**, 335.
[2] BOESEKEN and FELIX, *Ber.*, 1928, **56**, 1855.
[3] KUHN and BEIN, *Z. anorg. Chem.*, 1934, **216**, 321.

CHAPTER XXX.

ELECTRONIC THEORIES.

Drude's Theory of Spiral Vibrators.—The theory of optical rotatory power developed by Drude in 1896 depended on the supposition that in a *dissymmetric crystal* or in a *dissymmetrically isotropic medium* the paths of the ions are not straight lines, but helices (p. 120). Unlike the molecular theories of optical rotatory power described in Chapter XXIX, Drude's theory does not deal with individual molecules, and does not include any function that is based directly upon molecular dimensions; but this contrast is in fact only superficial, since the "pitch" p of the spiral (Fig. 153) provides a linear scale, which must obviously be related to the scale of the framework in which it is developed, and therefore to the linear dimensions and atomic separations of the crystal or of the molecule.

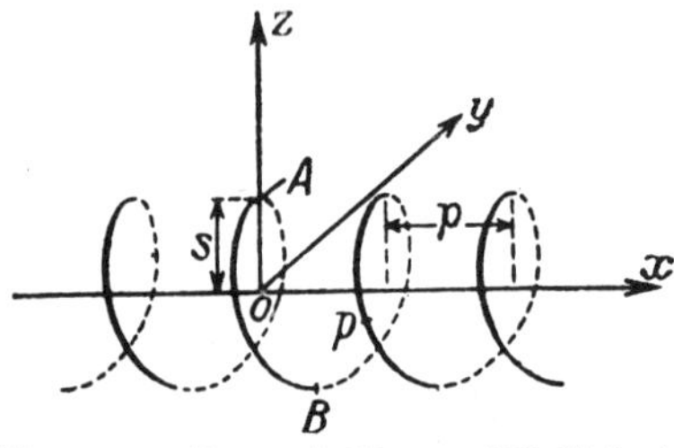

FIG. 153.—DRUDE'S MODEL (W. Kuhn's diagram).

Drude's theory was criticised by Born[1] in 1918 in a paper on *Elektronentheorie des natürlichen optischen Drehungsvermögens isotroper und anisotroper Flüssigkeiten*, in the following terms:

> "All theories which aim at elucidating the optical properties of matter have in common the assumption that there exist within it particles which can be set into vibration by the vibrations of light waves. In the explanation of dispersion and absorption, of magneto-optical and electro-optical phenomena, this conception has led to a uniform picture of the processes which occur; but one important class of optical properties is as yet only quite superficially related to this theory, namely, the power of certain substances to rotate the plane of polarisation, and one or two other phenomena associated with it, to which the somewhat obscure name 'natural optical activity' has become attached. The nature of these phenomena seemed to enforce the assumption that they must be bound up with a screw-like structure of the molecules; but such an assertion

[1] BORN, *Ann. Physik*, 1918 (iv), **55**, 177–178.

about the structure of the whole molecule has no place in ordinary dispersion theory, which deals only with individual, quasi-elastically bound electrons or ions. It has, therefore, generally been customary to link up optical rotatory power with the electromagnetic theory of light only in a formal way, in that new terms were included in the fundamental equations of dispersion theory, which express the screw-like, symmetryless character of the phenomenon, but which cannot be associated with any known forces. Drude appears to have been specially conscious of this avoidance of a deeper understanding of the molecular model, for he attempts, in his *Lehrbuch*, to make the use of these additional terms more plausible through the hypothesis that electrons in the molecules are constrained to move in screw-like orbits. This conception, however, seems to have found little support."

To this earlier criticism, Born now adds the following remarks :

"Drude's mathematical treatment does not correspond to his model. He works out the equations for a damped isotropic resonator and adds to the forces some terms expressing that the action of the electric field of the light depends not only on its value at the place of the resonator, but also on that at neighbouring points ; and these additional forces are assumed to have a rotational character." [1]

Born's original criticism of the arbitrary character of the screw terms, which were introduced by Drude into the electromagnetic equations in order to express the optical rotatory power of a dissymmetrically isotropic medium, was supplemented fifteen years later by W. Kuhn,[2] who showed that Drude had neglected a factor which would reduce the rotatory power of the model to zero. Nevertheless, the dispersion formulæ deduced by Drude and Natanson from this model (p. 425) are substantially identical, both in the region of transparency and in the region of absorption, with those which Kuhn himself deduced from his own model (p. 379). Indeed, the only important difference between the formulæ depends on the fact that in Drude's formula a linear scale is provided by the pitch of the spiral, whereas in Kuhn's formula it depends on the distance between two dissymmetrically coupled vibrators.

Stark's Valency-electron Theory.—In 1914 Stark [3] applied to the problem of optical activity a theory of valency which he had developed in 1908. Rotation of the plane of polarisation is explained in the following way. When the electric vector of a light-wave acts upon a valency electron it produces a displacement which is opposed by a restitutional force. In an isotropic field, this restoring force would be directed towards the position of rest of the

[1] Born, private communication.
[2] W. Kuhn, *Z. ph. C.*, 1933, B. **20**, 325.
[3] Stark, *Jahrb. Radioakt.*, 1914, **11**, 194; *Prinzipien der Atomdynamik*, **3**, 262.

electron; but in Stark's theory the field is not isotropic, since he supposes that the restoring force acts along the direction of the chemical bond. The combined action of the displacing and restoring forces acting on the electron then produces a displacement, which is not in general parallel to the electric vector of the wave field. The polarisation of the molecule is thus slightly rotated from the direction of the electric field. This rotation vanishes when averaged for all the random positions of the molecule, when the system contains *one*, *two* or *three* uncoupled valency electrons; but *four* valency electrons give a definite rotation, provided that they are arranged dissymmetrically. It is also shown that the effect of interchanging two of these four electrons is simply to reverse the direction of the rotation. Furthermore, if two of the valency electrons are identical, so that the molecule acquires a plane of symmetry passing through the other two electrons, the resultant rotation becomes zero.

Stark's theory was entirely of a descriptive nature, no mathematical development being attempted; but it appears to have been the first attempt to give an account of the chemist's picture, in which optical rotatory power is produced, not by a spiral vibrator, but by the association of four dissimilar radicals with an asymmetric carbon atom. Stark's mechanism for explaining optical rotation is not found, however, in modern theories, since it is no longer regarded as essential that the restitutional forces acting on the valency electrons shall be anisotropic. Conversely, the influence of one moving electron on the motion of the others, which forms the basis of these theories, does not play any part in Stark's mechanism.

Born's Theory of Coupled Vibrators.[1]—About 1915, Born, Oseen and Gray put forward independently and almost simultaneously the explanation of optical activity which forms the basis of most modern theories.

The general hypothesis may be expressed in the following way: A molecule is regarded as a system of discrete units, which are fixed more or less rigidly relative to one another. Each of these units possesses the property of assuming an induced polarisation under the action of an applied electric field. When a beam of plane-polarised light is incident upon such a molecule, the components become polarised under the action of the electric vector of the light wave. Each of these polarised units then produces a field of force which in its turn acts upon each of the other units. The resultant polarisation of each unit is determined by the combined influence of the applied external field and of the fields created by all the other units of the molecule. The phenomenon by which the state of one of the units of a molecule is thus influenced by the state of other units of the same molecule is described as COUPLING.

Theories based upon the idea of coupled systems have been developed in two different ways, as regards the polarisability of the

[1] BORN, *Physikal. Zeit.*, 1915, **16**, 251; *Ann. der Physik*, 1918, **55**, 177–240.

separate units of the molecule. Thus, in the molecular theories of Gray, de Mallemann and Boys (Chapter XXIX) no inquiry is made into the nature or form of the function of polarisation. These theories aim at producing a formula in which the rotation is an explicit function of the polarisabilities (and therefore of the molecular refractivities) of the component units of the molecule; they therefore have the merit of being directly applicable to available experimental data. On the other hand, in the electronic theories of Born and Oseen, the polarisability is a definite function of the frequency, and a complete mathematical analysis of the problem was undertaken on this basis. Thus in Born's theory each unit of the molecule was treated as a charged particle which could move slightly from its position of rest under the influence of an electric field; but this displacement was opposed by a force whose components were given the form of linear functions of the components of displacement not only of the one particle but of all. A displacement of one particle, therefore, gave rise to forces acting upon each of the other particles, and the coupling became general throughout the molecule.

Born showed that *three* coupled isotropic electrons could not give rise to optical rotatory power, but that *four* non-planar coupled isotropic electrons could produce this effect. His theory, like that of Stark, therefore confirmed the known relations between optical activity and molecular dissymmetry. In Born's theory, uncoupled electrons contribute nothing to the rotatory power of a molecule; but the small differences of phase in the incident light-wave on reaching the different vibrators in the molecule were shown to be an essential factor in the development of optical rotatory power, which is reduced to zero when the wave-length is very large compared with the linear dimensions of the molecule.

The final formula for the rotation does not depend explicitly upon the electric susceptibilities of the individual units, but upon a large number of constants, characteristic of the molecule, which it is difficult to link up with known data. Apart from the calculations of Hermann and of Hylleraas on sodium chlorate and bromate and on β-quartz (p. 346), the theory was therefore not immediately fertile so far as its applications to chemical problems was concerned.

A mathematical analysis on similar lines was made by Oseen[1] in 1915, by Landé[2] in 1918, and by Gans[3] in 1924 and 1926; and simplified formulæ derived from the general theory have recently been worked out by Born himself (p. 389).*

H. S. Allen's Magneton Theory.—A MAGNETON THEORY of optical rotation was developed by H. S. Allen[4] in 1920 from A. L.

[1] OSEEN, *Ann. der Physik*, 1915, **48**, 1. [2] LANDÉ, *ibid.*, 1918, **56**, 225.
[3] GANS, *Z. Physik*, 1923, **17**, 353; 1924, **27**, 164; *Ann. der Physik*, 1926, **79**, 548. [4] H. S. ALLEN, *Phil. Mag.*, 1920, [vi], **40**, 426–439.
* For a quantum-mechanical treatment see ROSENFELD, *Z. Physik*, 1928, **52**, 161; BORN and JORDAN, *Elementare Quantenmechanik*, 1930, § 47, 250; TEMPLE, *T.F.S.*, 1930, **26**, 293; EWALD, *ibid.*, p. 313.

Parson's magneton concept of atomic structure.[1] Instead of a point charge of negative electricity, the electron is considered as a ring electron, i.e. "as a charge of negative electricity distributed uniformly around a ring which rotates on its axis with high speed, and therefore behaves like a small magnet." Allen proposed to replace Drude's electron moving in a spiral path by a magneton vibrating to and fro along a straight line.

On the basis of this theory the rotation, δ, per unit length is given by

$$\delta = \frac{2\pi^2}{\lambda^2}\sum \frac{\theta f L}{1 - \left(\frac{\tau_h}{\tau}\right)^2}, \quad . \quad . \quad . \quad \text{(i)}$$

where L is the number of magnetons in unit volume,
f is the strength of the absorption band,

whilst τ and τ_h are the frequencies of the light and the natural frequency of the magneton respectively. This equation is identical with the expression given by Drude. In the particular case of a tetrahedral arrangement the magneton theory gives similar results to those deduced by Gray (p. 358). In place of the familiar tetrahedral model, Fig. 154 (*a*), of an asymmetric molecule, Allen represents such

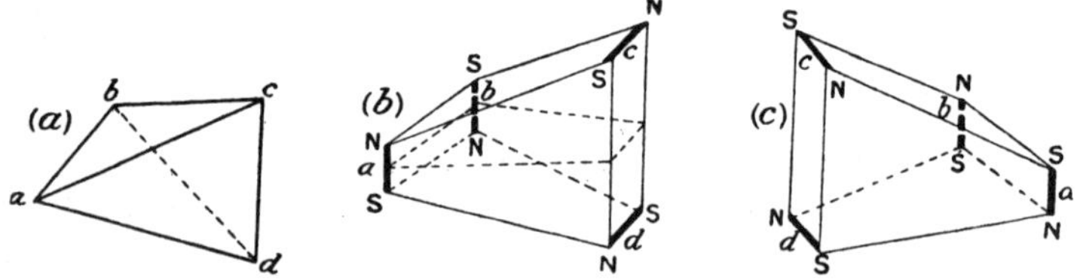

FIG. 154.—H. S. ALLEN'S MAGNETON MODELS.

a molecule and its mirror image as in Figs. 154 (*b*), (*c*), where each apex of the tetrahedron has been replaced by a magneton, N S. The pairs of magnetons associated with the groups *a* and *b* are under the influence of those associated with the groups *c* and *d*. If the two latter are identical their effects both on *a* and on *b* counterbalance one another, and optical activity disappears, since the molecule has a plane of symmetry, as shown by the dotted line.

J. J. Thomson's Tetrahedral Model.—A paper by J. J. Thomson,[2] *On Some Optical Effects including Refraction and Rotation of the Plane of Polarisation due to the Scattering of Light by Electrons*, was published in 1920. This included a theory to account for the optical activity of dissymmetric molecules, and was written independently of the work of Stark, Oseen and Born. Though by no means identical, the fundamental principles of this and the foregoing theories are similar, in that some form of coupling between the electrons is regarded as essential to the development of optical rotatory power. Thus Thomson states:

[1] PARSON, *Smithsonian Collection*, No. 2371, Nov. 1915.
[2] J. J. THOMSON, *Phil. Mag.*, 1920, [vi], **40**, 713.

"We cannot explain this rotation (of the plane of polarisation) if we consider isolated electrons in an atom, but we shall see that we could account for it by a system of electrons held so firmly in position that they act somewhat as a rigid body, a force acting on one electron displacing the whole system of electrons."

In the first instance it was assumed that, under the influence of a light wave, the whole molecule moves like a rigid body, the electronic and atomic systems remaining fixed relatively to one another. In general, this movement includes a rotation as well as a translational displacement of the framework. The electrons of the framework therefore scatter the light in a different manner from the scattering by free electrons and thus give rise to a rotation of plane-polarised light.

The theoretical rotations deduced from this hypothesis were, however, much smaller than those observed experimentally. Moreover, no rotatory dispersion was indicated. The model was therefore altered by postulating the existence of *two* tetrahedral frameworks, namely, (i) an immovable tetrahedron of four atoms surrounding the central asymmetric atom, (ii) a tetrahedron of valency electrons corresponding with the bonds between these atoms. The latter tetrahedron could be displaced by the action of the incident light, but was held in place by restoring forces, which tend to bring it back to its original configuration. Such a system was shown to give rise to a rotation of the right order of magnitude and to account also for the occurrence of rotatory dispersion.

On this theory, the geometrical form of the molecule is of vital importance, since optical activity vanishes if the electrons are at the corners of a regular tetrahedron. On the other hand, the atomic weights have very little effect in determining the magnitude of the rotation, since this is governed principally by the differences between the frequencies of the electrons which bind the atoms to the central carbon atom.

W. Kuhn's Review of the Phenomena of Optical Activity.—An interesting review of the fundamental phenomena of optical activity, which must be borne in mind in formulating any theory of optical rotatory power, was made by W. Kuhn[1] in 1929. It may be summarised as follows:

(*a*) Fresnel's equation for rotatory power

$$\phi = \frac{\pi}{\lambda_{vac.}}(n_l \sim n_r), \quad . \quad . \quad . \quad . \quad \text{(ii)}$$

shows that an extremely small difference between the refractive indices for left and right circularly polarised light will give rise to a relatively large rotation. Thus, for a substance of specific gravity 1, whose rotation per decimetre for blue-green light ($\lambda = 5000$ A.U.)

[1] W. Kuhn, *Z. ph. C.*, 1929, B. **4**, 14; *T.F.S.*, 1930, **26**, 293; *Ber.*, 1930, **63**, 190.

is 100°, the difference between the two refractive indices for circularly polarised light amounts to less than 3×10^{-6}. The actual values might thus be 1·410547 and 1·410544. Kuhn therefore states as *Fact I*: "*Even for substances of high rotatory power, the difference between the refractive indices for left and right circularly polarised light is of the order of* 1 *part in* 1,000,000."

(*b*) The refractive index, n, of a medium for unpolarised light at wave-lengths remote from the regions of absorption, is related to the position and intensity of the absorption bands by the equation

$$n^2 - 1 = \frac{Le^2}{\pi m} \sum \frac{f_1}{\nu_1^2 - \nu^2} \text{ for gases} \qquad \text{(iii)}$$

and from the theory of Lorenz and Lorentz by

$$\frac{n^2 - 1}{n^2 + 2} = \frac{Le^2}{3\pi m} \sum \frac{f_1}{\nu_1^2 - \nu^2} \text{ for liquids,} \qquad \text{(iv)}$$

where L = number of molecules per unit volume (Loschmidt number),
e and m = charge and mass of an electron,
ν_1 = frequency of maximum selective absorption,
f_1 = the "strength" of a given absorption band.

The value of f is related to the form of the band by the equation

$$f = \frac{mc}{\pi e^2} \frac{1}{L} \int \mu_\nu d\nu, \qquad \text{(v)}$$

where $\int \mu_\nu d\nu$ represents the area of the absorption curve, μ being defined by $I/I_0 = e^{-\mu l}$.* Only for extremely intense bands will the value of f approach unity. Thus, the value of f for the very intense band of *p-nitrosodimethylaniline*, log $\epsilon_{max.} = 4{\cdot}5$, is only of the order of 0·45.[1] Bands due to simple unsaturated groups usually have much smaller f-values, e.g. two ketones with $\epsilon = 20$ have $f = 0{\cdot}0007$.[2]

These weak bands in the near ultra-violet affect the refractive index of unpolarised light only to the extent of approximately 1 part in 10,000; the refractive index is therefore governed almost entirely by the very intense bands in the Schumann region. On the other hand, the weak ultra-violet bands, which contribute so little to the refractive indices of unpolarised light, contribute still less to the circular double refraction ($n_l \sim n_r$). As *Fact II* therefore he states:

* It is unfortunate that Kuhn, in all his papers, uses the symbols ϵ and κ for absorption coefficient and molecular extinction coefficient respectively. The symbols in common use in absorption spectroscopy have been defined in *The International Critical Tables* as follows:

Absorption index	κ defined by	$I/I_0 = e^{-\frac{4\pi\kappa l}{\lambda}}$
Absorption coefficient	μ ,, ,,	$I/I_0 = e^{-\mu l}$
Molecular extinction coefficient	ϵ ,, ,,	$I/I_0 = 10^{-\epsilon c l}$

where c = molar concentration. For this reason, in giving an account of Kuhn's work the symbols μ and ϵ have been used as defined above.

[1] Allsopp (not yet published). [2] Lishmund (not yet published).

"*For most bands of active substances lying in the nearer ultra-violet, the contribution to* $(n_l \sim n_r)$ *is in hundredths or thousandths of the contribution to the usual refractive index.*"

(*c*) It follows from *Facts I* and *II* that, if the contributions of the intense bands in the Schumann region to the circular double refraction $(n_l \sim n_r)$ were all of the same sign and proportional to the intensity of the bands, then the total rotatory power of a compound in the visible spectrum would be several thousand times larger than any rotations which have hitherto been observed. Observations of the rotatory power of strongly absorbing substances show, however, that the contribution of an optically active absorption band to $(n_l \sim n_r)$ is not proportional to the intensity of the absorption,* but is usually smaller for strong bands than for weak bands. The rotatory power in the visible spectrum is therefore often governed principally by the first weak absorption band. Kuhn's *Fact III* is therefore that: "*On approaching shorter wave-lengths, the rotation in general increases up to the point of passing through the first (weak) absorption band.*"

Facts I and *II* taken in conjunction with *Fact III* led Kuhn to the supposition that "*The relative difference in the behaviour towards right and left circularly polarised light in the strong bands lying in the outer ultra-violet,* $(f \simeq 1)$, *must often change sign in such a way that the rotation at a great distance from these bands disappears to a first approximation.*"

The principal conclusions from this review of the phenomenon of optical activity may be summarised as follows:

(*a*) The partial rotations associated with the intense absorption bands in the Schumann region are not all of the same sign and therefore annul one another to a large extent in a distant part of the spectrum.

(*b*) The rotatory power of a compound in the visible spectrum is chiefly governed by the nearest absorption bands, probably of weak intensity, situated in the visible and near ultra-violet.

(*c*) In comparison with their relative intensities, greater optical activity is associated with weak bands than with strong bands.

(*d*) A quantitative relationship exists between optical activity (circular double refraction) and circular dichroism.

Kuhn's Molecular Model.—A theoretical basis for these relationships was provided by working out a special case of Born's theory of coupled electronic vibrators. For this purpose a simplified molecular model was used containing two anisotropic *rectilinear oscillators* instead of four isotropic oscillators. Particle 1 (Fig. 155) is bound elastically to its position of rest in such a way that it is able to oscillate only in the direction Ox. Particle 2, at a distance d

* This is not altogether surprising, since unpublished experimental work makes it doubtful whether the partial refractions of an organic compound are proportional to the strengths of the corresponding bands, as is postulated in equations (iii) and (iv) (ALLSOPP, private communication).

from particle 1 measured along Oz, is similarly bound, and is able to oscillate only in the direction Oy. The particles have masses $m_1 m_2$ and charges $e_1 e_2$ which are subsequently connected with the charge and mass e and m of an electron by the relations

$$\frac{e_1^2}{m_1} = f_1 \frac{e^2}{m}, \quad \frac{e_2^2}{m_2} = f_2 \frac{e^2}{m}.$$

If the two particles vibrate quite independently of one another, i.e. if there is no coupling force between them, left or right circularly polarised light would have no preferential effect on the motions of the particles. This would correspond to equal refractive indices for the two types of circular vibration and therefore to optical inactivity.

If a coupling force is introduced between the two particles, their motions become more complicated. Kuhn first investigated the effect of the coupling on the free vibrations of the system. This is found to have two principal modes of vibration, in each of which both particles vibrate with a common frequency. When the coupling is weak, the characteristic frequencies of these modes of vibration are only slightly different from the natural (unperturbed) frequencies of the individual particles, the difference depending upon the strength of coupling involved. In one of the modes of vibration positive displacements of the one particle are accompanied by positive displacements of the second, and negative with negative, whilst in the other mode positive displacements of the one are associated with negative displacements of the other and negative with positive. It is not surprising, therefore, in view of the opposite nature of these vibrations, that the absorption bands to which they refer make contributions to the rotation which are opposite in sign.

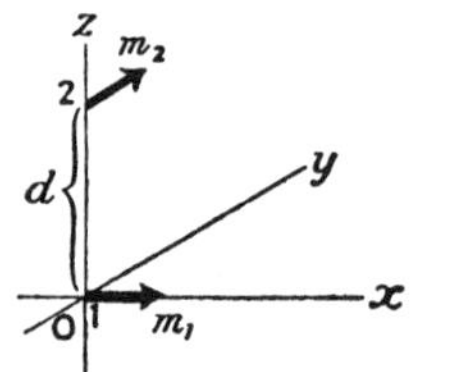

FIG. 155.—KUHN'S SIMPLIFIED MODEL OF A DISSYMMETRIC MOLECULE.

If a plane polarised vibration is incident on the coupled system from the negative Oz direction, and the distance d of separation of the particles is $\lambda/4$, i.e. the phase of the disturbance differs for the two vibrators by $\pi/2$, Kuhn showed that the right circular component would accelerate both particles, whereas the left circular component would accelerate one and retard the other. When the frequency of the incident vibration differs from the two frequencies of the system, this gives rise to a difference in the refractive indices of the two circular components, and therefore to optical activity. When the frequency of the incident disturbance is the same as one of the frequencies of the system (i.e. in the case of resonance) it corresponds to a difference in the absorptive power for the two circular components and thus to circular dichroism.

By considering the action of left and right circularly polarised light on a hypothetical molecule containing two such coupled vibrators, Kuhn derived an expression for the rotation per centimetre for light differing widely in frequency from the natural frequency of either vibrator. In the case of very weak coupling this was of the form

$$\phi = \text{constant}\left\{k_{12}d\frac{\nu^2}{\nu_1^{\,2} - \nu_2^{\,2}}\left[\frac{1}{\nu_1^{\,2} - \nu^2} - \frac{1}{\nu_2^{\,2} - \nu^2}\right]\right\}, \quad . \quad \text{(vi)}$$

where k_{12} = a constant characteristic of the coupling force between the oscillators,
d = the distance of separation,
ν_1 and ν_2 = the characteristic frequencies of the system,
ν = the frequency of the incident light.

If the vibrators are described as electrons, this equation leads to the following conclusions:

(i) A coupling force between the electrons is essential for optical activity, since if $k_{12} = 0$ then $\phi = 0$. The magnitude of the coupling required for this purpose depends on the relative intensities of the absorption bands associated with the two electrons. Thus, if the electrons both give rise to strong absorption bands, a strong coupling will be needed to give them a large circular dichroism and thus produce a large optical activity; but, if one of the bands is weak and the other strong, a weak coupling will suffice.

(ii) For optical activity to result, the two electrons must be separated by a finite distance, since if $d = 0$ then $\phi = 0$.

(iii) Equation (vi) is identical in form with a Drude equation *with two terms of opposite sign.* The partial rotations due to two coupled electrons will therefore be opposite in sign at frequencies which are less than that of either absorption band, i.e. when $\nu < \nu_1$ and $< \nu_2$.

(iv) Since the sum of the two numerators is zero, the partial rotations due to two coupled electrons will tend to annul one another at longer wave-lengths unless their frequencies are widely separated.

(v) For a dissymmetric system, containing an arbitrary number of electrons, bound in an arbitrary way to their positions of rest (including the case of isotropic binding), and coupled simultaneously with each other, Kuhn deduces the general equation

$$\phi = \nu^2 \sum_i \frac{a_i}{\nu_i^{\,2} - \nu^2}, \quad \text{where } \sum_i a_i = 0. \quad . \quad . \quad \text{(vii)}$$

This generalisation is, of course, a reversion in the direction of Born's original theory, which leads to a similar equation, with the same condition that *the sum of the numerators must be zero* (*Optik*, p. 528, equations 7 and 9).

Separation and Orientation of Coupled Vibrators.— If the vibrations of the coupled electrons are at right angles to one another, and the molecules of the substance are all orientated in the optimum manner, the circular dichroism is at a maximum when the distance of separation, d, of the low-frequency electron from the high-frequency electron to which it is coupled is given by

$$g_0 = 4\pi d/\lambda_0,^* \quad . \quad . \quad . \quad . \quad \text{(viii)}$$

where g_0 is the maximum value of the "dissymmetry factor" $g = (\epsilon_l \sim \epsilon_r)/\epsilon$ (p. 394) and λ_0 is the wave-length at which this maximum occurs. On the other hand, if the molecules are orientated chaotically, then

$$g_0 = 2\pi d/\lambda_0.^* \quad . \quad . \quad . \quad . \quad \text{(ix)}$$

Moreover, if the coupled vibrations do not take place at right angles, the value of d obtained from this expression will be too small, i.e. this equation gives only a *minimum* value for the distance of separation.

This formula was applied by Kuhn to some of the compounds which he had himself investigated, and led to the following *minimum* distances of separation of the coupled electrons:

Ethyl α-bromopropionate	$3{\cdot}4 \times 10^{-8}$ cm.
Methyl α-azidopropionate	$2{\cdot}9 \times 10^{-8}$ cm.
α-Azido-propionic dimethylamide	11×10^{-8} cm.
Alkaline copper tartrate	24×10^{-8} cm.
Potassium chromium tartrate	80×10^{-8} cm.

These distances appear to have no relation whatever to the linear dimensions of the molecule. Thus, in the case of the two azido compounds, the distance of the azido group from the asymmetric carbon atom and from the carbonyl radical of the carboxyl group must be substantially identical for the *ester* and the *amide;* but, although the *minimum* distance d has a normal value of 3 A.U. for the ester, it has a value four times greater for the amide; and the *minimum* value for potassium chromium tartrate is many times bigger than the largest value that can be assigned to the linear dimensions of the molecule.

In searching for the origin of this discrepancy, Kuhn investigated [1] a more general case of the two-vibrator model, in which the vibrations are no longer considered to be at right angles to one another, but are inclined at a small negative angle η (Fig. 156). He obtained the result that, although the maximum value of the *circular dichroism* $(\epsilon_l - \epsilon_r)$ corresponds to $\eta = \pi/2$, the maximum value of the *dissymmetry factor* $g = (\epsilon_l - \epsilon_r)/\epsilon$ is obtained when $\eta = 2\pi d/\lambda$, i.e. when η is very small, so that the two vibrations are almost antiparallel. This result applies only to a *particularly orientated* model, in which the light is propagated along the axis of z. When the treatment is extended to cover an assembly of such models *orientated at random*, an additional numerical

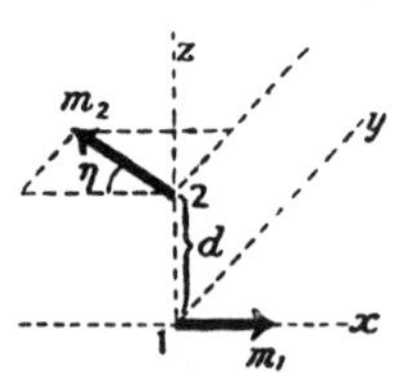

FIG. 156.—KUHN'S MODIFIED MOLECULAR MODEL.

* The relation $g\lambda = 4\pi d$ or $2\pi d$ is valid throughout the absorption band.

[1] KUHN and BEIN, *Z. ph. C.*, 1933, B. **22**, 406–422.

factor is deduced, whereby the value of η for maximum value of g becomes still smaller, namely, $\eta_m = 2\pi d/\sqrt{2\cdot5}\,\lambda$. The distance d is then given by $d \geqslant \frac{g\lambda}{2\pi}\frac{\sqrt{3}}{2}\sqrt{f}$, where f is a measure of the intensity of the optically-active absorption band, as on p. 378. The values thus deduced for three typical compounds are shown in Table 43.

TABLE 43.—SEPARATION AND ORIENTATION OF COUPLED VIBRATORS.

	λ_0.	f.	$\bar{g}$.	d.	η.
α-Azidopropionic dimethylamide	$2{\cdot}9 \times 10^{-5}$	5×10^{-4}	0·024	$> 0{\cdot}22 \times 10^{-8}$	2·2°
Camphor in hexane	$3{\cdot}0 \times 10^{-5}$	2×10^{-4}	0·1	$> 0{\cdot}60 \times 10^{-8}$	1·4°
Potassium chromium tartrate (aq.)	$5{\cdot}8 \times 10^{-5}$	7×10^{-4}	0·09	$> 1{\cdot}90 \times 10^{-8}$	2·6°

The distances thus deduced are now all smaller than the linear dimensions of the molecule, and are therefore physically possible; but no clear physical meaning can be assigned to them, since they are only *minimum* values, and cannot therefore be used to calculate the actual distance between the coupled vibrators which are responsible for optical rotatory power, even if the model were sufficiently realistic to justify such calculations.

The angle η between the directions in which the two electrons vibrate at these minimum distances is about 2°. Chemists will find it difficult to believe that the optical rotatory power of these compounds can be associated with so close an approach to planar symmetry, and will probably be justified in attributing this anomaly to a lack of correspondence between the structure of the molecules and the model on which the calculations were based. Thus it is obviously impossible to represent the *four* dissimilar radicals of an asymmetric carbon atom by *two* vibrators; and the contrast between the complexity of the molecule and the simplicity of the model is still greater in camphor, which contains an unsaturated radical and *two* asymmetric carbon atoms. The chromitartrate ion, which contains *six* asymmetric carbon atoms, *six* carboxyl-radicals and a coloured metallic atom, is in even more striking contrast with the two-vibrator model for which the data in Table 43 were calculated.

Kuhn's calculations indicate that there is one orientation of the molecule in which one of the absorption-coefficients, e.g. ϵ_r, becomes zero, whilst the other remains finite. If all the molecules were correctly oriented, the medium would then behave as a circular polariser, like tourmaline for plane-polarised light. This cannot occur in solution, but some "liquid crystals" show a close approximation to this effect (p. 347).

Optical Rotatory Power of Axially Symmetrical Molecules.—Kuhn's simplified molecular model has the advantage that optical rotatory power appears as a *first-order effect* of the interaction of *two linear vibrators* as contrasted with a *third-order effect* (p. 388) of the interaction of *four isotropic vibrators;* but the model presents a very incomplete picture of the molecule, since the dissymmetric coupling which he postulates cannot be realised except in a complex framework of nuclei and electrons, which is deleted from his simplified sketch. Thus, even the simplest type of optically-active molecule, with a single asymmetric carbon atom, appears to be beyond the scope of his hypothesis, since the four different radicals call inexorably for a system of four vibrators to represent it. Much more favourable conditions are found, however, in axially-symmetrical molecules, such as the symmetrical spiranes, and the highly symmetrical ions of the cobaltioxalate type, where the optical activity is due to a dissymmetric repetition of a relatively simple pattern and not to an increase in the number of dissimilar radicals in the molecule.

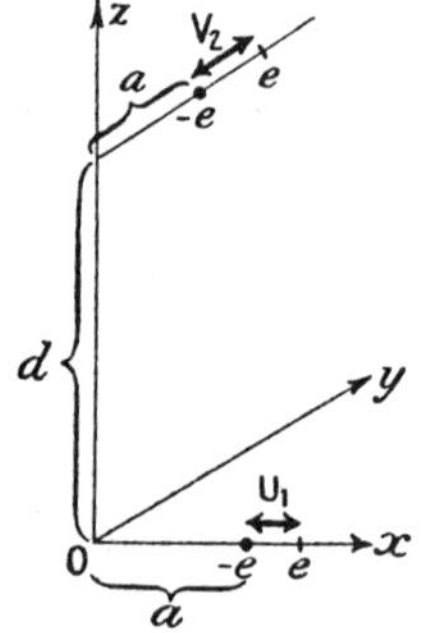

FIG. 157.—COUPLED VIBRATORS IN A SPIRO-COMPOUND.

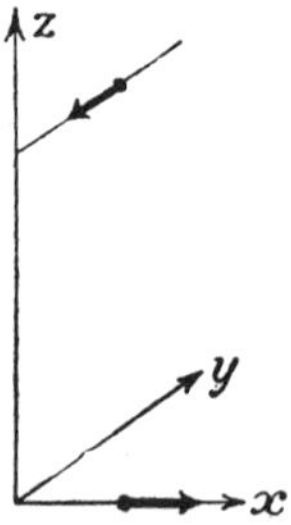

FIG. 158.—COUPLED VIBRATORS IN A DEXTROROTATORY MOLECULE.

(*a*) *Spiranes.*—Suitable conditions for investigation were found by Kuhn and Bein in Boeseken's *dipyruvic pentaerythritol* (Fig. 151, p. 370). For the purpose of analysis, the molecule of the spirane was "resolved" into a simple model as follows: The line joining the *lower* CH_3 and COOH groups is taken as the axis Ox, and the line through the centres of the two rings as the axis Oz. The line joining the *upper* CH_3 and COOH groups will then be parallel to the axis Oy. The COOH groups are replaced by two resonators of equal charge and mass, and similarly bound to their equilibrium positions. The resulting model (Fig. 157) differs from the original model (Fig. 155) only in that the resonators now vibrate about the points ($x=a$, $y=o$, $z=o$), ($x=o$, $y=a$, $z=d$) instead of about ($x=o$, $y=o$, $z=o$) and ($x=o$, $y=o$, $z=d$) respectively. The mathematical analysis remains exactly the same, and it is found that for *positive* coupling between the two resonators, their displacements must always have opposite signs (Fig. 158). In this case, the model has a greater absorption coefficient for left-circular light than for right-circular light, so that on the long wave-length side of the absorption band, it produces a dextrorotation. The molecular configuration of Fig. 151 will therefore correspond to *dextrorotation*.

(*b*) *Cobaltioxalates*.—The symmetry of the cobaltioxalate ion can be represented (p. 85) by placing a Co atom at the centre of a regular octahedron, with three acid groups arranged dissymmetrically on three edges (Fig. 152 (*b*), p. 371). Kuhn simplifies this molecule to a model (Fig. 152 (*a*), p. 370) consisting of an electrically charged resonator, charge e_1 and mass m_1, at the centre, and three equal resonators, charge e_2 and mass m_2, at the middle points of three edges, and constrained to vibrate only *along* the edges. These three lines correspond to the directions of the planes which contain the bivalent or "chelate" radicals

$$\begin{array}{c}\text{—O—C=O} \\ | \\ \text{—O—C=O}\end{array}$$

The peripheral particles are each coupled with the central particle and with each other.

The analysis predicts that this coupling should give rise to a splitting of the absorption bands into two components, with dissymmetry factors of different magnitude or even of different sign. This effect could not be detected in the spirane, since no data for the circular dichroism of the COOH band in such compounds are yet available; but a similar effect was easily observed in the two visible absorption bands of $K_3[Co(C_2O_4)_3]$ at 6000 and 4150 A.U. respectively, since measurements of circular dichroism (p. 336) showed that both bands were composite. As already stated (p. 371) the model illustrated in Fig. 152 (*a*) is identified with that form of the ion which is lævorotatory for sodium light but dextrorotatory for red light.

Present Position of the Electronic Theory of Optical Rotatory Power.—The following review of the present position forms the introduction to a paper by Professor Max Born, which has been submitted for publication in the *Proceedings* of the Royal Society:

> "The problem of correlating optical activity with other optical and chemical properties of a substance has been treated in recent times by several authors, using rather different methods. Kuhn [1] starts from the general theory, given by the author of this paper,[2] and simplifies it in such a way that he gets practical formulæ. Gray,[3] de Mallemann [4] and Boys [5] develop a theory of their own. Boys finds a rather simple expression for the rotatory power, which contains nothing else but the refractivities and the effective radii of the chemical groups involved.

[1] KUHN, *Z. ph. C.*, 1929, B. **4**, 14; *T.F.S.*, 1930, **26**, 293–308.
[2] BORN, *Phys. Z.*, 1915, **16**, 251, 437; *Ann. Physik*, 1917, [iv], **55**, 177.
[3] GRAY, *Phys. Rev.*, 1916, **7**, 472.
[4] DE MALLEMANN, *Rev. gen. sci.*, 1927, **38**, 453.
[5] BOYS, *P.R.S.*, 1934, A. **144**, 655–692.

"There is no question about the general theory. It is, in fact, a problem of quantum optics and has been treated from this standpoint.[1] The chief result is that the rotation effect is due to the same set of virtual oscillators which determine the dispersion. Therefore it is possible to replace the quantum mechanical system by a classical one, consisting of a set of harmonic resonators coupled by electric and magnetic forces.[2] The problem of finding a practical formula consists in choosing this resonator model in such a way that its shape and mechanical constants can be connected with known properties of the molecule and its constituent atoms. This is rather simple for crystals of inorganic substances containing only few atoms or ions, if the dissymmetry is due not to the shape of the single molecule, but to the lattice structure; for in these cases it is possible to attribute unambiguously oscillators to the atoms or ions. This has been done by my collaborators * for two types of crystals, namely, $NaClO_3$ and β-quartz; the results, especially for quartz, are a definite proof that the general theory is right.

"In the case of fluids or gases the problem is much more involved; for active molecules contain always a rather great number of different atoms, and it is not easy to attribute unambiguously oscillators to the atoms. In fact, experience seems to indicate that these oscillators are much influenced by the bonds between atoms or groups (radicals).

"The papers mentioned above follow two different ways. One single harmonic resonator, even if it is anisotropic, i.e. has different eigen-frequencies in the three dimensions of space, gives no optical rotation. The minimum number of anisotropic resonators giving rotation is two, and they have to fulfil two conditions: (1) they must not have a plane or a centre of symmetry, and (2) they must be coupled by some interaction forces. Kuhn's model belongs to this type; he takes the maximum anisotropy possible of a resonator confining its motion to a straight line, and he chooses the lines of the two resonators perpendicular to the line connecting their centres and to each other. Such pairs of resonators are possible only in molecules consisting of 4, or more than 4 atoms, because a system of 2 or 3 atoms has always a plane of symmetry (the plane of the 2 or 3 nuclei). This

[1] BORN and JORDAN, *Elementare Quantenmechanik* (Springer, Berlin, 1930), § 47, 250.

[2] BORN, *Optik* (Springer, Berlin, 1933).

* HERMANN (*Zeitschr. f. Phys.*, 1923, **16**, 103) has calculated the optical activity of $NaClO_3$ and $NaBrO_3$; his results give the right order of magnitude, but are not accurate because of a numerical mistake in the computations. HYLLERAAS (*Zeitschr. f. Phys.* 1927, **44**, 871) has calculated the mechanical and optical properties (lattice parameter, double refraction and optical activity) of quartz, and has found excellent agreement with observations. It is strange that in spite of these results most of the authors working on these problems, experimentally or empirically, used to say that my theory has never been applied to a practical case.

theory can be called semi-empirical; for it makes no attempt to reduce *a priori* the dimensions and location of the assumed resonators to other known properties of the molecule. In fact, the pair of resonators has only one advantage over Drude's helical orbits (with which it shares the property of depending on one geometrical constant, the height of the helix—the distance of the resonators), namely, that the derivation of the dispersion law is consistent with itself, whereas Drude's mathematical treatment does not really correspond to his model. But the consequences drawn from this model concerning the connection of rotation and circular dichroism have really nothing to do with the special model and had been derived from Drude's theory by Natanson [1] earlier.

"The other way consists in considering the molecule as built from several groups with given optical properties, and to calculate the optical activity as a consequence of the interaction of these groups. The simplest assumption possible about the single group is to attribute to it an isotropic resonator; then the minimum number of groups giving rotation is four. This is the idea of Boys' paper quoted above. But his treatment is insufficient for various reasons. Instead of using the well-prepared formulæ of the general theory, by which the effect of the forced vibrations under the action of light is reduced to the determination of the free vibrations of the system, he starts by calculating directly the forced vibrations; if one considers the complications of this procedure there is no wonder that his results do not agree with the rigorously proved formulæ of the general theory. Without going into details I mention only this point: Boys' expression for the rotatory power is proportional to the product of refractivities of the constituent groups; from this it follows that the rotatory power and dispersion (index of refraction) depend on wave-length in the same way. But the general theory, in accordance with experience, shows that this is not the case. In fact, both quantities are sums of terms of the form (ν frequency of light, ν_k eigen-frequency of the system)

$$\frac{a_k}{\nu_k^2 - \nu^2}.$$

Whereas, however, in the case of the refractive index the a_k are all positive (namely, the squares of the eigen-vectors of the vibration ν_k), in the case of rotatory power their sum is nought, so that some of them must be negative (they are scalar products of the eigen-vectors with other vectors connected with the location of the groups). Therefore Boys' theory cannot explain the fact that strong absorption bands may give a very small contribution to the optical activity and *vice versa*.

[1] NATANSON, *Bull. Akad. Sci., Krakow*, 1908, 764–783.

"For this reason it seemed to me desirable to derive the properties of a model, of the same type as that of Boys, from the general theory by the application of a method of successive approximations.[1] For purely technical reasons I have chosen, instead of a set of isotropic resonators, a system in which each resonator has an intrinsic, but weak, anisotropy. For an isotropic resonator with three exactly equal frequencies is a degenerate system; the effect of the interaction with other resonators of the same kind will be in the first instance (secular perturbation) a splitting of these equal frequencies into narrow triplets, transforming in this way each primarily isotropic resonator into an anisotropic one. The coupling of these, which gives rise to optical activity, is given by higher approximations. Therefore it is much more convenient to assume each oscillator anisotropic from the beginning, avoiding in this way the degeneracy. The principal axes and frequencies of the resonators are, of course, arbitrarily chosen; therefore the formula for the rotatory power cannot hold for frequencies in the neighbourhood of the absorption lines. But this is no objection against the theory, for there are two physical facts which prevent the theory from representing the behaviour of optical activity inside the region of absorption. Thus (i) in real molecules there are no absorption lines, but very complicated bands, and (ii) these bands are diffuse, by reason of the damping of the vibrations. Both these effects are entirely out of the power of our theories, and a formula trying to represent optical activity inside the absorption band can be nothing more than a half-empirical extrapolation. The difficulty of the application of the perturbation method to our system of resonators consists in the circumstance that the minimum number of (isotropic) oscillators having no plane of symmetry is four, so that the approximation has to be extended to the third order (the first order coupling two of the resonators, the second three, the third four). But it is possible to derive a rather simple formula, depending only on the frequencies and strengths of the resonators and on their spacial configuration.

"This formula holds only if there are no equal resonators in the system, otherwise there is another degeneracy, and in general the perturbation method will give infinite amplitudes. I have, however, found a special limiting case of the model where this does not occur. The geometrical shape of the model is a tetrahedron,

[1] I had the privilege of discussing the problems of optical activity with Professor Lowry, to whom I am much indebted for a great number of facts and hints. Remarks of his like that in his presidential address to the section of Chemistry at the Aberdeen Meeting of the British Association ("The real theory of optical rotatory power may be found by the mathematician, but is concealed from the chemist, in the papers of Born . . .") have induced me to a new effort to find a formula which could be understood and used by the chemist.

the surface planes of which are four congruent triangles, so that the figure has a binary axis of symmetry. For this case, the rotation remains finite, when two pairs of frequencies are tending to equality.

"The expression for the angle of rotation becomes very simple if the two pairs of resonators are perpendicular to one another and to the line connecting their centres. Molecules of such a shape seem to have been prepared rather frequently. I have therefore developed in detail the formula for this type of molecule, which is identical with a type that has been studied by Kuhn and Bein.[1]

"The constants of this formula are all well-defined properties of the molecule, but the data now available are not sufficient to give more than a confirmation of the order of magnitude. This is, however, sufficient for our purpose, since the theory itself cannot be expected to give more than the order of magnitude of the rotation. This is due to a limitation in the method of approximations employed, which actually breaks down when the resonance frequencies of any two atoms are close together (actually when $\nu_1^2 - \nu_2^2$ is less than $\dfrac{e^2}{4\pi^2 m r_{12}^3}$). When such a case arises, the interaction of these particular atoms needs to be calculated by an exact method, and only the interaction of this group with the more distant parts of the system may be treated by the perturbation method.

"Furthermore, since the whole argument is based on classical mechanics it cannot therefore be claimed to be completely rigorous. Actually quantum mechanics should be applied, and it is in fact proposed to treat the problem from this standpoint in subsequent papers."

Application of Born's Molecular Model.—In his paper Born considers the optical rotatory power of a system of four vibrators and shows that a great simplification results when these are situated at the corners of a tetrahedron formed from four congruent triangles, with identical lengths of sides, a, b, c, as in Fig. 159 (a). A still further simplification occurs if isosceles triangles are used as in Fig. 159 (b), so that $a = b$.

Born's formula for the rotation of his simplified tetrahedral model, Fig. 159 (c), in its complete form is

$$\chi = 3{\cdot}49 \times 10^{-11} \frac{\rho(n^2+2)}{M} f_{\mathrm{I}}^2 f_{\mathrm{II}}^2 \frac{\lambda_{\mathrm{I}}^6 \lambda_{\mathrm{II}}^6}{(\lambda_{\mathrm{II}}^2 - \lambda_{\mathrm{I}}^2)^3} \left\{\frac{\lambda_{\mathrm{I}}^2}{\lambda^2 - \lambda_{\mathrm{I}}^2} - \frac{\lambda_{\mathrm{II}}^2}{\lambda^2 - \lambda_{\mathrm{II}}^2}\right\}$$

$$\times \frac{1}{l^8} \frac{d}{l}\left(\frac{d^2}{l^2} - \frac{1}{2}\right)^2 \div \left(\frac{d^2}{l^2} + \frac{1}{2}\right)^6, \quad \text{(xi)}$$

[1] KUHN and BEIN, *Z. ph. C.*, 1933, B. **22**, 406; 1934, B. **24**, 335.

where χ = rotation in degrees per decimetre,
ρ = density of the medium,
n = its refractive index,
M = its molecular weight,
l = the distance AB in A.U.,
d = the distance OP in A.U.,
$\lambda_I\ \lambda_{II}$ are the characteristic wave-lengths of A and B, measured in A.U.,
$f_I f_{II}$ are the strengths of the vibrators A and B, and are given by

$$\frac{e_I^2}{m_I} = f_I \frac{e^2}{m}, \quad \frac{e_{II}^2}{m_{II}} = f_{II} \frac{e^2}{m},$$

where e_I, e_{II}, m_I, m_{II} are the charges and masses of the two vibrators and e, m the charge and mass of an electron.

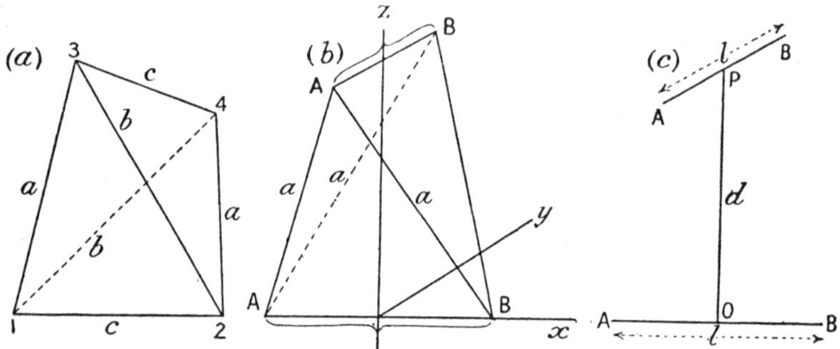

Fig. 159.—Born's Molecular Model.

The dispersion of χ is given solely by the term

$$\left\{\frac{\lambda_I^2}{\lambda^2 - \lambda_I^2} - \frac{\lambda_{II}^2}{\lambda^2 - \lambda_{II}^2}\right\}.$$

This is, of course, the same as the dispersion factor in Kuhn's equation. It can be simplified if we expand it for values of λ which are large compared with λ_I, λ_{II}. The dispersion factor then becomes

$$\frac{\lambda_I^2 - \lambda_{II}^2}{\lambda^2}$$

so that the dispersion in regions of long wave-lengths follows Biot's Law of Inverse Squares.

In order to apply his formula to a particular molecule Born makes the assumption that $f_I = f_{II} = 1$. Introducing this assumption, and also the above approximation for the dispersion factor, we get for the rotation of the model shown in Fig. 159(c) the value

$$\chi = 3{\cdot}49 \times 10^{-11} \frac{\rho(n^2+2)}{M\lambda^2} \left\{\frac{\lambda_I^3 \lambda_{II}^3}{(\lambda_{II}^2 - \lambda_I^2)}\right\}^2 \frac{1}{l^8} \frac{\frac{d}{l}\left(\frac{d^2}{l^2} - \frac{1}{2}\right)^2}{\left(\frac{d^2}{l^2} + \frac{1}{2}\right)^6}. \qquad \text{(xii)}$$

This model is almost identical with the structure of the spiro compound

$$\begin{array}{c} H_2N \\ H \end{array}\!\!>C\!<\!\begin{array}{c} CH_2 \\ CH_2 \end{array}\!\!>C\!<\!\begin{array}{c} CH_2 \\ CH_2 \end{array}\!\!>C\!<\!\begin{array}{c} NH_2 \\ H \end{array}$$

recently resolved by Pope and Jansen, in which the planes of symmetry of the two rings are destroyed by the unlike radicals H and NH_2 attached to the terminal carbon atoms.

The rotatory power given by the preceding formula, assuming

$$l = 1{\cdot}5 \text{ A.U.} \qquad d = 4{\cdot}0 \text{ A.U.},$$
$$\lambda_{I} = 1200 \text{ A.U.} \qquad \lambda_{II} = 1600 \text{ A.U.},$$

is $[M]_{5461}$ about 30°, in general agreement with the observed rotatory power, $[M]_{5461} = 18°$, of the salts; but only the order of magnitude need be considered, since the rotation depends on the *eighth* power of the linear dimensions, and the other data are merely guesses.

The sign of the rotation is not included in the formula. It is, however, an immediate deduction from the theory that the light must be twisted in the same direction as the twist of the line *AB* on the shortest route from one edge of the tetrahedron to the other. The model shown in Fig. 159 (*c*), and the formula set out above, must therefore give rise to a right screw and therefore to a lævorotation of the light. This conclusion is opposite to that reached tentatively by Kuhn and Bein (p. 371) in the case of Boeseken's spiro-acid.

The theory of coupled vibrators (p. 374), introduced by Born in 1915, had as one of its principal merits the elimination of Drude's highly artificial hypothesis of *spiral* vibrators, and of the complex fields of force which would be required to produce them. The development now recorded has the merit of making unnecessary the equally unreal hypothesis of *linear* vibrators, which formed the original basis of Kuhn's theory. Thus, even in the spiro-compounds (which have the same axial-symmetry as Kuhn's molecular model in its simplest possible form) Born's hypothesis of four isotropic vibrators, *A*, *B*, *A*, *B*, corresponding with the four dissymmetrically-arranged radicals of the preceding formula, has been shown to be at least as adequate as Kuhn's hypothesis of two linear vibrators at right angles to one another.

It is of interest to notice that, in seeking to deduce the optical rotatory power of dissymmetric molecules, de Mallemann, Boys and Born have all been obliged, by the limitations of the experimental data, as well as for the sake of simplicity of calculation, to make use of *isotropic* vibrators. Stark's hypothesis of *anisotropic* vibrators, depending on the directions of the chemical bonds, represents a refinement which may become of practical importance when the form of the ellipsoid or spheroid of polarisability of the electrons of a given radical or bond can either be predicted theoretically or determined experimentally; but, until this can be done, formulæ based upon this hypothesis must remain suspended in empty space, pending the development of new methods of investigation. When these conditions have been fulfilled, the anisotropy of the radicals may become an important additional factor in attempts to predict the absolute rotatory power of organic compounds. Reasons have already been given (p. 363) for believing that this factor may be of minor importance in the isosteric radicals F, OH, NH_2 and CH_3; and one may suspect that spacial dissymmetry, and not the anisotropy of the H and NH_2 radicals, will prove to be of dominant importance in the saturated spiro-base resolved by Pope and Jansen. On the other hand, the carboxyl-group of Boeseken's spiro-acid is unlikely to be equally polarisable in and perpendicular to the plane of the $-C\langle{}^{O}_{O}\rangle H$ group, although even in this case the degree of anisotropy may be reduced by free rotation of the carboxyl-group about the single bond which links it to the carbon-skeleton, thus reducing the ellipsoid of polarisability to an oblate or prolate spheroid.

In the absence of experimental data in reference to the anisotropy of the carboxyl group, Kuhn has gone to the opposite limit, by reducing the ellipsoid of polarisability to a straight line, at right angles to the principal axis of the molecule, instead of regarding it as approximately spherical. His hypothesis of linear vibrators then leads to the following formula for the rotation ρ per centimetre column of a compound of this type:

$$\rho = \pm \frac{L}{c^2} \frac{3d}{\pi} \frac{a^2}{(d^2 + 2a^2)^2} \frac{1}{\sqrt{(d^2 + 2a^2)}} f^2 \left(\frac{e^2}{m}\right)^2 \nu^2 \frac{1}{(\nu_0{}^2 - \nu^2)^2}$$

where d is the separation of the coupled vibrators, measured along the axis of the molecule,
a is their perpendicular distance from the axis,
ν_0 is their natural frequency, and
L is the number of pairs in 1 c.c.

Putting in the values $d=5 \times 10^{-8}$, $a=1{\cdot}5 \times 10^{-8}$, $f=1$, $\lambda_0=2000$ Å.U., this gives a specific rotation of about 400°.[1] The observed specific rotation of Backer and Schurinck's spiro-acid (p. 67) was about 1°.

Kuhn's formula differs primarily from that of Born in that it involves only a single pair of vibrators, and includes only a single natural frequency, whereas Born's formula is concerned with *two* pairs of vibrators, and includes the mutual interaction of the two vibrators of each pair with one another, as well as with those of the opposite pair. The fact that the two formulæ are not equivalent to one another is also emphasised by the fact that Kuhn's formula depends on the *inverse square* of the linear dimensions of the model, whereas Born's formula involves the *inverse eighth power* of the length l.

[1] Kuhn, private communication.

CHAPTER XXXI.

OPTICALLY-ACTIVE ABSORPTION BANDS.

Circular Dichroism and Circular Double Refraction.— Cotton's discovery of circular dichroism may be regarded as a sequel to Fresnel's discovery of circular double refraction. Thus, according to classical theory, *ordinary refraction* is related to absorption of light of shorter wave-length in the manner first postulated by Maxwell in 1869 (p. 118). It was to be expected that a similar relation would exist between CIRCULAR DOUBLE REFRACTION, $n_r \sim n_l$, which Fresnel had demonstrated as a cause of optical rotatory power in quartz, and the selective absorption $\epsilon_r \sim \epsilon_l$, of right and left circularly polarised light. This relation was realised by Cotton (p. 149) who discovered the effect in question, and described it as CIRCULAR DICHROISM. The analogy between the two phenomena is illustrated in Fig. 160.[1]

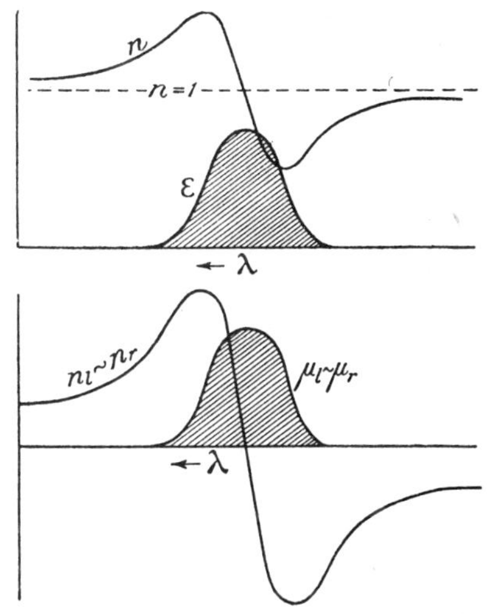

FIG. 160.—CIRCULAR DICHROISM AND CIRCULAR DOUBLE REFRACTION COMPARED WITH ORDINARY ABSORPTION AND REFRACTION.

Dissymmetry Factor of Optically-Active Absorption Bands.—From this analogy it appears that a close relation is likely to exist between the *absorption of ordinary light* and the *selective absorption of* r *and* l *circularly polarised light.* In other words, optical rotatory power must have its origin in OPTICALLY-ACTIVE ABSORPTION BANDS, in just the same way as refraction has its origin in ordinary absorption bands of shorter wave-length. Cotton showed, however, that the absorption bands in an optically-active medium are not necessarily active, since some of them, e.g. those of a dyestuff in a sugar solution (p. 155), contribute

[1] Compare W. KUHN, *Ber.*, 1930, **63**, 193, where, however, the axis of wave lengths is reversed.

nothing to the optical rotatory power of the medium, even in the region of the absorption bands themselves. Although, therefore, it is impossible for circular dichroism to appear except in a region of ordinary absorption, there is no quantitative relation between the two phenomena. The optical activity of an absorption band can, however, be measured very conveniently by means of the ratio of circular dichroism to ordinary absorption,

$$g = \frac{\epsilon_r \sim \epsilon_l}{\epsilon}. \quad . \quad . \quad . \quad . \quad . \quad \text{(i)}$$

This ratio has been described by W. Kuhn as the "anisotropy factor";[1] but this name is an unfortunate one, since the media to which it has been applied hitherto have all been *isotropic*. Moreover, there is no causal relationship between the *anisotropy* of an absorbing medium and its ability to rotate the plane of polarisation of light, which depends exclusively on its *dissymmetry*. The ratio in question may therefore be described more exactly as the DISSYMMETRY FACTOR.[2]

Relations between Circular Dichroism and Absorption.—The dissymmetry factor is an arbitrary variable which depends in some way on the dissymmetry of the molecular field in which the electronic transition takes place, since it vanishes when this field becomes symmetrical by acquiring a plane or centre of symmetry. It can vary in magnitude from zero in optically-inactive absorption bands to a maximum not much greater than $g = 0{\cdot}1$ in the weak absorption bands of unsaturated compounds; but its magnitude can only be predicted under the same conditions as the absolute magnitude of the optical rotatory power of the medium. The equations of Natanson, however, enable us to anticipate (p. 402) that, within the range covered by a single absorption band, the factor g will be proportional to the frequency ν of the incident light,

$$g = (\epsilon_l \sim \epsilon_r)/\epsilon \propto \nu. \quad . \quad . \quad . \quad . \quad . \quad \text{(ii)}$$

When this relation holds good, the form of a curve of circular dichroism can be deduced, with the help of a single arbitrary constant, from that of the absorption band in which it occurs, even when the dissymmetry factor is so small that direct measurements of circular dichroism are difficult or impossible. It is therefore both interesting and important to ascertain the mathematical form of the absorption curves of optically-active compounds, since the equations thus developed can then be used, not only to deduce the form of the curves of circular dichroism, but as a basis for deducing the more complex equations which represent the rotatory dispersion of the medium.

Mathematical Form of Absorption Curves.—(*a*) Helmholtz[3] supposed that absorption was influenced by a damping due to friction between atomic vibrators. The equations which he deduced from

[1] KUHN, *T.F.S.*, 1930, **26,** 299; *Z. ph. C.*, 1930, B. **8,** 286.
[2] LOWRY and HUDSON, *Phil. Trans.*, 1933, A. **232,** 119, footnote.
[3] HELMHOLTZ, *Pogg. Ann. Phys. Chem.*, 1875, **154,** 582.

this hypothesis were modified by Ketteler[1] to a form which was in harmony with the electro-magnetic theory of light. The KETTELER-HELMHOLTZ ABSORPTION EQUATION deduced in this way may be written in the form

$$2n\kappa = \frac{D\Gamma\lambda^3}{(\lambda^2 - \lambda_0^2)^2 + \Gamma^2\lambda^2}, \quad . \quad . \quad . \quad \text{(iii)}$$

where n is the refractive index and κ is the absorption index of the medium at wave-length λ, λ_0 is the wave-length of maximum absorption, Γ is a damping factor having the dimensions of a wave-length, and D is a constant.

The ABSORPTION INDEX, κ, is defined by $I/I_0 = e^{-\frac{4\pi\kappa d}{\lambda}}$, where I_0 and I are the intensities of the incident and transmitted light and d is the thickness of the absorbing medium.

If the band is narrow the refractive index n will remain approximately constant; the MOLECULAR EXTINCTION COEFFICIENT, ϵ, (defined by $I/I_0 = 10^{-\epsilon cd}$, where c is the concentration in gram-molecules per litre), will then be given by

$$\epsilon = \frac{\text{constant } \lambda^2}{(\lambda^2 - \lambda_0^2)^2 + \Gamma^2\lambda^2}. \quad . \quad . \quad . \quad \text{(iv)}$$

This equation represents an absorption curve which is unsymmetrical with respect both to frequencies and to wave-lengths, since it is steeper on the long wave-length side when plotted on a scale of frequencies and steeper on the short wave-length side when plotted on a scale of wave-lengths.

(*b*) The absorption band of *carbon disulphide*, which is much steeper on the side of longer wave-lengths, is too unsymmetrical to be represented by the preceding equation (iv). For the purpose of analysis, Bruhat and Pauthenier,[2] divided it into five narrow components, each of which could be represented separately by a term of the Ketteler-Helmholtz equation. The resultant curve was very similar to the experimental curve; but it included a number of tiny maxima and minima, instead of being perfectly smooth. This method of analysis suggests that an unsymmetrical absorption band is more composite in character than a symmetrical band; but this conclusion is devoid of any experimental justification, since the precise form of the absorption curve appears to depend merely on the way in which the rotation-vibrational fine structure of the band is distributed in relation to the fundamental electronic transition.

(*c*) Planck[3] in 1903 developed a new theory of absorption, based solely on an assumed re-emission of energy by the electrically charged vibrators, in the form of ether waves of the same period as

[1] KETTELER, *Phil. Mag.*, 1876, [v], **2**, 332, 414, 508; *Pogg. Ann. Phys. Chem.*, 1877, **160**, 466; *Wied. Ann. Phys. Chem.*, 1879, **7**, 658; 1881, **12**, 481.
[2] BRUHAT and PAUTHENIER, *Ann. de Phys.*, [x], 1926, **5**, 440.
[3] PLANCK, *Sitz. ber. Preuss. Akad. Wiss.*, 1904, **22**, 740–750.

those absorbed. Planck's results lead to the absorption equation

$$\kappa^2 = \frac{\sqrt{(\alpha^2+\beta^2-\alpha)^2+\beta^2}-(\alpha^2+\beta^2-\alpha)}{2(\alpha^2+\beta^2)}, \qquad \text{(v)}$$

where $\alpha = \frac{\lambda_0{}^2-(1-\gamma)\lambda^2}{3\gamma\lambda^2}$, $\beta = \frac{\sigma\lambda_0}{3\pi\gamma\lambda}$, $\gamma = \frac{\sigma L\lambda_0{}^3}{4\pi^3}$,

λ_0 = wave-length *in vacuo*, corresponding to the free period of the vibrators,
σ = logarithmic decrement of the vibrators,
L = number of vibrators per unit volume.

Planck showed that the form of the absorption curve will depend on the value of L. When L is large the curve is *steeper on the side of longer wave-lengths* and with its maximum on the long wave-length side of λ_0. When L is small the curve is *symmetrical on a scale of wave-lengths* and with its maximum at λ_0. *The curve for the molecular extinction coefficients, ϵ, will thus always be steeper on the side of longer wave-lengths* no matter what the value of L. This result was deduced for *line* spectra,† but nevertheless corresponds accurately with the observations of Hudson and others (p. 397) on the absorption *bands* of solutions.

(*d*) Bielecki and Henri[1] applied the Ketteler-Helmholtz equation to a number of experimental absorption curves, and showed that only for a limited range of wave-lengths near the centre of an absorption band was there a close agreement between the theoretical and experimental curves ; elsewhere the absorption decreased more rapidly than was accounted for on the theory of damped vibrators. Since the *vapours* of acetylene, benzene, etc., give complex band spectra, they supposed that the broad bands given by *liquids* and *solutions* are not due to the damping of a single electronic vibrator, but to the blurring of a spectrum which includes many individual lines. These individual lines were attributed to modifications of the frequency of the electron, which were subsequently shown to depend on the superposition of nuclear vibrations and molecular rotations on the original electronic transition. On the provisional hypothesis that these frequencies are distributed according to Gauss' law of probable errors, and that the absorption index is proportional to the number of electrons of each frequency, they deduced the exponential formula

$$\epsilon = \alpha\nu e^{-\beta(\nu_0-\nu)^2}, \quad . \quad . \quad . \quad . \quad \text{(vi)}$$

where $\alpha = \epsilon_{max.}/\nu_0$,
β = a constant embodying the " half-width " of the band, and
ϵ and $\epsilon_{max.}$ = the molecular extinction coefficients at frequencies ν and ν_0 (maximum of absorption), respectively.

[1] BIELECKI and HENRI, *Phys. Zeit.*, 1913, **14**, 516.

† See BORN, *Optik*, Berlin, 1933, pp. 477–486, for theoretical absorption and dispersion curves of gases under pressure.

This expression was found to agree with the experimental data for the absorption of solutions of *acetone* in alcohol and in water, on both sides of the maximum, of solutions of *acetic acid* in water, where only one arm of the curve can be observed, and of solutions of *acetaldehyde* in water, where one arm only of a second band is observed on the short wave-length side of the principal band. The authors therefore suggested that the exponential equation should be retained as an empirical relation, even if the assumptions on which it was based were proved to be incorrect. The curves calculated from this equation are unsymmetrical with respect to frequencies and wave-lengths in precisely the same way as those based on the Ketteler-Helmholtz equation.

(*e*) Kuhn and Braun[1] found a similar discrepancy between the observed absorption coefficients of *methyl α-azidopropionate* and theoretical values deduced by means of the Ketteler-Helmholtz equation. They therefore proposed to represent the absorption curve by the empirical equation

$$\mu = \mu_{\text{max.}} e^{-\left(\frac{\nu_0 - \nu}{\theta}\right)^2}, \qquad \text{(vii)}$$

where μ and $\mu_{\text{max.}}$ are the absorption coefficients, defined by $I/I_0 = e^{-\mu l}$, at frequencies ν and ν_0 respectively, ν_0 is the frequency corresponding to the maximum absorption at the head of the band, and θ is a constant related to the half-width ν' of the band by $\nu' = 1{\cdot}6651\theta$.* A similar equation

$$\epsilon = \epsilon_{\text{max.}} e^{-\left(\frac{\nu_0 - \nu}{\theta}\right)^2}, \qquad \text{(viii)}$$

may be used for molecular extinction coefficients.

This equation differs from that proposed by Bielecki and Henri only in the omission of the factor ν/ν_0. It represents a probability distribution of absorption intensities decreasing symmetrically on either side of the frequency of maximum absorption; it therefore gives a theoretical absorption curve which is symmetrical on a scale of frequencies. Kuhn and Braun showed that this equation represented their experimental data much more accurately than the Ketteler-Helmholtz equation.

(*f*) Absorption curves which are *steeper on the side of shorter wave-lengths* have been observed in diatomic xenon,[2] and in a few similar unstable molecules, whose component atoms are perhaps held together by weak forces of the van der Waals' type, instead of by electronic linkages; but experience has shown that the absorption curves of ordinary chemical molecules are either *symmetrical on a scale of wave-*

[1] KUHN and BRAUN, *Z. ph. C.*, 1930, B. **8**, 281–313.

[2] MCLENNAN and TURNBULL, *P.R.S.*, 1930, A. **129**, 266–283.

* The arbitrary constant, 1·6651, ensures that the theoretical curve based on the above equation shall coincide with the experimental curve at the two points where the absorption is half its maximum value as well as at the maximum.

lengths or are *steeper on the side of longer wave-lengths.* The equations of Ketteler and Helmholtz (iv), of Bielecki and Henri (vi) and of Kuhn and Braun (vii, viii), cannot therefore be used to represent either of these predominant types of absorption band, since they give rise to curves which are *steeper on the side of shorter wave-lengths.* Thus, of the six absorption equations cited above, equation (v) of Planck is the only one which agrees with the actual form of the absorption bands of normal molecules (p. 400), and even this is expressed in a form which does not lend itself to general use. An equation of a much simpler type has, however, been used by Lowry and Hudson[1] to represent the symmetrical absorption bands recorded by them in

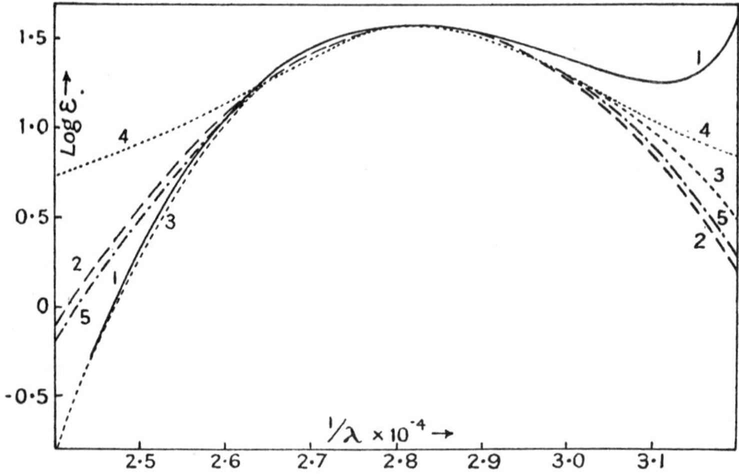

FIG. 161.—ANALYSIS OF THE FIRST ABSORPTION BAND OF METHYL *d*-BORNYL XANTHATE (Lowry and Hudson).

1. Experimental curve.
2. Theoretical curve (Kuhn and Braun).
3. ,, ,, (Lowry and Hudson).
4. ,, ,, (Ketteler and Helmholtz).
5. ,, ,, (Bielecki and Henri).

compounds containing the chromophoric radicals >C=O and >C=S.

This equation was devised in the first instance in order to represent the ultra-violet absorption bands of the optically-active xanthates which had already been studied by Tschugaeff (p. 317). Thus curve 1 in Fig. 161 shows the form of the first absorption band of *menthyl d-bornyl xanthate*, $C_{10}H_{17}$. O . CS . SCH_3. Curve 4 shows a series of values calculated from the Ketteler-Helmholtz equation, the constants of the equation being adjusted so that the calculated values coincide with the observed values, (i) at the maximum and (ii) at the

[1] LOWRY and HUDSON, *Phil. Trans.*, 1933, A. **232**, 117.

point (0·3 lower on the logarithmic scale) at which the absorption on the side of longer wave-lengths is half as great as at the maximum. The theoretical curve 4 (as Bielecki and Henri found in the case of acetone) bears no resemblance to the experimental curve 1, since it spreads out over it like an umbrella instead of dropping steeply towards the negative side of the axis. Curves 2 and 5 calculated from the equations of Kuhn and Braun and of Bielecki and Henri also fall off less rapidly than the experimental curve.

This deviation could not be accidental, since several other curves gave the same result, but were still not displaced far enough from those deduced from the Ketteler-Helmholtz equation. The *first* absorption band of the xanthates, however, is not well adapted to mathematical analysis, since it is approached so closely by a second absorption band that only a shallow minimum is developed between them. The form of the band can therefore only be studied on the side of longer wave-lengths. The absorption curves of the optically-active *methyl α-azidopropionate* and *α-azidopropionic dimethylamide*, studied by Kuhn, are equally unsuited for exact analysis, since the azido band is overlapped by the carboxylic band, leaving only a shallow minimum between them, so that only one arm of the absorption curve is developed; moreover, in the corresponding bromo- and chloro-compounds the absorption band is reduced to a mere step-out, on which it is impossible to identify with certainty even the wave-length of maximum selective absorption. The *second* absorption band of the xanthates, however, is much more fully developed; and repeated observations on different members of the series showed that it was *symmetrical on a scale of wave-lengths*, instead of being *symmetrical on a scale of frequencies* as Kuhn and Braun had postulated. The absorption bands of these compounds were therefore represented by the equation

$$\epsilon = \epsilon_{\text{max.}} e^{-\left(\frac{\lambda-\lambda_0}{\theta}\right)^2}, \qquad \text{(ix)}$$

where $\epsilon_{\text{max.}}$ and ϵ are the molecular extinction coefficients at wave-lengths λ_0 and λ, and $\theta = k\lambda'$ is a multiple of the half-width λ' of the band.

The new equation can also be written in terms of frequencies, in the form

$$\epsilon = \epsilon_{\text{max.}} e^{-\left[\frac{\nu_0}{\nu}\frac{\nu_0-\nu}{\theta}\right]^2} \qquad \text{(x)}$$

where $\theta = k\nu'$, and ν' (the half-width of the band expressed in frequencies) $= c\lambda'/\lambda_0^2$ approximately, whilst $c =$ velocity of light. In this form it can be compared with the equations (viii) of Kuhn and Braun and (vi) of Bielecki and Henri, when the latter is seen to be intermediate between the two former, since it is unsymmetrical both on a scale of wave-lengths and (by reason of the factor ν) also on a scale of frequencies. Hudson [1] has suggested that the unsymmetrical

[1] H. Hudson, *Thesis*, Cambridge, 1933, p. 162.

absorption curves of organic compounds might be represented by increasing the power of ν_0/ν in equation (x), but this modification has not yet been adequately tested, and there is at present no theoretical justification for it.

Verification of the Preceding Absorption Equation.—(*a*) Theoretical values for the molecular extinction coefficients of *menthyl* d-*bornyl xanthate*, calculated from the equation of Lowry and Hudson, are shown as curve 2 in Fig. 161. This curve is not strictly comparable with curves 2, 3 and 4, since it does not intersect the experimental curve at the point of half-width, 0·3 below the maximum on the logarithmic scale; but the agreement on the long-wave-length side of the maximum is almost within the limits of experimental error.

(*b*) Curve 1 in Fig. 162 shows the form of the *two* absorption bands of *methylene l-menthyl xanthate*,[1]

$$C_{10}H_{19}O\,.\,CS\,.\,S\,.\,CH_2\,.\,S\,.\,CS\,.\,O\,.\,C_{10}H_{19}.$$

Curve 2 is based on Kuhn and Braun's postulate of symmetry on a scale of frequencies, whilst curve 3 is symmetrical on a scale of wave-lengths. Curve 2 lies to the left of the experimental curve on *both* arms of the second absorption band, and any attempt to secure a closer agreement on one arm results in a greater deviation on the other arm. Curve 3, however, follows the course of the two absorption bands fairly closely throughout.

(*c*) A much more satisfactory analysis can be made of the absorption bands of aliphatic aldehydes and ketones, since some of these bands are isolated so completely that the minimum of absorption on the side of shorter wave-lengths has not yet been recorded. As an example, Fig. 163 shows the ultra-violet absorption band of *penta-acetyl-μ-arabinose*,[2] $H[CH\,.\,O\,.\,CO\,.\,CH_3]_5\,.\,CHO$. The form of the band can be represented very closely by the equation (ix) of Lowry and Hudson, and its symmetry on a scale of wave-lengths is clearly shown by this agreement; but at the shortest wave-lengths the experimental curve rises above the calculated curve (as in Figs. 163 and 167), perhaps on account of overlapping by the acetate band, which in other compounds begins to absorb strongly at about 2500 A.U.

(*d*) This limitation does not occur in the absorption curves of a series of ketones prepared and purified by Read and examined by Lishmund.[3] A few of these curves are strictly symmetrical on a scale of wave-lengths, and several others are steeper on the side of longer wave-lengths, but none of them is steeper on the side of shorter wave-lengths. The symmetry of the curves is remarkably exact, but the deviations from the values calculated from the equation of Lowry and Hudson are too regular to be attributed with confidence to experimental errors. In particular, the experimental curves are

[1] Lowry and Hudson, *Phil. Trans.*, 1933, A. **232**, 124.
[2] Hudson, Wolfrom and Lowry, *J.*, 1933, 1179.
[3] Unpublished results.

often slightly pinched at the top and expanded at the base. This effect may be attributed either to the inadequacy of the equation or perhaps to the superposition of two absorption bands, such as has already been demonstrated by measurements of circular dichroism in camphor and in camphor-β-sulphonic acid (pp. 406 and 407).

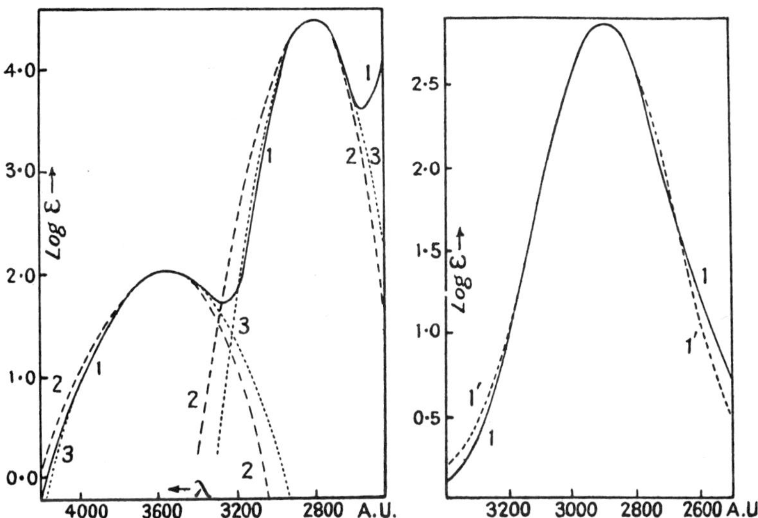

FIG. 162.—ABSORPTION CURVE OF METHYLENE *l*-MENTHYL XANTHATE.

1. Experimental curve.
2. Theoretical curve (Kuhn and Braun).
3. ,, ,, (Lowry and Hudson).

FIG. 163.—ABSORPTION CURVE OF PENTA-ACETYL-μ-GALACTOSE.

1. Experimental.
2. Theoretical (Lowry and Hudson).

(*e*) In a widely different field Mead [1] has pointed out that

> "*The spectra of the complex salts provide unique opportunities for testing the absorption equations, as the bands are well separated from one another and the half-widths can be correctly estimated.* They are obviously symmetrical on a wave-length scale, and . . . Lowry and Hudson's equation has been found to represent them with a high degree of accuracy in all cases."

This is illustrated in the case of the chromioxalate ion $[Cr(C_2O_4)_3]^{---}$ (Fig. 164), where the values calculated from the equations of Bielecki and Henri and of Kuhn and Braun are seen to deviate substantially from the experimental curve, which can be expressed, however, to a very close approximation by the equation of Lowry and Hudson. The small systematic deviations between the observed and calculated curves are of the same character as those referred to under (*d*) above.

[1] MEAD, *T.F.S.*, 1934, **30**, 1055.

Form of the Curves of Circular Dichroism.—Natanson,[1] by calculating the influence of a damping factor on optical rotatory power (p. 425), deduced for the form of the curve of circular dichroism the equation (xi)

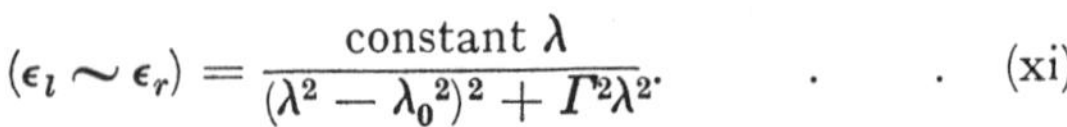

$$(\epsilon_l \sim \epsilon_r) = \frac{\text{constant } \lambda}{(\lambda^2 - \lambda_0^2)^2 + \Gamma^2\lambda^2}. \qquad \text{(xi)}$$

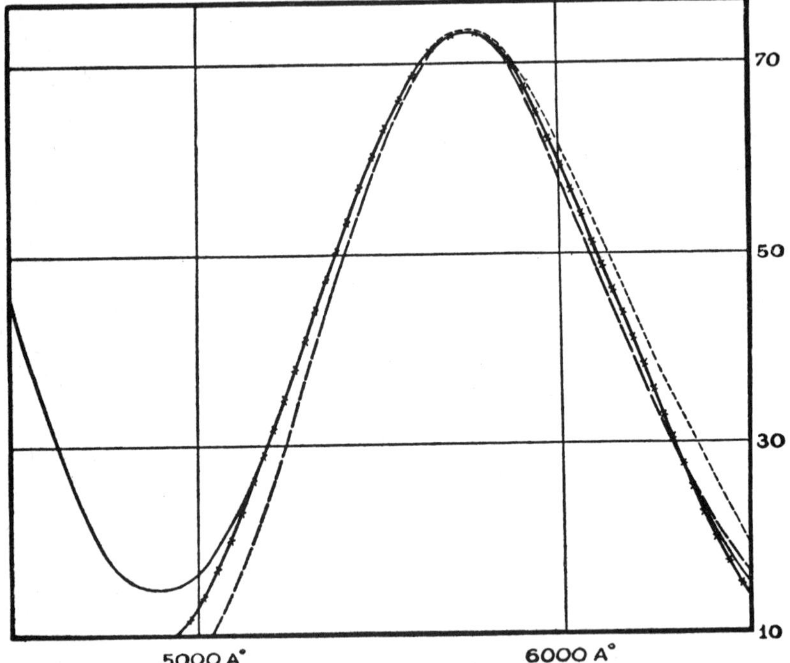

FIG. 164.—ABSORPTION BAND OF CHROMIOXALATE ION.

——————— Experimental - - - - - - - - - -Kuhn and Braun.
– – – – – – – Bielecki and Henri. × × × × × Lowry and Hudson.

This equation differs by a factor $1/\lambda$ from the Ketteler-Helmholtz equation (equation iv, p. 395) for ordinary absorption. *The dissymmetry factor, $g = (\epsilon_l \sim \epsilon_r)/\epsilon$, in a given absorption band must therefore be proportional to the frequency.* On this basis a series of equations can be deduced for the curve of circular dichroism, corresponding to the equations for ordinary absorption which have been cited above. Thus

(*a*) Assuming that the ratio $(\epsilon_l \sim \epsilon_r)/\epsilon$ is proportional to ν, Bielecki and Henri's exponential formula (vi) would give

$$(\epsilon_l \sim \epsilon_r) = (\epsilon_l \sim \epsilon_r)_{\text{max.}} \frac{\nu^2}{\nu_0^2} e^{-\beta(\nu_0 - \nu)^2}, \qquad \text{(xii)}$$

[1] NATANSON, *Bull. Acad. Sci., Krakow*, 1908, pp. 764–783.

where $(\epsilon_l \sim \epsilon_r)_{max.}$ is the maximum difference between the molecular extinction coefficients for the two circular components of plane polarised light at frequency ν_0. Both of the above equations represent curves of circular dichroism which are steeper on the side of longer wave-lengths when the latter are plotted as abscissæ.

(*b*) The absorption equation (viii) of Kuhn and Braun gives an equation for circular dichroism of the form

$$(\epsilon_l \sim \epsilon_r) = (\epsilon_l \sim \epsilon_r)_{max.} \frac{\nu}{\nu_0} e^{-\left(\frac{\nu_0 - \nu}{\theta}\right)^2} \quad . \quad . \quad \text{(xiii)}$$

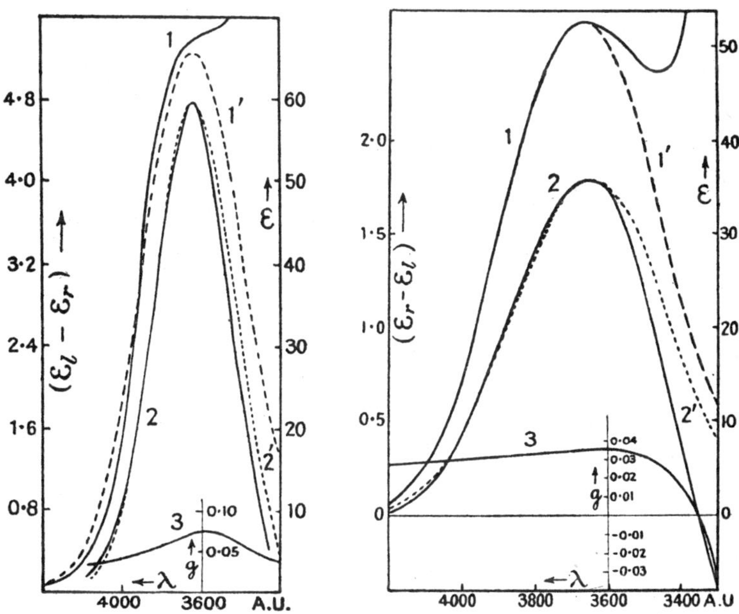

FIG. 165.—ABSORPTION AND CIRCULAR DICHROISM OF *l*-MENTHYL DIXANTHIDE.

FIG. 166.—ABSORPTION AND CIRCULAR DICHROISM OF *d*-BORNYL DIXANTHIDE

1. Experimental absorption.
1′. Theoretical absorption (L. and H.).
2. Experimental circular dichroism.
2′. Theoretical circular dichroism (L. and H.).
3. Dissymmetry factor *g*.

This equation is similar in form to the absorption equation of Bielecki and Henri; but in the case of *potassium chromium tartrate*, Kuhn and Szabo [1] chose to omit the factor ν/ν_0, and used an expression of the same form as their absorption equation, namely,

$$(\epsilon_l \sim \epsilon_r) = (\epsilon_l \sim \epsilon_r)_{max.} e^{-\left(\frac{\nu_0 - \nu}{\theta}\right)^2} \quad . \quad . \quad \text{(xiv)}$$

[1] KUHN and SZABO, *Z. ph. C.*, 1931, B. **15**, 59–73.

(c) The absorption equation (ix) of Lowry and Hudson leads to an equation for circular dichroism of the form

$$(\epsilon_l - \epsilon_r) = (\epsilon_l - \epsilon_r)_{\text{max.}} \frac{\lambda_0}{\lambda} e^{-\left(\frac{\lambda - \lambda_0}{\theta}\right)^2} \quad . \quad . \quad \text{(xv)}$$

or

$$= (\epsilon_l - \epsilon_r)_{\text{max.}} \frac{\nu}{\nu_0} e^{-\left[\frac{\nu_0}{\nu}\left(\frac{\nu_0 - \nu}{\theta}\right)\right]^2} \quad . \quad . \quad \text{(xvi)}$$

Applications of the Equation of Lowry and Hudson.—(a) *Xanthates*. Figs. 165 and 166 show the results obtained by applying the preceding equation (xv) of Lowry and Hudson to the experimental data for the circular dichroism of the weak ultra-violet absorption band of d-*bornyl dixanthide* and l-*menthyl dixanthide*.[1]

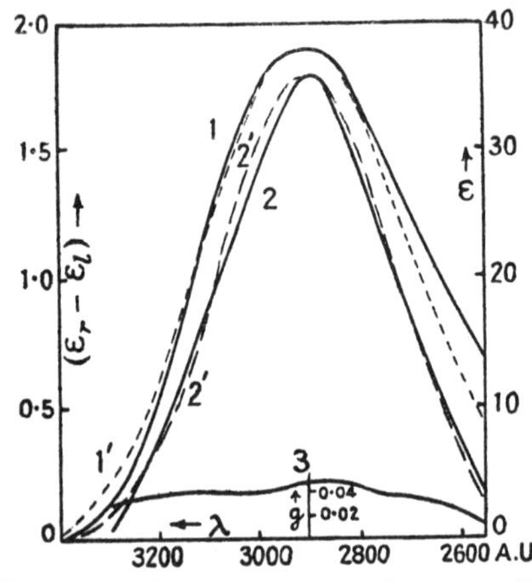

FIG. 167.—ABSORPTION AND CIRCULAR DICHROISM OF TETRA-ACETYL-μ-ARABINOSE.

1. Observed absorption.
1'. Theoretical absorption (L. and H.).
2. Circular dichroism.
2'. " " (L. and H.).
3. Dissymmetry factor.

Theoretical and experimental values are in close agreement except on the high-frequency side of the absorption band, where small deviations are shown in Fig. 165 and much larger ones in Fig. 166. The fact that the circular dichroism of *d*-bornyl dixanthide falls off much more rapidly on the side of shorter wave-lengths than on the side of longer wave-lengths is in direct contradiction of the general rule that these curves are steeper on the long wave-length side, even when the absorption curves are symmetrical on a scale of wave-lengths; but this anomaly can be proved to be due to the superposition of a second curve of circular dichroism, of opposite sign to the first, and depending on the optical activity of the second much stronger band at 2800 A.U. The resultant curve of observed circular dichroism therefore falls to zero at 3350 A.U., and then actually crosses the axis and gives values of opposite sign. The great intensity of the second absorption band made it impossible, however, to extend the measurements of circular dichroism into the region of absorption covered by this band.

[1] LOWRY and HUDSON, *Phil. Trans.*, 1933, A. **232**, 131 and 132.

(*b*) *Aldehydic Sugars.* Values for the absorption and circular dichroism of *tetra-acetyl-μ-arabinose*[1] are set out in Fig. 167. The curve 2 of circular dichroism conforms closely to the theoretical curve 2′ deduced from the equation (xv) of Lowry and Hudson, but the curve 3 representing the dissymmetry factor, *g*, instead of rising progressively in proportion to the frequency throughout the band, falls off on the side of shorter wave-lengths. This decrease is seen at once to be due to a separation of the experimental and theoretical absorption curves 1 and 1′, which has already been attributed to an overlapping of the aldehydic band by the absorption bands of the acetate groups (p. 400).

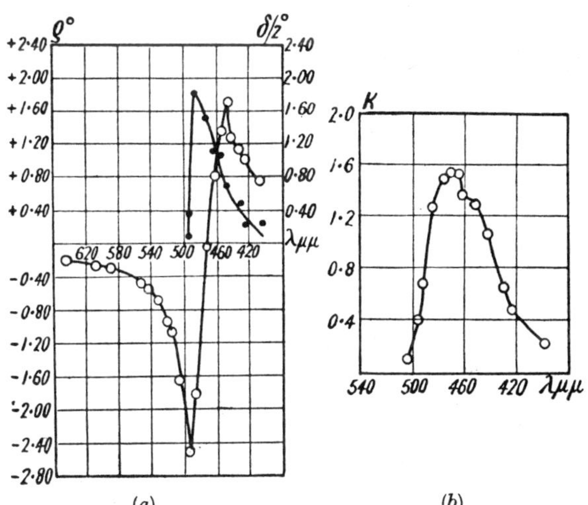

FIG. 168.—ROTATORY DISPERSION, CIRCULAR DICHROISM AND ABSORPTION OF CAMPHORQUINONE (Wedeneewa).

(*a*) Circular dichroism (black dots).
Rotatory dispersion (hollow circles).
(*b*) Absorption (hollow circles).

Circular Dichroism of Camphor, Camphorquinone and Camphor β-Sulphonic Acid.—The investigations of Wedeneewa[2] (p. 156) on the rotatory dispersion of solutions of camphorquinone in toluene showed that the maximum absorption was at 4700 A.U., but the maximum of circular dichroism was displaced to 4900 A.U. (Fig. 168). Similar displacements have been recorded by Kuhn and Gore[3] for solutions of camphor in hexane (Fig. 169), and by Lowry and French[4] for solutions of camphor-β-sulphonic acid in water (Fig. 170). The phenomenon observed in these two compounds is

[1] HUDSON, WOLFROM and LOWRY, *J.*, 1934, p. 1184.
[2] WEDENEEWA, *Ann. der Physik*, 1923, **72**, 122.
[3] KUHN and GORE, *Z. ph. C.*, 1931, B. **12**, 389.
[4] LOWRY and FRENCH, *J.*, 1932, p. 2655.

not, however, precisely the same as in camphorquinone, since the circular dichroism is concentrated in the long wave-length half of the >C=O band, whereas in camphorquinone it extends throughout the whole of the quinonoid band, although the maximum is not

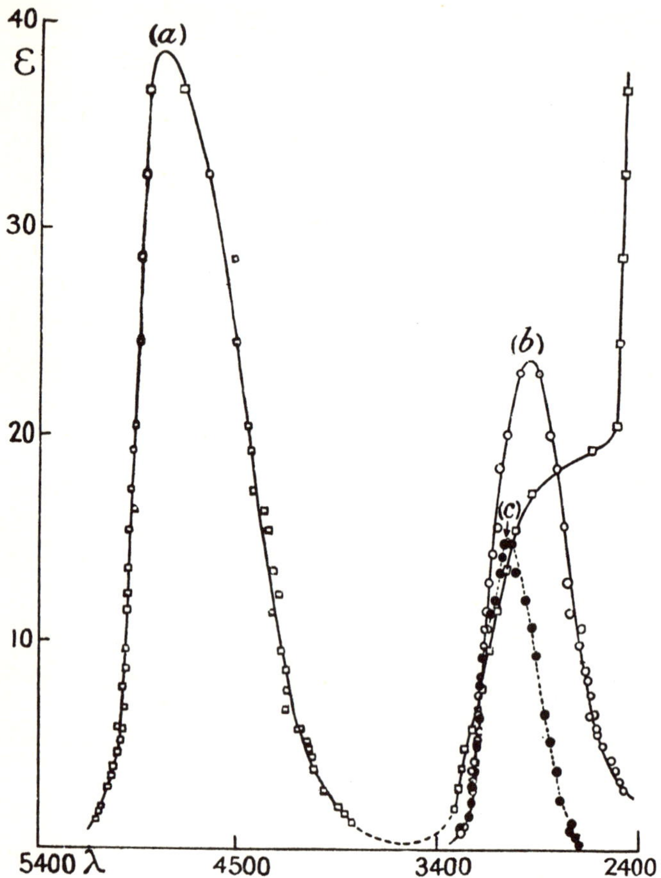

FIG. 169.—ABSORPTION AND CIRCULAR DICHROISM.
(*a*) Absorption of camphorquinone in *cyclo*-hexane (Lowry and Gore).
(*b*) ,, ,, camphor in hexane (Kuhn and Gore).
(*c*) Circular dichroism of camphor in hexane (Kuhn and Gore).

symmetrically placed. The dissymmetry factor $g = (\epsilon_l \sim \epsilon_r)/\epsilon$ therefore falls rapidly in passing through the ultra-violet ketonic band towards higher frequencies (Fig. 170), while it remains nearly constant in passing through the visual quinonoid band.

The concentration of circular dichroism in the long wave-length side of the >C=O band of camphor has been interpreted by Kuhn and Gore [1] and by Kuhn and Lehmann [2] as an indication that this band consists of two superposed components, having their origin in different electronic transitions, which cannot be resolved by

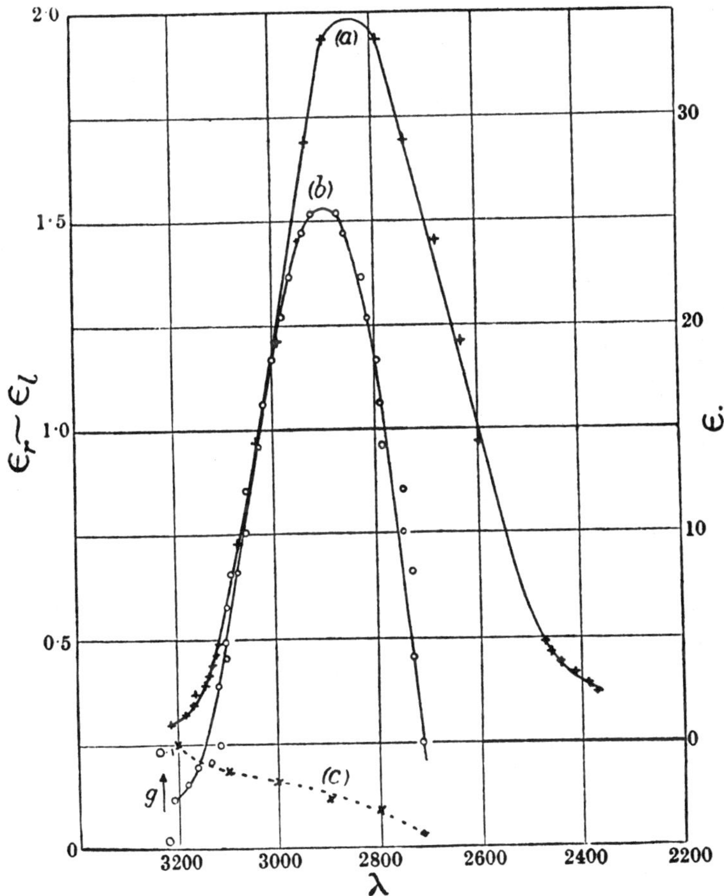

FIG. 170.—(a) ABSORPTION, (b) CIRCULAR DICHROISM, (c) DISSYMMETRY FACTOR OF CAMPHOR β SULPHONIC ACID IN WATER (Lowry and French).

measurements of absorption of unpolarised light. The low frequency "subsidiary band" contributes practically the whole of the circular dichroism and associated optical activity, whereas the high-frequency

[1] KUHN and GORE, *Z. ph. C.*, 1931, B. **12**, 389.

[2] KUHN and LEHMANN, *Z. Elektrochem.*, 1931, **37**, 549; *Z. ph. C.*, 1932, B. **18**, 32–48.

" principle band " is almost devoid of circular dichroism and optical activity.

Kuhn and Lehmann[1] have deduced the relative intensities of these components as follows. On the assumption that absorption as well as circular dichroism is an additive property, it follows that

$$g_{\text{obs.}} = \frac{(\epsilon_l - \epsilon_r)_{\text{obs.}}}{\epsilon_{\text{obs.}}} = \frac{(\epsilon_l - \epsilon_r)_{\text{subs.}} + (\epsilon_l - \epsilon_r)_{\text{prin.}}}{\epsilon_{\text{subs.}} + \epsilon_{\text{prin.}}}$$

$$= \frac{\epsilon_{\text{subs.}}\, g_{\text{subs.}} + \epsilon_{\text{prin.}}\, g_{\text{prin.}}}{\epsilon_{\text{subs.}} + \epsilon_{\text{prin.}}}.$$

The rapid decrease in the value of $g_{\text{obs.}}$ in passing from low to high frequencies shows that $g_{\text{prin.}}$ is very small, so that

$$g_{\text{obs.}} = \frac{\epsilon_{\text{subs.}}\, g_{\text{subs.}}}{\epsilon_{\text{subs.}} + \epsilon_{\text{prin.}}}$$

$$\therefore\ \epsilon_{\text{subs.}} = \frac{g_{\text{obs.}}(\epsilon_{\text{subs.}} + \epsilon_{\text{prin.}})}{g_{\text{subs.}}}$$

$$= \frac{(\epsilon_l - \epsilon_r)_{\text{obs.}}}{g_{\text{subs.}}}.$$

The maximum value of $g_{\text{obs.}}$ for camphor in hexane is 0·1; since $g_{\text{prin.}}$ is negligible, $g_{\text{subs.}} = 0{\cdot}1$ approximately. The corresponding maximum value of $(\epsilon_l - \epsilon_r)_{\text{obs.}}$ is about 1·4. The subsidiary active component band will therefore have a maximum intensity of $\epsilon_{\text{subs.}} = 14$ approximately, whereas the observed maximum intensity of the whole band is about $\epsilon_{\text{obs}} = 24$.

Although the ketonic or quinonoid >CO band in camphor, camphor-β-sulphonic acid and camphorquinone is composite in character from the point of view of optical activity, the optical activity of the —CHO band in the aldehydic sugar acetates appears to be relatively uniform throughout. Thus the maxima of circular dichroism and absorption of unpolarised light for tetraacetyl-μ-arabinose (Fig. 167) are separated by only 10 A.U. In the glucose compound, however, indications were observed of a small complexity at the extreme low frequency side, where a very weak circular dichroism of opposite sign, between 3100 A.U. and 3300 A.U., was postulated in order to represent the experimental data.

Induced Dissymmetry of Unsaturated and Chromophoric Groups.—The relationship between absorption and rotation in compounds which possess optically-active absorption bands was considered by Tschugaeff and Ogorodnikoff in 1913[2] in the case of a molecule containing two identical asymmetric centres linked together through

[1] Kuhn and Lehmann, *Z. Elektrochem.*, 1931, **37**, 549; *Z. ph. C.*, 1932, B. **18**, 32–48.

[2] Tschugaeff and Ogorodnikoff, *Z. ph. C.*, 1913, **85**, 507–508.

two identical chromophoric groups. Their views are illustrated in the following scheme:

$$\begin{array}{ccccccccc} A' & & \overbrace{\phantom{C\text{---}S}}^{B'} & & & & \overbrace{\phantom{S\text{---}C}}^{B''} & & A'' \\ M\text{---}O\text{---} & C & \text{---}S\text{---} & CH_2 & \text{---}S\text{---} & C & \text{---}O\text{---}M \\ & \| & & & & \| & \\ & S & & & & S & \end{array}$$

$[\alpha] = 2[\beta]$ where $[\beta]$ is the partial rotation of each half of the molecule.

$$\underset{\text{due to A}'}{[\beta]} = \underset{\text{B}'\text{ on A}'}{[\gamma']} + \underset{\text{B}''\text{ on A}'}{[\gamma'']} + [\gamma'''] + \text{etc.}$$

"In order to obtain a rough picture of the relationships in question, let us consider first the interactions of the radicals in the methylene ester,

$$M\text{---}OCS\text{---}S\text{---}CH_2\text{---}S\text{---}CSO\text{---}M.$$

Let A′ and A″ be the two centres of activity (the two menthyl radicals) which are identical by reason of the symmetry of the molecule. Also let B′ be the chromophore (CS_2) lying nearest to the radical A′, and B″ the second chromophore, identical with B′. The total rotation $[\alpha]$ produced by the substance in question can be regarded as the sum of two equal partial rotations $[\beta]$, of which one depends on the radical A′ and the second on A″: $[\alpha] = 2[\beta]$. Moreover, each of the partial rotations $[\beta]$ can be divided into several parts $[\gamma']$, $[\gamma'']$, $[\gamma''']$, etc., so that $[\beta] = [\gamma'] + [\gamma''] + [\gamma'''] + \ldots$ Let us consider more closely the relationships at the radical A′. Of the different partial rotations, which together make up the rotation $[\beta]$, one, say $[\gamma']$, is determined by the influence on A′ of the nearer chromophore B′, and the second, say $[\gamma'']$, by the influence of the more distant chromophore B″ on A′. We must assume that the value of $[\gamma'']$ is smaller than $[\gamma'] + [\gamma'''] + \ldots$, but that in the methylene ester it cannot be neglected in comparison with the other components of $[\beta]$. In the optical superposition of the separate $[\gamma]$-values, $[\gamma'']$ must therefore play a notable rôle.

"If we now go to the ethylene and trimethylene esters, we have obviously to deal with quite similar relationships, as regards the relative values of $[\gamma'']$, $[\gamma']$, $[\gamma''']$, etc. In the trimethylene ester the value of $[\gamma'']$, appears to be quite inconsiderable, and practically to disappear in comparison with $[\gamma'] + [\gamma''']$, etc. In the ethylene-compound it will be a little larger than in the trimethylene ester. In this way it is possible to explain the big jump, which occurs in the series

$$M\text{---}O\text{---}CS\text{---}S\text{---}(CH_2)_n\text{---}S\text{---}CS\text{---}O\text{---}M,$$

between the cases $n = 1$ and $n = 2$, as well as the quite insignificant difference between $n = 2$ and $n = 3$."

This passage contains two important features.

(1) The influence of the chromophoric group is assumed to decrease as its distance from the asymmetric carbon atom increases, and to become negligible when this distance is large.

(2) The two chromophoric groups are regarded as contributing independently to the partial rotations of the two asymmetric carbon atoms, in much the same way as in the later theories of de Mallemann and Boys, i.e. by introducing an unsaturated group into one of the four dissimilar radicals which provide the product of asymmetry of these atoms.

(1) The influence of the distance of the chromophoric group from an asymmetric centre on its contribution to the rotatory power of the latter can be interpreted in terms of Born's theory of coupled vibrators (p. 374), since the coupling is assumed to be electrical in origin, and must therefore decrease with increasing distance, e.g. in accordance with an inverse-cube law. This coupling of a chromophore with neighbouring radicals has been described by Kuhn [1] as VICINAL ACTION, but the phenomenon under consideration was clearly recognised by Tschugaeff, and is independent of any model that may be used to illustrate it.

(2) The two chromophores are regarded as having an "influence" on the two asymmetric carbon atoms, but no reciprocal effect is indicated. An alternative view was put forward by Lowry and Walker in 1924 [2] (p. 146), when a consideration of the experimental evidence led them to adopt the view that

> "Unsaturated chromophoric groups can be made asymmetric by induction from a fixed asymmetric centre, and thus develop optical activity."

This hypothesis was based upon the observation that

> "The dispersion equations for camphor and its derivatives are haunted by a low-frequency term *the period of which is definitely characteristic of the ketonic group*." [3]

The conditions under which this phenomenon may be expected to occur were indicated in the suggestion that

> "The chromophoric groups can exhibit an 'induced asymmetry' as a result of which they may themselves become optically-active *when coupled sufficiently closely to an asymmetric complex*."

A further limitation was made in the suggestion that

> "Since the experimental evidence shows that only a very modest optical rotatory power is developed in molecules which contain no unsaturated atom or group of atoms, it is clear that induced asymmetry can only give rise to important developments of optical rotatory power when applied to unsaturated or chromophoric groups."

[1] KUHN, FREUDENBERG and WOLF, *Ber.*, 1930, **63**, 2370; KUHN, *Ber.*, 1930, **63**, 190.

[2] LOWRY and WALKER, *Nature*, April 19, 1924, **113**, 565.

[3] LOWRY and CUTTER, *J.*, 1925, **127**, 613.

Thus the $>CH_2$ and $>CMe_2$ groups of camphor, although exposed to a dissymmetric intramolecular field of force, do not appear to make any important contribution to the rotatory power of camphor, whereas the dextrorotation of the ketone is due entirely to the fact that the lævorotation of the two fixed asymmetric centres is swamped by the dextrorotation induced in the polarisable $>C{=}O$ group by the internal field of the molecule.

The development of optical activity by induction in a polarisable radical has been described by the author as INDUCED DISSYMMETRY[1] (p. 147). An alternative view, also based upon Born's theory of coupled electrons, has been developed by Kuhn, who attributes the effect in question to the "induced anisotropy" of the absorption band. Born himself, however, does not regard anisotropy as essential to the development of optical activity in absorbing media, and his formula for the rotatory power of the spiro-compounds (p. 391) is just as valid for isotropic as for slightly anisotropic radicals.

Dissymmetry of the Ketonic Group in Optically-Active Ketones.—The view that the dissymetric internal field in optically-active ketones renders the carbon atom of the $>C{=}O$ group optically active by induction was supported[2] by citing the behaviour of these ketones when reduced. In this process, an unsymmetrical ketone of the type R′ . CO . R″ always yields an inactive or racemic secondary alcohol, since the two stereoisomeric products are formed in equal quantities,

e.g. $$CH_3 . CO . C_2H_5 \xrightarrow{+H_2} dl\ CH_3 . \mathbf{C}HOH . C_2H_5 ;$$

but this equality of yield is no longer maintained when the ketone is optically active.

> "Thus in the camphor series two stereoisomeric products may be formed in the ratio of 8 : 1 instead of 1 : 1. The hypothesis of the equality of the two links of a double bond is therefore not merely unproved, but, in the case of asymmetric compounds, is directly opposed to the experimental evidence."

The production of unequal yields of borneol and of *neo*-borneol by reduction of camphor may be due in part to steric effects, but, since the symmetry of the highly polarisable $>CO$ group is necessarily destroyed by a transverse electric field, the only question at issue is to determine whether the distortion thus produced is adequate to account for the chemical and physical effects which accompany this loss of symmetry. Additional evidence on this point has been provided by the "molecular theory" of optical rotatory power (Ch. XXIX, p. 369). This theory accounts satisfactorily for the rotations observed in some of the simplest optically-active molecules, containing only one asymmetric carbon atom and no unsaturated or chromophoric group, but, when extended to the region covered by

[1] LOWRY, *Nature*, April 22, 1933, **131**, 566.
[2] LOWRY and WALKER, *Nature*, April 19, 1924, **113**, 565.

the ultra-violet absorption band of an optically-active ketone, it only accounts for about 1 per cent. of the observed effects. In particular, the zero refraction, which is required in order to reproduce the Cotton effect in absorbing media, can only be developed by postulating the presence of an additional asymmetric centre *within the chromophoric group*. The extension of the molecular theory to absorbing media appears therefore to depend on postulating an induced dissymmetry in the chromophoric group, and to this extent therefore the theory lends support to this postulate.

Dissymmetric Photochemical Decomposition.—Circular dichroism implies that circularly polarised light of opposite signs is unequally absorbed by the optically active medium. Conversely, circularly polarised light of a given sign will be unequally absorbed by the *d* and *l* forms of the optically-active material, and, if photochemically active, will give rise to unequal decomposition.

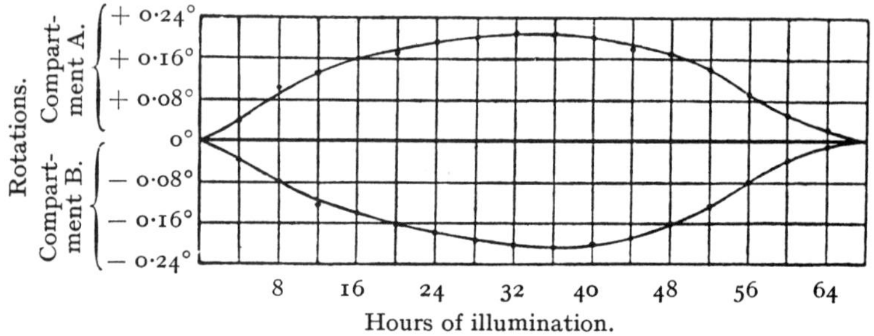

FIG. 171.—ROTATORY POWER OF HUMULENE NITROSITE IRRADIATED WITH CIRCULARLY POLARISED LIGHT.

This prediction, made by van't Hoff in 1894,[1] was verified in 1930 by Kuhn and Knopf,[2] who acted upon a solution in hexane of *d l- α-azidopropionic dimethylamide* with circularly polarised light of wave-length 3200 to 2800 A.U. from a mercury lamp with a benzene filter. The product showed a rotation $\alpha_{5780} = -1{\cdot}04°$ per decimetre when 37 per cent. had been decomposed by left-handed circularly polarised light, and $+0{\cdot}78°$ when 35 per cent. had been decomposed by a right-handed beam.

Similar experiments were made by Mitchell[3] with a 7 per cent. solution of blue humulene nitrosite in ethyl butyrate, in a double cell, of which one side was illuminated with right-handed and the other with left-handed circularly-polarised red light. The rotations of the two solutions increased in opposite directions to a maximum $\alpha_{5461} = \pm 0{\cdot}21°$ after about thirty-six hours of illumination, and then

[1] VAN'T HOFF, *Chemistry in Space*, English translation, 1898, p. 45.
[2] KUHN and KNOPF, *Z. ph. C.*, 1930, B. **7**, 292–310.
[3] MITCHELL, *J.*, 1930, pp. 1829–1834.

decreased again to zero after sixty-eight hours, by reason of further photochemical decomposition (Fig. 171).

Similar curves have been plotted by Karagunis and Drikos[1] in a paper, *Zur Stereochemie der freien Triarylmethylradikale.* These authors carried out the reaction

$$\left.\begin{array}{r} C_6H_5 \\ C_6H_5 \,.\, C_6H_4 \\ C_{10}H_7 \end{array}\right\rangle C + Cl_2 = 2\left.\begin{array}{r} C_6H_5 \\ C_6H_5 \,.\, C_6H_4 \\ C_{10}H_7 \end{array}\right\rangle CCl$$

in left- and right-circularly polarised light of wave-lengths Hg 4358 and Na 5890. When irradiated with violet light, selective chlorination of the *d* and *l* forms of the radical gave rise to curves diverging to about $\pm 0{\cdot}07°$ after about one hour and then converging again. Similarly, irradiation with yellow light gave curves diverging to $\pm 0{\cdot}08°$ and then converging again after about seventy-five minutes.

[1] Karagunis and Drikos, *Z. ph. C.*, 1934, B. **26**, 428–438.

CHAPTER XXXII.

ANALYSIS OF ROTATORY DISPERSION.

(*a*) *IN TRANSPARENT MEDIA.*

Graphic and Algebraic Methods.—In order to demonstrate his Law of Inverse Squares (p. 11) Biot in 1818 [1] plotted, against the square of the wave-length, the length of the column of quartz required to produce rotations of 180°, 360°, etc. (Fig. 37, p. 125). A similar process of plotting $1/\alpha$ against λ^2 was proposed by Lowry and Dickson in 1913 [2] (p. 125) as a method of testing the validity of Drude's equation $\alpha = k/(\lambda^2 - \lambda_0^2)$. In subsequent years it has been used extensively, since it provides a very convenient means of discriminating between dispersions which are approximately simple and those which are obviously complex (see especially Fig. 38, p. 126, and Figs. 39 to 42, pp. 128 and 129); but it is too inexact to be used as a precise method of analysis. The well-deserved criticisms of Hunter [3] were indeed only an elaboration of views which had already been expressed clearly by Lowry and Abram,[4] when they described this process as "the easiest *although perhaps the least exact* method" of analysing rotatory dispersions. The opinion, expressed by Lowry and Richards,[5] that "the only valid way of testing the character of a given dispersion curve is by the numerical method described below" is therefore entirely justified. It is, however, often worth while, before attempting a complete algebraic analysis of a dispersion curve, to plot (*a*) α against λ, (*b*) $1/\alpha$ against λ^2. Thus

(*a*) *α against λ*: (i) If α increases progressively as λ decreases, so that the dispersion curve is always convex towards the axis of wave-lengths, the rotatory dispersion is *normal;* but it is not necessarily *simple*, since more than one term of Drude's equation may be required to express the form of the curve. (ii) If, however, the curve exhibits a reversal of sign, a maximum, or an inflection, or if the curve is concave and bulges away from the axis of wave-lengths, the dispersion is *anomalous* and can only be represented by two terms of opposite sign.

[1] Biot, *Mem. Acad. Sci.*, 1818, **2,** 41–136.
[2] Lowry and Dickson, *J.*, 1913, **108,** 1075.
[3] Hunter, *J.*, 1924, **125,** 1198–1206.
[4] Lowry and Abram, *J.*, 1919, **115,** 303.
[5] Lowry and Richards, *J.*, 1924, **125,** 2514.

(*b*) $1/\alpha$ *against* λ^2: (i) If $1/\alpha$ against λ^2 gives a straight line, the rotatory dispersion is either *simple* or *pseudo*-simple, i.e. it can be represented at least approximately by one term of Drude's equation; but this is not always easily distinguished from cases (ii) and (iii) below. In particular, if the ratio $\alpha_{4358}/\alpha_{5461}$ is less than 1·63, it is almost certain that the dispersion is of type (iii).

(ii) If the curve is *concave* towards the axis of λ^2, i.e. if it droops towards the horizontal axis on which the square of the wave-lengths are plotted, then *two terms of similar sign* are required, and the rotatory dispersion is *complex* but *normal;* the curvature in question is, however, very slight, except in the immediate neighbourhood of the nearer absorption band, and may easily be overlooked.

(iii) If the curve is *convex* towards the axis of λ^2 then *two terms of opposite sign* are required, and the rotatory dispersion is *complex;* but it may be either *anomalous* or *quasi-anomalous* (p. 139), according to the relative magnitude of the two terms. If the dispersion is anomalous the curve will exhibit a minimum at the maximum of rotation, and will be interrupted by an asymptote at the reversal of sign.

Graphical Analysis of Complex and Anomalous Rotatory Dispersions.—In view of the very real difficulty of analysing complex and anomalous dispersion curves, which cannot be expressed by one term of Drude's equation, graphical methods have been used extensively in order to classify these complicated data. All these methods depend on the fact that *changes of conditions, which affect profoundly the rotation constant of a partial rotation, are often without influence on the dispersion constant.* Thus, the homologous series of secondary alcohols studied by Pickard and Kenyon exhibit remarkable changes of rotatory power as the length of the growing chain is increased (see, for instance, Fig. 124, p. 264); but the dispersion ratios are almost constant, with the exception of the initial compound, which sometimes gives a slightly higher value than the other members of the series (p. 265). Indeed, nearly all straight chain aliphatic secondary alcohols give dispersion ratios which differ but little from the mean value $\alpha_{4358}/\alpha_{5461} = 1{\cdot}64$ (approx.). An even more striking illustration of this rule is provided by Longchambon's observation (p. 341) that the dispersion ratios are often the same for solids and solutions, although the specific rotations may differ very widely.

This rule, which can be verified directly in cases of simple rotatory dispersion, may be expected to apply also to the partial rotations which combine to produce a complex rotatory dispersion. If the conditions are favourable, a family of complex dispersion curves may then be represented by an equation of the type

$$\alpha = \frac{k}{\lambda^2 - \lambda_1^2} + \frac{1 - k}{\lambda^2 - \lambda_2^2}$$

exactly as if they were produced by mixtures of two components with simple but unequal dispersions. Thus the families of curves which represent the rotatory dispersions of aqueous solutions of tartaric acid of different concentrations (Fig. 30, p. 96) or of ethyl tartrate in different solvents (Fig. 32, p. 98), show a close resemblance to the theoretical curves (Fig. 44, p. 142) plotted from the preceding equation. This result can be explained by the fact that these data can all be resolved by algebraic analysis into the sum of two partial rotations of opposite sign, of which the dispersion constants are approximately constant at 0·03 and 0·07 respectively, although the rotation constants vary very widely.*

The diagrams described below have all been designed to test whether a family of dispersion curves conforms to the mixture rule. They are therefore of no assistance in analysing any individual dispersion curve, but can be used to sort out those unruly members of a family which refuse to conform to a rule which the other members obey. Most of the data to which the test has been applied can be expressed as the sum of two simple partial rotations, as in the case of tartaric acid, where "fixed" derivatives have actually been prepared (p. 290), which exhibit simple rotatory dispersions, with dispersion constants corresponding with those of the two-terms of the Drude equation; † but the mixture rule could be obeyed equally well if each of the limiting cases included two partial rotations which could not be isolated experimentally. Indeed, the rotatory dispersions of the limiting cases need not be simple and need not even be related in any way to one another; nor is it legitimate to assume that, because the mixture rule is obeyed, the active medium actually contains two sorts of dissymmetric molecules, since the same effect may be produced by variations in the activity of two optically-active absorption bands in a single molecule, as the properties of the medium are altered (p. 112).

Darmois' Dispersion Diagram.—Darmois' method of analysis[1] is based on Biot's additive rule of optical rotations, according to which the specific rotatory power of a mixture is given by the equation

$$[\alpha]_p = p[\alpha]_1 + (1 - p)[\alpha]_2,$$

where $[\alpha]_1$, $[\alpha]_2$ and $[\alpha]_p$ are the specific rotations, at a given wave-length, of the two pure components and of a mixture containing a fraction p of one of them respectively. The corresponding dispersion curves are represented arbitrarily by *ab*, *cd*, *ef*, in Fig. 172. If the ordinates at two wave-lengths, λ_1 and λ_2, cut the three curves at A, B, C, and at A_1, B_1, C_1, respectively, then $\overline{CA} = [\alpha]_1 - [\alpha]_p$, and $\overline{CB} = [\alpha]_2 - [\alpha]_p$, so that

$$\frac{\overline{CA}}{\overline{CB}} = \frac{1 - p}{p} = \frac{\overline{C_1A_1}}{\overline{C_1B_1}}.$$

* It has already been pointed out (p. 143) that the curve for which the two dispersion constants are equal and opposite provides a boundary between normal and anomalous rotatory dispersion.

† The most remarkable of these is *solid tartaric acid*, which, as LONGCHAMBON has shown (p. 111), corresponds with the simple lævorotatory component of its solutions.

[1] DARMOIS, *A.C.P.*, 1911, [viii], **22**, 572.

The ratio of the intercepts on the two ordinates is therefore independent of the wave-length, and AA_1, BB_1, CC_1 must all pass through the same point O. Thus, the family of lines corresponding to CC_1 for all mixtures of the same pair of substances will be concurrent. Conversely, measurements of the rotatory powers of a series of mixtures at two wave-lengths (e.g. the violet and yellow lines of mercury) which yield a series of concurrent lines, obviously obey the mixture rule; and if, in addition, the specific rotations of the two pure components are known, the proportions in which they are present in any mixture can be deduced. From observations of this kind, Darmois concluded that the various fractions obtained in the dis-

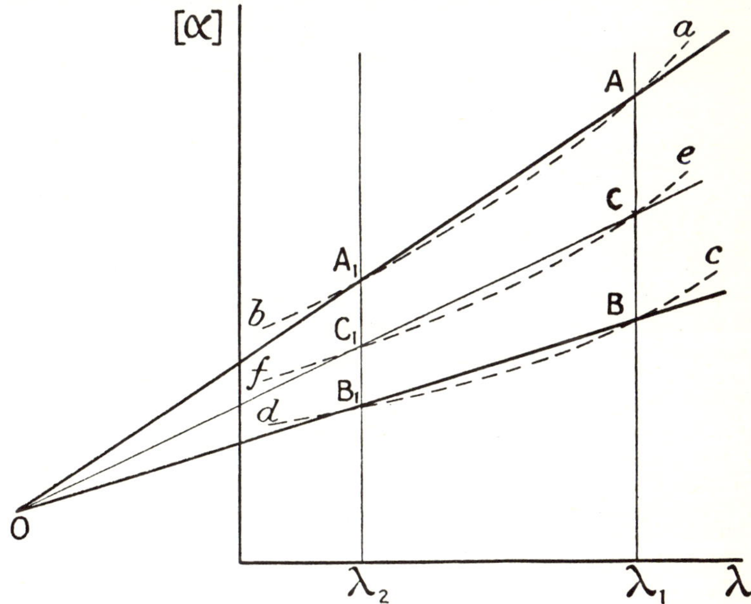

FIG. 172.—DARMOIS' DISPERSION DIAGRAM.

tillation of *turpentine* contain only α- and *β-pinene* (compare Fig. 36, p. 110). He also applied the method to the analysis of synthetic *borneol*, and subsequently to solutions containing *copper* and *malic acid* in various proportions.[1]

Darmois' relation was restated by de Mallemann [2] in a more general form, namely, that "*the tangents to the dispersion curves at points corresponding to the same abscissa pass through a point.*" He also showed that the specific rotations (for the yellow and blue lines of mercury) of aqueous solutions of tartaric acid containing various

[1] DARMOIS, *Journ. de Phys.*, 1924, [iv], **5**, 225.
[2] DE MALLEMANN, *ibid.*, 1923, [iv], **4**, p. 29.

salts can be fitted with great accuracy into a Darmois diagram, in harmony with the views of Arndtsen and of Longchambon that solutions of tartaric acid contain two types of molecule, which have been described by the latter author as α and β tartaric acid.

Characteristic Diagram of Armstrong and Walker.—Armstrong and Walker in 1913[1] plotted the specific rotations of a series of related compounds for a selected wave-length, e.g. Hg 5461, as abscissæ and the rotations for other selected wave-lengths as ordinates. The data for the first selected wave-length necessarily give a straight line inclined at an angle of 45°. Under favourable conditions the rotations for the other wave-lengths can also be represented by straight lines; but substances whose dispersions are not closely related to those of other members of the family may show marked deviations from these straight lines. The resulting system of intersecting straight lines was described as a CHARACTERISTIC DIAGRAM.

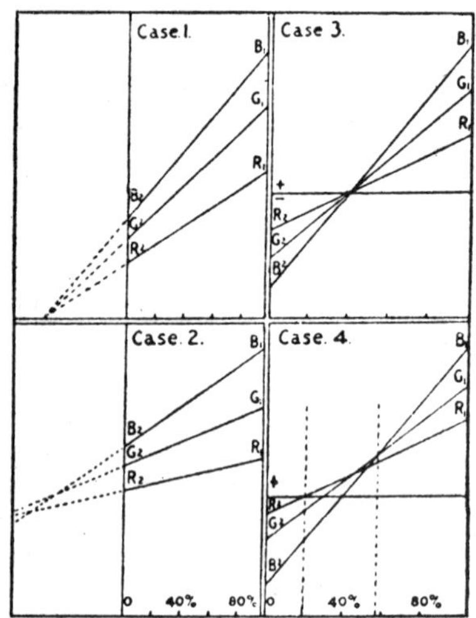

FIG. 173.—CHARACTERISTIC DIAGRAM OF ARMSTRONG AND WALKER (1913).

These diagrams (Fig. 173) may be of four types, corresponding to mixtures of the following four kinds:—

Case 1. Two substances having rotatory powers of the same sign and the same dispersive power.

Case 2. Two substances having rotatory powers of the same sign but different dispersive power.

Case 3. Two substances having rotatory powers of opposite sign but the same dispersive power.

Case 4. Two substances having rotatory powers of opposite sign and also of different dispersive power.

In Fig. 173 $R_1G_1B_1$ and $R_2G_2B_2$ are the specific rotations of the pure constituents for red, green and blue light. From the diagrams, it is clear that the abnormal sequence of colours, which is characteristic of anomalous rotatory dispersion, can only occur in case 4, and will then be observed only in the region between the two dotted lines in the diagram.

Patterson's Rational Zero and Rational Dispersion Coefficient.—The characteristic diagram of Armstrong and Walker was used by Pickard and Kenyon in their work on the rotatory

[1] H. E. ARMSTRONG and E. E. WALKER, *P.R.S.*, 1913, A. **88**, 388–403.

dispersion of alcohols and esters.[1] It was modified, however, by Patterson,[2] who plotted as ordinates, not rotations, but their differences from those at the standard wave-length, positive differences being plotted upwards and negative ones downwards. This method has the advantage of giving a wider scale of ordinates. When the dispersion is anomalous, the lines for the various wave-lengths do not converge to a point on the horizontal axis. In this case Patterson proposes to regard the point of intersection of a given line with the axis as a RATIONAL ZERO, and to calculate a RATIONAL DISPERSION COEFFICIENT by subtracting the rotation at this point from the observed molecular rotations before deducing a dispersion ratio. Thus the values of $[M]_{5780}$ for a series of derivatives of *benzoyl-l-tetrahydroquinaldine*[3] would become equal to the values of $[M]_{5461}$ if extrapolated to $+ 220°$. In a particular case in which $[M]_{5461} = 753°$, $[M]_{5780} = 648{\cdot}4°$, the "rational dispersion coefficient" for the green and yellow lines is therefore given by the ratio

$$(753° - 220°)/(648{\cdot}4° - 220°) = 1{\cdot}2454.$$

This ratio remains remarkably constant over the range of compounds which conform to the same straight-line law on the characteristic diagram, whereas the dispersion ratios deduced directly vary widely in this range. The "rational zero" must, however, be different for each wave-length, since it is not possible on an ordinary curve of rotatory dispersion for more than one other wave-length at a time to give the same rotation as the standard mercury-green line.

Application of Drude's Equation.—The general equation is of the form

$$[\alpha] = \sum \frac{k_n}{\lambda^2 - \lambda_n{}^2}.$$

Three cases may then be considered.[4]

(*a*) *Two-constant equation* $\alpha = k/(\lambda^2 - \lambda_0{}^2)$. When $1/\alpha$ is plotted against λ^2, a linear relation should be found, from which a rough value of $\lambda_0{}^2$ can be deduced; but this preliminary test should always be followed up by an algebraic analysis. The easiest method is to choose two wave-lengths, e.g. Hg 5461 and 4358 or Cd 5085 and 6438, separated rather widely from one another, but if possible, capable of being read without any change in the set-up of the apparatus; the two constants of the equation can then be deduced from the rotations for these two wave-lengths. The equation is of the form $y = p/(x - a)$ where $x = \lambda^2$ and y is the rotation for wave-length λ. Inserting two pairs of values for x and y, a and p can at once be deduced from the equation

$$x_1y_1 - ay_1 = x_2y_2 - ay_2 = p.$$

[1] PICKARD and KENYON, *J.*, 1914, **105**, 844, 845, 847, 1118, 2269, *et seq.*
[2] PATTERSON, *J.*, 1916, **109**, 1176–1203.
[3] POPE and WINMILL, *J.*, 1912, **101**, 2309.
[4] LOWRY and OWEN, *T.F.S.*, 1930, **26**, 371–376.

In order to check the validity of the formula thus deduced, rotations should be calculated for as long a series of observed wave-lengths as possible, and the sign and magnitude of the differences noted. It is then easy to see which of the readings are in harmony with their neighbours, and which readings exhibit abnormal differences. In many cases an improved equation can then be deduced by selecting two other wave-lengths which appear to give normal differences. After the final adjustment positive and negative differences should be distributed quite irregularly.

(*b*) *Three-constant equation* $\alpha = k_1/(\lambda^2 - \lambda_1^2) \pm k/\lambda^2$. An equation of this type is generally called for when a small regular curvature is seen on plotting $1/\alpha$ against λ^2. The equation employed is now of the form

$$y = \frac{p}{x - a} + \frac{q}{x},$$

where x and y have the same significance as before, and a, p, q, are the three arbitrary constants of the equation. The equation may be written

$$x^2y - axy - (p + q)x + aq = 0,$$

or

$$x^2y + Axy + Bx + D = 0,$$

where $A = -a, \quad B = -(p + q) \quad \text{and} \quad D = aq.$

Substituting two pairs of values x_1, y_1 and x_2, y_2, and eliminating D, we obtain

$$(x_1^2y_1 - x_2^2y_2) + A(x_1y_1 - x_2y_2) + B(x_1 - x_2) = 0,$$

and with a third set of values x_3, y_3, a second equation. The two may be written as

$$k_1A + l_1B + n_1 = 0$$
$$k_2A + l_2B + n_2 = 0$$

where

$$k_1 = x_1y_1 - x_2y_2, \qquad k_2 = x_1y_2 - x_3y_3$$
$$l_1 = x_1 - x_2, \qquad l_2 = x_1 - x_3$$
$$n_1 = x_1^2y_1 - x_2^2y_2, \qquad n_2 = x_1^2y_1 - x_3^2y_3.$$

The values of A and B will then be

$$A = \frac{n_2l_1 - n_1l_2}{k_1l_2 - k_2l_1}, \quad B = \frac{n_2k_1 - n_1k_2}{l_1k_2 - l_2k_1}.$$

The value of D and hence those of a, p, and q readily follow.

(*c*) *Four-constant equation* $\alpha = k_1/(\lambda^2 - \lambda_1^2) \pm k_2/(\lambda^2 - \lambda_2^2)$. The equation is of the form

$$y = \frac{p}{x - a} + \frac{q}{x - b},$$

where a, b, p, q, are the four arbitrary constants. By expansion the equation becomes

$$x^2y - (a + b)xy - (p + q)x + aby + bp + aq = 0,$$

or

$$x^2y + Axy + Bx + Cy + D = 0$$

where

$$A = -(a+b), \quad B = -(p+q), \quad C = ab, \quad D = bp + aq.$$

Eliminating D from

$$x_1^2y_1 + Ax_1y_1 + Bx_1 + Cy_1 + D = 0$$

and

$$x_2^2y_2 + Ax_2y_2 + Bx_2 + Cy_2 + D = 0,$$

$$(x_1^2y_1 - x_2^2y_2) + A(x_1y_1 - x_2y_2) + B(x_1 - x_2) + C(y_1 - y_2) = 0$$

is obtained, and with two other sets of values, x_3, y_3, x_4, y_4, two other equations result. The three equations may be conveniently written thus,

$$k_1A + l_1B + m_1C + n_1 = 0,$$
$$k_2A + l_2B + m_2C + n_2 = 0,$$
$$k_3A + l_3B + m_3C + n_3 = 0,$$

where

$$\begin{array}{lll} k_1 = x_1y_1 - x_2y_2, & k_2 = x_1y_1 - x_3y_3. & k_3 = x_1y_1 - x_4y_4, \\ l_1 = x_1 - x_2, & l_2 = x_1 - x_3, & l_3 = x_1 - x_4, \\ m_1 = y_1 - y_2, & m_2 = y_1 - y_3, & m_3 = y_1 - y_4, \\ n_1 = x_1^2y_1 - x_2^2y_2, & n_2 = x_1^2y_1 - x_3^2y_3, & n_3 = x_1^2y_1 - x_4^2y_4. \end{array}$$

The solution of these equations is given by the following determinants :

$$\frac{A}{\begin{vmatrix} l_1m_1n_1 \\ l_2m_2n_2 \\ l_3m_3n_3 \end{vmatrix}} = \frac{-B}{\begin{vmatrix} k_1m_1n_1 \\ k_2m_2n_2 \\ k_3m_3n_3 \end{vmatrix}} = \frac{C}{\begin{vmatrix} k_1l_1n_1 \\ k_2l_2n_2 \\ k_3l_3n_3 \end{vmatrix}} = \frac{-1}{\begin{vmatrix} k_1l_1m_1 \\ k_2l_2m_2 \\ k_3l_3m_3 \end{vmatrix}}$$

i.e.

$$A = -\left[\frac{l_1(m_2n_3 - m_3n_2) + m_1(n_2l_3 - n_3l_2) + n_1(l_2m_3 - l_3m_2)}{k_1(l_2m_3 - l_3m_2) + l_1(m_2k_3 - m_3k_2) + m_1(k_2l_3 - k_3l_2)}\right]$$

$$B = \left[\frac{k_1(m_2n_3 - m_3n_2) + m_1(n_2k_3 - n_3k_2) + n_1(k_2m_3 - k_3m_2)}{\text{same denominator}}\right]$$

$$C = -\left[\frac{k_1(l_2n_3 - l_3n_2) + l_1(n_2k_3 - n_3k_2) + n_1(k_2l_3 - k_3l_2)}{\text{same denominator}}\right]$$

From these the values of A, B, and C can be found which in turn give D. The values of the original constants a, b, p and q, then readily follow.

The dispersion curves are obviously very sensitive to small errors in the experimental data selected as a basis for the calculations. For this reason more than one equation must usually be tried before a complex rotatory dispersion can be expressed satisfactorily. The adjustment can be done by two methods. In the first, the preliminary equation is used to calculate a series of differences from the observed values, with the help of which it is possible to distinguish between " good " and " bad " readings, i.e. readings which are concordant with or diverge widely from their neighbours. A fresh calculation can then be made with the help of four fresh wave-lengths selected from the "good" readings, or by using idealised readings deduced from the average differences of a group of consecutive readings. The second method is to calculate the effect of a small variation in each of the four constants on the rotatory power for four selected wave-lengths. It is then possible to calculate, by methods similar to those described above, the *variations* that must be made in the four constants, in order to give the desired adjustments at the four wave-lengths selected for this purpose. In either case, the adjustment is not complete until the theoretical curve intersects the experimental curve (and therefore gives ± differences) at four points, between which the systematic errors, if any are left, should be opposite in sign in consecutive segments of the curve.

It will be seen that the squares and fourth powers of the wave-

TABLE 44.—WAVE-LENGTHS OF STANDARD LINES.

No.	Line.	Intensity.	Wave-Length $\lambda \times 10^4$.	λ^2 $\times 10^{-4}$ cm.	λ^4 $\times 10^{-4}$ cm.	$\log_{10} \lambda$.	$2 \log_{10} \lambda$.	$4 \log_{10} \lambda$.
			Å.U.					
1	Li Red .	10	6707·86	0·4499538	0·2024585	$\bar{1}$·8265840	$\bar{1}$·6531680	$\bar{1}$·3063360
2	Cd Red .	10	6438·470	0·4145391	0·1718426	$\bar{1}$·8087828	$\bar{1}$·6175656	$\bar{1}$·2351312
3	Zn Red .	10	6362·345	0·4047941	0·1638588	$\bar{1}$·8036172	$\bar{1}$·6072344	$\bar{1}$·2144688
4	Li Orange .	10	6103·6	0·3725393	0·1387856	$\bar{1}$·7855861	$\bar{1}$·5711722	$\bar{1}$·1423444
5	Na Yellow 1	8	5895·932	0·3476202	0·1208398	$\bar{1}$·7705525	$\bar{1}$·5411050	$\bar{1}$·0822100
6	Na Y (*O.M.*)	—	*5892·617*	0·3472295	0·1205683	$\bar{1}$·7703083	$\bar{1}$·5406166	$\bar{1}$·0812332
7	Na Yellow 2	10	5889·965	0·3469169	0·1203513	$\bar{1}$·7701127	$\bar{1}$·5402254	$\bar{1}$·0804508
8	Hg Yellow 1	10	5790·66	0·3353177	0·1124378	$\bar{1}$·7627281	$\bar{1}$·5254562	$\bar{1}$·0509124
9	Cu Yellow 1	6	5782·15	0·3343327	0·1117783	$\bar{1}$·7620894	$\bar{1}$·5241788	$\bar{1}$·0483576
10	Hg Y (*O.M.*)	—	*5780·13*	0·3340992	0·1116223	$\bar{1}$·7619377	$\bar{1}$·5238754	$\bar{1}$·0477508
11	Hg Yellow 2	10	5769·598	0·3328827	0·1108108	$\bar{1}$·7611456	$\bar{1}$·5222912	$\bar{1}$·0445824
12	Cu Yellow 2	5	5700·24	0·3249273	0·1055777	$\bar{1}$·7558931	$\bar{1}$·5117862	$\bar{1}$·0235724
13	Ag Green 1	6	5471·51	0·2993742	0·08962492	$\bar{1}$·7381072	$\bar{1}$·4762144	$\bar{2}$·9524288
14	Ag G (*O.M.*)	—	*5467·71*	0·2989585	0·08937622	$\bar{1}$·7378055	$\bar{1}$·4756110	$\bar{2}$·9512220
15	Ag Green 2	10	5465·43	0·2987093	0·08922726	$\bar{1}$·7376244	$\bar{1}$·4752488	$\bar{2}$·9504976
16	Hg Green .	10	5460·73	0·2981957	0·08892067	$\bar{1}$·7372507	$\bar{1}$·4745014	$\bar{2}$·9490028
17	Tl Green .	10	5350·47	0·2862754	0·08195359	$\bar{1}$·7283920	$\bar{1}$·4567840	$\bar{2}$·9135680
18	Cu Green 1	6	5220·06	0·2724903	0·07425095	$\bar{1}$·7176755	$\bar{1}$·4353510	$\bar{2}$·8707020
19	Cu G1 (*O.M.*)	—	*5218·90*	0·2723692	0·07418598	$\bar{1}$·7175790	$\bar{1}$·4351580	$\bar{2}$·8703160
20	Cu Green 1′	10	5218·203	0·2722964	0·07414535	$\bar{1}$·7175210	$\bar{1}$·4350420	$\bar{2}$·8700840
21	Ag Green 3	10	5209·04	0·2713410	0·07362593	$\bar{1}$·7167577	$\bar{1}$·4335154	$\bar{2}$·8670308
22	Cu Green 2	8	5153·26	0·2655608	0·07052256	$\bar{1}$·7120820	$\bar{1}$·4241640	$\bar{2}$·8483280
23	Cu Green 3	7	5105·545	0·2606660	0·06794677	$\bar{1}$·7080422	$\bar{1}$·4160844	$\bar{2}$·8321688
24	Cd Green .	10	5085·823	0·2586560	0·06690294	$\bar{1}$·7063613	$\bar{1}$·4127226	$\bar{2}$·8254452
25	Zn Blue 1 .	10	4810·534	0·2314123	0·05355164	$\bar{1}$·6821932	$\bar{1}$·3643864	$\bar{2}$·7287728
26	Cd Blue 1 .	10	4799·912	0·2303915	0·05308023	$\bar{1}$·6812332	$\bar{1}$·3624664	$\bar{2}$·7249328
27	Zn Blue 2 .	10	4722·162	0·2229882	0·04972372	$\bar{1}$·6741409	$\bar{1}$·3482818	$\bar{2}$·6965636
28	Zn Blue 3 .	10	4680·138	0·2190369	0·04797714	$\bar{1}$·6702586	$\bar{1}$·3405172	$\bar{2}$·6810344
29	Cd Blue 2 .	10	4678·151	0·2188510	0·04789575	$\bar{1}$·6700742	$\bar{1}$·3401485	$\bar{2}$·6802970
30	Li Blue 1 .	9	4603·0	0·2118761	0·04489148	$\bar{1}$·6630410	$\bar{1}$·3260820	$\bar{2}$·6521640
31	Li B (*O.M.*)	—	*4602·5*	0·2118301	0·04487198	$\bar{1}$·6629938	$\bar{1}$·3259876	$\bar{2}$·6519752
32	Li Blue 1′ .	9	4602·0	0·2117840	0·04485248	$\bar{1}$·6629466	$\bar{1}$·3258932	$\bar{2}$·6517864
33	Hg Violet .	10	4358·34	0·1899513	0·03608148	$\bar{1}$·6393211	$\bar{1}$·2786422	$\bar{2}$·5572844

lengths enter into the above equations. A list of the wave-lengths of the lines usually employed as monochromatic sources of light is given in Table 44, together with the squares and fourth powers of the wave-lengths and the corresponding logarithms. The wave-lengths have been taken from the *International Critical Tables* (Vol. V, 1929) and are expressed in International Ångströms; the squares and fourth powers are expressed in tenth metres. The names in the second column are those commonly used to describe the lines when used as light-sources in work on rotatory dispersion. The letters *O.M.* denote the Optical Mean of a doublet when used without separation: the wave-length shown in the table is the weighted mean of the components as deduced with the help of the intensity numbers.

CHAPTER XXXIII.

ANALYSIS OF ROTATORY DISPERSION.

(*b*) *IN ABSORBING MEDIA.*

EQUATIONS OF DRUDE, NATANSON AND KUHN.

Extension of Drude's Theory to the Region of Absorption. —In order to extend his theory of rotatory dispersion to include the region of absorption, Drude in 1906 [1] reintroduced a FRICTION CO-EFFICIENT which he had neglected in deducing his equation for rotatory dispersion in a region of complete transparency. A DAMPING FACTOR of this kind had already been postulated in 1875 by Helmholtz [2] (p. 394), in order to account for the phenomena of absorption and refractive dispersion. The omission of this coefficient from the equation leads to infinite values both for the rotatory power and for the absorption at a wave-length corresponding with the natural period of the optically-active vibrator. This result can be realised approximately in the case of an absorption *line*, as in R. W. Wood's experiments on the magnetic rotation of sodium vapour in the immediate neighbourhood of the D-line (p. 461); but it is only of theoretical significance in optically-active media, where the absorption *bands* are always of finite width and intensity, instead of being of zero width and infinite intensity.

When this frictional term was reintroduced, Drude's equation showed that the difference in velocity of propagation of the two circular components of plane polarised light would (i) increase to a maximum as an absorption band was penetrated, (ii) fall rapidly to zero at the centre of the band, (iii) increase again to a maximum of opposite sign on leaving the region of absorption, and finally (iv) decrease slowly on passing once more into a region of complete transparency. Since the rotatory power is dependent on the difference in velocity of the two circularly polarised beams, which can be expressed as a CIRCULAR DOUBLE REFRACTION, the rotatory dispersion will also behave in the same "anomalous" manner, as shown in the theoretical curves of Fig. 174 (p. 429), where an "anomalous" partial rotation is superposed on a "normal" partial rotation. This theoretical result is in strict agreement with the experimental data obtained by Cotton and described in Chapter XI.

[1] DRUDE, *Theory of Optics*, English trans., 1907, p. 415.
[2] HELMHOLTZ, *Ann. Phys. Chem.*, 1875, **154**, 582.

Drude's theory also indicated that the two circular components of plane polarised light would be absorbed to unequal extents and that the emergent light would therefore be elliptically polarised, again in agreement with Cotton's experimental observations. Since, however, his equation included an arbitrary constant, *k*, which may be equal to zero, Drude concluded that even in an optically-active medium these phenomena are limited to OPTICALLY-ACTIVE ABSORPTION BANDS, i.e. to bands due to resonators (ions or electrons) which vibrate in spiral instead of linear paths. Thus in Cotton's mixture of magenta and sugar (p. 155) the resonators which give rise to absorption and those which produce optical activity are quite distinct, and do not therefore give rise to the Cotton effect.

Natanson's Equations.—Drude's equation for the rotatory power of a *transparent* medium could be applied directly to existing experimental data, but his expression for rotatory dispersion in a region of selective *absorption* was not given in a correspondingly convenient form. In 1908 and 1909, however, Natanson,[1] with the help of a "damping factor" analogous to Drude's friction coefficient, deduced expressions for the difference between the refractive indices for left and right circularly polarised light in an absorbing electronic system and for the corresponding difference between the absorption coefficients. Natanson's equations may be written in the form

$$n_l \sim n_r = \frac{D\lambda(\lambda^2 - \lambda_0^2)}{(\lambda^2 - \lambda_0^2)^2 + \Gamma^2\lambda^2} \quad . \quad . \quad . \quad \text{(i)}$$

and

$$\kappa_l \sim \kappa_r = \frac{D\Gamma\lambda^2}{(\lambda^2 - \lambda_0^2)^2 + \Gamma^2\lambda^2}, \quad . \quad . \quad . \quad \text{(ii)}$$

where D = constant,

λ_0 = the wave-length corresponding to the head of an absorption band,

Γ = damping factor, having the dimensions of a wave-length,

n_l, n_r and κ_l, κ_r = refractive indices and absorption indices for left and right circularly polarised light respectively, κ being defined by $I/I_0 = e^{-\frac{4\pi\kappa d}{\lambda}}$.

If ρ and ϕ are respectively the rotation and ellipticity imposed on the plane polarised ray by unit length of column of the medium, then since

$$\rho = \frac{\pi}{\lambda}(n_l \sim n_r) \quad . \quad . \quad . \quad . \quad . \quad \text{(iii)}$$

and

$$\phi = \frac{\pi}{\lambda}(\kappa_l \sim \kappa_r) \text{ approximately,*} \quad . \quad . \quad \text{(iv)}$$

[1] NATANSON, *Bull. Acad. Sci., Krakow,* 1908, 764; *Journ. de Phys.*, 1909, [iv], 8, 321.

* This equation is deduced in the next paragraph.

the above equations can be written in the form

$$\rho = \frac{D'(\lambda^2 - \lambda_0^2)}{(\lambda^2 - \lambda_0^2)^2 + \Gamma^2\lambda^2} \quad . \quad . \quad . \quad \text{(v)}$$

and

$$\phi = \frac{D'\Gamma\lambda}{(\lambda^2 - \lambda_0^2)^2 + \Gamma^2\lambda^2}. \quad . \quad . \quad . \quad \text{(vi)}$$

Conversion of Ellipticities into Molecular Extinction Coefficients.—Let a_0 be the amplitude of the circular components incident on the medium and let a_l and a_r be the amplitudes of the left and right circular components respectively after traversing unit thickness of the medium.

Then

$$\frac{a_l}{a_0} = e^{-\frac{2\pi\kappa_l}{\lambda}} \quad \text{and} \quad \frac{a_r}{a_0} = e^{-\frac{2\pi\kappa_r}{\lambda}},$$

therefore

$$\tan\phi = \frac{a_r - a_l}{a_r + a_l} = \frac{e^{-\frac{2\pi\kappa_r}{\lambda}} - e^{-\frac{2\pi\kappa_l}{\lambda}}}{e^{-\frac{2\pi\kappa_r}{\lambda}} + e^{-\frac{2\pi\kappa_l}{\lambda}}} = \frac{1 - e^{-\frac{2\pi}{\lambda}(\kappa_l - \kappa_r)}}{1 + e^{-\frac{2\pi}{\lambda}(\kappa_l - \kappa_r)}}.$$

Since $(\kappa_l - \kappa_r)$ is always small, the exponentials can be replaced by the first two terms of the series only;

thus

$$\tan\phi = \frac{\frac{2\pi}{\lambda}(\kappa_l - \kappa_r)}{2 - \frac{2\pi}{\lambda}(\kappa_l - \kappa_r)};$$

but ϕ is small and therefore $\tan\phi \simeq \phi$ and $\frac{2\pi}{\lambda}(\kappa_l - \kappa_r)$ can be neglected in comparison with 2,

therefore

$$\phi = \frac{\pi}{\lambda}(\kappa_l - \kappa_r). \quad . \quad . \quad . \quad . \quad \text{(iv)}$$

The ellipticity produced by a column l cm. long is then

$$\phi = \frac{\pi l}{\lambda}(\kappa_l - \kappa_r).$$

If, in place of the absorption index κ, we use the molecular extinction coefficient ϵ, defined by

$$I/I_0 = 10^{-\epsilon cl} \text{ (where } c = \text{molar concentration)},$$

then

$$(\kappa_l - \kappa_r) = \frac{c\lambda}{4\pi \log_{10} e}(\epsilon_l - \epsilon_r).$$

The ellipticity ϕ is then related to the difference $(\epsilon_l - \epsilon_r)$ between the molecular extinction coefficients by

$$\phi = \frac{cl}{4 \log_{10} e}(\epsilon_l - \epsilon_r). \quad \text{(vii)}$$

In these equations the ellipticity is measured in radians.

Relation between the Signs and Magnitudes of Rotation and Circular Dichroism. Natanson's and Bruhat's Rules.—When a plane polarised ray traverses an optically-active medium, dextrorotation results if the right circularly polarised component is transmitted faster than the left, and *vice versa*. The sign of the elliptical vibration produced by an optically-active absorbing medium is the same as that of the circular component which is least absorbed. From the ratio of the rotation to the ellipticity,

$$\frac{\rho}{\phi} = \frac{(\lambda^2 - \lambda_0^2)}{\Gamma\lambda}. \quad \text{(viii)}$$

Natanson formulated a rule connecting the sign of the rotation and the circular dichroism. For when $\lambda > \lambda_0$, ρ and ϕ must be of the same sign since Γ is always positive, and when $\lambda < \lambda_0$, ρ and ϕ must be of opposite signs. Thus, *on the long wave-length side of an optically-active absorption band the sign of the elliptical vibration is the same as the sign of the rotation, while on the short wave-length side the ellipse is of opposite sign to the rotation.* Bruhat [1] has stated Natanson's rule in the following terms :

> " *On the red side of the absorption band, the ray which is less absorbed is propagated with the greater velocity ; on the violet side the reverse is the case.*"

These relationships may be summarised as follows :

(i) $v_r > v_l$ or $n_r < n_l$ produces dextrorotation.
$v_r < v_l$,, $n_r > n_l$,, lævorotation

(ii) On the long wave-length side of an optically-active absorption band, if

$v_r > v_l$, i.e. dextrorotation, then $\kappa_r < \kappa_l$, i.e. right ellipse,
$v_r < v_l$, ,, lævorotation, ,, $\kappa_r > \kappa_l$, ,, left ,,

On the short wave-length side, if

$v_r > v_l$, i.e. dextrorotation, then $\kappa_r > \kappa_l$, i.e. left ellipse,
$v_r < v_l$, ,, lævorotation, ,, $\kappa_r < \kappa_l$, ,, right ,,

where v_r and v_l are the velocities of the right and left circular components respectively;

where n_r and n_l are the refractive indices of the right and left circular components respectively;

where κ_r and κ_l are the absorption indices of the right and left circular components respectively.

[1] BRUHAT, *Ann. de Phys.*, 1915, [ix], **3**, 232 and 417 ; 1920, [ix], **13**, 25.

The relation between the magnitudes of the rotation and of the circular dichroism has been established by Bruhat as follows:

In the equations for the circular double refraction and circular dichroism, substitute, $\frac{\lambda}{\lambda_0} = x$ and $\frac{\Gamma}{\lambda_0} = K$,

then
$$n_l \sim n_r = \frac{D}{\Gamma}\left[\frac{Kx(x^2 - 1)}{(x^2 - 1)^2 + K^2x^2}\right] \quad . \quad . \quad \text{(ix)}$$

and
$$\kappa_l \sim \kappa_r = \frac{D}{\Gamma}\left[\frac{K^2x^2}{(x^2 - 1)^2 + K^2x^2}\right]. \quad . \quad . \quad \text{(x)}$$

The circular double refraction $(n_l \sim n_r)$ has a maximum value of $D/2\Gamma$ when $(x^2 - 1) = Kx$ and has a minimum value of $-D/2\Gamma$ when $(x^2 - 1) = -Kx$. The difference between the maximum and minimum values is therefore D/Γ. The circular dichroism $(\kappa_l \sim \kappa_r)$ has a maximum value of D/Γ when $x = 1$. Thus, *the difference between the maximum and minimum values of the circular double refraction* $(n_l \sim n_r)$ *is equal to the maximum value of the circular dichroism* $(\kappa_l \sim \kappa_r)$.

Let λ_1 and λ_2 be the wave-lengths at which the maximum and minimum values of the circular double refraction occur, i.e. when $(x^2 - 1) = \pm Kx$. Then

$$\lambda_1 = \tfrac{1}{2}(\sqrt{\Gamma^2 + 4\lambda_0^2} + \Gamma) \quad \text{and} \quad \lambda_2 = \tfrac{1}{2}(\sqrt{\Gamma^2 + 4\lambda_0^2} - \Gamma). \quad \text{(xi)}$$

The maximum and minimum thus occur at wave-lengths, one on either side of λ_0, such that $\lambda_1\lambda_2 = \lambda_0^2$.

The maximum value of the ellipticity $(\phi_{max.})$ which occurs at wave-length λ_0 is $D'/\Gamma\lambda_0$. The difference between the maximum and minimum values of the actual rotation $(\rho_{max.} - \rho_{min.})$ which occur at λ_1 and λ_2 respectively is $(D'/2\Gamma\lambda_1 + D'/2\Gamma\lambda_2)$. If the absorption band is narrow, λ_1 and λ_2 are not very different from λ_0, hence this difference is approximately $D'/\Gamma\lambda_0$. Thus, *for a narrow absorption band the difference between the rotation maximum and minimum is approximately equal to the maximum ellipticity.*

The maximum and minimum of circular double refraction occur when $(x^2 - 1) = \pm Kx$. Substituting this value in the expression for the circular dichroism, then $\kappa_l \sim \kappa_r = D/2\Gamma$. Thus, *the maximum and minimum of circular double refraction occur at those wave-lengths at which the circular dichroism has half its maximum value.*

If the values of λ_1 and λ_2 given above are subtracted, then $(\lambda_1 - \lambda_2) = \Gamma$. Thus the "damping factor" is equal to the difference between the wave-lengths either at which the circular double refraction has its maximum and minimum values or at which the circular dichroism $(\kappa_l \sim \kappa_r)$ has half its maximum value. If this latter is termed the HALF-WIDTH of the curve of circular dichroism, then *the "damping factor" is numerically equal to the half-width*, and *the maximum and minimum of circular double refraction occur at wave-lengths corresponding to the two ends of this half-width.*

Systems Containing more than one Group of Optically-active Absorbing Electrons.—The observed rotatory power of every dissymmetric molecule is the algebraic sum of a number of partial rotations, each of which has its origin in a particular electronic transition taking place within the molecule. Each partial rotation will be represented by a sinuous curve of rotatory dispersion as given

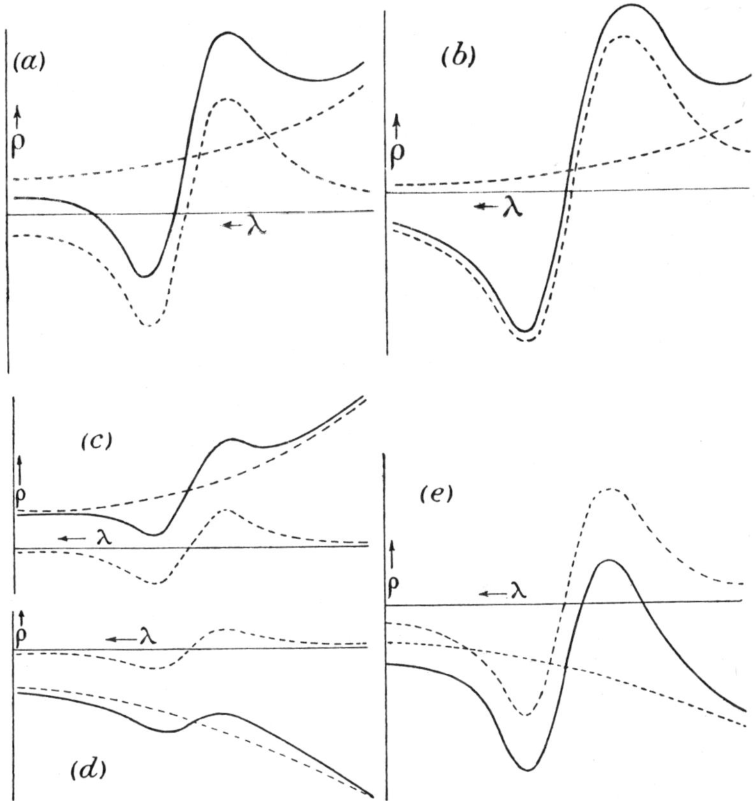

FIG. 174.—SUPERPOSITION OF PARTIAL ROTATIONS.

The broken lines represent the separate partial rotations; the continuous lines, which are the algebraic sum of the partial rotations, represent the total observed rotations.

by Natanson's equation, and will appear to have its origin in that part of the spectrum where the absorption band, caused by the electronic transition, is located. It is, however, the exception rather than the rule for more than one optically-active absorption band to be situated in a region of the spectrum accessible to polarimetric observation. Most of the data for the rotatory dispersion of absorbing media are therefore the algebraic sum of (i) a partial rotation associated with an absorption band in the region under investigation, and (ii) a partial rotation associated with an inaccessible absorption band, usually situated in the Schumann region. The former gives rise to a sinuous curve of rotatory dispersion, and the latter to a quasi-hyperbolic curve. Variations in the relative sign and magnitude of these partial rotations can give rise to a variety of curves of total rotatory dispersion, some of which are represented in Fig. 174 (*a*) to (*e*).

Applications of Natanson's Equations.—(*a*) Since Γ is numerically equal to the half-width of the curve of circular dichroism $(\kappa_l \sim \kappa_r)$, and $D = \Gamma(\kappa_l \sim \kappa_r)_{\text{max.}}$, Natanson's equations can be applied directly to experimental data. Thus, Natanson himself calculated the rotatory dispersion of potassium chromium tartrate from Cotton's measurements of circular dichroism, and showed that there was a very close agreement in the region of maximum absorption. Larger deviations at other wave-lengths were attributed to the influence of neighbouring absorption bands.

(*b*) Equations (i) to (iv), together with the rules regarding the relative sign and magnitude of circular dichroism and circular double refraction, give the relation

$$\rho_1\lambda_1 - \rho_2\lambda_2 = \phi_0\lambda_0 \qquad \text{(xii)}$$

where λ_1 and λ_2 are the wave-lengths at which the rotation has its maximum and minimum values of ρ_1 and ρ_2 respectively, and λ_0 is the wave-length of maximum ellipticity ϕ_0. Bruhat tested this expression by applying it to the experimental data for the solutions of optically-active absorbing compounds investigated by Cotton, Olmstead and himself. In the majority of cases there was a remarkably close agreement, the average discrepancy being of the order of 15 per cent. Marked discrepancies were always associated with the existence of neighbouring optically-active absorption bands, which would have a very appreciable influence on the observed rotation while affecting the maximum value of the circular dichroism very little. Since the rules of Natanson and Bruhat refer to a single optically-active absorption band, they would then clearly be no longer valid. For eleven other solutions the experimental data were insufficient to verify this relation with regard to magnitude, but the sign of the rotation and ellipticity agreed with Natanson's rule. This rule is also in harmony with the observations of Giesel and Stumpf on rotatory dispersion in cholesteric liquids (Chapter XXVIII, p. 349), which showed that the rotation is *negative* on the long wave-length side and *positive* on the short wave-length side of the point of maximum selective reflection when the *right* circular component is selectively reflected and the *left* circular component is transmitted.

(*c*) Bruhat made a much more thorough test of the validity of Natanson's equations by deducing from measurements of circular dichroism the value of the constants D, Γ and λ_0, and using these in order to calculate theoretical values of the rotatory power which were then plotted side by side with the experimental data. The rotatory dispersion of *potassium chromium tartrate* could thus be represented by a single term of Natanson's equation, but the data for copper saccharate required, in addition, an undamped term of Drude's equation to represent the simple partial rotatory dispersion of the remainder of the molecule. Tschugaeff's *l-bornyl dithiourethane*, on the other hand, required three terms as follows :

$$\rho = \frac{D(\lambda^2 - \lambda_0^2)}{(\lambda^2 - \lambda_0^2)^2 + \Gamma^2\lambda^2} + \frac{A}{(\lambda^2 - \lambda_1^2)} + \frac{B}{\lambda^2}. \quad \text{(xiii)}$$

In this equation, the first term, of the Natanson type, represents the sinuous partial rotation associated with the absorption band in the visible spectrum. The second term, of the Drude type, represents the simple partial rotation associated with a second absorption band of characteristic frequency, c/λ_1 in the near ultra-violet. The third term, also of the Drude type, represents another simple partial rotation associated with a more remote frequency in the Schumann region.

(*d*) Further evidence of the correctness of the rules formulated by Natanson and Bruhat was provided in 1923 by Wedeneewa's analysis of her own data for the rotatory dispersion and circular dichroism of camphorquinone (Fig. 168, p. 405). She found that a single term of Natanson's equations did not give an adequate representation either of the circular dichroism or of the rotatory dispersion of camphorquinone. In particular, the curve of circular dichroism was so unsymmetrical it had already fallen almost to zero at 80 A.U. from the maximum on the long wave-length side of the band, but was still appreciable at 800 A.U. from the maximum on the short wave-length side. An approximate representation was obtained, however, by subdividing the curve of circular dichroism into eight narrow components, each 40 A.U. wide. Each component was expressed by a term of Natanson's equation, but the maxima of the four components of low frequency were assumed to be twice as great as those of the four components of high frequency. A summation of the eight corresponding partial rotations gave theoretical curves approximating to those observed experimentally.

Circular Dichroism and the Absorption of Unpolarised Light.—An obvious analogy exists between Natanson's equation for circular dichroism

$$\kappa_l \sim \kappa_r = \frac{D\Gamma\lambda^2}{(\lambda^2 - \lambda_0^2)^2 + \Gamma^2\lambda^2} \quad \text{(11)}$$

and the Ketteler-Helmholtz equation for the absorption of unpolarised light

$$2n\kappa = \frac{\delta\Gamma\lambda^3}{(\lambda^2 - \lambda_0^2)^2 + \Gamma^2\lambda^2}. \quad \text{(xiv)}$$

Since the refractive index n will remain approximately constant inside a narrow band, the ratio of these two quantities can be expressed by the equation

$$g = \frac{\kappa_l \sim \kappa_r}{\kappa} \propto \frac{1}{\lambda}. \qquad \text{(xv)}$$

From this equation it follows that the DISSYMMETRY FACTOR, g (p. 394), in a given absorption band is proportional to the frequency $\nu \propto 1/\lambda$, provided that λ_0 and Γ have the same values in the two expressions, i.e. that the maxima of circular dichroism and of absorption occur at the same wave-length and that the half-widths of the two curves are the same.

Kuhn's Deduction of Dispersion Equations for Absorbing Media.—Like Natanson, Kuhn extended his calculations (p. 381) to the case in which the frequency of the incident light approaches one of the resonance frequencies of his system of coupled vibrators. As in the theories of Drude and Natanson, a "friction term" was introduced, to limit the amplitude of oscillation of the particles. This calculation gave an expression for the rotation per centimetre which, in the case of very weak coupling, was of the form

$$\phi = \text{constant}\left\{k_{12}d\frac{\nu^2}{\nu_1^2-\nu_2^2}\left[\frac{\nu_1^2-\nu^2}{(\nu_1^2-\nu^2)^2+\nu^2\nu_1'^2}-\frac{\nu_2^2-\nu^2}{(\nu_2^2-\nu^2)^2+\nu^2\nu_2'^2}\right]\right\} \qquad \text{(xvi)}$$

where ν'_1 and ν'_2 are friction coefficients and are to be taken as the mean width of the absorption band or line. This equation, for a single characteristic frequency, is identical with that given by Natanson. For values of ν far removed from ν_1 and ν_2 it reduces to a two-term Drude equation, with the supplementary condition that the two-terms are of opposite sign.

Corresponding to the above expression for rotatory power Kuhn obtained, for the circular dichroism associated with the first vibration (ξ_1) of the system, an equation of the form

$$(\mu_l - \mu_r)_{\xi_1} = \text{constant}\left[\frac{\nu^3\nu_1'}{(\nu_1^2 - \nu^2)^2 + \nu^2\nu_1'^2}\right]. \qquad \text{(xvii)}$$

Allowing for the difference between κ and μ, this is equivalent to the equation (ii) for circular dichroism deduced in 1908 by Natanson, but expressed in frequencies instead of in wave-lengths. For a narrow absorption band, for which $\nu_1^2 - \nu^2$ can be put equal to $2\nu(\nu_1 - \nu)$, Kuhn approximated this equation to the form

$$(\mu_l - \mu_r)_{\xi_1} = \text{constant} \,.\, \nu\left[\frac{\nu_1'}{4(\nu_1 - \nu)^2 + \nu_1'^2}\right]. \qquad \text{(xviii)}$$

The circular dichroism associated with the second vibration (ξ_2) of the system was shown to be of opposite sign, so that

$$(\mu_l - \mu_r)_{\xi_2} = -\,\text{constant} \,.\, \nu\left[\frac{\nu_2'}{4(\nu_2 - \nu)^2 + \nu_2'^2}\right]. \qquad \text{(xix)}$$

The integral of the terms inside the brackets of these two equations is equal to a constant $\pi/2$ so that

$$\int_{\xi_1+\xi_2}(\mu_l - \mu_r)\lambda d\nu = 0. \quad . \quad . \quad . \quad \text{(xx)}$$

This equation, which is analogous to the expression

$$\phi = \nu^2 \sum_i \frac{a_i}{\nu_i^2 - \nu^2}, \quad \text{where } \sum_i a_i = 0 \quad . \quad . \quad \text{(xxi)}$$

for rotatory power as given previously (p. 381), shows that the circular dichroism must change sign in the various regions of the absorption spectrum of a substance in such a way that $\int(\mu_l - \mu_r)\lambda d\nu = 0$.

The circular dichroism is quantitatively related to the rotatory power (e.g. for the absorption band ξ_1) by

$$(n_l - n_r)_{\xi_1} = \text{constant}\frac{1}{\nu_1^2 - \nu^2} \cdot \frac{\nu}{\nu_1}\int_{\xi_1}(\mu_l - \mu_r)d\nu. \quad . \quad \text{(xxii)}$$

It follows that when $\mu_l > \mu_r$ and $\nu_1 > \nu$ then $n_l > n_r$, i.e. on the long wave-length side of an absorption band dextrorotation results if the left circular component of plane polarised light is absorbed more than the right, in agreement with the rule already formulated by Natanson (p. 427).

The constants in these equations are not entirely arbitrary, e.g. the constant in equation (xxii) has the value $c/2\pi^2$.

Observations of the Cotton Effect in Colourless Compounds.—The observations of the "Cotton effect" described in Chapter XI were confined to coloured compounds, in which the phenomena in question can be studied within the range of the visible spectrum. The earliest observations of the rotatory dispersion of colourless compounds, at wave-lengths beyond the ordinary limits of transparency, appear to have been made in 1911 by Darmois,[1] who studied the rotatory dispersion of solutions of *camphor* on either side of the ultra-violet absorption band, and of α- and *β-pinene* to within a short distance of their absorption band. Servant[2] in 1932 repeated the latter measurements and extended them into the region of absorption of these compounds, where both show anomalies characteristic of the Cotton effect.

Measurements of rotatory dispersion throughout an ultra-violet absorption band were first described by W. Kuhn[3] in 1930 in the case of

I. *α-Azidopropionic dimethylamide* $N_3 . CH(CH_3) . CO . N(CH_3)_2$

which has a maximum of selective absorption at 2900 A.U. His results are reproduced in Fig. 175. Curve 1 represents the specific rotation, whilst curve 2 shows the molecular extinction coefficients plotted as log ϵ, but labelled log κ by the author.

The partial rotations due to the absorption band could not be

[1] DARMOIS, *A.C.P.*, 1911, [viii], **22,** 577–589.
[2] SERVANT, *C.R.*, 1932, **194,** 368.
[3] W. KUHN, *T.F.S.*, 1930, **26,** 293–308.

deduced from its optical activity, since no measurements of circular dichroism were made, and the dissymmetry factor was therefore unknown; but by assuming that the curve of circular dichroism has the same maximum and the same half-width as the absorption band, equation (xvi) can be expressed in a form which contains only one arbitrary constant,[1] namely,

$$[\alpha] = \frac{a\lambda_0^2(\lambda^2 - \lambda_0^2)}{(\lambda^2 - \lambda_0^2)^2 + \lambda^2\lambda'^2}, \quad . \quad . \quad \text{(xxiii)}$$

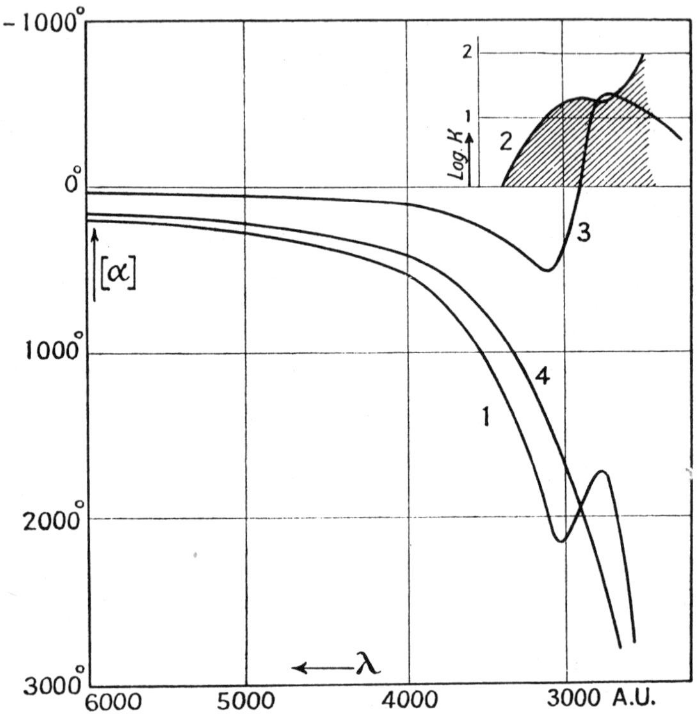

FIG. 175.—ABSORPTION AND ANOMALOUS ROTATORY DISPERSION OF α-AZIDOPROPIONIC DIMETHYLAMIDE $N_3 . \mathbf{C}H(CH_3) . CO . N(CH_3)_2$.

1. Observed rotations [α].
2. Absorption, log ε.
3. Partial rotation of azido group.
4. Residual rotation.

where λ_0 is the maximum and λ' the half-width of the band. After assuming a suitable value for *a*, the calculated partial rotations of the band were plotted as curve 3, and the difference between 1 and 3 was plotted as curve 4. This residual curve represents the sum of the partial rotations due to the more distant absorption bands in the

[1] KUHN, *Z. ph. C.*, 1929, B. **4**, 26; 1930, B. **8**, 284.

further ultra-violet and Schumann regions of the spectrum, and should therefore show no trace of the anomaly due to the azido band.

Since this result was realised, at least to a first approximation, it could be claimed that the partial rotation due to the azido band had been correctly segregated. Kuhn was therefore able to draw the following conclusions:

(i) The partial rotation of the N_3 group, in the visible spectrum at $\lambda = 6000$ A.U. amounts to 20 per cent. of the total rotatory power; the N_3 band, which provides only about one hundred thousandth part of the total optical absorption of the molecule, must therefore be relatively very active.

(ii) A large part of the residual rotation (curve 4) is probably associated with the absorption band next nearest to the visible, due to the dimethylamide group, —CO . $N(CH_3)_2$.

(iii) The partial rotations associated with the N_3— and —CO . $N(CH_3)_2$ bands are of the same sign (both negative), so that these two groups cannot be directly coupled; it is therefore probable that these groups are both coupled with other radicals which have intense absorption bands in the Schumann region and which contribute a positive partial rotation.

CHAPTER XXXIV.

ANALYSIS OF ROTATORY DISPERSION.

(*c*) *IN ABSORBING MEDIA* (*continued*).

EQUATIONS OF KUHN AND BRAUN AND OF LOWRY AND HUDSON.

Inadequacy of Equations based upon a Damping Factor. It was shown by Bruhat (p. 430) that the anomalous rotatory dispersion of *coloured* compounds in the region of absorption can be expressed, at least to a first approximation, by Natanson's equation; and a similar proof in reference to the ultra-violet absorption band of a *colourless* compound was given by Kuhn (p. 433). The relation between absorption, circular dichroism and rotatory dispersion, which was thus established, has not been modified in any fundamental way, but has been expressed in a general form by Kuhn.[1] It follows that, since a damping factor is inadequate to express the absorption of unsaturated organic compounds (p. 396), the equations for circular dichroism and for rotatory dispersion in the region of absorption must be modified in the same way as the absorption-equations already discussed in Chapter XXXI. From this point of view, the equations of Kuhn and Braun and of Lowry and Hudson, which are described in the present chapter, may be regarded as intermediate steps in the direction of an ideal equation, which will express within the limits of experimental error the rotatory dispersion of those compounds whose optical properties conform most closely to a simple mathematical law. Even in the most favourable cases, however, the process of approximation is not yet quite complete, since small but systematic deviations between the observed and calculated rotations can still be detected; and no successful attempt has yet been made to express in an accurate quantitative way the more complicated types of dispersion curve.

Equation of Kuhn and Braun.—The equation of Kuhn and Braun [2] is based on the assumption that the relation between absorption and rotation, which is described in the preceding chapter, is valid for a *narrow* absorption band, but cannot be applied to the *wide* absorption bands of organic compounds, which often cover a

[1] KUHN and FREUDENBERG, *Handbuch der chemischen Physik*, Leipzig, 1932, **8**, III, p. 77, equation 82.

[2] KUHN and BRAUN, *Z. ph. C.*, 1930, B. **8**, 281–313.

range of several hundred Ångström units. They therefore integrated the effects produced by an absorption band depending on a probability distribution of frequencies, and thus deduced an equation for rotatory dispersion which may be regarded as a second approximation towards the ideal.

From Kuhn's model, the partial rotation α_{ξ_1} associated with an absorption line of frequency ν_1 is given by

$$\alpha_{\xi_1} = \frac{1}{2\pi}\frac{\nu^2}{\nu_1}\frac{\nu_1^2 - \nu^2}{(\nu_1^2 - \nu^2)^2 + \nu^2\nu'^2}\int_{\xi_1}(\mu_l - \mu_r)_{\nu_1} d\nu_1. \quad \text{(i)}$$

The dissymmetry factor is given by $g_{\nu_1} = (\mu_l - \mu_r)_{\nu_1}/\mu_{\nu_1}$ . . (ii)

The partial rotation $d\alpha$ corresponding to a strip of the absorption band with a frequency range $d\nu_1$ is therefore given by

$$d\alpha = \frac{1}{2\pi}\frac{\nu^2}{\nu_1}\frac{\nu_1^2 - \nu^2}{(\nu_1^2 - \nu^2)^2 + \nu^2\nu'^2} g_{\nu_1}\mu_{\nu_1} d\nu_1. \quad \text{(iii)}$$

Substituting for μ_{ν_1} the value $\mu_{\max.} e^{-\left(\frac{\nu_0 - \nu_1}{\theta}\right)^2}$ given by the exponential absorption equation (vii) (p. 397), and integrating over the range of frequencies covered by the absorption band, the partial rotation α is given by

$$\alpha = \int_0^\infty \frac{1}{2\pi}\frac{\nu^2}{\nu_1}\frac{\nu_1^2 - \nu^2}{(\nu_1^2 - \nu^2)^2 + \nu^2\nu'^2} g_{\nu_1}\mu_{\max.} e^{-\left(\frac{\nu_0 - \nu_1}{\theta}\right)^2} d\nu_1. \quad \text{(iv)}$$

But since the dissymmetry factor g is proportional to the frequency i.e.

$$g_{\nu_1} = g_0\nu_1/\nu_0,$$

then

$$\alpha = \frac{1}{2\pi}\frac{\nu^2}{\nu_0} g_0\mu_{\max.}\int_0^\infty e^{-\left(\frac{\nu_0 - \nu_1}{\theta}\right)^2}\frac{\nu_1^2 - \nu^2}{(\nu_1^2 - \nu^2)^2 + \nu^2\nu'^2} d\nu_1. \quad \text{(v)}$$

By making the substitution $\dfrac{\nu_0 - \nu_1}{\theta} = x$, Kuhn and Braun showed that the partial rotation α per cm. column of the medium can be expressed by the equation

$$\alpha = \frac{1}{2\sqrt{\pi}}\frac{\nu}{\nu_0} g_0\mu_{\max.}\left[e^{-\left(\frac{\nu_0 - \nu}{\theta}\right)^2}\int_0^{\frac{\nu_0 - \nu}{\theta}} e^{x^2} dx - \frac{\theta}{2(\nu_0 + \nu)}\right]. \quad \text{(vi)}$$

If [M] is the molecule partial rotation associated with the absorption band and c is the molar concentration, then $\alpha = [M]c/100$. Also, if ϵ_l and ϵ_r are the molecular extinction coefficients for left and right circularly polarised light, defined by $I/I_0 = 10^{-\epsilon c l}$ then

$$g_0\mu_{\max.} = \frac{(\epsilon_l - \epsilon_r)_{\max.} c}{\log_{10} e}. \quad \text{(vii)}$$

Substituting these values of α and $g_0\mu_{max.}$ the molecular partial rotation is given by

$$[M] = \frac{100}{2\sqrt{\pi}} \cdot \frac{(\epsilon_l - \epsilon_r)_{max.}}{\log_{10} e} \frac{\nu}{\nu_0}\left[e^{-\left(\frac{\nu_0-\nu}{\theta}\right)^2}\int_0^{\frac{\nu_0-\nu}{\theta}} e^{x^2}dx - \frac{\theta}{2(\nu_0+\nu)}\right]. \quad \text{(viii)}$$

The maximum value of the term $e^{-\left(\frac{\nu_0-\nu}{\theta}\right)^2}\int_0^{\frac{\nu_0-\nu}{\theta}} e^{x^2}dx$ is 0·54073 approximately, when $\frac{\nu_0 - \nu}{\theta} = 0{\cdot}9$. The second term $\frac{\theta}{2(\nu_0+\nu)}$ is small and changes very little with ν so that the value of [M] is also a maximum when $\frac{\nu_0-\nu}{\theta} = 0{\cdot}9$. Let $[\phi]$ be this maximum value of [M] which occurs at frequency ν_ϕ given by $\frac{\nu_0 - \nu_\phi}{\theta} = 0{\cdot}9$

then $$[\phi]^{*} = \frac{100}{2\sqrt{\pi}} \cdot \frac{(\epsilon_l - \epsilon_r)_{max.}}{\log_{10} e} \frac{\nu_\phi}{\nu_0} 0{\cdot}54073. \quad \text{(ix)}$$

The molecular partial rotation [M] is then given by

$$[M] = \frac{[\phi]}{0{\cdot}541} \frac{\nu}{\nu_\phi}\left[e^{-\left(\frac{\nu_0-\nu}{\theta}\right)^2}\int_0^{\frac{\nu_0-\nu}{\theta}} e^{x^2}dx - \frac{\theta}{2(\nu_0+\nu)}\right] \text{approximately.} \quad \text{(x)}$$

In deducing partial rotations with the help of this equation the values of ν_0 and θ should be taken from the curve of circular dichroism and not from the curve of absorption of unpolarised light; but, when the circular dichroism has not been measured directly, the above equation can be applied by assigning an arbitrary value to $[\phi]$ and using the values of ν_0 and θ given by absorption data (compare pp. 394 and 434).

If the frequency ν of the incident light is widely removed from the frequency ν_0 corresponding to the maximum of circular dichroism, i.e. in regions of the spectrum at great distances from an absorption band, $\frac{\nu_0 - \nu}{\theta}$ is large, and

$$e^{-\left(\frac{\nu_0-\nu}{\theta}\right)^2}\int_0^{\frac{\nu_0-\nu}{\theta}} e^{x^2}dx = \frac{\theta}{2(\nu_0-\nu)} \text{ approximately.} \quad \text{(xi)}$$

For these regions of the spectrum

$$\begin{aligned}[M] &= \frac{[\phi]}{0{\cdot}541} \frac{\nu}{\nu_\phi}\left[\frac{\theta}{2(\nu_0-\nu)} - \frac{\theta}{2(\nu_0+\nu)}\right] \\ &= \frac{[\phi]}{0{\cdot}541} \frac{\theta}{\nu_\phi}\left[\frac{\nu^2}{\nu_0^2-\nu^2}\right] = \text{constant}\left[\frac{\nu^2}{\nu_0^2-\nu^2}\right] \\ &= \frac{\text{constant}}{\lambda^2 - \lambda_0^2}.\end{aligned}$$

* The value of $[\phi]$ obtained from the maximum circular dichroism by means of this equation is in radians, and must be multiplied by 57·296 if $[\phi]$ is required in degrees.

Thus in regions of complete transparency Kuhn and Braun's equation reduces to Drude's equation.

Applications of the Equations of Kuhn and Braun.—

(*a*) The relative merits of the old and new equations were tested in the case of

II. *Methyl α-azidopropionate*, N_3 . **C**$H(CH_3)$. CO . OCH_3.

Thus, in Fig. 176, curve 1 represents the *total* observed molecular rotation. The partial rotation associated with the N_3-absorption band (curve 2) deduced by means of the new equation is represented by 3. The *residual rotation* (curve 1 minus curve 3) is then represented by curve 4. If the partial rotation associated with the N_3-absorption band is deduced on the basis of the Drude-Natanson equation, the residual rotation is represented by the dotted curve 4′.

Neither of the curves of residual rotation, 4 and 4′, is entirely free from anomalies in the region covered by the azido absorption band, but curve 4 is much the more satisfactory of the two. Kuhn and Braun attributed the small remaining anomaly in the residual rotation curve 4, partly to experimental error and partly to the fact that the form of their exponential absorption equation was not in exact agreement with experimental observation.

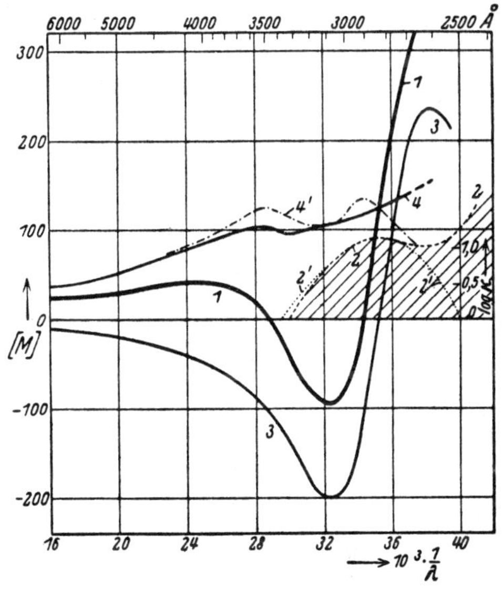

FIG. 176.—ABSORPTION AND ROTATORY DISPERSION OF METHYL α-AZIDOPROPIONATE.
II. N_3 . **C**$H(CH_3)$. CO . OCH_3.

1. Total rotation [M].
2. Absorption (experimental).
2′. ,, (K. and B.).
3. Partial rotation (K. and B.).
4. Residual ,, ,, ,,
4′. ,, ,, (D. and N.).

(*b*) In the same paper Kuhn and Braun described a similar analysis by means of their new equation of the rotatory dispersion of

I. *α-Azidopropionic dimethylamide*, N_3 . **C**$H(CH_3)$. CO . $N(CH_3)_2$

(p. 433) and

III. *Ethyl α-bromopropionate*, Br . **C**$H(CH_3)$. CO . OC_2H_5.

The corresponding observations on

IV. *α-Bromopropionic dimethylamide*, Br . **C**H(CH_3) . CO . N(CH_3)$_2$ could not be extended far into the region covered by the absorption band on account of its great intensity.

(*c*) Kuhn, Freudenberg and Wolf [1] investigated

V. *Methyl α-chloropropionate*, Cl . **C**H(CH_3) . CO . OCH_3 and

VI. *α-Chloropropionic dimethylamide*, Cl . **C**H(CH_3) . CO . N(CH_3)$_2$.

They detected in the curve of total rotatory power of the chlorodimethylamide, VI, an anomaly which could not be observed in the bromodimethylamide, IV.

These observations led to the following conclusions:

(i) The three amides, I, III, V are *lævorotatory*, but the three esters, II, IV, VI are *dextrorotatory*. On the other hand, the *anomalous partial rotation* in the absorption band is of the same sign throughout, since in every case it is *lævorotatory* at longer wave-lengths. The *induced activity* of the ultra-violet absorption band is therefore of the same sign for the amide and for the ester, whether the α-substituent is N_3, Br or Cl.

(ii) The *total rotations* are *positive in the esters* where a positive normal rotation is superposed on a weaker negative anomalous rotation, but *negative in the amides*, where a negative anomalous rotation is enhanced by the superposition of a stronger normal rotation. The change of sign on passing from the esters to the amides is therefore due to a reversal of the *normal* rotation, when one of the substituents on the asymmetric carbon atom is changed from — CO . OCH_3 to — CO . N(CH_3)$_2$.

(iii) The fact that the normal and anomalous partial rotations in the amides are of similar sign suggests that the absorption bands associated with these two partial rotations are not coupled with one another, but are coupled independently with absorption bands of opposite sign in the Schumann region (compare p. 435). If this conclusion is accepted, the reversal of sign of the normal partial rotation is due to a change in the *induced activity* of the — CO . R radical, and not to a reversal of the sign of the fixed dissymmetry of the asymmetric carbon atom, such as Boys assumes to be responsible for the change of sign on passing from amyl alcohol to amyl chloride (p. 365).

Anomalous Rotatory Dispersion of β-Octyl Nitrite.—Kuhn and Lehmann [2] investigated the optical activity in the region of absorption of *β-octyl nitrite*, O : N . O . **C**H(CH_3) . C_6H_{13}, which had already been prepared and investigated in a more preliminary way by Pickard and Hunter.[3] This ester has a strong absorption band at 2300 A.U. in which measurements of rotatory dispersion could not be made, and a weaker band at 3650 A.U. which extends over a range of nearly 1000 A.U. This includes four subsidiary

[1] Kuhn, Freudenberg and Wolf, *Ber.*, 1930, **63**, 2367–2379.

[2] Kuhn and Lehmann, *Z. Elektrochem.*, 1931, **37**, 549; *Z. ph. C.*, 1932, B. **18**, 32–48.

[3] Pickard and Hunter, *J.*, 1923, **123**, 434–444.

maxima and a step-out (curve 1, Fig. 177) and gives rise to an equally complex curve 2 of rotatory dispersion.

The *circular dichroism* in the weak absorption band included four strong equally spaced *negative* components at 3870, 3710, 3570 and 3455 A.U. (curves 1 to 4, Fig. 178 *a*) corresponding with the four subsidiary maxima of absorption, together with a weak but wide *negative* component 5 at 3350 A.U., corresponding with the step-out in the absorption curve, and finally a fairly strong and wide *positive* component 6 at 3950 A.U. which differed too little in wave-length from the negative component at 3870 A.U. to be shown on the curve of general absorption. The component of opposite sign was attributed to a second electronic transition (compare the ketonic band of camphor).

The superposition of the six components of circular dichroism gave the dotted theoretical curve 2 of Fig. 178 (*b*), which agrees well with the experimental curve 1. The partial rotations corresponding

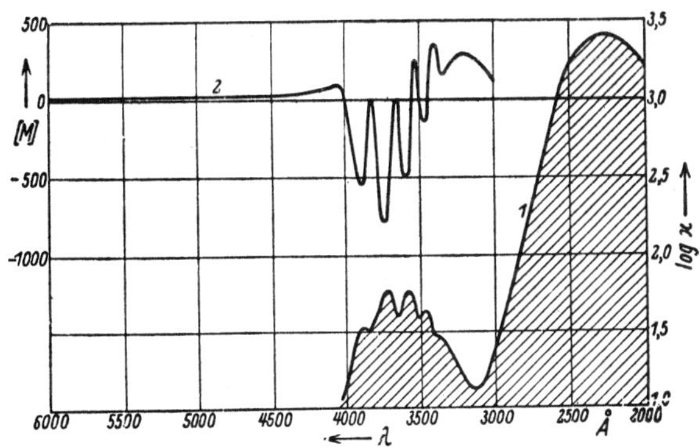

FIG. 177.—ABSORPTION AND ROTATORY DISPERSION OF OCTYL NITRITE.
1. Absorption. 2. Rotation.

with the six components of circular dichroism were calculated from the equation of Kuhn and Braun and are plotted in Fig. 179. Their superposition gave the theoretical curve 2 of total rotations shown in Fig. 180, in close agreement with the experimental curve 1.

Equation of Lowry and Hudson.—The relation between absorption and rotation can be investigated very advantageously in the optically-active xanthates prepared, and studied in the visual region of the spectrum, by Tschugaeff (pp. 317 and 398), since the ultra-violet absorption bands are more clearly defined than in Kuhn's optically-active propionates. It has already been shown (p. 399) that the form of the absorption bands cannot be expressed by the equation of Kuhn and Braun, since they are symmetrical on a scale of wave-lengths instead of on a scale of frequencies. Since the circular dichroism appeared to be proportional to the product of the absorption and the frequency, it was obvious that the equations for circular dichroism and for rotatory dispersion must be altered in the same way as the equation for absorption. A modified equation for

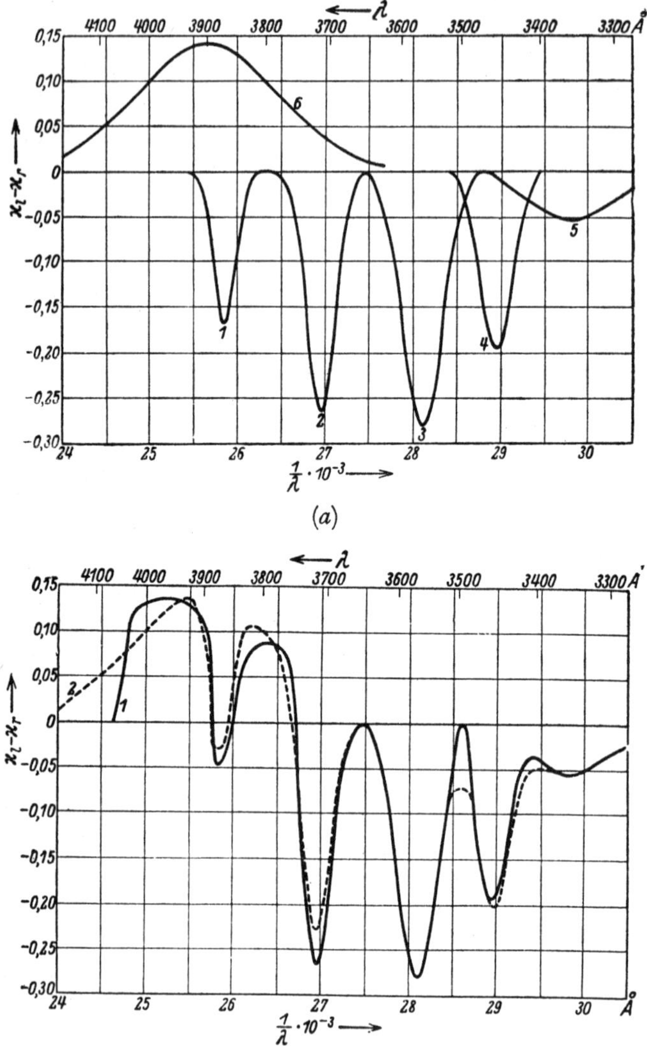

(a)

Curve 1 observed. Curve 2 calculated.

(b)

FIG. 178.—ANALYSIS AND SYNTHESIS OF THE CIRCULAR DICHROISM OF OCTYL NITRITE (Kuhn and Lehmann, 1931).

rotatory dispersion in the region of absorption was therefore developed on exactly the same lines as that of Kuhn and Braun.

According to this equation the rotation is given by

$$\alpha = \int_0^\infty \frac{1}{2\pi}\frac{\nu^2}{\nu_1}\frac{\nu_1{}^2 - \nu^2}{(\nu_1{}^2 - \nu^2)^2 + \nu^2\nu'^2} g_{\nu_1}\mu_{\max.} e^{-\left[\frac{\nu_0}{\nu}\left(\frac{\nu_0 - \nu}{\theta}\right)\right]^2} d\nu_1 . \quad . \quad \text{(xii)}$$

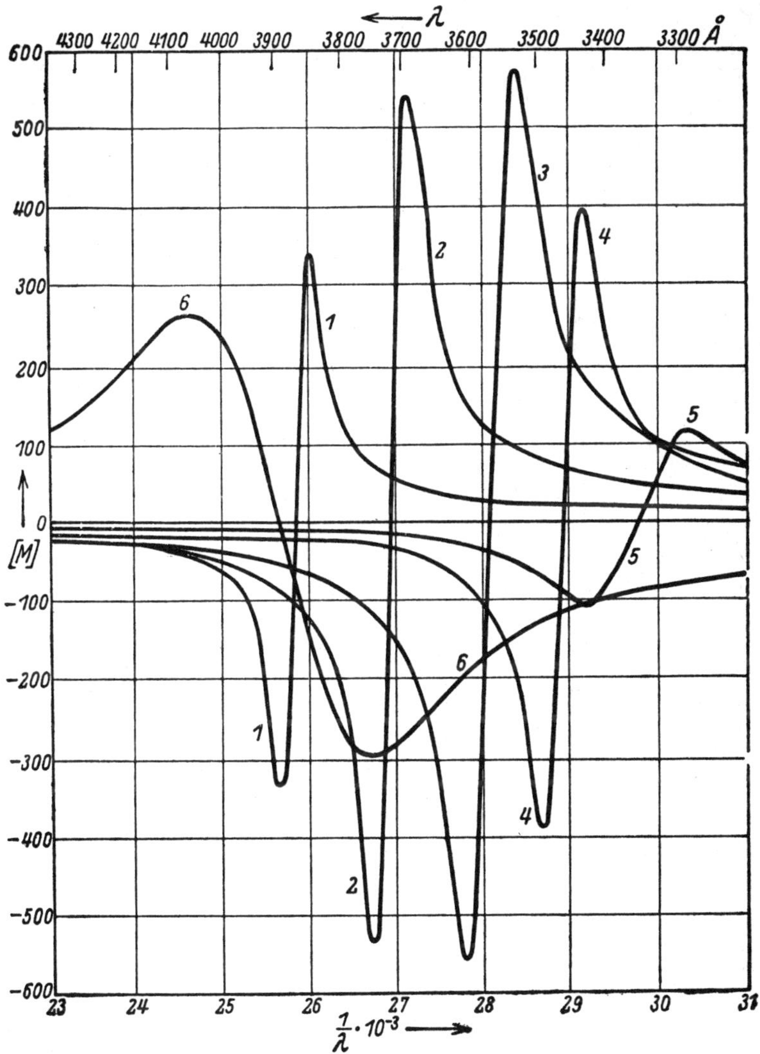

FIG. 179.—ANALYSIS OF THE ROTATORY DISPERSION OF OCTYL NITRITE (Kuhn and Lehmann, 1931).

This gives for the molecular rotation

$$[\mathrm{M}] = \frac{[\phi]}{m}\frac{\nu}{\nu_\phi}\left[e^{-\left[\frac{\nu_0}{\nu}\left(\frac{\nu_0 - \nu}{\theta}\right)\right]^2}\int_0^{\frac{\nu_0}{\nu}\left(\frac{\nu_0-\nu}{\theta}\right)} e^{x^2}dx + \frac{\nu\theta}{2\nu_0(\nu_0+\nu)}\right] \quad \text{(xiii)}$$

or in wave-lengths

$$[\mathrm{M}] = \frac{[\phi]}{m}\frac{\lambda_\phi}{\lambda}\left[e^{-\left(\frac{\lambda-\lambda_0}{\theta}\right)^2}\int_0^{\frac{\lambda-\lambda_0}{\theta}} e^{x^2}dx + \frac{\theta}{2(\lambda+\lambda_0)}\right] \quad . \text{(xiv)}$$

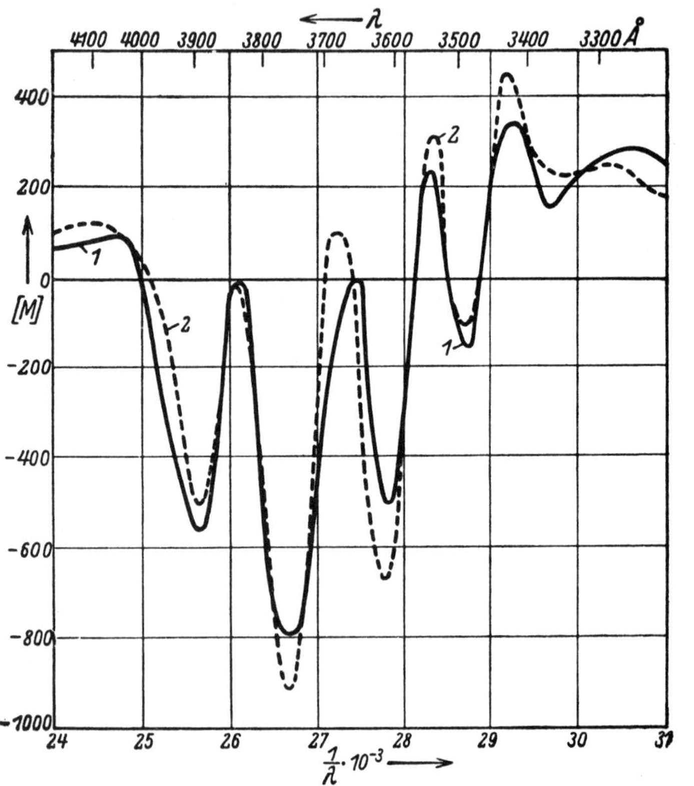

Curve 1 observed. Curve 2 calculated.

FIG. 180.—ROTATORY DISPERSION OF OCTYL NITRITE (Kuhn and Lehmann, 1931).

where m is the maximum value of the sum of the terms inside the bracket, and the other symbols have their usual significance. Like the corresponding absorption equations, this new equation for rotatory dispersion expresses the data for the xanthates more accurately than the equation of Kuhn and Braun. It also fulfils the essential condition that it shall reduce to Drude's equation for wave-lengths far removed from the absorption band. Thus for values of λ which are substantially larger than λ_0

$$[M] = \frac{[\phi]}{m}\frac{\lambda_\phi}{\lambda}\left[\frac{\theta}{2(\lambda - \lambda_0)} + \frac{\theta}{2(\lambda + \lambda_0)}\right]$$

$$= \frac{[\phi]}{m}\lambda_\phi\theta\left[\frac{1}{\lambda^2 - \lambda_0^2}\right] = \frac{\text{constant}}{\lambda^2 - \lambda_0^2}$$

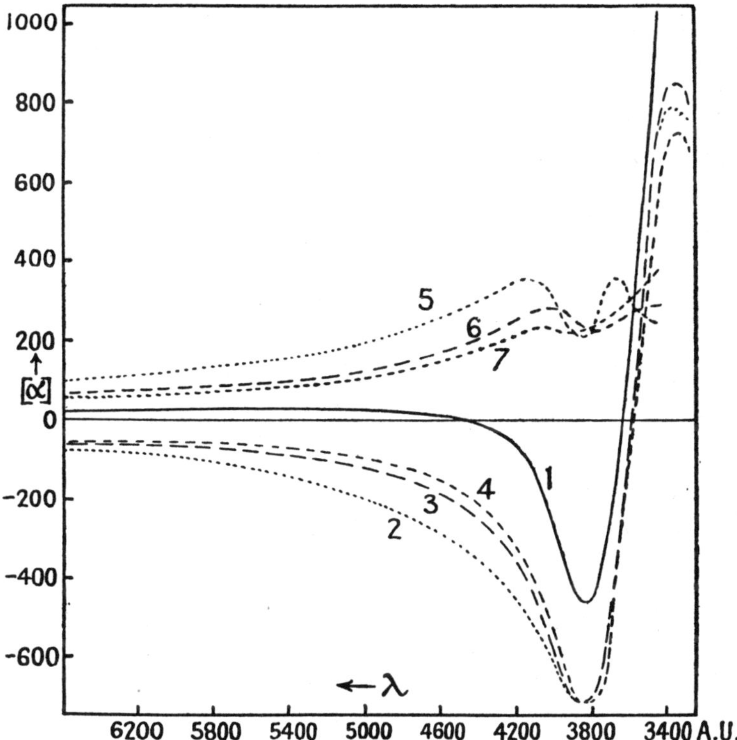

FIG. 181.—ROTATORY DISPERSION OF ETHYL *d*-BORNYL XANTHATE.

1. Observed rotation.
2. Partial rotation of CS . S band (D. and N.).
3. ,, ,, ,, ,, ,, (K. and B.).
4. ,, ,, ,, ,, ,, (L. and H.).
5. Residual rotation (1–2) (D. and N.).
6. ,, ,, (1–3) (K. and B.).
7. ,, ,, (1–4) (L. and H.).

Verification of the Equation of Lowry and Hudson.—The merits of this equation can be tested by the method used by Kuhn and Braun, namely, by calculating the partial rotations due to the nearest absorption band and plotting the residual rotations due to optically-active absorption bands of shorter wave-length. In every case in which a comparison has been made the anomalies in the residual rotations are suppressed more completely when the anomalous partial rotations of the absorption band are calculated

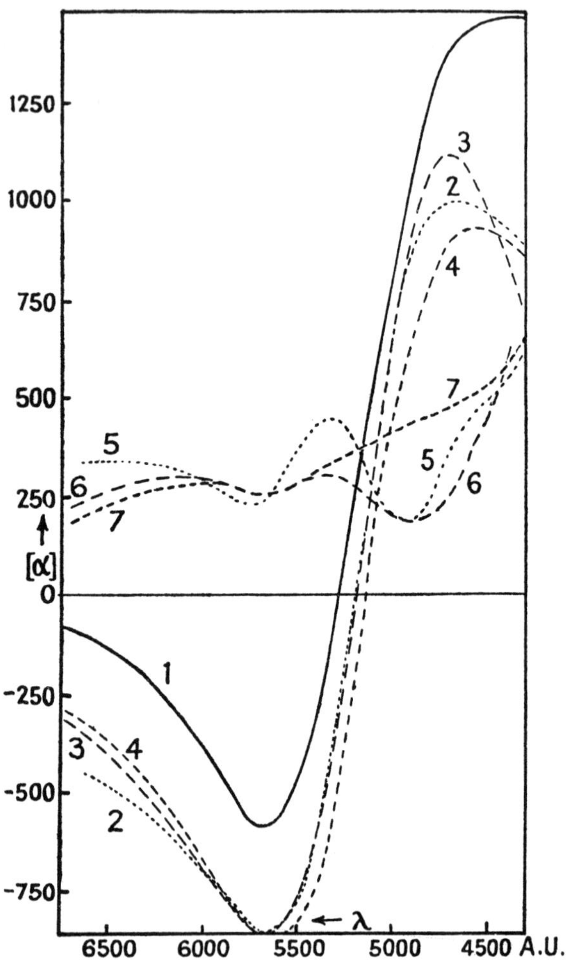

FIG. 182.—ROTATORY DISPERSION OF DIPHENYL-*l*-MENTHYL DITHIOURETHANE.

1. Observed rotation.
2. Partial rotation of $N(CS)_2$ band (D. and N.).
3. ,, ,, ,, ,, ,, (K. and B.).
4. ,, ,, ,, ,, ,, (L. and H.).
5. Residual rotation (1–2) (D. and N.).
6. ,, ,, (1–3) (K. and B.).
7. ,, ,, (1–4) (L. and H.).

from the new equation than when they are calculated from the equation of Kuhn and Braun.

(a) *Xanthates.* Fig. 181 shows the data for the anomalous rotatory dispersion of

I. *Ethyl* d-*bornyl xanthate,* $C_{10}H_{17}O . CS . S . C_2H_5$.

Fig. 182 shows the corresponding data for the anomalous rotatory dispersion in the visible spectrum of the red compound,

II. *Diphenyl*-l-*menthyl dithiourethane,*

$$C_{10}H_{19}O . CS . N(C_6H_5) . CS . C_6H_5.$$

In each figure, curve 1 shows the observed total rotations. Curves 2, 3, 4 show the partial rotations of the first absorption band as

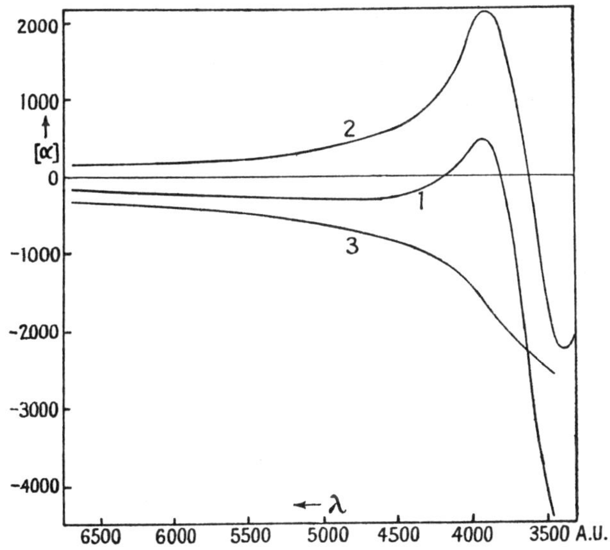

FIG. 183.—ROTATORY DISPERSION OF *l*-MENTHYL DIXANTHATE.
1. Observed rotation.
2. Partial rotation of band (L. and H.).
3. Residual rotation (L. and H.).

calculated from the equations of (i) Drude and Natanson, (ii) Kuhn and Braun, (iii) Lowry and Hudson.

Curves 5, 6, 7 show the corresponding residual rotations. The progressive elimination of the anomalies in curves 5, 6 and 7 provides evidence of the progressive approximation of the three equations (*D.* and *N.*, *K.* and *B.*, *L.* and *H.*) to the experimental data. The slight ripple remaining in curve 7 is perhaps due to the overlapping of two optically-active absorption bands, or to an imperfect adjustment of the constants of the equation ; but it may also be cited as evidence that the process of approximation to the ideal equation is not yet complete.

The absorption band in the diphenyl-dithiourethane is about 2500 A.U. wide; it therefore provides exceptionally good material for discriminating between the three formulæ, since the conditions are very unfavourable for eliminating the anomalies. In the more favourable case of

III. l-*Menthyl dixanthate*, $(C_{10}H_{19}O \, . \, CS \, . \, S)_2$ (Fig. 183),

the elimination of the anomalies in curve 3 is more nearly complete than in curve 7 of Figs. 181 and 182, but it still leaves room for possible improvement.

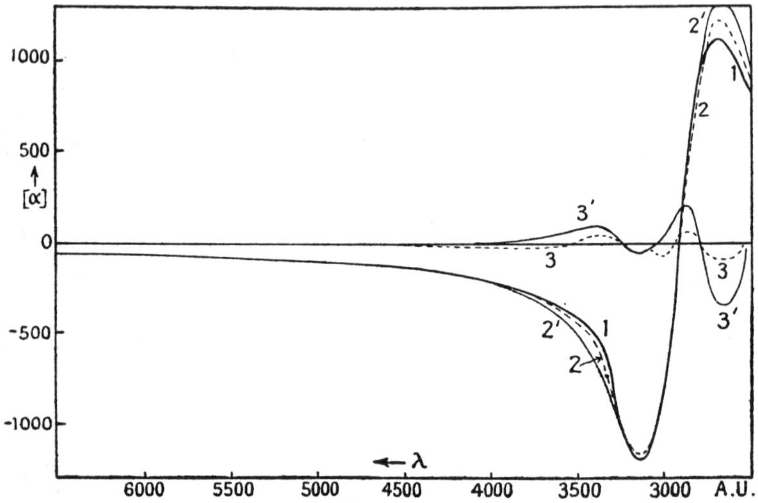

FIG. 184.—ROTATORY DISPERSION OF TETRA-ACETYL-μ-ARABINOSE.

1. Observed rotation.
2. Partial rotation of CHO (L. and H.).
2′. ,, ,, ,, ,, (K. and B.).
3. Residual rotation (L. and H.).
3′. ,, ,, (K. and B.).

(*b*) *Aldehydic Sugars*.—A particularly interesting case has been found in

IV. *Tetra-acetyl-μ-arabinose*, $H[CHOAc]_4 \, . \, CHO$ (Fig. 184),

since the curves 3 and 3′ representing the residual rotation cut the axis of zero rotation in five places. The variations of positive and negative residual rotations, which are greater in curve 3′ (K. and B.) than in 3 (L. and H.), can be attributed partly to experimental error and partly to the inadequacy of the equations; but they are so evenly distributed on either side of the axis that it is impossible to say with certainty whether the residual normal rotation is positive or negative. It therefore appears that the whole of the observed rotatory power

TABLE 45.—SUMMARY OF EQUATIONS FOR ABSORPTION, CIRCULAR DICHROISM AND ROTATORY DISPERSION.

Absorption.	Circular Dichroism.	Rotatory Dispersion.	Authors.
$2n\kappa = \dfrac{D'\Gamma\lambda^3}{(\lambda^2 - \lambda_0^2)^2 + \Gamma^2\lambda^2}$	$(\kappa_l - \kappa_r) = \dfrac{D''\Gamma\lambda^2}{(\lambda^2 - \lambda_0^2)^2 + \Gamma^2\lambda^2}$	$\alpha = \dfrac{D'''(\lambda^2 - \lambda_0^2)}{(\lambda^2 - \lambda_0^2)^2 + \Gamma^2\lambda^2}$	Ketteler, Helmholtz (absorption). Natanson (circular dichroism). Drude, Natanson (rotatory dispersion).
$\epsilon = \epsilon_{\text{max.}} {}^{\nu}/_{\nu_0}\, e^{-\beta(\nu_0-\nu)^2}$			Bielecki and Henri.
$\epsilon = \epsilon_{\text{max.}} \dfrac{\nu^2\nu'^2}{(\nu_0^2 - \nu^2)^2 + \nu^2\nu'^2}$ $\simeq \epsilon_{\text{max.}} \dfrac{\nu'^2}{4(\nu_0 - \nu)^2 + \nu'^2}$		$\alpha = \text{const.} \dfrac{\nu^2(\nu_0^2 - \nu^2)}{(\nu_0^2 - \nu^2)^2 + \nu^2\nu'^2}$ $= \text{const.} \dfrac{\lambda^2 - \lambda_0^2}{(\lambda^2 - \lambda_0^2)^2 + \lambda^2\lambda'^2}$	Kuhn.
$\epsilon = \epsilon_{\text{max.}} e^{-\left(\frac{\nu_0 - \nu}{\theta}\right)^2}$		$[M] =$ $\dfrac{[\phi]}{0{\cdot}541}\dfrac{\nu}{\nu_\phi}\left[e^{-\left(\frac{\nu_0-\nu}{\theta}\right)^2}\int_0^{\frac{\nu_0-\nu}{\theta}} e^{x^2}dx - \dfrac{\theta}{2(\nu_0+\nu)}\right]$	Kuhn and Braun.
	$(\epsilon_l \sim \epsilon_r)$ $= (\epsilon_l \sim \epsilon_r)_{\text{max.}} e^{-\left(\frac{\nu_0 - \nu}{\theta}\right)^2}$		Kuhn and Szabo.
$\epsilon = \epsilon_{\text{max.}} e^{-\left[\frac{\nu_0}{\nu}\left(\frac{\nu_0 - \nu}{\theta}\right)\right]^2}$ $= \epsilon_{\text{max.}} e^{-\left(\frac{\lambda - \lambda_0}{\theta}\right)^2}$	$(\epsilon_l \sim \epsilon_r)$ $= (\epsilon_l \sim \epsilon_r)_{\text{max.}} {}^{\nu}/_{\nu_0}\, e^{-\left[\frac{\nu_0}{\nu}\left(\frac{\nu_0 - \nu}{\theta}\right)\right]^2}$ $= (\epsilon_l \sim \epsilon_r)_{\text{max.}} {}^{\lambda_0}/_{\lambda}\, e^{-\left(\frac{\lambda - \lambda_0}{\theta}\right)^2}$	$[M] =$ $\dfrac{[\phi]}{m}\dfrac{\nu}{\nu_\phi}\left[e^{-\left[\frac{\nu_0}{\nu}\left(\frac{\nu_0-\nu}{\theta}\right)\right]^2}\int_0^{\frac{\nu_0}{\nu}\left(\frac{\nu_0-\nu}{\theta}\right)} e^{x^2}dx + \dfrac{\nu\theta}{2\nu_0(\nu_0+\nu)}\right]$ $= \dfrac{[\phi]}{m}\dfrac{\lambda\phi}{\lambda}\left[e^{-\left(\frac{\lambda-\lambda_0}{\theta}\right)^2}\int_0^{\frac{\lambda-\lambda_0}{\theta}} e^{x^2}dx + \dfrac{\theta}{2(\lambda+\lambda_0)}\right]$	Lowry and Hudson.

of the molecule is contributed by the induced dissymmetry of the aldehydic group, and that the partial rotations due to the fixed dissymmetry of the asymmetric carbon atoms have cancelled out. In these circumstances the anomalous partial rotation of the absorption band has automatically isolated itself without requiring any *ad hoc* mathematical analysis for this purpose. The conditions are thus uniquely favourable for determining the correct geometrical form of the curve. It is therefore of interest to notice that whereas the equation of Kuhn and Braun gives a difference of 200° between the magnitude of the two maxima, and a maximum deviation of 200° from the observed values, these differences are reduced to 30° by using the equation of Lowry and Hudson, whilst in the experimental curve the difference between the two maxima disappears altogether.

Complete equality of the two maxima has also been observed in

V. *Penta-acetyl-μ-fructose*, $H[CHOAc]_4 . CO . CH_2 . O . CO . CH_3$,

where the asymmetric atoms have the same relative configuration as in IV, but the H of the CHO group has been replaced by the radical — $CH_2 . O . CO . CH_3$. It is therefore possible that equality of the positive and negative maxima is a characteristic feature of the anomalous partial rotation of a homogeneous absorption band; but the evidence for optical cancellation of the fixed dissymmetry of the asymmetric carbon atoms is not yet complete enough to prove the complete equality of the two maxima.

Application of the Equation of Lowry and Hudson.—

$$[M] = \frac{[\phi]}{m}\frac{\lambda_\phi}{\lambda}\left[e^{-\left(\frac{\lambda-\lambda_0}{\theta}\right)^2}\int_0^{\frac{\lambda-\lambda_0}{\theta}} e^{x^2}dx + \frac{\theta}{2(\lambda+\lambda_0)}\right].$$

TABLE 46.—KUHN'S INTEGRALS.

c.	$e^{-c^2}\int_0^c e^{x^2}dx$.	c.	$e^{-c^2}\int_0^c e^{x^2}dx$.
0·0	0·00000	1·2	0·50726
0·1	0·09903	1·3	0·48339
0·2	0·19465	1·4	0·4565
0·3	0·27620	1·5	0·4282
0·4	0·35943	1·6	0·4000
0·5	0·42444	1·8	0·3468
0·6	0·47477	2·0	0·30135
0·7	0·51050	2·2	0·2629
0·8	0·53210	2·4	0·2353
0·9	0·54073	2·6	0·2122
1·0	0·53806	2·8	0·1936
1·1	0·52620	3·1623	0·1667

The calculation of rotatory dispersion by means of the equation of Lowry and Hudson or of the equation of Kuhn and Braun is greatly simplified by making use of the table of values of the integral $e^{-c^2}\int_0^c e^{x^2}dx$ for various values of c, as given by Kuhn [1] in Table 46.

For values of c greater than 3·2 the value of the integral $e^{-c^2}\int_0^c e^{x^2}dx$ is approximately equal to $1/2c$.

For other values of c, the integral can, if necessary, be evaluated from the convergent series :

$$e^{-c^2}\int_0^c e^{x^2}dx = c - \frac{2c^3}{1.3} + \frac{2^2c^5}{1.3.5} - \frac{2^3c^7}{1.3.5.7} + \frac{2^4c^9}{1.3.5.7.9} - \ldots + \ldots$$

The values given in the above table, and $1/2c$ for larger values of c, are, however, sufficient for calculating rotatory power over the whole spectral region, and the series need only be used when it is required to calculate the rotatory power for some *particular* wave-length near the region of absorption.

The calculation of rotatory dispersion by means of the above equation is less laborious if advantage is taken of the facts that (i) the function $e^{-\left(\frac{\lambda-\lambda_0}{\theta}\right)^2}\int_0^{\frac{\lambda-\lambda_0}{\theta}} e^{x^2}dx$, which principally governs the value of the partial rotation, is symmetrical with respect to λ_0, so that λ increases or decreases uniformly for uniform increments or decrements of $c\left(c = \frac{\lambda - \lambda_0}{\theta}\right)$, and (ii) that the value of the term $\frac{\theta}{2(\lambda + \lambda_0)}$ varies very little over the spectral region which is usually open to experimental observation.

The method of calculation is most conveniently illustrated by taking a hypothetical example :

Let the maximum of circular dichroism occur at wave-length $\lambda_0 = 3000$ A.U.

" " half-width * of the band of circular dichroism be $\lambda' = 500$ A.U.

" " maximum value of the circular dichroism be $(\epsilon_r - \epsilon_l)_{\text{max.}} = 2{\cdot}0$.

[1] W. Kuhn, Freudenberg's *Stereochemie*, p. 381.

* The value of λ' here used is the value which, when substituted in the equation

$$(\epsilon_r - \epsilon_l) = (\epsilon_r - \epsilon_l)_{\text{max.}}\lambda_0/\lambda\, e^{-\left(\frac{\lambda-\lambda_0}{\theta}\right)^2},$$

where $\lambda' = 1{\cdot}6651\theta$, gives the best representation of the experimental curve of circular dichroism.

Since $$\lambda' = 1{\cdot}6651\theta$$

$$c = \frac{\lambda - \lambda_0}{\theta} = \frac{(\lambda - \lambda_0)\ 1{\cdot}6651}{\lambda'}$$

$$\therefore\ \lambda = \lambda_0 + \frac{c\lambda'}{1{\cdot}6651}.$$

From the above table $e^{-c^2}\int_0^c e^{x^2}dx$ is a maximum when $c = 0{\cdot}9$ approximately. The wave-length λ_ϕ, at which [M] is a maximum is therefore given by

$$\lambda_\phi = 3000 + \frac{0{\cdot}9 \times 500}{1{\cdot}6651}$$

$$= 3270{\cdot}3 \text{ A.U.},$$

and, when $c = 0$, $\lambda = \lambda_0 = 3000$ A.U.

These values are shown in heavy type in column I of Table 47 below.

As c increases by 0·9, λ increases by 270·3 A.U. throughout, or as c changes by 0·1 λ changes by 30·03 A.U.

In column II are tabulated values of c, from Kuhn's table and beyond, at convenient intervals, from $c = 13{\cdot}0$ to $c = -3{\cdot}5$. In column III are tabulated the corresponding values of the integral $e^{-c^2}\int_0^c e^{x^2}dx$, as given by Kuhn, or as given by $1/2c$ for large values of c. These two columns are the first to be drawn up. In column I are inserted the values of λ_0 and λ_ϕ (3000 and 3270 A.U. respectively) opposite $c = 0$ and $c = 0{\cdot}9$. The remaining values of λ in column I are obtained by adding to, or subtracting from, the two wave-lengths already tabulated, the following differences:

30·03 A.U. for every change of c by 0·1
60·06 ,, ,, ,, c ,, 0·2.
150·16 ,, ,, ,, c ,, 0·5
300·3 ,, ,, ,, c ,, 1·0

(For all practical purposes fractions of an Ångström unit may be neglected in making the additions and subtractions.)

In column IV are tabulated the values of the term

$$\frac{\theta}{2(\lambda + \lambda_0)} = \frac{\lambda'}{3{\cdot}3302(\lambda + \lambda_0)}$$

corresponding to the values of λ thus obtained. Since this term increases very slowly with decreasing wave-length, it is only necessary to calculate it for some of the values of λ; the remaining values can then be written in by interpolation. (In column IV the calculated values are shown in heavy type, and the interpolated values in ordinary type.)

The addition of the numbers in columns III and IV gives those in column V, which represent the values of the square bracket in the original equation. The maximum value (0·5646) in this column, corresponding to $\lambda_\phi = 3270$ A.U., is the value of m in the original equation.

Now the value of $[\phi]$ is given by

$$[\phi] = -\frac{100}{2\sqrt{\pi}}\frac{(\epsilon_r - \epsilon_l)_{\text{max.}}}{\log_{10} e}\frac{\lambda_0}{\lambda_\phi}m \text{ radians.}$$

(The negative sign is introduced because the partial rotations are negative on the long wave-length side of λ_0 when $\epsilon_r > \epsilon_l$; see Natanson's rules, p. 427.)

$$\begin{aligned} \therefore [\phi] &= -\frac{100 \times 57{\cdot}296}{2\sqrt{\pi}\log_{10} e}\left[\frac{(\epsilon_r - \epsilon_l)_{\text{max.}}\lambda_0 m}{\lambda_\phi}\right] \text{ degrees} \\ &= -3723\left[\frac{(\epsilon_r - \epsilon_l)_{\text{max.}}\lambda_0 m}{\lambda_\phi}\right] \\ &= -\frac{3723 \times 2 \times 3000 \times 0{\cdot}5646}{3270} \\ &= -3858^\circ. \end{aligned}$$

This is the maximum value of [M] on the long wave-length side of λ_0 and is written in heavy type in column VI opposite $\lambda = \lambda_\phi = 3270$ A.U. The remaining values of [M] in column VI are obtained by multiplying the numbers in column V by $\frac{[\phi]}{m} \cdot \frac{\lambda_\phi}{\lambda}$ for the corresponding values of λ.*

Columns I and VI thus give the wave-lengths and corresponding molecular partial rotations associated with the absorption band in question.

In the majority of cases it is unnecessary to calculate the rotatory power for as many wave-lengths as are given in the above example. A conveniently spaced range of wave-lengths is obtained by using only alternate values of c ($c = 0{\cdot}0$, $0{\cdot}2$, $0{\cdot}4$, etc.). The value 3·1623 for c given in Kuhn's table of integrals has no special significance,

* It is not essential, either to deduce $[\phi]$, or to use λ_ϕ and m, since [M] can be obtained directly by substituting $(\epsilon_r - \epsilon_l)_{\text{max.}}$ in the equation in its original form:

$$[M] = -\frac{100}{2\sqrt{\pi}} \cdot \frac{(\epsilon_r - \epsilon_l)_{\text{max.}}}{\log_{10} e} \cdot \frac{\lambda_0}{\lambda}\left[e^{-\left(\frac{\lambda - \lambda_0}{\theta}\right)^2}\int_0^{\frac{\lambda - \lambda_0}{\theta}} e^{x^2}dx + \frac{\theta}{2(\lambda + \lambda_0)}\right].$$

The above method is, however, described, because in the absence of measurements of circular dichroism, it is often possible to assign an approximate value to $[\phi]$, i.e. $[M]_{\text{max.}}$, by an examination of the observed rotatory power in the region of absorption. In this case the values of λ_0 and λ' must refer to the point of maximum absorption and the half-width of the absorption band respectively.

TABLE 47.—HUDSON'S EQUATION.

$\lambda_\theta = 3000$ A.U., $\lambda' = 500$ A.U., $(\epsilon_r - \epsilon_l)$max. $= 2{\cdot}0$.

I.	II.	III.	IV.	V.	VI.
λ.	c.	$e^{-c^2}\int_0^c e^{x^2}dx.$	$\frac{\theta}{2(\lambda+\lambda_0)}.$	$\left[e^{-c^2}\int_0^c e^{x^2}dx + \frac{\theta}{2(\lambda+\lambda_0)}\right]$	[M].
6904	13·0	0·0385	0·0152	0·0537	− 174
6604	12·0	0·0417	0·0156	0·0573	− 194
6304	11·0	0·0454	0·0161	0·0615	− 218
6003	10·0	0·0500	0·0167	0·0667	− 248
5703	9·0	0·0566	0·0173	0·0729	− 285
5403	8·0	0·0625	0·0179	0·0805	− 333
5102	7·0	0·0714	0·0185	0·0890	− 389
4802	6·0	0·0833	0·0192	0·1025	− 477
4502	5·0	0·1000	0·0200	0·1200	− 595
4351	4·5	0·1111	0·0204	0·1315	− 675
4201	4·0	0·1250	0·0208	0·1458	− 775
4051	3·5	0·1429	0·0213	0·1642	− 905
3950	3·1623	0·1667	0·0216	0·1883	− 1065
3841	2·8	0·1936	0·0219	0·2155	− 1253
3781	2·6	0·2122	0·0221	0·2343	− 1385
3721	2·4	0·2353	0·0223	0·2576	− 1546
3661	2·2	0·2629	0·0225	0·2854	− 1741
3601	2·0	0·3013	0·0227	0·3240	− 2009
3541	1·8	0·3468	0·0230	0·3698	− 2333
3481	1·6	0·4000	0·0232	0·4232	− 2715
3450	1·5	0·4282	0·0233	0·4515	− 2923
3420	1·4	0·4565	0·0234	0·4799	− 3134
3390	1·3	0·4834	0·0235	0·5069	− 3340
3360	1·2	0·5073	0·0236	0·5309	− 3530
3330	1·1	0·5262	0·0237	0·5499	− 3689
3300	1·0	0·5381	0·0238	0·5619	− 3804
3270 = $\lambda\phi$	0·9	0·5407	0·0239	0·5646 = m	− 3858 = [ϕ]
3240	0·8	0·5321	0·0241	0·5562	− 3835
3210	0·7	0·5105	0·0242	0·5347	− 3720
3180	0·6	0·4748	0·0243	0·4991	− 3505
3150	0·5	0·4244	0·0244	0·4488	− 3183
3120	0·4	0·3594	0·0245	0·3839	− 2748
3090	0·3	0·2762	0·0246	0·3008	− 2174
3060	0·2	0·1946	0·0247	0·2193	− 1601
3030	0·1	0·0990	0·0248	0·1238	− 913
3000 = λ_0	0·0	0·0000	0·0250	0·0250	− 186
2970	− 0·1	− 0·0990	0·0251	− 0·0739	556
2940	− 0·2	− 0·1946	0·0252	− 0·1694	1287
2910	− 0·3	− 0·2762	0·0254	− 0·2508	1925
2880	− 0·4	− 0·3594	0·0255	− 0·3339	2589
2850	− 0·5	− 0·4244	0·0256	− 0·3988	3127
2820	− 0·6	− 0·4748	0·0258	− 0·4490	3556
2790	− 0·7	− 0·5105	0·0259	− 0·4846	3879
2760	− 0·8	− 0·5321	0·0261	− 0·5060	4096
2730	− 0·9	− 0·5407	0·0262	− 0·5145	4209
2700	− 1·0	− 0·5381	0·0263	− 0·5118	4234
2670	− 1·1	− 0·5262	0·0265	− 0·4997	4180
2640	− 1·2	− 0·5073	0·0266	− 0·4807	4066
2610	− 1·3	− 0·4834	0·0267	− 0·4567	3909
2580	− 1·4	− 0·4565	0·0269	− 0·4296	3719
2550	− 1·5	− 0·4282	0·0270	− 0·4012	3514
2519	− 1·6	− 0·4000	0·0272	− 0·3728	3306
2459	− 1·8	− 0·3468	0·0274	− 0·3194	2901
2399	− 2·0	− 0·3013	0·0277	− 0·2736	2547
2339	− 2·2	− 0·2629	0·0281	− 0·2348	2242
2279	− 2·4	− 0·2353	0·0284	− 0·2069	2028
2219	− 2·6	− 0·2122	0·0287	− 0·1835	1848
2159	− 2·8	− 0·1936	0·0291	− 0·1645	1702
2050	− 3·1623	− 0·1667	0·0297	− 0·1370	1492
1949	− 3·5	− 0·1429	0·0303	− 0·1126	1290

and is inconvenient in deducing the corresponding value of λ by the above additive method, so may be omitted. The error involved in writing $1/2c$ for $e^{-c^2}\int_0^c e^{x^2}dx$ for large values of c amounts to approximately 5 per cent. when $c = 3$, so that a slight discontinuity in the curve of partial rotations results in this region unless the true values of the integral are extended beyond this point. The slight discontinuity is, however, negligible in comparison with the experimental errors involved in measuring circular dichroism on which the whole calculation is based.

The above method of calculation may be summarised as follows:

(i) Six columns headed as shown are drawn up.

(ii) In column II are tabulated values of c for which the corresponding integral is known or is easily obtained from $1/2c$.

(iii) In column III are tabulated the corresponding values of the integral.

(iv) From $\lambda_\phi = \lambda_0 + \frac{c\lambda'}{1\cdot6651}$, where $c = 0\cdot9$, is calculated the wave-length λ_ϕ corresponding to the maximum value of [M] on the long wave-length side of λ_0. Then $\frac{\lambda_\phi - \lambda_0}{0\cdot9}$ is the increment of λ for every increment of c by $0\cdot1$.

(v) By addition to or subtraction from λ_0 or λ_ϕ the wave-lengths corresponding to all the values of c are tabulated in column I.

(vi) Values of $\frac{\lambda'}{3\cdot3302(\lambda + \lambda_0)}$ for wave-lengths at convenient intervals are deduced and entered in column IV, the remainder being obtained by interpolation.

(vii) Columns III and IV are summed to give column V. The maximum value in this column (corresponding to $\lambda = \lambda_\phi$ and $c = 0\cdot9$) is m.

(viii) $[\phi]$ is deduced from the maximum value of the circular dichroism by

$$[\phi] = -3723\left[\frac{(\epsilon_r - \epsilon_l)_{\text{max.}}\lambda_0 m}{\lambda_\phi}\right].$$

(ix) The partial molecular rotations in column VI are obtained by multiplying the numbers in column V by $\frac{[\phi]}{m} \cdot \frac{\lambda_\phi}{\lambda}$.

The partial molecular rotations associated with an optically-active absorption band can be calculated by this method in but a small fraction of the time required for the evaluation of a two-term Drude equation for the region of transparency only.

Application of the Equation of Kuhn and Braun.—

$$[\text{M}] = \frac{[\phi]}{0\cdot541}\frac{\nu}{\nu_\phi}\left[e^{-\left(\frac{\nu_0-\nu}{\theta}\right)^2}\int_0^{\frac{\nu_0-\nu}{\theta}} e^{x^2}dx - \frac{\theta}{2(\nu_0+\nu)}\right].$$

(*a*) This equation can be evaluated in a precisely analogous manner to that described above. Frequencies are, however, used throughout instead of wave-lengths, and since the function $e^{-\left(\frac{\nu_0-\nu}{\theta}\right)^2}\int_0^{\frac{\nu_0-\nu}{\theta}} e^{x^2}dx$ is symmetrical with respect to ν_0, uniform increments of c $\left(c = \frac{\nu_0 - \nu}{\theta}\right)$ correspond to uniform increments of ν. Thus for known values of c the corresponding values of ν may be obtained by addition to, or subtraction from, ν_0 or ν_ϕ.

(*b*) An alternative method, suggested by Mitchell,[1] consists in plotting c against Kuhn's values of the integral, and reading off the values of the integral for those values of c which correspond to particular frequencies or wave-lengths.

[1] MITCHELL, *The Cotton Effect* (G. Bell & Sons, 1933), p. 40.

CHAPTER XXXV.

THEORIES OF MAGNETIC ROTATORY POWER.

Applications of Fresnel's Theory.—(*a*) Shortly after Faraday's discovery of magnetic rotatory power, Airy [1] in 1846 suggested that the phenomenon might be explained in the same way in which Fresnel had accounted for natural rotatory power, namely that, under the influence of the magnetic field, the plane polarised beam is resolved into two circularly polarised beams of opposite sense which are propagated through the medium with different velocities. The same idea was put forward independently by Verdet in 1855.[2]

(*b*) A direct demonstration of this hypothesis as applied to the origin of magnetic rotatory power was given almost simultaneously by Righi [3] and by H. Becquerel [4] in 1878. In Becquerel's experiments, plane polarised light was rendered circularly polarised by passage through a quarter-wave plate. It was then brought into an interferometer system, similar to that employed by Rayleigh, in which the initial beam is divided into two beams which are allowed to interfere. A plate of glass was placed in one of these beams, between the poles of an electromagnet.* When the magnet was excited, the interference fringes were seen to move; and the direction of motion was reversed when the field was reversed or when the quarter-wave plate was turned through a right-angle. The change of phase thus produced was obviously due to a difference in the velocities with which right and left circularly polarised beams traversed the glass plate when under the influence of the magnetic field.

(*c*) The two circularly polarised beams were actually separated for the first time in 1901 by Brace,[5] who made use of the fact that two beams travelling with different velocities under the influence of the magnetic field are reflected at different angles at a glass-air interface. A further difference in the angle of reflection can be produced by reflecting the beams for a second time whilst moving through the field *in the reverse direction*, provided that the sense of each rotation has

[1] Airy, *Phil. Mag.*, 1846, [iii], **28,** 469.
[2] Verdet, *A.C.P.*, 1858, [iii], **52,** 129.
[3] Righi, *Nuovo Cimento*, 1878, **3,** 212.
[4] H. Becquerel, *C.R.*, 1879, **88,** 334.
[5] Brace, *Phil. Mag.*, 1901, [vi], **1,** 464.
* A second plate, outside the poles, was used to compensate certain possible secondary effects of the magnet on the glass.

been reversed. In order to achieve this result Brace employed the apparatus shown in Fig. 185. Two long right-angled glass prisms, P_1 and P_2, were cemented together with a half-wave plate of mica, M, between them, and were placed in a magnetic field, H, whose direction was parallel to the plane of the mica. A beam, L, of plane polarised light, from a vertical slit illuminated with a powerful flame from sodium burning in hydrogen, entered the main prisms through a small auxiliary prism, P_3, at one end of the system. Since its path was perpendicular to the field, no effect was produced thereby, but by passing through the half-wave plate, M, the plane of vibration was turned through a right angle. After total reflection at A, however,

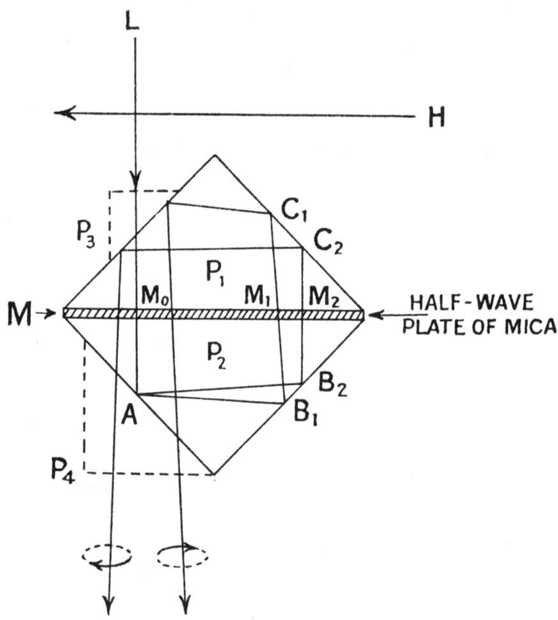

FIG. 185.—SEPARATION OF CIRCULARLY POLARISED BEAMS BY A MAGNETIC FIELD (Brace, 1901).

the path of the light was parallel to the magnetic field; the beam was therefore broken up into two components reflected at slightly different angles. These components were again reflected at B_1, B_2, and, on passing through the mica for the second time at M_1, M_2, the directions of rotations of the two circularly polarised beams were reversed by the action of the half-wave plate. As a result of this reversal, the angular separation of the two beams was doubled by the third reflection at C_1, C_2, in spite of the fact that the reflected beams were now travelling through the magnetic field in the opposite direction to that of their path when first reflected at A. In this way the light was passed round the system five times in a quasi-spiral path, undergoing twenty internal reflections, and was finally inspected through a second auxiliary prism, P_4, at the other end of the system. On turning on the magnetic field, the image of the slit in the observing telescope was doubled, and the opposite polarisations of the beams could also be demonstrated.

(*d*) The difference in the velocities of the two components was first measured by Mills,[1] who employed a Michelson interferometer in which one half of the field was illuminated with right and the other with left circularly polarised light, the dividing line being perpendicular to the interference fringes. Both beams passed through a tube of CS_2 which was placed along the axis of a solenoid. When the current in the solenoid was started, the fringes in the two halves of the field were displaced in opposite directions, and it was found that *the faster circular component was the one whose sense of rotation was the same as that of the exciting current.* The difference in velocities could be deduced from the displacement of the fringes.

(*e*) With the help of the powerful field of the Bellevue electromagnet, Cotton and Schérer [2] have been able to observe directly the MAGNETIC CIRCULAR DICHROISM inside an absorption band in solutions of cobalt salts and of thiobenzophenone.

Relationship between Magnetic Rotatory Power and Refractive Index.—Airy,[3] Verdet,[4] and de la Rive [5] sought to explain the Faraday phenomenon in terms of a mechanical effect on the medium when placed in a magnetic field. These explanations were justified by experiments such as those of Matteucci [6] and Wertheim [7] on the effect of mechanical strain on the optical properties of glass; but they need not now be considered in detail, since they did not lead to any satisfactory mathematical analysis of the phenomenon. A concrete result was obtained, however, when Verdet [8] in 1865 deduced two alternative formulæ for magnetic rotatory dispersion in terms of refractive dispersion, viz.

$$\delta = \frac{A}{\lambda^2}\left(n - \lambda\frac{dn}{d\lambda}\right) \text{ or } \delta = B\frac{n^2}{\lambda^2}\left(n - \lambda\frac{dn}{d\lambda}\right), \qquad \text{(i) and (ii)}$$

where n is the refractive index of the medium, and A and B are arbitrary constants. He calculated $dn/d\lambda$ with the help of a Cauchy formula, and then tested the equations against the data for CS_2 and for creosote. The second expression gave the better agreement, but the divergence was still greater than the experimental error.

[1] J. MILLS, *Phys. Rev.*, 1904, **18**, 65.

[2] COTTON, *C.R.*, 1932, **195**, 561; SCHÉRER, *ibid.*, p. 950; COTTON and SCHÉRER, *ibid.*, p. 1342; SCHÉRER and CORBONNIER, *C.R.*, 1933, **196**, 1724 and 1932; see also R. W. WOOD, *Phil. Mag.*, 1905, **9**, 725.

[3] AIRY, *ibid.*, 1846, [iii], **28**, 469.

[4] VERDET, *A.C.P.*, 1854, [iii], **41**, 370.

[5] DE LA RIVE, *Traité d'Electricité*, 1853, **1**, 555; *A.C.P.*, 1868, [iv], **15**, 57.

[6] MATTEUCCI, *A.C.P.*, 1848, [iii], **24**, 354; *P.R.S.*, 1848, **5**, 741.

[7] WERTHEIM, *C.R.*, 1851, **32**, 289.

[8] VERDET, *A.C.P.*, 1863, [iii], **69**, 415.

A more satisfactory relationship between magnetic rotatory power and refractive index was established by H. Becquerel between 1875 and 1899.[1] He first deduced, from measurements on a large number of liquids, an empirical formula relating the magnetic rotation δ to the refractive index n for a given wave-length, namely,

$$\delta = An^2(n^2 - 1) \quad . \quad . \quad . \quad . \quad \text{(iii)}$$

where A is a constant characteristic of the substance. He then improved the technique for obtaining monochromatic light, by employing flames impregnated with *sodium*, *lithium* and *thallium* salts, and made a more extensive study of the relationship between rotation and refraction in a series of fifty-six liquid and solid compounds. The results, referred to carbon disulphide as a standard, were again in close accord with his formula; moreover, with the help of a careful comparison of the rotatory powers of water and CS_2, he was able to show that Verdet's experimental results also satisfied the formula. In this connection he recorded the interesting observation that compounds of similar composition give comparable values for the constant A. He then investigated the formulæ for magnetic rotatory dispersion, and found that Verdet's formula (ii) could be replaced by the empirical expression[1]

$$\delta = A\frac{n^2(n^2 - 1)}{\lambda^2} \quad . \quad . \quad . \quad . \quad . \quad \text{(iv)}$$

Attempts to put these empirical formulæ on a theoretical basis met with little success, until Zeeman discovered the effect now known by his name, and recognised its relationship with the Faraday effect. Becquerel then developed in 1897[1] a theory in which a rotatory motion under the action of a magnetic field was attributed to the ether, and concluded that the rotation δ for 1 cm. of substance in a field of 1 Gauss should be given by

$$\delta = \frac{2\pi}{c\tau}\lambda\frac{dn}{d\lambda} \quad . \quad . \quad . \quad . \quad \text{(v)}$$

where τ is the natural period of the rotations. This expression did not agree well with the experimental results for the magnetic rotatory dispersion of CS_2, but a better result was obtained with creosote—the only other substance for which data were then available.

Magnetic Rotatory Dispersion in Metallic Vapours.—In the following year,[1] Becquerel demonstrated the rapid increase in the magnetic rotatory power of sodium vapour which takes place in the immediate vicinity of the D-line, and called attention to the similarity of this phenomenon to the anomalous refractive dispersion of the vapour in the same region. The relationship between the two phenomena could be expressed quantitatively with fair accuracy

[1] H. Becquerel, *C.R.*, 1875, **80**, 1375; *A.C.P.*, 1885, [vi], **6**, 145; *C.R.*, 1897, **125**, 679; 1898, **127**, 647 and 899; 1899, **128**, 145.

by the expression (iv), which certainly gave a better agreement with experiment than did Verdet's formula (i) or (ii).

The magnetic rotatory dispersion of sodium vapour was also investigated, almost simultaneously, by Macaluso and Corbino,[1] who observed that *the sign of the rotation is the same on both sides of the D-lines.* Their experiments were repeated on a larger scale by R. W. Wood,[2] with an apparatus closely resembling that employed by him in his observations of the refractive dispersion of sodium vapour in the same region (p. 119). A piece of sodium was introduced into a steel tube, closed with glass end-plates inclined to one another, so that it formed a prism. The tube was evacuated, and the sodium was then vaporised by heating with a suitably placed flame. The tube was enclosed in an electromagnet and set up between the Nicol prisms of a spectropolarimeter, with an arc lamp as source of light. The Nicol prisms were first set parallel, so that the complete spectrum appeared, crossed by the absorption bands at the D-lines. With the magnetic field turned on, a dark band, corresponding to a rotation of $\pi/2$, appeared on either side of these, and as the analyser was turned, these two bands moved outwards along the spectrum. In this way the extinction wave-lengths for any given setting of the Nicols could be determined.

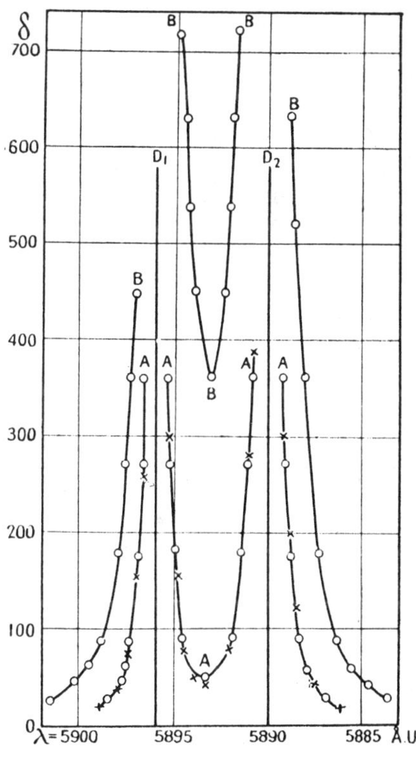

FIG. 186.—MAGNETIC ROTATORY DISPERSION OF SODIUM VAPOUR (R. W. Wood, 1905).

By this method a complete curve of rotatory dispersion on either side of the D-lines could be drawn. Two such curves, AAAAA, BBBBB, corresponding to two different pressures of vapour, are shown in Fig. 186. The rapid increase in rotation as the D-lines are approached, and the minimum in the curve between them, show clearly that the sign of the rotation is the same on both sides of the

[1] MACALUSO and CORBINO, *Rend. Acad. Lincei*, 1898, [v], **7**, 293.
[2] R. W. WOOD, *Phil. Mag.*, 1905, [vi], **10**, 408.

absorption, whilst the numerical values, coupled with the corresponding data for the refractive dispersion, were found to agree very closely with a formula relating the two quantities which had been proposed by Drude, as described below.

Observations of the magnetic rotations of metallic vapours have also been made by Geiger,[1] using a sodium flame, by Hansen,[2] with sodium and lithium flames, by Senftleben,[3] also with a sodium flame, by Minkowski [4] in non-luminous sodium vapour, and by Kuhn [5] in non-luminous thallium and cadmium vapours. The object of all these experiments was to establish the relationship between the strength f of an absorption line and its contribution to the refractive index, which is expressed in the equation for refractive dispersion, as derived on both classical and quantum bases, namely,

$$n^2 - 1 = \frac{Le^2}{\pi m} \sum \frac{f_1}{\nu_1{}^2 - \nu^2} \quad . \quad . \quad . \quad \text{(vi)}$$

(see equations iii and iv, p. 378). This relationship appears to be true to a high degree of accuracy; but it must be emphasised that these results apply only to absorption *lines* in metallic vapours, and that it does not therefore follow that it would be equally valid when applied to the partial refractions contributed by the wide diffuse bands of organic compounds,[6] which may extend over as much as 1000 A.U.

Electronic Theories of Magnetic Rotatory Power.—(*a*) The first electronic theory, based on the hypothesis that electrically charged particles exist in the magnetically active medium, was propounded in 1900 by Drude,[7] who adopted the idea put forward by Ampère and Weber that paramagnetism is due to "molecular currents," depending on the orbital movement of ionic charges. Drude assumed that similar currents might be induced in diamagnetic substances by a magnetic field. These ionic charges can be used to account for refractive dispersion; thus the refractive index n of the medium at wave-length λ can be expressed by the formula

$$n^2 = a + \frac{b}{\lambda^2 - \lambda_0{}^2} \quad . \quad . \quad . \quad \text{(vii)}$$

where λ_0 is the wave-length corresponding with the characteristic frequency of oscillation of the ions and a and b are constants. Drude concluded that the magnetic rotation δ would be given by

$$\delta = n\left(\frac{a'}{\lambda^2} + \frac{b'}{\lambda^2 - \lambda_0{}^2}\right), \quad . \quad . \quad . \quad \text{(viii)}$$

[1] L. GEIGER, *Ann. Physik*, 1907, **23**, 758; **24**, 597.

[2] H. M. HANSEN, *ibid.*, 1914, **43**, 169; W. VOIGT and H. M. HANSEN, *Phys. Zeits.*, 1912, **13**, 217.

[3] H. SENFTLEBEN, *Ann. Physik*, 1915, **47**, 949.

[4] R. MINKOWSKI, *ibid.*, 1921, **66**, 206; R. LADENBURG and R. MINKOWSKI, *Zeits. Phys.*, 1921, **6**, 153.

[5] W. KUHN, *Danske Vidensk. Selskab., Kopenhagen*, 1926, **7**, 12.

[6] See footnote, p. 379; cf. ALLSOPP, *P.R.S.*, 1934, A. **146**, 310.

[7] DRUDE, *Lehrbuch der Optik*, 1900.

where a' and b' are a different pair of constants. This expression agreed with the experimental data for the magnetic rotation and refractive dispersion of CS_2; but it requires that the rotation shall change sign at wave-length $\lambda = \lambda_0$, i.e. on passing through the characteristic frequency of an absorption band. The experiments of Macaluso and Corbino (p. 461) showed, however, that this is not the case. Drude therefore sought an alternative basis for his theory, and found it in the "Hall effect." Thus, if the ions (without having the orbital motion of "molecular currents") are merely free to vibrate as the light-wave passes over them, they will be acted on in a magnetic field by a force which is at right-angles both to the direction of motion and to the direction of the field; and this force must be added to those already effective in the system. This hypothesis led to an expression for δ in a slightly different form, namely,

$$\delta = \frac{1}{n}\left(\frac{a'}{\lambda^2} + \frac{b'\lambda^2}{(\lambda^2 - \lambda_0{}^2)^2}\right) \quad . \quad . \quad . \quad \text{(ix)}$$

The agreement with the data for CS_2 in this case is not quite so satisfactory; but the new equation is capable of accounting for the similarity of sign of the rotation on either side of the absorption band. It received confirmation from the experiments of R. W. Wood (p. 461), the results of which could be expressed by the modified equation

$$\delta = \frac{B\lambda^2}{(\lambda^2 - \lambda_0{}^2)^2} \quad . \quad . \quad . \quad . \quad \text{(x)}$$

in which the refractive index is taken as constant throughout the region of absorption.

(*b*) A simple theory, explaining both the Faraday and the Zeeman effects, was developed by Larmor[1] in 1900. This theory accounts for the two phenomena on the basis of the "Larmor precession," namely, the effect that in a magnetised medium the electrons move with a definite frequency in circular orbits around the direction of the field as axis. Suppose that circularly polarised light, whose sense of rotation is the same as that of the electrons, passes through the medium in the direction of the field. In order that it may continue to travel with the same velocity as it would have done in the unmagnetised medium, it will be necessary to increase its frequency by the frequency of rotation of the electrons, in order that the relative phase of light vibrations and electron vibrations may remain the same. On the other hand, circularly polarised light of the opposite sense of rotation would need to have its frequency reduced by the same amount. It follows that, if the frequency of the circular vibrations remains the same, the velocities (or refractive indices) for the two senses of rotation must be changed, one becoming accelerated and the other retarded. In the magnetic field, therefore, the double refraction postulated by Airy will occur.

[1] LARMOR, *Ether and Matter*, p. 352.

This reasoning leads to the result that the magnetic rotation δ at wave-length λ, for unit length and unit field, is

$$\delta = \frac{e}{2mc^2}\lambda\frac{dn}{d\lambda}, \quad . \quad . \quad . \quad . \quad \text{(xi)}$$

where c is the velocity of light and e/m is the ratio of charge to mass of an electron. This expression is identical in form with the formula (v) deduced on quite other grounds by Becquerel.

(*c*) Richardson [1] has modified Larmor's result by substituting for $dn/d\lambda$ the value obtained by differentiating a dispersion equation of the Ketteler-Helmholtz or Drude type. Using the same notation as above, he obtained the result

$$n\delta\lambda^2 = \sum K\left(\frac{\lambda^2}{\lambda^2 - \lambda_0^2}\right)^2, \quad . \quad . \quad . \quad . \quad \text{(xii)}$$

where K is a constant, and the summation is carried over all the absorption bands. If both sides of this equation are divided by the factor $n\lambda^2$, it becomes identical with Drude's equation (ix), since a' was shown by Drude to be negligibly small and the a'/λ^2 term can therefore be omitted. In this simplified form, Drude's equation has been shown by Evans and others [2] to be valid for a large number of simple organic compounds.

(*d*) Despite the confirmation of Drude's equation (ix) by R. W. Wood's experiments on sodium vapour, closer investigation has shown that the phenomenon is more complicated than he supposed. Thus, Wood himself found that an aqueous solution of *praesodymium chloride* gives a magnetic rotation which has opposite signs on either side of the absorption band,[3] and can so be termed ANOMALOUS MAGNETIC ROTATORY DISPERSION, whilst that of *neodymium nitrate* has the same sign throughout.[4] Early experiments on organic compounds led to similar results, although many of the conclusions then reached were afterwards found to be untenable on account of the low intensities of the magnetic fields which had been used.[5] The work of Roberts [6] on the magnetic rotatory dispersions of the paramagnetic salts of Fe, Ni and Co, however, definitely established the existence of this "anomalous" type of dispersion in the region of absorption. Roberts' observations on cobalt salts have been con-

[1] S. S. RICHARDSON, *Phil. Mag.*, 1916, [vi], **31**, 232 and 454.
[2] EVANS and STEPHENS, *ibid.*, 1927, [vii], **3**, 546, and subsequent papers.
[3] R. W. WOOD, *ibid.*, 1905, [vi], **9**, 725.
[4] *Ibid.*, 1908, [vi], **15**, 270.
[5] See, for instance, SCHMAUSS, *Ann. Physik*, 1900, **2**, 280; 1902, **8**, 842; 1903, **10**, 853; BATES, *ibid.*, 1903, **12**, 1091; ELIAS, *ibid.*, 1911, **35**, 299; PFLEIDERER, *Zeits. Phys.*, 1926, **39**, 663; KRETHLOW, *ibid.*, 1927, **42**, 840.
[6] ROBERTS, SMITH and RICHARDSON, *Phil. Mag.*, 1922, [vi], **44**, 912; ROBERTS, *ibid.*, 1925, [vi], **49**, 397; 1930, [vii], **9**, 361.

firmed by Miescher.[1] Miescher also studied the magnetic rotatory dispersions through the absorption bands of the *dithiourethane* of *l-borneol*, of *benzylidene-camphoryledone-3-acetone*, and of *camphorquinone*, but found no anomaly. Cotton and Schérer[2] found that *cobalt chloride* gives curves for magnetic circular dichroism and magnetic rotatory dispersion in the region of absorption which are exactly analogous to those for the natural circular dichroism and rotatory dispersion of optically-active compounds. On the other hand, a diamagnetic organic compound, *thiobenzophenone*, in the fused state, gave a similar curve for the magnetic circular dichroism, but a rotation with *the same sign on either side of the absorption band.*

There are thus two types of magnetic rotatory dispersion, namely (i) a DIAMAGNETIC TYPE OF ROTATORY DISPERSION, which can be represented by Drude's formula (ix); this includes the diamagnetic organic compounds and some metallic vapours, etc., (ii) a PARAMAGNETIC TYPE, which does not conform to Drude's formula on account of the presence of paramagnetic atoms. The rotations of the latter type, which includes the salts of cobalt, are obviously composite; but Roberts, Miescher and others have been able to explain them on the basis of a theory of magnetic rotatory power developed by Ladenburg.[3] This theory, which has been brilliantly confirmed by the work of Becquerel and de Haas on the magnetic rotations of crystals of salts of the rare earths at low temperatures,[4] accounts for the existence of both diamagnetic and paramagnetic rotations by means of simple quantum theory.

The Molecular Theory of Magnetic Rotatory Power.—The electronic theories outlined above take no account of the structure of the molecules in which the electrons are situated. They assume only that the interaction between the electrons and the light wave can be expressed in terms of the ordinary refractive properties of the medium, and that the effect of the magnetic field can be represented by appropriate modifications of the equations. A chemical molecule, however, has a discrete structure which determines the distribution and polarisability of the electrons in it. The simplest arbitrary representation of such a molecule is an ellipsoid with three different axes corresponding to the three principal polarisabilities. The magneto-optical properties of such a molecule will vary with the direction of propagation of the light wave relative to these axes; the observed rotation will therefore be an average value for a large number of molecules distributed in random orientations.

J. J. Thomson[5] in 1920, in a paper dealing with the scattering of light by the electrons of an *unsymmetrical* molecule, deduced expressions for both natural and magnetic rotatory powers of such a

[1] MIESCHER, *Helv. Phys. Acta,* 1930, **3**, 93.
[2] COTTON and SCHÉRER, *loc. cit.*, p. 459.
[3] LADENBURG, *Zeits. Phys.*, 1925, **34**, 898; 1927, **46**, 168.
[4] BECQUEREL and DE HAAS, *ibid.*, 1929, **52**, 678; **57**, 11.
[5] J. J. THOMSON, *Phil. Mag.*, 1920, [vi], **40**, 713.

molecule. Thus Verdet's constant δ at a wave-length λ remote from an absorption frequency at λ_0 was given by

$$\delta = \frac{A(n-1)}{\lambda^2 - \lambda_0^2} \quad . \quad . \quad . \quad . \quad \text{(xiii)}$$

where A is a constant and n is the refractive index.

A more detailed molecular theory has been developed on similar lines by de Mallemann.[1] He defines SPECIFIC and MOLECULAR MAGNETIC ROTATORY POWERS by absolute terms as follows :

$$\text{Specific magnetic rotatory power} \quad [\Lambda] = \frac{\Lambda}{d}$$

$$\text{Molecular magnetic rotatory power} \quad [\Lambda]_M = \frac{M\Lambda}{d}$$

where Λ is Verdet's constant in radians, M is the molecular weight and d the density of the substance.

The MOLECULAR ROTATORY POWER $[\Lambda]_M$ of a *gas* at wave-length λ can then be expressed as the sum of the ATOMIC ROTATIONS of the constituent atoms or radicals as follows :

$$[\Lambda]_M = \frac{\pi}{2Ne\lambda^2}\sum_k \frac{[R_k]_A{}^2}{p_A} = \sum_k [\Lambda_k]_A$$

where $[R_k]_A$ is the atomic refractivity for an atom containing p_k optical electrons, N is Avogadro's number and e is the electronic charge.

The atomic rotations can be deduced from measurements of Verdet's constants (for the vapours) of homologous organic series in exactly the same way as the atomic refractions, and the following values have been obtained :

TABLE 46.—ATOMIC MAGNETIC ROTATIONS ($\lambda = 5780$ A.U.).

Atom or Bond .	H	C	Cl	Br	I	CN	=	≡
$[\Lambda]_A \times 10^5$.	1·80	2·20	9·0	19·5	41·0	7·2	8·5	13·0

The calculated values for H and C are $1{\cdot}79 \times 10^{-5}$ and $2{\cdot}18 \times 10^{-5}$ respectively. The agreement with experiment is remarkable.

The specific rotation of a molecule depends on its physical state, but on the basis of the Lorenz theory of refraction, de Mallemann's theory leads to the conclusion that the ratio $[\Lambda]_{gas}/[\Lambda]_{liq.}$ should be equal to $9n/(n^2+2)^2$, n being the refractive index of the liquid. This again is confirmed by experiment, as can be seen by comparing the values set out in columns 5 and 6 of Table 47.

[1] DE MALLEMANN, *Journ. de Phys.*, 1926, **7**, 295; *Congrès International d'Electricité*, Section 1, Report No. 31, 1932.

TABLE 47.—MAGNETIC ROTATIONS OF LIQUIDS AND GASES.

(λ = 5780 A.U.).

	$n_{liq.}$	$[\Lambda]_{liq.} \times 10^2$.	$[\Lambda]_{gas} \times 10^2$.	$\frac{[\Lambda]_{gas}}{[\Lambda]_{liq.}}$.	$\frac{9n}{(n^2+2)^2}$.	$\frac{9n[\Lambda]_{liq.}}{(n^2+2)^2}$.
Pentane . .	1·357	1·90	1·61	0·85	0·83	1·58
Ethyl chloride .	1·379	1·49	1·26	0·85	0·815	1·20
Ethyl bromide .	1·425	1·31	1·07	0·82	0·79	1·03
Ethyl iodide .	1·514	1·56	1·21	0·77	0·74	1·15
Chloroform .	1·446	1·13	0·93	0·82	0·78	0·88
Benzene . .	1·502	3·53	2·72	0·77	0·75	2·65
Carbon bisulphide	1·627	3·43	2·30	0·67	0·68	2·33

Since the molecular rotation of the liquid is equal to the sum of the atomic rotations, multiplied by the factor $(n^2 + 2)^2/9n$, the additive relationship predicted for the atomic rotations of gases can be extended to liquids. Moreover, since this factor approximates very closely to 1 in all gases, it follows that the molecular rotation or the corresponding specific rotation becomes invariant when multiplied by $9n/(n^2 + 2)^2$, just as the theory predicts.

De Mallemann's theory is not yet completely developed, and the results which have so far been obtained apply only to simple molecules at wave-lengths remote from an absorption band. In particular, the theory will not predict the magnetic rotatory dispersion satisfactorily, even in the visible spectrum. Nevertheless, the agreement between theory and experiment which has been described seems to indicate the correctness of the fundamental hypotheses, and to suggest that the Faraday effect (as de la Rive and Perkin believed) is capable of becoming a powerful means of elucidating problems of molecular structure.

AUTHOR INDEX

SUBJECT INDEX

CATALOG OF DOVER BOOKS

CHEMISTRY AND PHYSICAL CHEMISTRY

ORGANIC CHEMISTRY, F. C. Whitmore. The entire subject of organic chemistry for the practicing chemist and the advanced student. Storehouse of facts, theories, processes found elsewhere only in specialized journals. Covers aliphatic compounds (500 pages on the properties and synthetic preparation of hydrocarbons, halides, proteins, ketones, etc.), alicyclic compounds, aromatic compounds, heterocyclic compounds, organophosphorus and organometallic compounds. Methods of synthetic preparation analyzed critically throughout. Includes much of biochemical interest. "The scope of this volume is astonishing," INDUSTRIAL AND ENGINEERING CHEMISTRY. 12,000-reference index. 2387-item bibliography. Total of x + 1005pp. 5⅜ x 8. Two volume set.
S700 Vol I Paperbound **$2.00**
S701 Vol II Paperbound **$2.00**
The set **$4.00**

THE PRINCIPLES OF ELECTROCHEMISTRY, D. A. MacInnes. Basic equations for almost every subfield of electrochemistry from first principles, referring at all times to the soundest and most recent theories and results; unusually useful as text or as reference. Covers coulometers and Faraday's Law, electrolytic conductance, the Debye-Hueckel method for the theoretical calculation of activity coefficients, concentration cells, standard electrode potentials, thermodynamic ionization constants, pH, potentiometric titrations, irreversible phenomena, Planck's equation, and much more. "Excellent treatise," AMERICAN CHEMICAL SOCIETY JOURNAL. "Highly recommended," CHEMICAL AND METALLURGICAL ENGINEERING. 2 Indices. Appendix. 585-item bibliography. 137 figures. 94 tables. ii + 478pp. 5⅝ x 8⅜.
S52 Paperbound **$2.35**

THE CHEMISTRY OF URANIUM: THE ELEMENT, ITS BINARY AND RELATED COMPOUNDS, J. J. Katz and E. Rabinowitch. Vast post-World War II collection and correlation of thousands of AEC reports and published papers in a useful and easily accessible form, still the most complete and up-to-date compilation. Treats "dry uranium chemistry," occurrences, preparation, properties, simple compounds, isotopic composition, extraction from ores, spectra, alloys, etc. Much material available only here. Index. Thousands of evaluated bibliographical references. 324 tables, charts, figures. xxi + 609pp. 5⅜ x 8. S757 Paperbound **$2.95**

KINETIC THEORY OF LIQUIDS, J. Frenkel. Regarding the kinetic theory of liquids as a generalization and extension of the theory of solid bodies, this volume covers all types of arrangements of solids, thermal displacements of atoms, interstitial atoms and ions, orientational and rotational motion of molecules, and transition between states of matter. Mathematical theory is developed close to the physical subject matter. 216 bibliographical footnotes. 55 figures. xi + 485pp. 5⅜ x 8. S94 Clothbound **$3.95**
S95 Paperbound **$2.45**

POLAR MOLECULES, Pieter Debye. This work by Nobel laureate Debye offers a complete guide to fundamental electrostatic field relations, polarizability, molecular structure. Partial contents: electric intensity, displacement and force, polarization by orientation, molar polarization and molar refraction, halogen-hydrides, polar liquids, ionic saturation, dielectric constant, etc. Special chapter considers quantum theory. Indexed. 172pp. 5⅜ x 8.
S64 Paperbound **$1.50**

ELASTICITY, PLASTICITY AND STRUCTURE OF MATTER, R. Houwink. Standard treatise on rheological aspects of different technically important solids such as crystals, resins, textiles, rubber, clay, many others. Investigates general laws for deformations; determines divergences from these laws for certain substances. Covers general physical and mathematical aspects of plasticity, elasticity, viscosity. Detailed examination of deformations, internal structure of matter in relation to elastic and plastic behavior, formation of solid matter from a fluid, conditions for elastic and plastic behavior of matter. Treats glass, asphalt, gutta percha, balata, proteins, baker's dough, lacquers, sulphur, others. 2nd revised, enlarged edition. Extensive revised bibliography in over 500 footnotes. Index. Table of symbols. 214 figures. xviii + 368pp. 6 x 9¼. S385 Paperbound **$2.45**

THE PHASE RULE AND ITS APPLICATION, Alexander Findlay. Covering chemical phenomena of 1, 2, 3, 4, and multiple component systems, this "standard work on the subject" (NATURE, London), has been completely revised and brought up to date by A. N. Campbell and N. O. Smith. Brand new material has been added on such matters as binary, tertiary liquid equilibria, solid solutions in ternary systems, quinary systems of salts and water. Completely revised to triangular coordinates in ternary systems, clarified graphic representation, solid models, etc. 9th revised edition. Author, subject indexes. 236 figures. 505 footnotes, mostly bibliographic. xii + 494pp. 5⅜ x 8. S91 Paperbound **$2.45**

Relativity, quantum theory, nuclear physics

THE PRINCIPLE OF RELATIVITY, A. Einstein, H. Lorentz, M. Minkowski, H. Weyl. These are the 11 basic papers that founded the general and special theories of relativity, all translated into English. Two papers by Lorentz on the Michelson experiment, electromagnetic phenomena. Minkowski's SPACE & TIME, and Weyl's GRAVITATION & ELECTRICITY. 7 epoch-making papers by Einstein: ELECTROMAGNETICS OF MOVING BODIES, INFLUENCE OF GRAVITATION IN PROPAGATION OF LIGHT, COSMOLOGICAL CONSIDERATIONS, GENERAL THEORY, and 3 others. 7 diagrams. Special notes by A. Sommerfeld. 224pp. 5⅜ x 8.
S81 Paperbound **$1.75**

SPACE TIME MATTER, Hermann Weyl. "The standard treatise on the general theory of relativity," (Nature), written by a world-renowned scientist, provides a deep clear discussion of the logical coherence of the general theory, with introduction to all the mathematical tools needed: Maxwell, analytical geometry, non-Euclidean geometry, tensor calculus, etc. Basis is classical space-time, before absorption of relativity. Partial contents: Euclidean space, mathematical form, metrical continuum, relativity of time and space, general theory. 15 diagrams. Bibliography. New preface for this edition. xviii + 330pp. 5⅜ x 8.
S267 Paperbound **$1.85**

PRINCIPLES OF QUANTUM MECHANICS, W. V. Houston. Enables student with working knowledge of elementary mathematical physics to develop facility in use of quantum mechanics, understand published work in field. Formulates quantum mechanics in terms of Schroedinger's wave mechanics. Studies evidence for quantum theory, for inadequacy of classical mechanics, 2 postulates of quantum mechanics; numerous important, fruitful applications of quantum mechanics in spectroscopy, collision problems, electrons in solids; other topics. "One of the most rewarding features . . . is the interlacing of problems with text," Amer. J. of Physics. Corrected edition. 21 illus. Index. 296pp. 5⅜ x 8. S524 Paperbound **$1.85**

PHYSICAL PRINCIPLES OF THE QUANTUM THEORY, Werner Heisenberg. A Nobel laureate discusses quantum theory; Heisenberg's own work, Compton, Schroedinger, Wilson, Einstein, many others. Written for physicists, chemists who are not specialists in quantum theory, only elementary formulae are considered in the text; there is a mathematical appendix for specialists. Profound without sacrifice of clarity. Translated by C. Eckart, F. Hoyt. 18 figures. 192pp. 5⅜ x 8. S113 Paperbound **$1.25**

SELECTED PAPERS ON QUANTUM ELECTRODYNAMICS, edited by **J. Schwinger.** Facsimiles of papers which established quantum electrodynamics, from initial successes through today's position as part of the larger theory of elementary particles. First book publication in any language of these collected papers of Bethe, Bloch, Dirac, Dyson, Fermi, Feynman, Heisenberg, Kusch, Lamb, Oppenheimer, Pauli, Schwinger, Tomonoga, Weisskopf, Wigner, etc. 34 papers in all, 29 in English, 1 in French, 3 in German, 1 in Italian. Preface and historical commentary by the editor. xvii + 423pp. 6⅛ x 9¼. S444 Paperbound **$2.45**

THE FUNDAMENTAL PRINCIPLES OF QUANTUM MECHANICS, WITH ELEMENTARY APPLICATIONS, E. C. Kemble. An inductive presentation, for the graduate student or specialist in some other branch of physics. Assumes some acquaintance with advanced math; apparatus necessary beyond differential equations and advanced calculus is developed as needed. Although a general exposition of principles, hundreds of individual problems are fully treated, with applications of theory being interwoven with development of the mathematical structure. The author is the Professor of Physics at Harvard Univ. "This excellent book would be of great value to every student . . . a rigorous and detailed mathematical discussion of all of the principal quantum-mechanical methods . . . has succeeded in keeping his presentations clear and understandable," Dr. Linus Pauling, J. of the American Chemical Society. Appendices: calculus of variations, math. notes, etc. Indexes. 611pp. 5⅜ x 8.
S472 Paperbound **$2.95**

ATOMIC SPECTRA AND ATOMIC STRUCTURE, G. Herzberg. Excellent general survey for chemists, physicists specializing in other fields. Partial contents: simplest line spectra and elements of atomic theory, building-up principle and periodic system of elements, hyperfine structure of spectral lines, some experiments and applications. Bibilography. 80 figures. Index. xii + 257pp. 5⅜ x 8. S115 Paperbound **$1.95**

THE THEORY AND THE PROPERTIES OF METALS AND ALLOYS, N. F. Mott, H. Jones. Quantum methods used to develop mathematical models which show interrelationship of basic chemical phenomena with crystal structure, magnetic susceptibility, electrical, optical properties. Examines thermal properties of crystal lattice, electron motion in applied field, cohesion, electrical resistance, noble metals, para-, dia-, and ferromagnetism, etc. "Exposition . . . clear . . . mathematical treatment . . . simple," Nature. 138 figures. Bibliography. Index. xiii + 320pp. 5⅜ x 8. S456 Paperbound **$1.85**

FOUNDATIONS OF NUCLEAR PHYSICS, edited by **R. T. Beyer.** 13 of the most important papers on nuclear physics reproduced in facsimile in the original languages of their authors: the papers most often cited in footnotes, bibliographies. Anderson, Curie, Joliot, Chadwick, Fermi, Lawrence, Cockcroft, Hahn, Yukawa. UNPARALLELED BIBLIOGRAPHY. 122 double-columned pages, over 4,000 articles, books classified. 57 figures. 288pp. 6⅛ x 9¼.
S19 Paperbound **$1.75**

MESON PHYSICS, R. E. Marshak. Traces the basic theory, and explicity presents results of experiments with particular emphasis on theoretical significance. Phenomena involving mesons as virtual transitions are avoided, eliminating some of the least satisfactory predictions of meson theory. Includes production and study of π mesons at nonrelativistic nucleon energies, contrasts between π and μ mesons, phenomena associated with nuclear interaction of π mesons, etc. Presents early evidence for new classes of particles and indicates theoretical difficulties created by discovery of heavy mesons and hyperons. Name and subject indices. Unabridged reprint. viii + 378pp. 5⅜ x 8. S500 Paperbound **$1.95**

See also: **STRANGE STORY OF THE QUANTUM, B. Hoffmann; FROM EUCLID TO EDDINGTON, E. Whittaker; MATTER AND LIGHT, THE NEW PHYSICS, L. de Broglie; THE EVOLUTION OF SCIENTIFIC THOUGHT FROM NEWTON TO EINSTEIN, A. d'Abro; THE RISE OF THE NEW PHYSICS, A. d'Abro; THE THEORY OF GROUPS AND QUANTUM MECHANICS, H. Weyl; SUBSTANCE AND FUNCTION, & EINSTEIN'S THEORY OF RELATIVITY, E. Cassirer; FUNDAMENTAL FORMULAS OF PHYSICS, D. H. Menzel.**

Hydrodynamics

HYDRODYNAMICS, H. Dryden, F. Murnaghan, Harry Bateman. Published by the National Research Council in 1932 this enormous volume offers a complete coverage of classical hydrodynamics. Encyclopedic in quality. Partial contents: physics of fluids, motion, turbulent flow, compressible fluids, motion in 1, 2, 3 dimensions; viscous fluids rotating, laminar motion, resistance of motion through viscous fluid, eddy viscosity, hydraulic flow in channels of various shapes, discharge of gases, flow past obstacles, etc. Bibliography of over 2,900 items. Indexes. 23 figures. 634pp. 5⅜ x 8. S303 Paperbound **$2.75**

A TREATISE ON HYDRODYNAMICS, A. B. Basset. Favorite text on hydrodynamics for 2 generations of physicists, hydrodynamical engineers, oceanographers, ship designers, etc. Clear enough for the beginning student, and thorough source for graduate students and engineers on the work of d'Alembert, Euler, Laplace, Lagrange, Poisson, Green, Clebsch, Stokes, Cauchy, Helmholtz, J. J. Thomson, Love, Hicks, Greenhill, Besant, Lamb, etc. Great amount of documentation on entire theory of classical hydrodynamics. Vol I: theory of motion of frictionless liquids, vortex, and cyclic irrotational motion, etc. 132 exercises. Bibliography. 3 Appendixes. xii + 264pp. Vol II: motion in viscous liquids, harmonic analysis, theory of tides, etc. 112 exercises. Bibliography. 4 Appendixes. xv + 328pp. Two volume set. 5⅜ x 8.
S724 Vol I Paperbound **$1.75**
S725 Vol II Paperbound **$1.75**
The set **$3.50**

HYDRODYNAMICS, Horace Lamb. Internationally famous complete coverage of standard reference work on dynamics of liquids & gases. Fundamental theorems, equations, methods, solutions, background, for classical hydrodynamics. Chapters include Equations of Motion, Integration of Equations in Special Gases, Irrotational Motion, Motion of Liquid in 2 Dimensions, Motion of Solids through Liquid-Dynamical Theory, Vortex Motion, Tidal Waves, Surface Waves, Waves of Expansion, Viscosity, Rotating Masses of liquids. Excellently planned, arranged; clear, lucid presentation. 6th enlarged, revised edition. Index. Over 900 footnotes, mostly bibliographical. 119 figures. xv + 738pp. 6⅛ x 9¼. S256 Paperbound **$2.95**

See also: **FUNDAMENTAL FORMULAS OF PHYSICS, D. H. Menzel; THEORY OF FLIGHT, R. von Mises; FUNDAMENTALS OF HYDRO- AND AEROMECHANICS, L. Prandtl and O. G. Tietjens; APPLIED HYDRO- AND AEROMECHANICS, L. Prandtl and O. G. Tietjens; HYDRAULICS AND ITS APPLICATIONS, A. H. Gibson; FLUID MECHANICS FOR HYDRAULIC ENGINEERS, H. Rouse.**

Acoustics, optics, electromagnetics

ON THE SENSATIONS OF TONE, Hermann Helmholtz. This is an unmatched coordination of such fields as acoustical physics, physiology, experiment, history of music. It covers the entire gamut of musical tone. Partial contents: relation of musical science to acoustics, physical vs. physiological acoustics, composition of vibration, resonance, analysis of tones by sympathetic resonance, beats, chords, tonality, consonant chords, discords, progression of parts, etc. 33 appendixes discuss various aspects of sound, physics, acoustics, music, etc. Translated by A. J. Ellis. New introduction by Prof. Henry Margenau of Yale. 68 figures. 43 musical passages analyzed. Over 100 tables. Index. xix + 576pp. 6⅛ x 9¼.
S114 Paperbound **$2.95**

THE THEORY OF SOUND, Lord Rayleigh. Most vibrating systems likely to be encountered in practice can be tackled successfully by the methods set forth by the great Nobel laureate, Lord Rayleigh. Complete coverage of experimental, mathematical aspects of sound theory. Partial contents: Harmonic motions, vibrating systems in general, lateral vibrations of bars, curved plates or shells, applications of Laplace's functions to acoustical problems, fluid friction, plane vortex-sheet, vibrations of solid bodies, etc. This is the first inexpensive edition of this great reference and study work. Bibliography. Historical introduction by R. B. Lindsay. Total of 1040pp. 97 figures. $5\frac{3}{8}$ x 8.
S292, S293, Two volume set, paperbound, **$4.00**

THE DYNAMICAL THEORY OF SOUND, H. Lamb. Comprehensive mathematical treatment of the physical aspects of sound, covering the theory of vibrations, the general theory of sound, and the equations of motion of strings, bars, membranes, pipes, and resonators. Includes chapters on plane, spherical, and simple harmonic waves, and the Helmholtz Theory of Audition. Complete and self-contained development for student and specialist; all fundamental differential equations solved completely. Specific mathematical details for such important phenomena as harmonics, normal modes, forced vibrations of strings, theory of reed pipes, etc. Index. Bibliography. 86 diagrams. viii + 307pp. $5\frac{3}{8}$ x 8. S655 Paperbound **$1.50**

WAVE PROPAGATION IN PERIODIC STRUCTURES, L. Brillouin. A general method and application to different problems: pure physics, such as scattering of X-rays of crystals, thermal vibration in crystal lattices, electronic motion in metals; and also problems of electrical engineering. Partial contents: elastic waves in 1-dimensional lattices of point masses. Propagation of waves along 1-dimensional lattices. Energy flow. 2 dimensional, 3 dimensional lattices. Mathieu's equation. Matrices and propagation of waves along an electric line. Continuous electric lines. 131 illustrations. Bibliography. Index. xii + 253pp. $5\frac{3}{8}$ x 8.
S34 Paperbound **$1.85**

THEORY OF VIBRATIONS, N. W. McLachlan. Based on an exceptionally successful graduate course given at Brown University, this discusses linear systems having 1 degree of freedom, forced vibrations of simple linear systems, vibration of flexible strings, transverse vibrations of bars and tubes, transverse vibration of circular plate, sound waves of finite amplitude, etc. Index. 99 diagrams. 160pp. $5\frac{3}{8}$ x 8. S190 Paperbound **$1.35**

LOUD SPEAKERS: THEORY, PERFORMANCE, TESTING AND DESIGN, N. W. McLachlan. Most comprehensive coverage of theory, practice of loud speaker design, testing; classic reference, study manual in field. First 12 chapters deal with theory, for readers mainly concerned with math. aspects; last 7 chapters will interest reader concerned with testing, design. Partial contents: principles of sound propagation, fluid pressure on vibrators, theory of moving-coil principle, transients, driving mechanisms, response curves, design of horn type moving coil speakers, electrostatic speakers, much more. Appendix. Bibliography. Index. 165 illustrations, charts. 411pp. $5\frac{3}{8}$ x 8. S588 Paperbound **$2.25**

MICROWAVE TRANSMISSION, J. S. Slater. First text dealing exclusively with microwaves, brings together points of view of field, circuit theory, for graduate student in physics, electrical engineering, microwave technician. Offers valuable point of view not in most later studies. Uses Maxwell's equations to study electromagnetic field, important in this area. Partial contents: infinite line with distributed parameters, impedance of terminated line, plane waves, reflections, wave guides, coaxial line, composite transmission lines, impedance matching, etc. Introduction. Index. 76 illus. 319pp. $5\frac{3}{8}$ x 8.
S564 Paperbound **$1.50**

THE ANALYSIS OF SENSATIONS, Ernst Mach. Great study of physiology, psychology of perception, shows Mach's ability to see material freshly, his "incorruptible skepticism and independence." (Einstein). Relation of problems of psychological perception to classical physics, supposed dualism of physical and mental, principle of continuity, evolution of senses, will as organic manifestation, scores of experiments, observations in optics, acoustics, music, graphics, etc. New introduction by T. S. Szasz, M. D. 58 illus. 300-item bibliography. Index. 404pp. $5\frac{3}{8}$ x 8. S525 Paperbound **$1.75**

APPLIED OPTICS AND OPTICAL DESIGN, A. E. Conrady. With publication of vol. 2, standard work for designers in optics is now complete for first time. Only work of its kind in English; only detailed work for practical designer and self-taught. Requires, for bulk of work, no math above trig. Step-by-step exposition, from fundamental concepts of geometrical, physical optics, to systematic study, design, of almost all types of optical systems. Vol. 1: all ordinary ray-tracing methods; primary aberrations; necessary higher aberration for design of telescopes, low-power microscopes, photographic equipment. Vol. 2: (Completed from author's notes by R. Kingslake, Dir. Optical Design, Eastman Kodak.) Special attention to high-power microscope, anastigmatic photographic objectives. "An indispensable work," J., Optical Soc. of Amer. "As a practical guide this book has no rival," Transactions, Optical Soc. Index. Bibliography. 193 diagrams. 852pp. $6\frac{1}{8}$ x $9\frac{1}{4}$. Vol. 1 T611 Paperbound **$2.95**
Vol. 2 T612 Paperbound **$2.95**

THE THEORY OF OPTICS, Paul Drude. One of finest fundamental texts in physical optics, classic offers thorough coverage, complete mathematical treatment of basic ideas. Includes fullest treatment of application of thermodynamics to optics; sine law in formation of images, transparent crystals, magnetically active substances, velocity of light, apertures, effects depending upon them, polarization, optical instruments, etc. Introduction by A. A. Michelson. Index. 110 illus. 567pp. $5\frac{3}{8}$ x 8. S532 Paperbound **$2.45**

OPTICKS, Sir Isaac Newton. In its discussions of light, reflection, color, refraction, theories of wave and corpuscular theories of light, this work is packed with scores of insights and discoveries. In its precise and practical discussion of construction of optical apparatus, contemporary understandings of phenomena it is truly fascinating to modern physicists, astronomers, mathematicians. Foreword by Albert Einstein. Preface by I. B. Cohen of Harvard University. 7 pages of portraits, facsimile pages, letters, etc. cxvi + 414pp. 5⅜ x 8. S205 Paperbound **$2.00**

OPTICS AND OPTICAL INSTRUMENTS: AN INTRODUCTION WITH SPECIAL REFERENCE TO PRACTICAL APPLICATIONS, B. K. Johnson. An invaluable guide to basic practical applications of optical principles, which shows how to set up inexpensive working models of each of the four main types of optical instruments—telescopes, microscopes, photographic lenses, optical projecting systems. Explains in detail the most important experiments for determining their accuracy, resolving power, angular field of view, amounts of aberration, all other necessary facts about the instruments. Formerly "Practical Optics." Index. 234 diagrams. Appendix. 224pp. 5⅜ x 8. S642 Paperbound **$1.65**

PRINCIPLES OF PHYSICAL OPTICS, Ernst Mach. This classical examination of the propagation of light, color, polarization, etc. offers an historical and philosophical treatment that has never been surpassed for breadth and easy readability. Contents: Rectilinear propagation of light. Reflection, refraction. Early knowledge of vision. Dioptrics. Composition of light. Theory of color and dispersion. Periodicity. Theory of interference. Polarization. Mathematical representation of properties of light. Propagation of waves, etc. 279 illustrations, 10 portraits. Appendix. Indexes. 324pp. 5⅜ x 8. S178 Paperbound **$1.75**

FUNDAMENTALS OF ELECTRICITY AND MAGNETISM, L. B. Loeb. For students of physics, chemistry, or engineering who want an introduction to electricity and magnetism on a higher level and in more detail than general elementary physics texts provide. Only elementary differential and integral calculus is assumed. Physical laws developed logically, from magnetism to electric currents, Ohm's law, electrolysis, and on to static electricity, induction, etc. Covers an unusual amount of material; one third of book on modern material: solution of wave equation, photoelectric and thermionic effects, etc. Complete statement of the various electrical systems of units and interrelations. 2 Indexes. 75 pages of problems with answers stated. Over 300 figures and diagrams. xix +669pp. 5⅜ x 8. S745 Paperbound **$2.75**

THE ELECTROMAGNETIC FIELD, Max Mason & Warren Weaver. Used constantly by graduate engineers. Vector methods exclusively: detailed treatment of electrostatics, expansion methods, with tables converting any quantity into absolute electromagnetic, absolute electrostatic, practical units. Discrete charges, ponderable bodies, Maxwell field equations, etc. Introduction. Indexes. 416pp. 5⅜ x 8. S185 Paperbound **$2.00**

ELECTRICAL THEORY ON THE GIORGI SYSTEM, P. Cornelius. A new clarification of the fundamental concepts of electricity and magnetism, advocating the convenient m.k.s. system of units that is steadily gaining followers in the sciences. Illustrating the use and effectiveness of his terminology with numerous applications to concrete technical problems, the author here expounds the famous Giorgi system of electrical physics. His lucid presentation and well-reasoned, cogent argument for the universal adoption of this system form one of the finest pieces of scientific exposition in recent years. 28 figures. Index. Conversion tables for translating earlier data into modern units. Translated from 3rd Dutch edition by L. J. Jolley. x + 187pp. 5½ x 8¾. S909 Clothbound **$6.00**

THEORY OF ELECTRONS AND ITS APPLICATION TO THE PHENOMENA OF LIGHT AND RADIANT HEAT, H. Lorentz. Lectures delivered at Columbia University by Nobel laureate Lorentz. Unabridged, they form a historical coverage of the theory of free electrons, motion, absorption of heat, Zeeman effect, propagation of light in molecular bodies, inverse Zeeman effect, optical phenomena in moving bodies, etc. 109 pages of notes explain the more advanced sections. Index. 9 figures. 352pp. 5⅜ x 8. S173 Paperbound **$1.85**

TREATISE ON ELECTRICITY AND MAGNETISM, James Clerk Maxwell. For more than 80 years a seemingly inexhaustible source of leads for physicists, mathematicians, engineers. Total of 1082pp. on such topics as Measurement of Quantities, Electrostatics, Elementary Mathematical Theory of Electricity, Electrical Work and Energy in a System of Conductors, General Theorems, Theory of Electrical Images, Electrolysis, Conduction, Polarization, Dielectrics, Resistance, etc. "The greatest mathematical physicist since Newton," Sir James Jeans. 3rd edition. 107 figures, 21 plates. 1082pp. 5⅜ x 8. S636-7, 2 volume set, paperbound **$4.00**

See also: **FUNDAMENTAL FORMULAS OF PHYSICS, D. H. Menzel; MATHEMATICAL ANALYSIS OF ELECTRICAL & OPTICAL WAVE MOTION, H. Bateman.**

Mechanics, dynamics, thermodynamics, elasticity

MECHANICS VIA THE CALCULUS, P. W. Norris, W. S. Legge. Covers almost everything, from linear motion to vector analysis: equations determining motion, linear methods, compounding of simple harmonic motions, Newton's laws of motion, Hooke's law, the simple pendulum, motion of a particle in 1 plane, centers of gravity, virtual work, friction, kinetic energy of rotating bodies, equilibrium of strings, hydrostatics, sheering stresses, elasticity, etc. 550 problems. 3rd revised edition. xii + 367pp. 6 x 9. S207 Clothbound **$3.95**

MECHANICS, J. P. Den Hartog. Already a classic among introductory texts, the M.I.T. professor's lively and discursive presentation is equally valuable as a beginner's text, an engineering student's refresher, or a practicing engineer's reference. Emphasis in this highly readable text is on illuminating fundamental principles and showing how they are embodied in a great number of real engineering and design problems: trusses, loaded cables, beams, jacks, hoists, etc. Provides advanced material on relative motion and gyroscopes not usual in introductory texts. "Very thoroughly recommended to all those anxious to improve their real understanding of the principles of mechanics." MECHANICAL WORLD. Index. List of equations. 334 problems, all with answers. Over 550 diagrams and drawings. ix + 462pp. 5⅜ x 8.
S754 Paperbound **$2.00**

THEORETICAL MECHANICS: AN INTRODUCTION TO MATHEMATICAL PHYSICS, J. S. Ames, F. D. Murnaghan. A mathematically rigorous development of theoretical mechanics for the advanced student, with constant practical applications. Used in hundreds of advanced courses. An unusually thorough coverage of gyroscopic and baryscopic material, detailed analyses of the Corilis acceleration, applications of Lagrange's equations, motion of the double pendulum, Hamilton-Jacobi partial differential equations, group velocity and dispersion, etc. Special relativity is also included. 159 problems. 44 figures. ix + 462pp. 5⅜ x 8.
S461 Paperbound **$2.00**

THEORETICAL MECHANICS: STATICS AND THE DYNAMICS OF A PARTICLE, W. D. MacMillan. Used for over 3 decades as a self-contained and extremely comprehensive advanced undergraduate text in mathematical physics, physics, astronomy, and deeper foundations of engineering. Early sections require only a knowledge of geometry; later, a working knowledge of calculus. Hundreds of basic problems, including projectiles to the moon, escape velocity, harmonic motion, ballistics, falling bodies, transmission of power, stress and strain, elasticity, astronomical problems. 340 practice problems plus many fully worked out examples make it possible to test and extend principles developed in the text. 200 figures. xvii + 430pp. 5⅜ x 8. S467 Paperbound **$2.00**

THEORETICAL MECHANICS: THE THEORY OF THE POTENTIAL, W. D. MacMillan. A comprehensive, well balanced presentation of potential theory, serving both as an introduction and a reference work with regard to specific problems, for physicists and mathematicians. No prior knowledge of integral relations is assumed, and all mathematical material is developed as it becomes necessary. Includes: Attraction of Finite Bodies; Newtonian Potential Function; Vector Fields, Green and Gauss Theorems; Attractions of Surfaces and Lines; Surface Distribution of Matter; Two-Layer Surfaces; Spherical Harmonics; Ellipsoidal Harmonics; etc. "The great number of particular cases . . . should make the book valuable to geophysicists and others actively engaged in practical applications of the potential theory," Review of Scientific Instruments. Index. Bibliography. xiii + 469pp. 5⅜ x 8. S486 Paperbound **$2.25**

THEORETICAL MECHANICS: DYNAMICS OF RIGID BODIES, W. D. MacMillan. Theory of dynamics of a rigid body is developed, using both the geometrical and analytical methods of instruction. Begins with exposition of algebra of vectors, it goes through momentum principles, motion in space, use of differential equations and infinite series to solve more sophisticated dynamics problems. Partial contents: moments of inertia, systems of free particles, motion parallel to a fixed plane, rolling motion, method of periodic solutions, much more. 82 figs. 199 problems. Bibliography. Indexes. xii + 476pp. 5⅜ x 8. S641 Paperbound **$2.00**

MATHEMATICAL FOUNDATIONS OF STATISTICAL MECHANICS, A. I. Khinchin. Offering a precise and rigorous formulation of problems, this book supplies a thorough and up-to-date exposition. It provides analytical tools needed to replace cumbersome concepts, and furnishes for the first time a logical step-by-step introduction to the subject. Partial contents: geometry & kinematics of the phase space, ergodic problem, reduction to theory of probability, application of central limit problem, ideal monatomic gas, foundation of thermo-dynamics, dispersion and distribution of sum functions. Key to notations. Index. viii + 179pp. 5⅜ x 8.
S147 Paperbound **$1.35**

ELEMENTARY PRINCIPLES IN STATISTICAL MECHANICS, J. W. Gibbs. Last work of the great Yale mathematical physicist, still one of the most fundamental treatments available for advanced students and workers in the field. Covers the basic principle of conservation of probability of phase, theory of errors in the calculated phases of a system, the contributions of Clausius, Maxwell, Boltzmann, and Gibbs himself, and much more. Includes valuable comparison of statistical mechanics with thermodynamics: Carnot's cycle, mechanical definitions of entropy, etc. xvi + 208pp. 5⅜ x 8. S707 Paperbound **$1.45**

THE DYNAMICS OF PARTICLES AND OF RIGID, ELASTIC, AND FLUID BODIES; BEING LECTURES ON MATHEMATICAL PHYSICS, A. G. Webster. The reissuing of this classic fills the need for a comprehensive work on dynamics. A wide range of topics is covered in unusually great depth, applying ordinary and partial differential equations. Part I considers laws of motion and methods applicable to systems of all sorts; oscillation, resonance, cyclic systems, etc. Part 2 is a detailed study of the dynamics of rigid bodies. Part 3 introduces the theory of potential; stress and strain, Newtonian potential functions, gyrostatics, wave and vortex motion, etc. Further contents: Kinematics of a point; Lagrange's equations; Hamilton's principle; Systems of vectors; Statics and dynamics of deformable bodies; much more, not easily found together in one volume. Unabridged reprinting of 2nd edition. 20 pages of notes on differential equations and the higher analysis. 203 illustrations. Selected bibliography. Index. xi + 588pp. 5⅜ x 8. S522 Paperbound **$2.35**

A TREATISE ON DYNAMICS OF A PARTICLE, E. J. Routh. Elementary text on dynamics for beginning mathematics or physics student. Unusually detailed treatment from elementary definitions to motion in 3 dimensions, emphasizing concrete aspects. Much unique material important in recent applications. Covers impulsive forces, rectilinear and constrained motion in 2 dimensions, harmonic and parabolic motion, degrees of freedom, closed orbits, the conical pendulum, the principle of least action, Jacobi's method, and much more. Index. 559 problems, many fully worked out, incorporated into text. xiii + 418pp. 5⅜ x 8.
S696 Paperbound **$2.25**

DYNAMICS OF A SYSTEM OF RIGID BODIES (Elementary Section), E. J. Routh. Revised 7th edition of this standard reference. This volume covers the dynamical principles of the subject, and its more elementary applications: finding moments of inertia by integration, foci of inertia, d'Alembert's principle, impulsive forces, motion in 2 and 3 dimensions, Lagrange's equations, relative indicatrix, Euler's theorem, large tautochronous motions, etc. Index. 55 figures. Scores of problems. xv + 443pp. 5⅜ x 8. S664 Paperbound **$2.35**

DYNAMICS OF A SYSTEM OF RIGID BODIES (Advanced Section), E. J. Routh. Revised 6th edition of a classic reference aid. Much of its material remains unique. Partial contents: moving axes, relative motion, oscillations about equilibrium, motion. Motion of a body under no forces, any forces. Nature of motion given by linear equations and conditions of stability. Free, forced vibrations, constants of integration, calculus of finite differences, variations, precession and nutation, motion of the moon, motion of string, chain, membranes. 64 figures. 498pp. 5⅜ x 8. S229 Paperbound **$2.35**

DYNAMICAL THEORY OF GASES, James Jeans. Divided into mathematical and physical chapters for the convenience of those not expert in mathematics, this volume discusses the mathematical theory of gas in a steady state, thermodynamics, Boltzmann and Maxwell, kinetic theory, quantum theory, exponentials, etc. 4th enlarged edition, with new material on quantum theory, quantum dynamics, etc. Indexes. 28 figures. 444pp. 6⅛ x 9¼.
S136 Paperbound **$2.45**

FOUNDATIONS OF POTENTIAL THEORY, O. D. Kellogg. Based on courses given at Harvard this is suitable for both advanced and beginning mathematicians. Proofs are rigorous, and much material not generally avialable elsewhere is included. Partial contents: forces of gravity, fields of force, divergence theorem, properties of Newtonian potentials at points of free space, potentials as solutions of Laplace's equations, harmonic functions, electrostatics, electric images, logarithmic potential, etc. One of Grundlehren Series. ix + 384pp. 5⅜ x 8.
S144 Paperbound **$1.98**

THERMODYNAMICS, Enrico Fermi. Unabridged reproduction of 1937 edition. Elementary in treatment; remarkable for clarity, organization. Requires no knowledge of advanced math beyond calculus, only familiarity with fundamentals of thermometry, calorimetry. Partial Contents: Thermodynamic systems; First & Second laws of thermodynamics; Entropy; Thermodynamic potentials: phase rule, reversible electric cell; Gaseous reactions: van't Hoff reaction box, principle of LeChatelier; Thermodynamics of dilute solutions: osmotic & vapor pressures, boiling & freezing points; Entropy constant. Index. 25 problems. 24 illustrations. x + 160pp. 5⅜ x 8. S361 Paperbound **$1.75**

THE THERMODYNAMICS OF ELECTRICAL PHENOMENA IN METALS and A CONDENSED COLLECTION OF THERMODYNAMIC FORMULAS, P. W. Bridgman. Major work by the Nobel Prizewinner: stimulating conceptual introduction to aspects of the electron theory of metals, giving an intuitive understanding of fundamental relationships concealed by the formal systems of Onsager and others. Elementary mathematical formulations show clearly the fundamental thermodynamical relationships of the electric field, and a complete phenomenological theory of metals is created. This is the work in which Bridgman announced his famous "thermomotive force" and his distinction between "driving" and "working" electromotive force. We have added in this Dover edition the author's long unavailable tables of thermodynamic formulas, extremely valuable for the speed of reference they allow. Two works bound as one. Index. 33 figures. Bibliography. xviii + 256pp. 5⅜ x 8. S723 Paperbound **$1.65**

REFLECTIONS ON THE MOTIVE POWER OF FIRE, by Sadi Carnot, and other papers on the 2nd law of thermodynamics by E. Clapeyron and R. Clausius. Carnot's "Reflections" laid the groundwork of modern thermodynamics. Its non-technical, mostly verbal statements examine the relations between heat and the work done by heat in engines, establishing conditions for the economical working of these engines. The papers by Clapeyron and Clausius here reprinted added further refinements to Carnot's work, and led to its final acceptance by physicists. Selections from posthumous manuscripts of Carnot are also included. All papers in English. New introduction by E. Mendoza. 12 illustrations. xxii + 152pp. 5⅜ x 8.
S661 Paperbound **$1.50**

TREATISE ON THERMODYNAMICS, Max Planck. Based on Planck's original papers this offers a uniform point of view for the entire field and has been used as an introduction for students who have studied elementary chemistry, physics, and calculus. Rejecting the earlier approaches of Helmholtz and Maxwell, the author makes no assumptions regarding the nature of heat, but begins with a few empirical facts, and from these deduces new physical and chemical laws. 3rd English edition of this standard text by a Nobel laureate. xvi + 297pp. 5⅜ x 8. S219 Paperbound **$1.75**

THE THEORY OF HEAT RADIATION, Max Planck. A pioneering work in thermodynamics, providing basis for most later work. Nobel Laureate Planck writes on Deductions from Electrodynamics and Thermodynamics, Entropy and Probability, Irreversible Radiation Processes, etc. Starts with simple experimental laws of optics, advances to problems of spectral distribution of energy and irreversibility. Bibliography. 7 illustrations, xiv + 224pp. 5⅜ x 8.
S546 Paperbound **$1.50**

A HISTORY OF THE THEORY OF ELASTICITY AND THE STRENGTH OF MATERIALS, I. Todhunter and K. Pearson. For over 60 years a basic reference, unsurpassed in scope or authority. Both a history of the mathematical theory of elasticity from Galileo, Hooke, and Mariotte to Saint Venant, Kirchhoff, Clebsch, and Lord Kelvin and a detailed presentation of every important mathematical contribution during this period. Presents proofs of thousands of theorems and laws, summarizes every relevant treatise, many unavailable elsewhere. Practically a book apiece is devoted to modern founders: Saint Venant, Lame, Boussinesq, Rankine, Lord Kelvin, F. Neumann, Kirchhoff, Clebsch. Hundreds of pages of technical and physical treatises on specific applications of elasticity to particular materials. Indispensable for the mathematician, physicist, or engineer working with elasticity. Unabridged, corrected reprint of original 3-volume 1886-1893 edition. Three volume set. Two indexes. Appendix to Vol. I. Total of 2344pp. 5⅜ x 8⅜.
S914–916 The set, Clothbound **$12.50**

THE MATHEMATICAL THEORY OF ELASTICITY, A. E. H. Love. A wealth of practical illustration combined with thorough discussion of fundamentals—theory, application, special problems and solutions. Partial Contents: Analysis of Strain & Stress, Elasticity of Solid Bodies, Elasticity of Crystals, Vibration of Spheres, Cylinders, Propagation of Waves in Elastic Solid Media, Torsion, Theory of Continuous Beams, Plates. Rigorous treatment of Volterra's theory of dislocations, 2-dimensional elastic systems, other topics of modern interest. "For years the standard treatise on elasticity," AMERICAN MATHEMATICAL MONTHLY. 4th revised edition. Index. 76 figures. xviii + 643pp. 6⅛ x 9¼.
S174 Paperbound **$2.95**

RAYLEIGH'S PRINCIPLE AND ITS APPLICATIONS TO ENGINEERING, G. Temple & W. Bickley. Rayleigh's principle developed to provide upper and lower estimates of true value of fundamental period of a vibrating system, or condition of stability of elastic systems. Illustrative examples; rigorous proofs in special chapters. Partial contents: Energy method of discussing vibrations, stability. Perturbation theory, whirling of uniform shafts. Criteria of elastic stability. Application of energy method. Vibrating systems. Proof, accuracy, successive approximations, application of Rayleigh's principle. Synthetic theorems. Numerical, graphical methods. Equilibrium configurations, Ritz's method. Bibliography. Index. 22 figures. ix + 156pp. 5⅜ x 8.
S307 Paperbound **$1.50**

INVESTIGATIONS ON THE THEORY OF THE BROWNIAN MOVEMENT, Albert Einstein. Reprints from rare European journals. 5 basic papers, including the Elementary Theory of the Brownian Movement, written at the request of Lorentz to provide a simple explanation. Translated by A. D. Cowper. Annotated, edited by R. Fürth. 33pp. of notes elucidate, give history of previous investigations. Author, subject indexes. 62 footnotes. 124pp. 5⅜ x 8.
S304 Paperbound **$1.25**

Dover publishes books on art, music, philosophy, literature, languages, history, social sciences, psychology, handcrafts, orientalia, puzzles and entertainments, chess, pets and gardens, books explaining science, intermediate and higher mathematics, mathematical physics, engineering, biological sciences, earth sciences, classics of science, etc. Write to:

Dept. catrr.
Dover Publications, Inc.
180 Varick Street, N. Y. 14, N. Y.